APPLICATIONS OF FRACTIONAL CALCULUS IN PHYSICS

APPLICATIONS OF FRACTIONAL CALCULUS IN PHYSICS

Editor

R. Hilfer

Universität Mainz and Universität Stuttgart
Germany

Published by

World Scientific Publishing Co. Pte. Ltd.

P O Box 128, Farrer Road, Singapore 912805

USA office: Suite 1B, 1060 Main Street, River Edge, NJ 07661

UK office: 57 Shelton Street, Covent Garden, London WC2H 9HE

Library of Congress Cataloging-in-Publication Data
Applications of fractional calculus in physics / edited by Rudolf Hilfer.
p. cm.
Includes bibliographical references and index.
ISBN 9810234570 (alk. paper)
1. Fractional calculus. 2. Mathematical physics. I. Hilfer, Rudolf.

QC20.7.F75 A67 2000
530.15'583--dc21 99-088745

British Library Cataloguing-in-Publication Data
A catalogue record for this book is available from the British Library.

Printed in Singapore by Uto-Print.

ποταμοῖς τοῖς αὐτοῖς ἐμβαίνομέν τε καὶ οὐκ ἐμβαίνομεν,
εἶμέν τε καὶ οὐκ εἶμεν.
Ἡράκλειτος

Preface

Although fractional calculus is a natural generalization of calculus, and although its mathematical history is equally long, it has, until recently, played a negligible role in physics. One reason could be that, until recently, the basic facts were not readily accessible even in the mathematical literature. This book intends to increase the accessibility of fractional calculus by combining an introduction to the mathematics with a review of selected recent applications in physics.

Many applications of fractional calculus amount to replacing the time derivative in an evolution equation with a derivative of fractional order. This is not merely a phenomenological procedure providing an additional fit parameter. Rather the chapters of this book illustrate that fractional derivatives seem to arise generally and universally, and for deep mathematical reasons. One central theme of this book is the fact that fractional derivatives arise as the infinitesimal generators of a class of translation invariant convolution semigroups. These semigroups appear universally as attractors for coarse graining procedures or scale changes. They are parametrized by a number in the unit interval corresponding to the order of the fractional derivative.

Despite their common theme all chapters are self contained and can be read independently of the rest. Editing has been kept to a minimum in order to preserve the diverse style and levels of formalization in the different areas of application. Its diversity shows that the field is still evolving and workers have not even agreed on a common notation for fractional integrals and derivatives.

Given the long mathematical history of fractional calculus it is appropriate that the book begins with a mathematical introduction to fractional calculus. Chapter I provides such an introduction, and reviews also mathematical applications to special functions, Euler, Bernoulli, and Stirling numbers. Chapter II discusses fractional evolution equations and their emergence from coarse graining. It stresses the general importance of fractional semigroups for applications in physics, and gives explicit solutions for some fractional differential equations. Chapter III continues the mathematical discussion of fractional semigroups and their infinitesimal generators from a functional analytic point of view. Chapters IV and V review phenomenological and physical arguments for the general importance of fractional derivatives. The arguments are based mainly on the ubiquity of long time memory in nonequilibrium processes and on the behaviour of trajectories in chaotic Hamiltonian systems. Polymer science applications of fractional calculus are discussed in Chapters VI and VII . Chapter VI focusses on surface interacting polymers and the decimation transformation of random walk models. Chapter VII discusses the Rouse model and rheological constitutive modelling. Applications to relaxation and diffusion models for biophysical phenomena are presented in Chapter VIII. Finally the last chapter (IX) reviews a somewhat unorthodox application in which fractional calculus is used to generalize the Ehrenfest classification of phase transitions in equilibrium thermodynamics.

Let me conclude this preface by wishing all readers the joy and excitement that I felt many times when wandering and wondering in the fields of fractional calculus and its applications. Last but not least it is a pleasant task to thank Marc Lätzel, Martin Ottmann and Marlies Parsons for their help with typesetting the manuscript.

R. Hilfer
May 1999
Stuttgart

Contents

CHAPTER I

AN INTRODUCTION TO FRACTIONAL CALCULUS

P.L. BUTZER
Lehrstuhl A für Mathematik, RWTH Aachen, Templergraben 55, D-52062 Aachen, Germany

U. WESTPHAL
Institut für Mathematik, Universität Hannover, Welfengarten 1, D-30167 Hannover, Germany

Contents

1 Various Approaches to the Fractional Calculus

1.1 Some history

As to the history of fractional calculus, already in 1695 L'Hospital raised the question as to the meaning of $d^n y/dx^n$ if $n = 1/2$, that is "what if n is fractional?". "This is an apparent paradox from which, one day, useful consequences will be drawn", Leibniz replied, together with "$d^{1/2}x$ will be equal to $x\sqrt{dx : x}$". S. F. Lacroix [100] was the first to mention in some two pages a derivative of arbitrary order in a 700 page text book of 1819. Thus for $y = x^a$, $a \in \mathbb{R}_+$, he showed that

$$\frac{d^{1/2}y}{dx^{1/2}} = \frac{\Gamma(a+1)}{\Gamma(a+1/2)} x^{a-1/2}. \tag{1.1}$$

In particular he had $(d/dx)^{1/2}x = 2\sqrt{x/\pi}$ (the same result as by the present day Riemann–Liouville definition below).

Although the name "fractional calculus" is actually a misnomer, the designation "integration and differentiation of arbitrary order" being more appropriate, one usually sticks to "fractional calculus", a terminology in use since the days of L'Hospital.

J. B. J. Fourier, who in 1822 derived an integral representation for $f(x)$,

$$f(x) = \frac{1}{2\pi} \int_{\mathbb{R}} f(\alpha)\, d\alpha \int_{\mathbb{R}} \cos p(x-\alpha)\, dp,$$

obtained (formally) the derivative version

$$\frac{d^\nu}{dx^\nu} f(x) = \frac{1}{2\pi} \int_{\mathbb{R}} f(\alpha)\, d\alpha \int_{\mathbb{R}} p^\nu \cos\{p(x-\alpha) + \frac{\nu\pi}{2}\}\, dp,$$

where "the number ν will be regarded as any quantity whatever, positive or negative".

It is usually claimed that Abel resolved in 1823 the integral equation arising from the brachistochrone problem, namely

$$\frac{1}{\Gamma(\alpha)} \int_0^x \frac{g(u)}{(x-u)^{1-\alpha}}\, du = f(x), \qquad 0 < \alpha < 1 \tag{1.2}$$

with the solution

$$g(x) = \frac{1}{\Gamma(1-\alpha)} \frac{d}{dx} \int_0^x \frac{f(u)}{(x-u)^\alpha} du. \tag{1.3}$$

As J. Lützen [107, p. 314] first showed, Abel never solved the problem by fractional calculus but merely showed how the solution, found by other means, could be written as a fractional derivative. Lützen also briefly summarized what Abel actually did. Liouville [103], however, did solve the integral equation (1.2) in 1832.

Perhaps the first serious attempt to give a logical definition of a fractional derivative is due to Liouville; he published nine papers on the subject between 1832 and 1837, the last in the field in 1855. They grew out of Liouville's early work on electromagnetism. There is further work of George Peacock (1833), D. F. Gregory (1841), Augustus de Morgan (1842), P. Kelland (1846), William Center (1848). Especially basic is Riemann's student paper of 1847 [139].

Liouville[a] started in 1832 with the well known result $D^n e^{ax} = a^n e^{ax}$ where $D = d/dx$, $n \in \mathbb{N}$, and extended it at first in the particular case $\nu = 1/2$, $a = 2$, and then to arbitrary order $\nu \in \mathbb{R}_+$ by

$$D^\nu e^{ax} = a^\nu e^{ax}. \tag{1.4}$$

He assumed the series representation for $f(x)$ as $f(x) = \sum_{k=0}^\infty c_k e^{a_k x}$ and defined the derivative of arbitrary order ν by

$$D^\nu f(x) = \sum_{k=0}^\infty c_k a_k^\nu e^{a_k x}. \tag{1.5}$$

Whereas this was Liouville's first approach, his second method was applied to the explicit function x^{-a}. He considered the integral $I = \int_0^\infty u^{a-1} e^{-xu}\, du$. Substituting $xu = t$ gives the result $I = x^{-a} \int_0^\infty t^{a-1} e^{-t}\, dt = x^{-a}\Gamma(a)$ (for $\Re e\ a > 0$). Operating on both sides of $x^{-a} = I/\Gamma(a)$ with D^ν, he obtained, using $D^\nu(e^{-xu}) = (-1)^\nu u^\nu e^{-xu}$,

$$D^\nu x^{-a} = (-1)^\nu \frac{\Gamma(a+\nu)}{\Gamma(a)} x^{-a-\nu}. \tag{1.6}$$

Liouville used the latter in his investigations of potential theory.

---

[a]The most serious and detailed examination of the work of Liouville is that presented by J. Lützen [107, pp. 303–349]

Since the ordinary differential equation $d^n y/dx^n = 0$ of n–th order has the complementary (general) solution $y = c_0 + c_1 x + \ldots + c_{n-1}x^{n-1}$, Liouville considered that the fractional order equation $d^\alpha y/dx^\alpha = 0$, $\alpha \in \mathbb{R}_+$, should have a suitable corresponding complementary solution, too. In this respect Riemann added a $\psi(x)$ to (1.8) below as the complementary function (shown to be of indeterminate nature by Cayley in 1880). For details on complementary functions see [127].

As partly indicated above, among the mathematicians spearheading research in the broad area of fractional calculus until 1941 were S.F. Lacroix, J.B.J. Fourier, N.H. Abel, J. Liouville, A. De Morgan, B. Riemann, Hj. Holmgren, K. Grünwald, A.V. Letnikov, N.Ya. Sonine, J. Hadamard, G.H. Hardy, H. Weyl, M. Riesz, H.T. Davis, A. Marchaud, J.E. Littlewood, E.L. Post, E.R. Love, B.Sz.-Nagy, A. Erdélyi and H. Kober.

Fractional calculus has developed especially intensively since 1974 when the first international conference in the field took place. It was organized by Bertram Ross [144] and took place at the University of New Haven, Connecticut in 1974. It had an exceptional turnout of 94 mathematicians; the proceedings contain 26 papers by the experts of the time. It was followed by the conferences conducted by Adam Mc Bride and Garry Roach [115] (University of Strathclyde, Glasgow, Scotland) of 1984, by Katsuyuki Nishimoto [125] (Nihon University, Tokyo, Japan) of 1989, and by Peter Rusev, Ivan Dimovski and Virginia Kiryakova [150] (Varna, Bulgaria) of 1996. In the period 1975 to the present, about 600 papers have been published relating to fractional calculus.

Samko et al in their encyclopedic volume [153, p. xxxvi] state and we cite: "We pay tribute to investigators of recent decades by citing the names of mathematicians who have made a valuable scientific contribution to fractional calculus development from 1941 until the present [1990]. These are M.A. Al-Bassam, L.S. Bosanquet, P.L. Butzer, M.M. Dzherbashyan, A. Erdélyi, T.M. Flett, Ch. Fox, S.G. Gindikin, S.L. Kalla, I.A. Kipriyanov, H. Kober, P.I. Lizorkin, E.R. Love, A.C. McBride, M. Mikolás, S.M. Nikol'skii, K. Nishimoto, I.I. Ogievetskii, R.O. O'Neil, T.J. Osler, S. Owa, B. Ross, M. Saigo, I.N. Sneddon, H.M. Srivastava, A.F. Timan, U. Westphal, A. Zygmund and others". To this list must of course be added the names of the authors of Samko et al [153] and many other mathematicians, particularly those of the younger generation.

Books especially devoted to fractional calculus include K.B. Oldham and J. Spanier [133], S.G. Samko, A.A. Kilbas and O.I. Marichev [153], V.S. Kiryakova [91], K.S. Miller and B. Ross [121], B. Rubin [147].

Books containing a chapter or sections dealing with certain aspects of fractional calculus include H.T. Davis [37], A. Zygmund [181], M.M.Dzherbashyan [45], I.N. Sneddon [159], P.L. Butzer and R.J. Nessel [25], P.L. Butzer and W. Trebels [28], G.O. Okikiolu [132], S. Fenyö and H.W. Stolle [55], H.M. Srivastava and H.L. Manocha [162], R. Gorenflo and S. Vessella [65].

There also exist two journals devoted especially to fractional calculus, namely the one edited by K. Nishimoto [126], and the one recently founded by V. Kiryakova [92].

For an historical survey of the field until 1975 one may consult Oldham-Spanier [133, pp. 1–15] as well as Ross [145].

1.2 Basic definitions; Riemann–Liouville and Weyl approaches

Let us consider some of the starting points for a discussion of classical fractional calculus; we will also introduce notations.

One development begins with a generalization of repeated integration. Thus if f is locally integrable on (a, ∞), then the n–fold iterated integral is given by

$$
\begin{aligned}
{}_aI_x^n f(x) &:= \int_a^x du_1 \int_a^{u_1} du_2 \ldots \int_a^{u_{n-1}} f(u_n)\, du_n \\
&= \frac{1}{(n-1)!} \int_a^x (x-u)^{n-1} f(u)\, du
\end{aligned}
\tag{1.7}
$$

for almost all x with $-\infty \leq a < x < \infty$ and $n \in \mathbb{N}$. Writing $(n-1)! = \Gamma(n)$, an immediate generalization is the *integral* of f of *fractional order* $\alpha > 0$,

$$
{}_aI_x^\alpha f(x) = \frac{1}{\Gamma(\alpha)} \int_a^x (x-u)^{\alpha-1} f(u)\, du \quad \text{(right hand),} \tag{1.8}
$$

and similarly for $-\infty < x < b \leq \infty$

$$
{}_xI_b^\alpha f(x) = \frac{1}{\Gamma(\alpha)} \int_x^b (u-x)^{\alpha-1} f(u)\, du \quad \text{(left hand),} \tag{1.9}
$$

both being defined for suitable f. The subscripts in I denote the terminals of integration (in the given order). Note the kernel $(u-x)^{\alpha-1}$ for (1.9).

Observe that (1.8) for $\alpha = n$ can be shown to be (see e. g. [145, pp. 7–10]) the unique solution of the initial value problem

$$
y^{(n)}(x) = f(x),\; y(a) = y'(a) = \ldots = y^{(n-1)}(a) = 0. \tag{1.10}
$$

When $a=-\infty$ equation (1.8) is equivalent to Liouville's definition, and when $a=0$ we have Riemann's definition (without the complementary function). One generally speaks of ${}_aI_x^\alpha f$ as the *Riemann–Liouville* fractional integral of order α of f, a terminology introduced by Holmgren (1863/64). On the other hand, one usually refers to

$$ {}_xW_\infty^\alpha f(x) = {}_xI_\infty^\alpha f(x) = \frac{1}{\Gamma(\alpha)} \int_x^\infty (u-x)^{\alpha-1} f(u)\, du \tag{1.11}$$

$$ {}_{-\infty}W_x^\alpha f(x) = {}_{-\infty}I_x^\alpha f(x) = \frac{1}{\Gamma(\alpha)} \int_{-\infty}^x (x-u)^{\alpha-1} f(u) du \tag{1.12}$$

as *Weyl* fractional integrals of order α, they being defined for suitable f.

The right and left hand fractional integrals ${}_aI_x^\alpha f(x)$ and ${}_xI_b^\alpha f(x)$ are related via the Parseval equality (fractional integration by parts) which we give for convenience for $a=0$ and $b=\infty$:

$$\int_0^\infty f(x)\, ({}_0I_x^\alpha g)\,(x)\, dx = \int_0^\infty ({}_xW_\infty^\alpha f)\,(x) g(x)\, dx. \tag{1.13}$$

The following properties are stated for right handed fractional integrals (with obvious changes in the case of left handed integrals).

Concerning existence of fractional integrals, let $f \in L^1_{\text{loc}}(a,\infty)$. Then, if $a > -\infty$, ${}_aI_x^\alpha f(x)$ is finite almost everywhere on (a,∞) and belongs to $L^1_{\text{loc}}(a,\infty)$. If $a=-\infty$, it is assumed that f behaves at $-\infty$ such that the integral (1.8) converges. Under these assumptions the fractional integrals satisfy the additive index law (or semigroup property)

$$ {}_aI_x^\alpha\ {}_aI_x^\beta f = {}_aI_x^{\alpha+\beta} f \quad (\alpha,\beta>0). \tag{1.14}$$

Indeed, by Dirichlet's formula concerning the change of the order of integration, we have

$$\begin{aligned} {}_aI_x^\alpha\ {}_aI_x^\beta f(x) &= \frac{1}{\Gamma(\alpha)} \int_a^x (x-u)^{\alpha-1}\, du \frac{1}{\Gamma(\beta)} \int_a^u (u-t)^{\beta-1} f(t)\, dt \\ &= \frac{1}{\Gamma(\alpha)\Gamma(\beta)} \int_a^x f(t)\, dt \int_t^x (x-u)^{\alpha-1} (u-t)^{\beta-1}\, du. \end{aligned}$$

The second integral on the right equals, under the substitution $y = \frac{u-t}{x-t}$,

$$(x-t)^{\alpha+\beta-1} \int_0^1 (1-y)^{\alpha-1} y^{\beta-1}\, dy = B(\alpha,\beta)(x-t)^{\alpha+\beta-1} = \frac{\Gamma(\alpha)\Gamma(\beta)}{\Gamma(\alpha+\beta)}(x-t)^{\alpha+\beta-1}, \quad (1.15)$$

$B(\alpha,\beta)$ being the Beta–function (see (1.29)). When this is substituted into the above the result follows. In particular, we have

$${}_aI_x^{n+\alpha} f = {}_aI_x^n \; {}_aI_x^\alpha f \quad (n \in \mathbb{N}, \alpha > 0) \qquad (1.16)$$

which implies by n-fold differentiation

$$\frac{d^n}{dx^n} \; {}_aI_x^{n+\alpha} f(x) = {}_aI_x^\alpha f(x) \quad (n \in \mathbb{N}, \alpha > 0)$$

for almost all x.

The above results also hold for complex parameters α, if the condition $\alpha > 0$ is replaced by $\Re e\ \alpha > 0$. Then the operation ${}_aI_x^\alpha$ may be considered as a holomorphic function of α for $\Re e\ \alpha > 0$ which can be extended to the whole complex plane by analytic continuation, if f is sufficiently smooth.

To understand and establish this fact, we assume, for convenience, that f is an infinitely differentiable function defined on $\mathbb{R}$ with compact support contained in $[a, \infty)$, if $a > -\infty$, implying that $f^{(n)}(a) = 0$ for $n = 0, 1, 2, \ldots$. Then for any fixed $x > a$ the integral in (1.8) is a holomorphic function of α for $\Re e\ \alpha > 0$. Now, integration by parts n-times yields

$${}_aI_x^\alpha f(x) = {}_aI_x^{n+\alpha} f^{(n)}(x) \quad (\Re e\ \alpha > 0, n \in \mathbb{N}). \qquad (1.17)$$

Applying the semigroup property (1.16) to the expression on the right in (1.17) and differentiating the result n-times with respect to x, we obtain

$$\frac{d^n}{dx^n} \; {}_aI_x^\alpha f(x) = {}_aI_x^\alpha f^{(n)}(x) \quad (\Re e\ \alpha > 0, n \in \mathbb{N}), \qquad (1.18)$$

showing that under the hypotheses assumed on f the operations of integration of fractional order α and differentiation of integral order n commute.

Returning to formula (1.17) we now realize that its right-hand side is a holomorphic function of α in the wider domain $\{\alpha \in \mathbb{C}; \Re e\ \alpha > -n\}$ and even equals there $\frac{d^n}{dx^n} \; {}_aI_x^{n+\alpha} f(x)$, by (1.18). Thus we can extend ${}_aI_x^\alpha f(x)$ to the domain $\{\alpha \in \mathbb{C}; \Re e\ \alpha \le 0\}$ analytically, defining for $\alpha \in \mathbb{C}$ with $\Re e\ \alpha \le 0$

$${}_aI_x^\alpha f(x) := {}_aI_x^{n+\alpha} f^{(n)}(x) = \frac{d^n}{dx^n} \; {}_aI_x^{n+\alpha} f(x) \qquad (1.19)$$

with any integer $n > -\Re e\ \alpha$. In particular, we obtain

$$ {}_aI_x^0 f(x) = f(x), \ {}_aI_x^{-n} f(x) = f^{(n)}(x) \quad (n \in \mathbb{N}). $$

Moreover, it is clear by complex function theory arguments that the semigroup property (1.14) remains valid for all $\alpha, \beta \in \mathbb{C}$.

The very elegant method of analytic continuation developed by M. Riesz and his school is restricted to a rather small class of functions (even if the assumptions are weakened somewhat which is, indeed, possible). But the expressions occurring in formula (1.19) are meaningful for much more general classes of functions and thus give rise to the following definitions of fractional derivatives which go back to Liouville.

Let α be a complex number with $\Re e\ \alpha > 0$ and $n = [\Re e\ \alpha] + 1$, where $[\Re e\ \alpha]$ denotes the integral part of $\Re e\ \alpha$. Then the right-handed fractional derivative of order α is defined by

$$ {}_aD_x^\alpha f(x) = \frac{d^n}{dx^n}\ {}_aI_x^{n-\alpha} f(x) \quad (n = [\Re e\ \alpha] + 1) \tag{1.20} $$

for any $f \in L^1_{\text{loc}}(a, \infty)$ for which the expression on the right exists.

One can unify the definitions of integrals and derivatives of arbitrary order α, $\Re e\ \alpha \neq 0$, (equivalently) by, for $n \in \mathbb{N}$,

$$ {}_aD_x^\alpha f(x) = \begin{cases} \frac{1}{\Gamma(-\alpha)} \int_a^x (x-u)^{-\alpha-1} f(u)\, du & (\Re e\ \alpha < 0) \\ \left(\frac{d}{dx}\right)^n {}_aI_x^{n-\alpha} f(x) & (\Re e\ \alpha > 0; n-1 \leq \Re e\ \alpha < n); \end{cases} \tag{1.21} $$

one often speaks of a *differintegral* of f of order α in this respect. This process is also referred to as *fractional integro-differentiation.*

Note that the left-handed fractional derivative of order α is defined by

$$ {}_xD_b^\alpha f(x) = (-1)^n \frac{d^n}{dx^n}\ {}_xI_b^{n-\alpha} f(x) \quad (n = [\Re e\ \alpha] + 1). \tag{1.22} $$

The fractional derivative of *purely imaginary order* $\alpha = i\theta$, $\theta \neq 0$, is defined by

$$ {}_aD_x^{i\theta} f(x) = \frac{1}{\Gamma(1-i\theta)} \frac{d}{dx} \int_a^x \frac{f(u)}{(x-u)^{i\theta}}\, du, $$

and the associated integral of order $\alpha = i\theta$ by

$$ {}_aI_x^{i\theta} f(x) = \frac{d}{dx}\, {}_aI_x^{1+i\theta} f(x) = \frac{1}{\Gamma(1+i\theta)} \frac{d}{dx} \int_a^x (x-u)^{i\theta} f(u)\, du $$

(since the fractional integral (1.8) diverges for $\alpha = i\theta$). Then the definition of fractional integro-differentiation for all $\alpha \in \mathbb{C}$ is completed by introducing the identity operator ${}_aD_x^0 f := {}_aI_x^0 f = f$ for $\alpha = 0$.

Let us also observe that the fractional operators are linear:

$$ {}_aD_x^\alpha [c_1 f_1(x) + c_2 f_2(x)] = c_1\, {}_aD_x^\alpha f_1(x) + c_2\, {}_aD_x^\alpha f_2(x), $$

c_1, c_2 being constants.

Concerning sufficient conditions for the existence of the fractional derivatives (1.20) and their relation to (1.26), let us consider the case $0 < \Re e\, \alpha < 1$, if $a > -\infty$. Suppose that f is absolutely continuous on the finite interval $[a, b]$, in notation $f \in AC[a, b]$, meaning that f is differentiable almost everywhere on (a, b) with $f' \in L^1(a, b)$ and has the representation on $[a, b]$

$$ f(x) = \int_a^x f'(u) du + f(a) = {}_aI_x^1 f'(x) + f(a). $$

Substituting this in ${}_aI_x^{1-\alpha} f(x)$ and noting that, by the semigroup property, the operators $I^{1-\alpha}$ and I^1 commute, we obtain

$$ {}_aI_x^{1-\alpha} f(x) = {}_aI_x^1\, {}_aI_x^{1-\alpha} f'(x) + \frac{f(a)}{\Gamma(2-\alpha)} (x-a)^{1-\alpha}. $$

By differentiating with respect to x this implies

$$ {}_aD_x^\alpha f(x) = \frac{d}{dx}\, {}_aI_x^{1-\alpha} f(x) = {}_aI_x^{1-\alpha} f'(x) + \frac{f(a)}{\Gamma(1-\alpha)} (x-a)^{-\alpha} \tag{1.23} $$

which shows that, in general, the operators ${}_aI_x^{1-\alpha}$ and $\frac{d}{dx}$ do not commute.

By Hölder's inequality one easily derives from (1.23) that ${}_aD_x^\alpha f \in L^r(a, b)$ for $1 \le r < 1/\Re e\, \alpha$.

Formula (1.23) may be extended to complex α with $\Re e\, \alpha \ge 1$. The results are summarized in Proposition 1.1 below. Beforehand we introduce the following notation: For $n \in \mathbb{N}$, $AC^{n-1}[a, b]$ denotes the set of $(n-1)$-times differentiable functions f on $[a, b]$ such that $f, f', \ldots, f^{(n-1)}$ are absolutely continuous on $[a, b]$. Note that $AC^0[a, b]$ equals $AC[a, b]$.

Proposition 1.1. *a) If $f \in AC[a,b]$, f being given just in the (finite) interval $[a,b]$, then ${}_aD_x^\alpha f$, ${}_xD_b^\alpha f$ exist a. e. for $0 < \Re e\, \alpha < 1$. Moreover, ${}_aD_x^\alpha f \in L^r(a,b)$ for $1 \le r < 1/\Re e\, \alpha$ with*

$$ {}_aD_x^\alpha f(x) = \frac{1}{\Gamma(1-\alpha)} \left\{ \frac{f(a)}{(x-a)^\alpha} + \int_a^x \frac{f'(u)}{(x-u)^\alpha}\, du \right\}. \tag{1.24}$$

Correspondingly for ${}_xD_b^\alpha f$.

b) If $f \in AC^{n-1}[a,b]$, $n = [\Re e\, \alpha] + 1$, then ${}_xD_a^\alpha f$ exists a. e. for $\Re e\, \alpha \ge 0$ and has the representation

$$ {}_aD_x^\alpha f(x) = \sum_{k=0}^{n-1} \frac{f^{(k)}(a)}{\Gamma(1+k-\alpha)} (x-a)^{k-\alpha} + \frac{1}{\Gamma(n-\alpha)} \int_a^x \frac{f^{(n)}(u)}{(x-u)^{\alpha-n+1}}\, du. \tag{1.25}$$

An alternative way to define a fractional derivative of order α, also due to Liouville, is

$$ {}_a\overline{D}_x^\alpha f(x) := {}_aI_x^{n-\alpha} f^{(n)}(x) \quad (n = [\Re e\, \alpha] + 1). \tag{1.26}$$

Obviously, f has to be n-times differentiable in order that the right-hand side of (1.26) exists. The relation between the two fractional derivatives (1.20) and (1.26) is given by formula (1.25) above, namely

$$ {}_aD_x^\alpha f(x) = {}_a\overline{D}_x^\alpha f(x) + \sum_{k=0}^{n-1} \frac{f^{(k)}(a)}{\Gamma(1+k-\alpha)} (x-a)^{k-\alpha}, $$

holding under the assumptions stated in Proposition 1.1 b). Definition (1.20) is commonly used in mathematical circles, while definition (1.26) is often preferable in problems of physical interest when initial conditions are expressed in terms of integer derivatives and Laplace transform methods are applied; see (5.27) for the Laplace transform of ${}_0\overline{D}_x^\alpha f$. The use of definition (1.26) in this context goes back to Caputo [32].

For a new real function space C_α, $0 \le \alpha \le 1$, larger than the space of continuous functions, enabling one to study fractional order continuity, derivability and integrability, see Bonilla et al [10].

There are many results concerned with two-weight problems for various types of fractional integrals (and transforms). Many arose especially after the appearance of the monographs Kokilashvili-Krbec [95], Genebashvili et al [60]. Let us indicate two of these. Let $L^p_w(0,\infty)$ be the class of Lebesgue measurable functions on $\mathbb{R}_+$ for which $\|f\|_{L^p_w} = \left(\int_0^\infty |f(u)|^p w(u)du\right)^{1/p} < +\infty$, where $w(x) > 0$ a.e. and is locally integrable on $\mathbb{R}_+$. For simplicity take $J^\alpha f(x) = \int_0^x (x-u)^{\alpha-1} f(u)du$. There holds:

Let $1 < p \le q < \infty$, $\frac{1}{p} < \alpha < 1$ or $\alpha > 1$, and $p' = p/(p-1)$. Then the operator J^α is bounded from $L^p(0,\infty)$ to $L^q_w(0,\infty)$, i.e.,

$$\|J^\alpha f\|_{L^q_w} \le A\|f\|_{L^p} \tag{1.27}$$

for a constant A independent of f, iff

$$B = \sup_{t>0} B(t) = \sup_{t>0} \left(\int_t^\infty u^{(\alpha-1)q} w(u)du \right)^{1/q} t^{1/p'} < \infty. \tag{1.28}$$

Moreover, if A is the best constant in (1.27), then $A \approx B$. Further, this operator J^α is compact from $L^p(0,\infty)$ to $L^q_w(0,\infty)$ iff condition (1.28) holds as well as the condition $\lim_{t\to 0} B(t) = \lim_{t\to\infty} B(t) = 0$.

The more general problem of boundedness and compactness of J^α from $L^p_v(0,\infty)$ (with $v \not\equiv 1$) to $L^q_w(0,\infty)$ has also been solved, the formulation being more difficult. The two-weight problem for the Weyl operator ${}_xW^\alpha_\infty f(x)$ is also a closed problem. There exist similar results for the Erdélyi-Kober operators, operators with power-logarithmic kernels, for Riesz potentials, etc. For this growing field see e.g. A. Meshki [118], H.P. Heinig [70] and [96], [97], and the extensive literature cited there.

1.3 Three examples

Let us compare the foregoing with the classical calculus. For this purpose let us first evaluate the fractional derivative ${}_0D^\alpha_x f(x)$ for the function $f(x) = x^b$ for $b > -1$. In view of the evaluation of the beta integral, valid for all $p > 0$ and $q > 0$ (or $\Re e\, p > 0$, $\Re e\, q > 0$), namely

$$B(p,q) := \int_0^1 u^{p-1}(1-u)^{q-1}\, du = \frac{\Gamma(p)\Gamma(q)}{\Gamma(p+q)}, \tag{1.29}$$

we have for $m-1 \le \alpha < m$, $m \in \mathbb{N}$,

$$\begin{aligned}
{}_0D_x^\alpha x^b &= \left(\frac{d}{dx}\right)^m \left\{ \frac{1}{\Gamma(m-\alpha)} \int_0^x (x-u)^{m-\alpha-1} u^b \, du \right\} \\
&= \frac{1}{\Gamma(m-\alpha)} \left(\frac{d}{dx}\right)^m \left\{ x^{m-\alpha+b} \int_0^1 (1-v)^{m-\alpha-1} v^b \, dv \right\} \\
&= \frac{1}{\Gamma(m-\alpha)} \frac{\Gamma(b+1)\Gamma(m-\alpha)}{\Gamma(b+1+m-\alpha)} \left(\frac{d}{dx}\right)^m x^{m-\alpha+b} \\
&= \frac{\Gamma(b+1)}{\Gamma(b+1+m-\alpha)} x^{b-\alpha} \frac{\Gamma(m-\alpha+b+1)}{\Gamma(m-\alpha+b-m+1)} \\
&= \frac{\Gamma(b+1)}{\Gamma(b-\alpha+1)} x^{b-\alpha},
\end{aligned} \tag{1.30}$$

where we made use of, with $p \in \mathbb{R}\backslash\{-1,-2,-3,\dots\}$,

$$\left(\frac{d}{dx}\right)^m x^p = \frac{\Gamma(p+1)}{\Gamma(p-m+1)} x^{p-m} = p(p-1)\dots(p-m+1)x^{p-m} \quad (m \in \mathbb{N}). \tag{1.31}$$

For $b = 0$ we obtain ${}_0D_x^\alpha c = cx^{-\alpha}/\Gamma(1-\alpha)$ for any $\alpha > 0$. Thus fractional differentiation of a constant c is zero only for positive integral values of $\alpha = n \in \mathbb{N}$ (recall $\Gamma(1-n) = \infty$). On the other hand, for any $\alpha > 0$, ${}_0D_x^\alpha f(x) \equiv 0$ if $f(x) = x^{\alpha-k}$, $k = 1, 2, \dots, 1+[\alpha]$.

Let us consider a second example, namely ${}_0I_x^\alpha f$ for $f(x) = \log x$. Indeed, under the substitution $u = x(1-v)$,

$$\begin{aligned}
{}_0I_x^\alpha f(x) &= \frac{1}{\Gamma(\alpha)} \int_0^x (x-u)^{\alpha-1} \log u \, du \qquad (1.32) \\
&= \frac{(\log x)x^\alpha}{\Gamma(\alpha)} \int_0^1 v^{\alpha-1} \, dv + \frac{x^\alpha}{\Gamma(\alpha)} \int_0^1 v^{\alpha-1} \log(1-v) \, dv \\
&= \frac{x^\alpha}{\Gamma(\alpha+1)} \log x - \frac{x^\alpha}{\Gamma(\alpha)\alpha} \int_0^1 \log(1-v) \, d(1-v^\alpha) \\
&= \frac{x^\alpha}{\Gamma(\alpha+1)} \left\{ \log x - (1-v^\alpha)\log(1-v)\Big|_0^1 - \int_0^1 \frac{1-v^\alpha}{1-v} \, dv \right\}
\end{aligned}$$

by integration by parts. Noting

$$\int_0^1 \frac{v^x - v^y}{1-v}\, dv = \psi(y+1) - \psi(x+1) \quad (\Re e\, x, \Re e\, y > -1),$$

where the psi function, defined by $\psi(x) = [\Gamma(x)]^{-1}(d/dx)\Gamma(x)$, obeys the recursion $\psi(x+1) - \psi(x) = x^{-1}$ with $-\psi(1) = \gamma = 0.5772157\ldots$, this yields

$$ {}_0I_x^\alpha \log x = \frac{x^\alpha}{\Gamma(\alpha+1)}\{\log x - \psi(\alpha+1) + \psi(1)\}. \tag{1.33}$$

Thus, for $m-1 \le \Re e\, \alpha < m$,

$$\begin{aligned} {}_0D_x^\alpha \log x &= \left(\frac{d}{dx}\right)^m {}_0I_x^{m-\alpha} \log x \qquad (1.34)\\ &= \left(\frac{d}{dx}\right)^m \frac{x^{m-\alpha}}{\Gamma(m-\alpha+1)}\left[\log x - \psi(m-\alpha+1) + \psi(1)\right], \end{aligned}$$

where classical termwise differentiation is to be applied. Combining the result with (1.33) we obtain for all $\alpha \in \mathbb{C}$,

$$ {}_0D_x^\alpha \log x = \frac{x^{-\alpha}}{\Gamma(1-\alpha)}\left\{\log x - \psi(1-\alpha) - \gamma\right\}, \tag{1.35}$$

where in case $\alpha = n \in \mathbb{N}$ the expression on the right has to be interpreted as the limit for $\alpha \to n$. In fact, from the limit $\lim_{\alpha\to n} \psi(1-\alpha)/\Gamma(1-\alpha) = (-1)^{-n}\Gamma(n)$, the rule $(d/dx)^n \log x = -\Gamma(n)(-x)^{-n}$ readily follows. However for $\alpha = -n \in \mathbb{N}$ there holds the classical result

$$ {}_0D_x^{-n} \log x = {}_0I_x^n \log x = \frac{x^n}{n!}\left\{\log x - \sum_{j=1}^n \frac{1}{j}\right\},$$

also observing that $\psi(n+1) - \psi(1) = \sum_{j=1}^n j^{-1}$, which follows from the recursion $\psi(x+1) - \psi(x) = x^{-1}$.

As to our third example, take Weyl's definition, for $m-1 \le \alpha < m$, $m \in \mathbb{N}$,

$$ {}_xD_\infty^\alpha f(x) = (-1)^m \left(\frac{d}{dx}\right)^m {}_xW_\infty^{m-\alpha} f(x) \tag{1.36}$$

for which, for the example $f(x) = e^{-px}$, $p > 0$, under the substitution $u - x = y/p$,

$$
\begin{aligned}
{}_xW_\infty^\alpha e^{-px} &= \frac{1}{\Gamma(\alpha)} \int_x^\infty (u-x)^{\alpha-1} e^{-pu}\, du \\
&= \frac{e^{-px}}{\Gamma(\alpha)p^\alpha} \int_0^\infty y^{\alpha-1} e^{-y}\, dy = \frac{e^{-px}}{p^\alpha} \qquad (\alpha > 0). \qquad (1.37)
\end{aligned}
$$

This yields for $p > 0$,

$$
{}_xD_\infty^\alpha e^{-px} = (-1)^m \left(\frac{d}{dx}\right)^m p^{-(m-\alpha)} e^{-px} = p^\alpha e^{-px} \quad (m - 1 \leq \alpha < m). \qquad (1.38)
$$

1.4 Approaches by Hadamard, by contour integration and other methods

The approach, due to J. Hadamard (1892), is to consider fractional differentiation of an analytic function via differentiation of its Taylor series $f(z) = \sum_{k=0}^\infty c_k (z - z_0)^k$ formally "α times", thus

$$
{}_{z_0}D_z^\alpha f(z) = \sum_{k=0}^\infty \frac{\Gamma(k+1)}{\Gamma(k+1-\alpha)} c_k (z - z_0)^{k-\alpha} \quad (c_k = f^{(k)}(z_0)/k!). \qquad (1.39)
$$

This approach is discussed in articles by Gaer-Rubel [59], Lavoie, Tremblay and Osler [101], Ross [145].

If z is complex, the natural extension of the Riemann-Liouville definition of the fractional integral is

$$
{}_aI_z^\alpha f(z) = \frac{1}{\Gamma(\alpha)} \int_a^z (z - \zeta)^{\alpha-1} f(\zeta) d\zeta \qquad (1.40)
$$

where the path of integration is along the line from $\zeta = a$ to $\zeta = z$ in the complex ζ-plane. The multivalued function $(z-\zeta)^{\alpha-1}$ is defined by $(z-\zeta)^{\alpha-1} = \exp[(\alpha - 1)\log(z - \zeta)]$ with $\log(z - \zeta) = \log|z - \zeta| + i \arg(z - \zeta)$. To obtain a single-valued branch for the integrand of (1.40) it is appropriate to choose $\arg(z - \zeta) = \arg(z - a)$ for all ζ on the line from a to z, where $\arg(z - a)$ is fixed by $-\pi < \arg(z - a) \leq \pi$.

An important method of defining a derivative of arbitrary order for holomorphic functions is given through a generalization of Cauchy's integral formula

$$f^{(n)}(z) = \frac{n!}{2\pi i} \int_C \frac{f(\zeta)}{(\zeta - z)^{n+1}} d\zeta \qquad (z \in \text{Int } C; n \in \mathbb{N}_0),$$

where f is holomorphic in a simply connected region $\mathcal{D}$ in the complex ζ-plane, and C is a piecewise smooth Jordan curve in $\mathcal{D}$. N.Y. Sonine (1872), P.A. Nekrassov (1888) et al. and more recently T.J. Osler [134] defined a fractional derivative of order α of $f(z)$ by

$$_aD_z^\alpha f(z) = \frac{\Gamma(\alpha+1)}{2\pi i} \int_{C(a,z+)} \frac{f(\zeta)}{(\zeta - z)^{\alpha+1}} d\zeta. \tag{1.41}$$

Here the integral curve $C(a, z+)$ in $\mathcal{D}$ is a closed contour starting at $\zeta = a$, encircling $\zeta = z$ once in the positive sense and returning to $\zeta = a$. For nonintegral α, the integrand has a branch line which is chosen as the ray beginning at $\zeta = z$ and passing through $\zeta = a$. Thus, this branch line cuts the integration contour in its beginning and ending point, but nowhere else. For the function $(\zeta - z)^{-\alpha-1}$ its principal branch is taken; it is that continuous range of the function for which $\arg(\zeta - z)$ is zero when $\zeta - z$ is real and positive.

Note that the generalized Cauchy integral formula (1.41) is defined for all $\alpha \in \mathbb{C}$; for negative integers $-n\,(n \in \mathbb{N})$ it is to be understood as the limit for $\alpha \to -n$. If $\Re e\ \alpha < 0$, then (1.41) coincides with the Riemann-Liouville integral $_aI_z^{-\alpha} f(z)$.

The Cauchy type complex contour integral approach is especially useful in the study of special functions. See e.g. Lavoie-Osler-Tremblay [102].

There exist further approaches to fractional integration by Erdélyi and Kober [51,93], R.K. Saxena [154], G.O. Okikiolu [131], Kalla-Saxena [90], K. Nishimoto [124], J. Cossar [36], S. Ruscheweyh [149], etc.

2 Leibniz Rule and Applications; Semigroups of Operators

2.1 Fractional Leibniz rule for functions

The classical Leibniz rule for the n-fold derivative of a product of two functions f, g as a sum of products of operations performed on each function is given by, provided f and g are n-times differentiable at z,

$$D_z^n[f(z)g(z)] = \sum_{k=0}^{n} \binom{n}{k} D_z^{n-k} f(z) \cdot D_z^k g(z), \tag{2.1}$$

where $D_z^n = (d/dz)^n$. This rule can be extended to fractional values of α: replacing n by α we have for holomorphic functions f, g

$$_aD_z^\alpha[f(z)g(z)] = \sum_{k=0}^{\infty} \binom{\alpha}{k} {}_aD_z^{\alpha-k} f(z) \cdot D_z^k g(z) \quad (\alpha \in \mathbb{C}) \tag{2.2}$$

a result which basically goes back to Liouville (1832), where ${}_aD_z^\alpha$ is now understood in the sense of Osler [134]. This formula suffers from the apparent drawback that the interchange of $f(z)$ and $g(z)$ on the right side of (2.2) is not obvious. An interesting generalization of this rule without the drawback, due to Y.Watanabe (1931) [170] and Osler [134] (1970), is

$$_aD_z^\alpha[f(z)g(z)] = \sum_{k=-\infty}^{\infty} \binom{\alpha}{k+\mu} {}_aD_z^{\alpha-k-\mu} f(z) \cdot {}_aD_z^{k+\mu} g(z), \tag{2.3}$$

where μ is arbitrary, rational, irrational, or a complex number. The special case $\mu = 0$ reduces to (2.2).

Observe that if $\Re e\ \alpha < 0$, then formula (2.2) is actually the counterpart of Leibniz rule for fractional integrals. There also exists a *symmetrical* fractional Leibniz rule in the form (due to Osler [137])

$$_aD_z^\alpha[f(z)g(z)] = \sum_{k=-\infty}^{\infty} c \binom{\alpha}{ck+\mu} {}_aD_z^{\alpha-ck-\mu} f(z) \cdot {}_aD_z^{ck+\mu} g(z), \tag{2.4}$$

where $\alpha, \mu \in \mathbb{C}$ for which $\binom{\alpha}{ck+\mu}$ is well-defined, and $0 < c \leq 1$; it yields (2.3) for $c = 1$, and reduces to (2.2) for $c = 1$ and $\mu = 0$.

As special cases of (2.3), there holds for $f(z) = 1$, and $g(z)$ renamed as $f(z)$,

$$_aD_z^\alpha f(z) = \frac{\Gamma(\alpha+1)\sin((\alpha-\mu)\pi)}{\pi} \sum_{k=-\infty}^{\infty} (-1)^k \frac{(z-a)^{k+\mu-\alpha}\ {}_aD_z^{\mu+k} f(z)}{(\alpha-\mu-k)\Gamma(\mu+k+1)} \tag{2.5}$$

for $\Re e\ \alpha > -1$, $\alpha - \mu \notin \mathbb{Z}$, noting $\Gamma(z)\Gamma(1-z) = \pi/\sin \pi z$ for $z \notin \mathbb{Z}$.

A further result, also due to Osler [134], is the case $\alpha = 0$ of (2.3), namely

$$f(z)g(z) = \frac{\sin \mu\pi}{\pi} \sum_{k=-\infty}^{\infty} (-1)^k \frac{{}_aD_z^{-(\mu+k)} f(z) \cdot {}_aD_z^{\mu+k} g(z)}{\mu + k} \qquad (\mu \notin \mathbb{Z}). \quad (2.6)$$

Note that (2.3) has under suitable conditions the interesting integral analogue

$${}_aD_z^{\alpha}[f(z)g(z)] = \int_{-\infty}^{\infty} \binom{\alpha}{\tau + \mu} {}_aD_z^{\alpha-\mu-\tau} f(z) \cdot {}_aD_z^{\tau+\mu} g(z)\, d\tau, \qquad (2.7)$$

where $\alpha, \mu \in \mathbb{C} \backslash \mathbb{Z}^-$. It assumes an elegant form for $\mu = 0$ (see Osler [136]). By setting $\alpha = 0$ formula (2.7) readily reduces to (see [87, p. 16])

$$f(z)g(z) = \frac{1}{\pi} \int_{-\infty}^{\infty} \frac{\sin \pi(\mu + \tau)}{\mu + \tau} {}_aD_z^{-\mu-\tau} f(z) \cdot {}_aD_z^{\mu+\tau} g(z)\, d\tau \qquad (\mu \notin \mathbb{Z}^-). \quad (2.8)$$

Recently Kalia and his coworkers deduced from the Leibniz rule (2.4) a number of interesting expansion formulae associated with the Gamma function $\Gamma(z)$, the *Psi function* $\psi(z) := \Gamma'(z)/\Gamma(z)$, with the *incomplete Gamma function* $\gamma(a, z)$, defined by

$$\gamma(a, z) = \int_0^a t^{z-1} e^{-t}\, dt \qquad (\Re e\; z > 0), \qquad (2.9)$$

as well as with the *entire incomplete Gamma function* $\gamma^*(a, z)$, defined by

$$a^z \gamma^*(a, z) = \frac{\gamma(a, z)}{\Gamma(z)} \qquad (|\arg(z)| \le \pi - \epsilon; 0 < \epsilon < \pi). \qquad (2.10)$$

One is the well-known result (see e.g. Magnus, Oberhettinger and Soni [108]) concerning the psi function (recall (1.33))

$$\psi(\beta - \alpha + 1) - \psi(\beta + 1) = \sum_{k=1}^{\infty} \frac{(-\alpha)_k}{k(\beta - \alpha + 1)_k} \qquad (2.11)$$

valid for $\Re e\; \beta > -1$ and $\beta - \alpha + 1 \notin \mathbb{Z}^-$. Here $(\alpha)_k = \alpha(\alpha + 1) \dots (\alpha + k - 1)$ for $k \in \mathbb{N}$.

A newer expansion, one for the entire incomplete Gamma function, is given for $\Re e\ \beta > -1; \alpha \notin \mathbb{Z}^-, \alpha - \nu \notin \mathbb{Z}$ by

$$\gamma^*(\beta-\alpha,z) = \frac{\sin\pi(\alpha-\nu)}{\pi}\Gamma(\alpha+1)\sum_{k=-\infty}^{\infty}\frac{(-1)^k\gamma^*(\beta-\nu-k,z)}{(\alpha-\nu-k)\Gamma(\nu+k+1)}. \quad (2.12)$$

The special case of (2.12) when $\beta = 0$, which easily reduces to

$$\gamma(\alpha,z) = -\frac{\sin\pi(\alpha+\nu)\cdot\sin(\pi\nu)}{\pi\sin\pi\alpha}\sum_{k=-\infty}^{\infty}\frac{z^{\alpha+\nu+k}}{\alpha+\nu+k}\gamma(-\nu-k,z) \quad (2.13)$$

with $\alpha, \alpha+\nu \notin \mathbb{Z}$, was given earlier by [134, p. 671, Entry 18]. The particular case $\nu = 0$ of (2.12) gives

$$\gamma^*(\beta-\alpha,z) = \frac{\sin\pi\alpha}{\pi}\Gamma(\alpha+1)\sum_{k=0}^{\infty}\frac{(-1)^k}{(\alpha-k)\,k!}\gamma^*(\beta-k,z) \quad (2.14)$$

valid for $\Re e\ \beta > -1; \alpha \notin \mathbb{Z}$.

An interesting application of the Leibniz rule to hypergeometric functions is given by the identity

$${}_2F_1(a,b,c;1) = \frac{\Gamma(c)\Gamma(c-a-b)}{\Gamma(c-a)\Gamma(c-b)} \quad (2.15)$$

valid for $\Re e\ c > \Re e(a+b)$, $c \notin Z_0^-$, which follows directly from the integral representation (which may be established by the Leibniz rule – see [121, p. 114])

$${}_2F_1(a,b,c;z) = \frac{\Gamma(c)}{\Gamma(a)\Gamma(c-a)}\int_0^1 u^{a-1}(1-u)^{c-a-1}\cdot(1-uz)^{-b}du \quad (2.16)$$

$$(\Re e\ c > \Re e\ a > 0,\ |z| < 1),$$

or by applying the Leibniz rule for fractional integrals to the product of $f(x) = x^\mu$ and $g(x) = x^\lambda$, $\lambda, \mu \geq 0$, yielding

$$\frac{\Gamma(\lambda+\mu+1)}{\Gamma(\lambda+\mu+\nu+1)} = \frac{\Gamma(\mu+1)}{\Gamma(\mu+\nu+1)}{}_2F_1(-\lambda,\nu,\mu+\nu+1;1).$$

The more conventional notation $a = -\lambda$, $b = \nu$, $c = \mu+\nu+1$ then gives (2.15).

Observe that formula (2.15) is actually a particular case of the Shannon sampling theorem of signal analysis (see [19]). A further one is given by ([121, p. 77]), namely

$$ {}_2F_1(a,b,c;z) = (1-z)^{-b}{}_2F_1(c-a,b,c;\frac{z}{z-1}). $$

Let us finally add an unusual integral analogue of the fractional version of Taylor's theorem (see Osler [135], [87, p. 297–8])

$$ f(z) = c\sum_{k=-\infty}^{\infty} \frac{{}_aD_w^{ck+\mu}f(w)}{\Gamma(ck+\mu+1)}(z-w)^{ck+\mu} \qquad (0 < c \leq 1; \mu \in \mathbb{C}) $$

in the form

$$ f(z) = \int_{-\infty}^{\infty} \frac{{}_aD_w^{t+\mu}f(w)}{\Gamma(t+\mu+1)}(z-w)^{t+\mu}\,dt \qquad (\mu \in \mathbb{C}; |z-w| = |w-a|). \quad (2.17) $$

For fractional versions of Taylor's formula with integral remainder terms as well as with those of Lagrange type one may consult M.M. Dzherbashyan et al. [47] and J. Trujillo et al. [168] and the literature cited there.

2.2 Fractional Landau-Kallman-Rota-Hille inequalities for operators

There exists an interesting inequality due to E. Landau (1913) which has been generalized in recent years in many directions. It is customary to write it in the form

$$ |f'(x)|^2 \leq 4\ \max_{x\in[0,1]}|f(x)| \cdot \max_{x\in[0,1]}|f''(x)| $$

provided f, f' and f'' are continuous on $[0,1]$. The inequality is also valid for the spaces $C[0,\infty]$ and $L^p(0,\infty)$, $L^p(-\infty,\infty)$ for any $1 \leq p \leq \infty$. In some cases the constant "4" may be different.

Now E. Hille [82] generalized this inequality by replacing the differential operators (d/dx) and $(d/dx)^2$ by semi-group generators A and their powers of arbitrary integral orders in the form

$$ ||A^k f||^n \leq C_{n,k}^n ||f||^{n-k} ||A^n f||^k, \qquad (2.18) $$

where n and k are integers with $1 \leq k < n$, $C_{n,k}$ being a constant independent of f.

As to the concepts involved above and below, a family $\{T(t); t \geq 0\}$ of operators mapping a Banach space X (with norm $\|\cdot\|_X$) into itself, satisfying the functional equation $T(t+s) = T(t)T(s)$ for all $s, t \geq 0$, $T(0) = I$ (identity operator), together with the strong continuity property

$$\lim_{t \to 0+} \|T(t)f - f\|_X = 0 \quad (f \in X),$$

is said to be a (C_0)-*semigroup* of operators. Its *infinitesimal generator* A is defined by $Af = s - \lim_{h\to 0+} h^{-1}[T(h) - I]f = T'(0+)f$ for all $f \in X$ for which this limit exists, so for all $f \in D(A)$.

The fractional power $(-A)^\alpha$ of order $\alpha > 0$ can be defined by

$$(-A)^\alpha f = \lim_{\varepsilon \to 0+} \frac{1}{C_{\alpha,m}} \int_\varepsilon^\infty \frac{[I - T(t)]^m f}{t^{\alpha+1}} dt \tag{2.19}$$

for each $f \in X$ for which the limit exists in the X-norm. Here $0 < \alpha < m$, $C_{\alpha,m}$ is the constant of (3.24), and so (2.19) can be regarded as an abstract generalization of the Marchaud fractional derivative of (3.22). Alternatively this power may be defined by

$$(-A)^\alpha f = \lim_{t \to 0+} \frac{[I - T(t)]^\alpha f}{t^\alpha} \tag{2.20}$$

which is motivated by the Liouville-Grünwald fractional derivative defined in (3.1).

The fractional version of (2.18), which is due to Trebels and Westphal [167], reads: *Let* $\{T(t);\ t \geq 0\}$ *be a uniformly bounded* ($\|T(t)\| \leq M$)(C_0)-*semigroup of operators,* A *its infinitesimal generator. If* $f \in D((-A)^\gamma)$ *and* $0 < \alpha < \gamma$, *then*

$$\|(-A)^\alpha f\|^\gamma \leq C^\gamma_{\gamma,\alpha} \|f\|^{\gamma-\alpha} \|(-A)^\gamma f\|^\alpha, \tag{2.21}$$

where $C_{\gamma,\alpha}$ *is a certain constant depending on* γ *and* α. Applying (2.21) with $\alpha = 1$, $\gamma = 2$ and $M = 1$ yields

$$\|Af\|^2 \leq \left(\frac{2}{\log 2}\right)^2 \|f\| \|A^2 f\|$$

with $C_{2,1} = (2/\log 2) \sim 2.885$ instead of the constant 2 (which is better).

As a further application one may deduce Bernstein's inequality for *fractional* derivatives from the classical one for integers. The latter reads

$$\| \sum_{k=-n}^{n} (ik)^r c_k e^{ikx} \| \leq n^r \|t_n(x)\| \qquad (r = 1, 2, \dots), \tag{2.22}$$

where $t_n(x) = \sum_{k=-n}^{n} c_k e^{ikx}$ is a trigonometric polynomial of degree n, the norm being either $C_{2\pi}$ or $L^p_{2\pi}$, $1 \le p < \infty$.

As to the fractional version, take the particular *translation* semigroup $T(t)f(x) = f(x-t)$ which has infinitesimal generator $A = (-d/dx)$. Then for fractional α

$$(-A)^\alpha t_n(x) = t_n^{(\alpha)}(x) = \sum_{k=-n}^{n} (ik)^\alpha c_k e^{ikx}$$

which is just the Riemann-Liouville or Liouville-Grünwald derivative of order α. Thus, by the foregoing theorem and (2.22) we have for $0 < \alpha < r$,

$$\|t_n^{(\alpha)}\|^r \le C^r_{r,\alpha} \|t_n\|^{r-\alpha} \|t_n^{(r)}\|^\alpha \le C^r_{r,\alpha} n^{r\alpha} \|t_n\|^r,$$

which gives

$$\|t_n^{(\alpha)}\| \le C_{r,\alpha} n^\alpha \|t_n\|.$$

Replacing $L^p_{2\pi}$ by $L^p(-\infty, +\infty)$ and t_n by entire functions of exponential type $\le n$, one arrives at results of Lizorkin [104]; see also Junggeburth-Scherer-Trebels [86], as well as Section 3.4.

Let us finally remark that the above argument may be extended to groups of operators using the analysis of Westphal [173, II]. Inequalities estimating intermediate derivatives via a higher derivative and the function itself are also referred to as Hadamard (1914) - Kolmogorov (1939) type inequalities (also in the more general instance when the terms of the inequalities are taken with respect to different norms). The literature in this respect is quite large; see e.g. [153, p. 313] and the literature cited there. It includes papers by Hardy, Landau and Littlewood (1935), S.P. Geisberg (1965), R.J. Hughes (1977), G.G. Magaril-Il'yaev and V.M. Tikhomirov (1981). The case for Marchaud-type derivatives is quite popular. For a recent and excellent account of all aspects of the classical Landau-Kolmogorov inequality see S. Bagdasarov [4], also M.K. Kwong and A. Zettl [99].

2.3 The behaviour of semigroup operators at zero and infinity with rates

Fractional integration may be used to investigate methods of summation of series, integrals and functions (see e.g. [153, pp. 276 ff, 304, 314 f] and the extended literature cited there). Thus for a function $\varphi : \mathbb{R}_+ \to \mathbb{C}$ one says $\varphi(t) \to L$ as $t \to 0+$ in the sense of Cesàro (C, α)-summability provided $(C^\alpha \varphi)(t) := \frac{\alpha}{t^\alpha} \int_0^t (t-u)^{\alpha-1} \varphi(u) du \to L$ as $t \to 0+$.

For semigroup operators, apart from the strong limit s-$\lim_{t\to 0+} T(t)f = f$ for $f \in X$ one is interested in the limit of the Cesàro means

$$C_T^\alpha(t)f := \frac{\alpha}{t^\alpha}\int_0^t (t-u)^{\alpha-1}T(u)f\,du \quad (\alpha > 0) \tag{2.23}$$

for $t \to 0+$ as well as for $t \to \infty$.

Concerning the *approximation* theoretical behaviour of $\{T(t); t \geq 0\}$ and $\{C_T^\alpha(t); t \geq 0\}$ for $t \to 0+$, it is connected as follows:

$$\begin{aligned} &\|T(t)f - f\|_X = \begin{cases} \mathcal{O}(t) \\ o(t) \end{cases} \\ \Longleftrightarrow\quad &\|C_T^\alpha(t)f - f\|_X = \begin{cases} \mathcal{O}(t) \\ o(t) \end{cases} \\ \Longleftrightarrow\quad &\begin{cases} f \in D(A) \\ f \in N(A) \end{cases} \end{aligned} \tag{2.24}$$

provided the Banach space X is reflexive. Otherwise $D(A)$ above has to be replaced by the completion of $D(A)$ relative to X, namely $\widetilde{D(A)}^X$ (for this concept see [15] and the literature cited there). Above, $N(A) = \{f \in D(A); Af = \theta\}$ is the null space of A.

As to the *ergodic* theoretical behaviour of $\{T(t); t \geq 0\}$ for $t \to \infty$,

$$s - \lim_{t\to\infty} C_T^\alpha(t)f = Pf \tag{2.25}$$

for each $f \in X$ and $\alpha \geq 1$, where P is the linear, bounded projection of X onto the kernel $N(A)$ parallel to $\overline{R(A)}$, the closure of the range $R(A)$. The result is true provided $\{T(t); t \geq 0\}$ is uniformly bounded and X is reflexive.

As to the *rate* of approximation of $C_T^\alpha(t)f$ to Pf for $t \to \infty$, the study of which was initiated in [29], there holds for $\alpha \geq 1$,

$$\begin{aligned} &\|t^{-1}\int_0^t T(u)f\,du - Pf\|_X = \begin{cases} \mathcal{O}(1/t) \\ o(1/t) \end{cases} \\ \Longleftrightarrow\quad &\|\alpha t^{-\alpha}\int_0^t (t-u)^{\alpha-1}T(u)f - Pf\| = \begin{cases} \mathcal{O}(1/t) \\ o(1/t) \end{cases} \\ \Longleftrightarrow\quad &\begin{cases} f \in D(B) \\ f \in N(B) = N(A), \text{i.e.}, Pf = f. \end{cases} \end{aligned} \tag{2.26}$$

Above, the operator B is an appropriate extension of the inverse of the generator, A^{-1}, restricted to $\overline{R(A)}$. [More precisely, B is the closed, densely defined operator mapping $f = Ag + Pf \in R(A) \oplus N(A)$ onto $g \in D(A) \cap N(P)$ with $Pg = \theta$.]

There is a most surprising connection between the approximation and ergodic theoretical behaviour of the semigroup $\{T(t); t \geq 0\}$, thus between the approximation of $T(t)f$ to f for $t \to 0+$ and of $(1/t)\int_0^t T_B(u)f du$ to Pf for $t \to \infty$, where $\{T_B(t);\, t \geq 0\}$ is the semigroup generated by B.

It was raised as an open problem in a different setting (astronomy) on the occasion of a colloquium talk given at Aachen in the early sixties by Prof. R. Kurth and established in the doctoral thesis of A. Gessinger [62].

In this respect, first note that the resolvent $R(\lambda; A) = (\lambda I - A)^{-1}$ of A can be represented as the Laplace transform (see Section 5.3) of $\{T(t); t \geq 0\}$ for each $\Re e\, \lambda > 0$ and $f \in X$ in the form

$$R(\lambda; A)f = \int_0^\infty e^{-\lambda t} T(t) f dt.$$

The Abelian mean ergodic theorem in this frame states that for each $f \in X$ the strong Abel limit

$$s - \lim_{\lambda \to \infty} \lambda R(\lambda; A) f = f.$$

The fundamental connection between $R(\lambda; A)$ and the resolvent $R(\lambda; B)$ of B is

$$\lambda R(\lambda; A)f - Pf = f - \lambda^{-1} R(\lambda^{-1}; B) f \quad (\Re e\, \lambda > 0; f \in X).$$

Further, if the operator A generates a holomorphic semigroup $\{T_A(t); t \geq 0\}$ (see [12, p. 15] or [15] for this concept) so does the operator B, to be denoted by $\{T_B(t); t \geq 0\}$, and conversely. On top, the latter semigroup associated with B can be given explicitly via

$$T_B(t) = f + Pf - \sqrt{t} \int_0^\infty \frac{J_1(2\sqrt{tu})}{\sqrt{u}} T_A(u) f du,$$

$J_1(u)$ being the Bessel function of first kind.

The fundamental connection between $T_A(t)$ and $T_B(t)$ announced above, now reads:

Theorem 2.1. *If $\{T_A(t); t \geq 0\}$ is a holomorphic semigroup with generator A acting on the reflexive Banach space X, then:*

$$\|T_A(t)f - f\|_X = \begin{cases} \mathcal{O}(t) \\ o(t) \end{cases} \quad (t \to 0+)$$

$$\iff \|t^{-1}\int_0^t T_B(u)f du - Pf\|_X = \begin{cases} \mathcal{O}(t^{-1}) \\ o(t^{-1}) \end{cases} \quad (t \to \infty)$$

$$\iff \|\lambda R(\lambda; A)f - f\|_X = \begin{cases} \mathcal{O}(\lambda^{-1}) \\ o(\lambda^{-1}) \end{cases} \quad (\lambda \to \infty)$$

$$\iff \|\lambda R(\lambda; B)f - Pf\|_X = \begin{cases} \mathcal{O}(\lambda) \\ o(\lambda) \end{cases} \quad (\lambda \to 0+)$$

$$\iff \begin{cases} f \in D(A) \\ f \in N(A) = N(B). \end{cases}$$

The foregoing matter should be of special interest in the physical applications sketched in Section 4.7 since often semigroup and ergodic theory methods are applied.

As an application to Theorem 2.1 consider the Weierstrass semigroup

$$[W_A(t)f](x) = \sum_{k=-\infty}^{\infty} e^{-tk^2} f^\wedge(k)e^{ikx} = \frac{1}{2\pi}\int_{-\pi}^{\pi} f(x-u)\vartheta_3(u,t)du \tag{2.27}$$

with $x \in \mathbb{R}$ and $t > 0$ where $\vartheta_3(x,t) = \sum_{k=-\infty}^{\infty} e^{-tk^2}e^{ikx}$ is Jacobi's theta function, and

$$f^\wedge(k) = \frac{1}{2\pi}\int_{-\pi}^{\pi} f(u)e^{-iku}du$$

are the Fourier coefficients of $f \in L^2_{2\pi}$. It is well-known that the infinitesimal generator is given by $[Af](x) = f''(x)$ with domain $D(A) = \{f \in L^2_{2\pi}; f \in AC^1_{2\pi}, f'' \in L^2_{2\pi}\}$. The associated operator B turns out to be (see [15])

$$[Bf](x) = -\sum_{\substack{k=-\infty \\ k\neq 0}}^{\infty} k^{-2} f^\wedge(k)e^{ikx} \quad (x \in \mathbb{R}) \tag{2.28}$$

for which $\|Bf\|_{L^2_{2\pi}} \le \|f\|_{L^2_{2\pi}}$, so that B is bounded. Since $\{W_A(t); t \ge 0\}$ is holomorphic, the operator B also generates a holomorphic semigroup, namely $\{W_B(t); t \ge 0\}$, given by

$$[W_B(t)f](x) = f^\wedge(0) + \sum_{\substack{k=-\infty \\ k\neq 0}}^{\infty} f^\wedge(k)e^{-tk^{-2}}e^{ikx} \quad (x, t \in \mathbb{R}), \tag{2.29}$$

where the exponent $-tk^2$ of (2.27) is now $-tk^{-2}$. The projector is given by $Pf = f^\wedge(0)$, and the resolvent by (see [16])

$$\begin{aligned}\lambda[R(\lambda, B)f](x) &= f(x) + f^\wedge(0) - \lambda^{-1}[R(\lambda^{-1}; A)f](x) \\ &= f^\wedge(0) + \sum_{\substack{k=-\infty \\ k\neq 0}}^{\infty} \lambda(\lambda + k^{-2})^{-1} f^\wedge(k)e^{ikx}.\end{aligned}$$

Corollary 2.2. *The following assertions are equivalent for $f \in L^2_{2\pi}$ and any $\alpha > 0$, $\beta > 0$:*

(i) $\|W_A(t)f - f\|_{L^2_{2\pi}} = \begin{cases} \mathcal{O}(t) \\ o(t) \end{cases} \quad (t \to 0+),$

(ii) $\|C^{\alpha}_{W_A}(t)f - f\|_{L^2_{2\pi}} = \begin{cases} \mathcal{O}(t) \\ o(t) \end{cases} \quad (t \to 0+),$

(iii) $\|t^{-1}\int_0^t W_B(u)f du - f^\wedge(0)\|_{L^2_{2\pi}} = \begin{cases} \mathcal{O}(t^{-1}) \\ o(t^{-1}) \end{cases} \quad (t \to \infty),$

(iv) $\|C^{1+\beta}_{W_B}(t)f - f^\wedge(0)\|_{L^2_{2\pi}} = \begin{cases} \mathcal{O}(t^{-1}) \\ o(t^{-1}) \end{cases} \quad (t \to \infty),$

(v) $\|\lambda R(\lambda; A)f - f\|_{L^2_{2\pi}} = \begin{cases} \mathcal{O}(\lambda^{-1}) \\ o(\lambda^{-1}) \end{cases} \quad (\lambda \to \infty),$

(vi) $\|\lambda R(\lambda; B)f - f^\wedge(0)\|_{L^2_{2\pi}} = \begin{cases} \mathcal{O}(\lambda) \\ o(\lambda) \end{cases} \quad (\lambda \to 0+),$

(vii) $\begin{cases} f \in D(A) = \{f \in L^2_{2\pi}; f \in AC^1_{2\pi}, f'' \in L^2_{2\pi}\} \\ f(x) = f^\wedge(0) \text{ a.e.} \end{cases}$

Observe that $w(x,t) := [W_A(t)f](x)$ of (2.27) is the solution of the heat equation $(\partial/\partial t)w(x,t) = (\partial^2/\partial x^2)w(x,t)$, $-\pi \le x \le \pi$, $t > 0$ with the boundary conditions $w(-\pi,t) = w(\pi,t)$, $(\partial/\partial x)w(-\pi,t) = (\partial/\partial x)w(\pi,t)$

and the initial condition $\lim_{t\to 0+} w(x,t) = f(x)$. Then assertions (i), (ii) and (iii) of Corollary 2.2 state that either $W_A(t)f(x)$ or the arithmetic mean $t^{-1}\int_0^t W_A(u)f du$ of the solution $W_A(t)f$ (having generator A) tends to the initial value $f(x)$ with the rate $\mathcal{O}(t)$ as $t \to 0+$ iff the mean $t^{-1}\int_0^t W_B(u)f du$ associated with the generator B (namely (2.29), (2.28)) tends to $f^\wedge(0) = (1/2\pi)\int_{-\pi}^{\pi} f(u)du$ with the rate $\mathcal{O}(t^{-1})$ as $t \to \infty$.

Let us consider now Abel summability of a uniformly bounded (C_0)-semigroup $\{T(t);\, t \geq 0\}$ with generator A acting on a Banach space X, the integral order powers of the resolvent $R(\lambda; A)$ now being extended to the fractional case $\gamma > 0$, thus

$$R^\gamma(\lambda; A)f = \frac{1}{\Gamma(\gamma)}\int_0^\infty t^{\gamma-1}e^{-\lambda t}T(t)f dt \quad (f \in X). \tag{2.30}$$

In this respect H. Komatsu [98] showed that

$$s - \lim_{\lambda\to 0+} \lambda^\gamma R^\gamma(\lambda; A)f \text{ exists } \Leftrightarrow f \in X_0,$$

where $X_0 = N(A) \oplus \overline{R(A)}$. If so, the limit is equal to Pf.

The mean ergodic theorem for $\lambda^\gamma R^\gamma(\lambda; A)f$ with rates, first established by U. Westphal [176], states that

Theorem 2.3. *There holds for $f \in X_0$, $\gamma > 0$ and $\lambda \to 0+$*

$$\|\lambda^\gamma R^\gamma(\lambda; A)f - Pf\|_X = \begin{cases} \mathcal{O}(\lambda^\gamma) \\ o(\lambda^\gamma) \end{cases}$$

$$\Longleftrightarrow \begin{cases} f \in \widetilde{D(B^\gamma)}^{X_0} \quad \textit{or } f \in D(B^\gamma) \textit{ if } X \textit{ is reflexive} \\ f \in N(A) \end{cases} \tag{2.31}$$

As to the above, $(-A)^\gamma$ is defined either by (2.19) or (2.20). Further, observe that $R^\gamma(\lambda; A)f \in D((-A)^\gamma)$ for $f \in X$, and

$$(-A)^\gamma R^\gamma(\lambda; A)f = [I - \lambda R(\lambda; A)]^\gamma f.$$

Then the operator B^γ is defined by $B^\gamma f = g$ where g is the unique element in $D((-A)^\gamma) \cap \overline{R(A)}$ satisfying $(-A)^\gamma g = f - Pf$. Note that B^γ is closed with $\overline{D(B^\gamma)} = X_0$, and $N(B^\gamma) = N(A)$.

It is interesting to observe that the rate of approximation in (2.31) increases with the exponent γ of the resolvent $R^\gamma(\lambda; A)$. This is not the case with the order α of the Cesàro means of (2.23).

Let us now apply Theorem 2.3 to the (C_0)-semigroup of translations on $L^2(0,\infty)$, namely $[T_A(t)f](x) = f(x-t)$ if $x > t \geq 0$, and $[T_A(t)f](x) = 0$, if $0 < x < t$ with $[Af](x) = -f'(x)$, domain $D(A) = \{f \in L^2(0,\infty); f \in AC_{\text{loc}}, f' \in L^2(0,\infty)\}$, and $\|T_A(t)\| \leq 1$, $N(A) = \{0\}$, $X_0 = L^2(0,\infty)$, the projector P being the null operator. Now for $f \in L^2(0,\infty)$, $\gamma > 0$,

$$[[I - T(t)]^\gamma f](x) = \sum_{j=0}^{[x/t]} (-1)^j \binom{\gamma}{j} f(x - jt),$$

so that $(-A)^\gamma f$ is the strong fractional Grünwald-Liouville derivative of f of order γ. In terms of the Laplace transform $(-A)^\gamma$ is characterized by (see Section 5.3) $D((-A)^\gamma) = \{f \in L^2(0,\infty); \exists g \in L^2(0,\infty)$ such that $s^\gamma f_{\mathcal{L}}^\wedge(s) = g_{\mathcal{L}}^\wedge(s), \Re e\ s > 0\}$, $[(-A)^\gamma f]_{\mathcal{L}}^\wedge(s) = s^\gamma f_{\mathcal{L}}^\wedge(s) \quad, \Re e\ s > 0$.

The operators $R^\gamma(\lambda; A)$ and their Laplace transforms are given for $f \in L^2(0,\infty)$, $\Re e\ s > 0$ by

$$\begin{aligned} [R^\gamma(\lambda; A)f](x) &= \frac{1}{\Gamma(\gamma)} \int_0^x t^{\gamma-1} e^{-\lambda t} f(x-t)dt \\ [R^\gamma(\lambda; A)f]_{\mathcal{L}}^\wedge(s) &= f_{\mathcal{L}}^\wedge(s)/(\lambda+s)^\gamma. \end{aligned}$$

Again, using the Laplace transform (cf. Section 5.3), it follows that $B^\gamma f = {}_0I_x^\gamma f$ and $D(B^\gamma) = \{f \in L^2(0,\infty);\ {}_0I_x^\gamma f \in L^2(0,\infty)\}$. This leads to the following application involving fractional integrals of the type ${}_0I_x^\gamma e^{\lambda x} f(x)$ with the exponential weight $e^{\lambda x}$, studied e.g. in [153, pp. 195–199].

Corollary 2.4. *Let $\gamma > 0$ and $f \in L^2(0,\infty)$. One has for $\lambda \to 0+$:*

a) $\left\| \frac{1}{\Gamma(\gamma)} \int_0^x t^{\gamma-1} e^{-\lambda t} f(x-t)dt \right\|_{L^2} = o(1) \quad \text{iff } f = 0;$

b) $\left\| \frac{1}{\Gamma(\gamma)} \int_0^x t^{\gamma-1} e^{-\lambda t} f(x-t)dt \right\|_{L^2} = \mathcal{O}(1)$

$\Leftrightarrow f \in D(B^\gamma) \Leftrightarrow {}_0I_x^\gamma f \in L^2(0,\infty).$

Observe that $(1/\Gamma(\gamma)) \int_0^x t^{\gamma-1} e^{-\lambda t} f(x-t)dt = e^{-\lambda x}\ {}_0I_x^\gamma e^{\lambda x} f(x)$.

3 Liouville-Grünwald, Marchaud and Riesz Fractional Derivatives

3.1 Liouville-Grünwald derivatives and their chief properties

A. K. Grünwald[b] (1867) [66] and A.V. Letnikov (1868) developed an approach to fractional differentiation for which the definition of the fractional derivative $D^\alpha f(x)$ is the limit of a fractional difference quotient, thus

$$D^\alpha f(x) = \lim_{h\to 0} \frac{\Delta_h^\alpha f(x)}{h^\alpha}, \tag{3.1}$$

$\Delta_h^\alpha f(x)$ being the right-handed difference of fractional order

$$\Delta_h^\alpha f(x) = \sum_{j=0}^{\infty} (-1)^j \binom{\alpha}{j} f(x - jh) \tag{3.2}$$

which coincides with the classical r-th right handed difference $\Delta_h^r f(x)$ for $\alpha = r \in \mathbb{N}$ (noting that $\binom{r}{j} = 0$ for $j \geq r+1$). While the arguments of Grünwald were rather formal, Letnikov's were rigorous. They showed in particular that $D^{-\alpha} f$ coincides with Liouville's expression (1.8) for $\alpha > 0$ and sufficiently good functions f. In fact, Liouville (1832) already formulated a definition of the form (3.1), however only for functions which are expressible as sums of exponentials. Although the foregoing approach was also developed further by E. L. Post (1930), it does not seem to have received general attention until the work of Westphal (1974) and Butzer and Westphal (1975) (see however also the earlier work of K. F. Moppert (1953), N. Stuloff (1951), and M. Mikolás (1963) — see [120] 1975). Observe that the difference (3.2) is generally not defined for $\alpha < 0$. Thus for the example $f(x) \equiv 1$, the sum

$$\sum_{j=0}^{r} (-1)^j \binom{\alpha}{j} = \frac{1}{\Gamma(1-\alpha)} \frac{\Gamma(r+1-\alpha)}{\Gamma(r+1)} \tag{3.3}$$

diverges for $r \to \infty$ if $\alpha < 0$. In this section we confine the matter to periodic functions, treating definition (3.1) in the norm-topologies of $C_{2\pi}$ and $L_{2\pi}^p$, $1 \leq p < \infty$. (Note that the series (3.2) always converges with respect to these norms.) Thus this approach is a global and not a pointwise one; it enables one

[b] According to Lavoie-Osler-Tremblay [101], the Grünwald approach is "the most difficult, yet in some ways the most natural approach to a representation for fractional differentiation". This approach is carried out in the frame of 2π-periodic functions below. The Butzer-Westphal treatment [30] via fractional differences was also incorporated into Samko et al [153, pp. 371–381].

to present a fully rigorous and thorough approach to fractional calculus using elementary means of Fourier analysis. It extends the matter in the integral case (i. e., $\alpha = r$) as considered in [25] (Chapter 10); see also [11] (1961).

The background to this approach is best seen by transfering definition (3.1) into the setting of Fourier expansions of 2π-periodic functions. If

$$f(x) \sim \sum_{k=-\infty}^{\infty} f^\wedge(k) e^{ikx},$$

$$f^\wedge(k) = [f(\cdot)]^\wedge(k) := \frac{1}{2\pi} \int_{-\pi}^{\pi} e^{-iku} f(u)\, du \qquad (k \in \mathbb{Z})$$

defining the finite Fourier transform of f (or Fourier-coefficients of f), one is led to introduce the fractional derivative $D^\alpha f$ for $f \in C_{2\pi}$ or $L^p_{2\pi}$ by

$$D^\alpha f(x) \sim \sum_{k=-\infty}^{\infty} f^\wedge(k)(ik)^\alpha e^{ikx} \tag{3.4}$$

(noting that $(d/dx)^r e^{ikx} = (ik)^r e^{ikx}$ for $\alpha = r$) or, equivalently, by

$$[D^\alpha f(\cdot)]^\wedge(k) = (ik)^\alpha f^\wedge(k) \qquad (k \in \mathbb{Z}). \tag{3.5}$$

To connect definition (3.1) with this set-up, instead of inserting the factor $(ik)^\alpha$ one could also take an approximation for it, i. e. ,

$$(ik)^\alpha = \lim_{h \to 0} \left(\frac{1 - e^{-ikh}}{h} \right)^\alpha, \tag{3.6}$$

suggesting as a definition

$$D^\alpha f(x) \sim \lim_{h \to 0} \sum_{k=-\infty}^{\infty} \left(\frac{1 - e^{-ikh}}{ikh} \right)^\alpha f^\wedge(k)(ik)^\alpha e^{ikx}. \tag{3.7}$$

This is indeed reasonable since (see [30, p. 126])

$$[\Delta_h^\alpha f(\cdot)]^\wedge(k) = (1 - e^{-ikh})^\alpha f^\wedge(k) \quad (k \in \mathbb{Z}). \tag{3.8}$$

Returning to the classical instance $\alpha = r \in \mathbb{N}$, it was Riemann who (in the case $r = 2$) first defined $D^r f$ by

$$\lim_{h \to 0} \frac{\Delta_h^r f(x)}{h^r} = D^r f(x)$$

in the pointwise sense. If the r-th ordinary derivative, defined recursively by $f^{(r)}(x) = \lim_{h\to 0}[f^{(r-1)}(x+h) - f^{(r-1)}(x)]/h$, exists at a point $x = x_0$, so does the pointwise (Riemann) derivative $D^r f$ and both are equal. But the converse is not necessarily valid. However, the existence of $D^r f$ in the uniform norm is equivalent to the fact that the r-th ordinary derivative $f^{(r)}(x)$ exists for *all* x and is continuous. A corresponding result holds with respect to the $L^p_{2\pi}$-norm. This again suggests that the fractional derivative $D^\alpha f$ can be handled by considering the limit in norm. If this is the case, then

$$\lim_{h\to 0} h^{-\alpha}[\Delta_h^\alpha f(\cdot)]^\wedge(k) = [D^\alpha f(\cdot)]^\wedge(k) \qquad (k \in \mathbb{Z}),$$

or $(ik)^\alpha f^\wedge(k) = [D^\alpha f(\cdot)]^\wedge(k)$ by (3.8) and (3.6), giving the connection with (3.5).

Thus, the α-th strong Liouville-Grünwald derivative $D^\alpha f$ of f will be that function g in $C_{2\pi}$ or $L^p_{2\pi}$, $1 \le p < \infty$, respectively, for which the limit

$$\lim_{h\to 0+} \|h^{-\alpha}\Delta_h^\alpha f - g\| = 0 \tag{3.9}$$

exists. For simplicity we shall briefly write $D^\alpha f \in C_{2\pi}$ or $D^\alpha f \in L^p_{2\pi}$, respectively.

To show how definition (3.9) is correctly related to formula (3.4) we shall prove that $((1 - e^{-ikh})/(ikh))^\alpha$ is the finite Fourier transform of some function $\chi_\alpha(x; h)$ which is 2π-periodic, integrable, and behaves properly as h approaches zero. These results are stated in Props. 3.1 through 3.3 below.

As to the fractional integral, since an r-fold indefinite integration of e^{ikx} yields $e^{ikx}/(ik)^r$, one is led to introduce it by, following H. Weyl [177],

$$I^\alpha f(x) \sim \sum_{k=-\infty}^{\infty}{}' f^\wedge(k) e^{ikx}/(ik)^\alpha \qquad (\alpha > 0), \tag{3.10}$$

(the dash indicating that the term $k = 0$ is omitted), or equivalently by

$$[I^\alpha f(\cdot)]^\wedge(k) = \begin{cases} (ik)^{-\alpha} f^\wedge(k), & k = \pm 1, \pm 2, \ldots \\ 0, & k = 0. \end{cases} \tag{3.11}$$

To give the formal definition (3.10) a more correct interpretation, one customarily defines for $f \in L^p_{2\pi}, 1 \le p < \infty$ (or $C_{2\pi}$) and $\alpha > 0$,

$$I^\alpha f(x) = (f * \psi_\alpha)(x) := \frac{1}{2\pi}\int_0^{2\pi} f(x-u)\psi_\alpha(u)\, du \tag{3.12}$$

$$\psi_\alpha(x) := \sum_{k=-\infty}^{\infty}{}' \frac{e^{ikx}}{(ik)^\alpha} = 2\sum_{k=1}^{\infty} \frac{\cos(kx - \alpha\pi/2)}{k^\alpha} \tag{3.13}$$

the series being convergent for all $x \in (0, 2\pi)$, $\alpha > 0$, uniformly so on $\epsilon \leq x \leq 2\pi - \epsilon$, $\epsilon > 0$, so that it represents the Fourier series of ψ_α. Thus

$$[\psi_\alpha(\cdot)]^\wedge(k) = \begin{cases} (ik)^{-\alpha}, & k = \pm 1, \pm 2, \ldots \\ 0, & k = 0, \end{cases}$$

where $(ik)^\alpha = |k|^\alpha \exp(\frac{\alpha\pi i}{2} \text{sign } k)$ for $k \in \mathbb{Z}$.

Hence the convolution and uniqueness theorems give that definition (3.12) is consistent with (3.10) for any $\alpha > 0$, noting again that (3.11) holds. Since $\psi_\alpha \in L^1_{2\pi}$ (see [25, p. 426]), $I^\alpha f$ belongs to $L^p_{2\pi}$ or $C_{2\pi}$ if f does so, since the right side of (3.12), which is called a *Weyl*[c] *fractional integral* of order α, is a convolution product.

3.2 A crucial proposition; basic theorems

The function $\chi_\alpha(x; h)$ mentioned after Definition (3.9) will depend upon a basic function, namely

$$p_\alpha(x) := \Delta_1^\alpha k_\alpha(x) = \begin{cases} [\Gamma(\alpha)]^{-1} \sum_{0 \leq j < x} (-1)^j \binom{\alpha}{j} (x - j)^{\alpha - 1}, & 0 < x < \infty \\ 0, & -\infty < x < 0, \end{cases} \tag{3.14}$$

where $k_\alpha(x) = [\Gamma(\alpha)]^{-1} x_+^{\alpha-1}$. It will be needed several times in this chapter. Some of its properties are collected in the following proposition, $\mathcal{F}[p_\alpha(\cdot)](v)$ denoting its Fourier transform $\int_{-\infty}^{\infty} e^{-ivu} p_\alpha(u)\, du$, $v \in \mathbb{R}$ (see Section 5.1). For references and proof see [172].

Proposition 3.1. *The function*[d] $p_\alpha(x)$ *has for* $\alpha > 0$ *the properties*

a) $p_\alpha \in L^1(\mathbb{R})$, $\int_{-\infty}^{\infty} p_\alpha(u)\, du = 1$;

b) $\mathcal{F}[p_\alpha(\cdot)](v) = \begin{cases} (iv)^{-\alpha}(1 - e^{-iv})^\alpha, & v \neq 0 \\ 1, & v = 0; \end{cases}$

c) $p_\alpha(x) \equiv 0$ *for* $x > \alpha$ *in the cases* $\alpha = 1, 2, \ldots$.

---

[c] In fact, Weyl only considered fractional integration in this form.

[d] The most difficult part of this proposition is the proof of p_α belonging to $L^1(\mathbb{R})$, a fact treated in Westphal [172] (see Section 3.2 of [175]).

If $\alpha > 1$, *then*

d) $p_\alpha(x) = \mathcal{O}(x^{-\alpha-1}) \qquad (x \to \infty)$,

e) p'_α *and* xp'_α *belong to* $L^1(\mathbb{R})$.

Now the function $\chi_\alpha(x;h)$ is defined by

$$\chi_\alpha(x;h) = \frac{2\pi}{h} \sum_{-(x/2\pi)<j<\infty} p_\alpha\left(\frac{x+2\pi j}{h}\right) \qquad (h > 0). \tag{3.15}$$

It has the properties, established in the following crucial proposition.

Proposition 3.2. *For* $\alpha > 0$ *and* $h > 0$ *one has*

a) $\chi_\alpha(\cdot;h) \in L^1_{2\pi}, \int_0^{2\pi} \chi_\alpha(u;h)\,du = 2\pi$;

b) $\|\chi_\alpha(\cdot;h)\|_{L^1_{2\pi}} \leq M$, *uniformly in* h;

c) $\lim_{h\to 0+} \int_\delta^\pi |\chi_\alpha(u;h)|\,du = 0$ *for* $\delta > 0$;

d) $[\chi_\alpha(\cdot;h)]^\wedge(k) = \begin{cases} (ikh)^{-\alpha}(1-e^{-ikh})^\alpha, & k \neq 0 \\ 1, & k = 0; \end{cases}$

e) $\chi_\alpha(x;h) = \frac{\Delta_h^\alpha \psi_\alpha(x)}{h^\alpha} + 1$.

Relation e) clarifies the need for $\chi_\alpha(x;h)$. Now the latter function is a so called approximate identity. This follows especially from

Proposition 3.3. *If* f *belongs to* $L^p_{2\pi}$, $1 \leq p < \infty$ *or to* $C_{2\pi}$, *then for any* $\alpha > 0$,

$$\lim_{h\to 0+} \|J_{h,\alpha} f - f\| = 0,$$

the norm being taken with respect to $L^p_{2\pi}$ *or* $C_{2\pi}$, *where*

$$J_{h,\alpha} f(x) := \frac{1}{2\pi} \int_0^{2\pi} \chi_\alpha(x-u;h) f(u)\,du \quad (h > 0, x \in \mathbb{R}). \tag{3.16}$$

Now to the basic results, including the monotonicity and additivity laws, and the fundamental theorem of the fractional calculus. Part a) expresses definition (3.9) in terms of the finite Fourier transform.

Theorem 3.4. *a) The following three assertions are equivalent for* $f \in L^p_{2\pi}$, $1 \le p < \infty$ *and* $\alpha > 0$:

(i) $D^\alpha f \in L^p_{2\pi}$;

(ii) $\exists g \in L^p_{2\pi} : (ik)^\alpha f^\wedge(k) = g^\wedge(k),\ k \in \mathbb{Z}$;

(iii) $\exists g \in L^p_{2\pi} : f(x) - f^\wedge(0) = I^\alpha g(x)$ *a. e.*

If (i) or (ii) are satisfied, then $D^\alpha f = g$.

b) There holds for $f \in L^p_{2\pi}, \alpha, \beta > 0$:

(i) *(Monotonicity law)*
If $D^\alpha f \in L^p_{2\pi}$, *then* $D^\beta f \in L^p_{2\pi}$ *for any* $0 < \beta < \alpha$;

(ii) *(Additivity law)*
$D^\alpha D^\beta f = D^{\alpha+\beta} f$ *whenever one of the two sides is meaningful;*

(iii) *(Fundamental theorem of fractional calculus)*

$$D^\alpha (I^\alpha) f = f - f^\wedge(0) = I^\alpha (D^\alpha f), \tag{3.17}$$

the latter equality holding if $D^\alpha f \in L^p_{2\pi}$.

c) If $f \in L^p_{2\pi}$ *for* $1 < p < \infty$, *then* $D^\alpha f \in L^p_{2\pi}$ *if and only if*

$$\|\Delta^\alpha_h f\|_{L^p_{2\pi}} = \mathcal{O}(h^\alpha) \qquad (h > 0).$$

d) If $f \in L^1_{2\pi}$, *then the following statements are equivalent for* $\alpha > 0$:

(i) $$\begin{cases} D^{\alpha-1} f \in BV_{2\pi} \cap L^1_{2\pi}, & \alpha > 1 \\ f \in BV_{2\pi}, & \alpha = 1 \\ I^{1-\alpha} f \in BV_{2\pi}, & \alpha < 1; \end{cases}$$

(ii) $\exists \mu \in BV_{2\pi}$ *such that* $(ik)^\alpha f^\wedge(k) = \mu^\vee(k) := \frac{1}{2\pi} \int_{-\pi}^{\pi} e^{-iku} d\mu(u)$, $k \in \mathbb{Z}$

(iii) $\|\Delta^\alpha_h f\|_{L^1_{2\pi}} = \mathcal{O}(h^\alpha) \qquad (h > 0)$;

(iv) $\exists \mu \in BV_{2\pi}$ *such that for every* $s \in C_{2\pi}$

$$\lim_{h \to 0+} \int_0^{2\pi} s(u)[h^{-\alpha} \Delta^\alpha_h f(u)]\, du = \int_0^{2\pi} s(u)\, d\mu(u).$$

Above $BV_{2\pi}$ is the space of all functions μ which are of bounded variation on every finite interval, are normalized by $\mu(x) = [\mu(x+0) + \mu(x-0)]/2$, and satisfy $\mu(x + 2\pi) = \mu(x) + \mu(2\pi) - \mu(0)$ for all x.

In connection with part b) (iii) let us remark that the formulae

$$h^{-\alpha}\Delta_h^\alpha(I^\alpha f) = J_{h,\alpha}f - f^\wedge(0) = I^\alpha(h^{-\alpha}\Delta_h^\alpha f),$$

being valid for $f \in L_{2\pi}^p$, $1 \le p < \infty$, may be regarded as precursors to (3.17). Theorem 3.4 b)(iii) is actually the counterpart of the fundamental theorem of the differential and integral calculus for the fractional case, in the sense that D^α and I^α, introduced independently of each other in this section, are precisely inverse operators provided $f^\wedge(0) = 0$. Therefore, it is justified to set $D^\alpha = I^{-\alpha}$ if $\alpha < 0$. Then one has as a generalization of Part b)(ii) that if α, β are any reals, $f^\wedge(0) = 0$ and $D^\alpha D^\beta f$ exists, then $D^\alpha D^\beta f = D^{\alpha+\beta} f$.

The foregoing results were considered for $L_{2\pi}^p$-functions. Their chief counterparts are now stated for $f \in C_{2\pi}$ all together in one theorem.

Theorem 3.5. *a) If $f \in C_{2\pi}$, then $D^\alpha f \in C_{2\pi}$ if and only if $\exists g \in C_{2\pi} : (ik)^\alpha f^\wedge(k) = g^\wedge(k)$, $k \in \mathbb{Z}$.*

b) The following statements are equivalent for $f \in C_{2\pi}$, $\alpha > 0$:

(i) $\exists g \in L_{2\pi}^\infty : (ik)^\alpha f^\wedge(k) = g^\wedge(k), \quad k \in \mathbb{Z}$,

(ii) $\exists g \in L_{2\pi}^\infty : f(x) - f^\wedge(0) = I^\alpha g(x)$ for all x,

(iii) $\|\Delta_h^\alpha f\|_{C_{2\pi}} = \mathcal{O}(h^\alpha), \quad (h \to 0+)$,

(iv) $\exists g \in L_{2\pi}^\infty$ such that for every $s \in L_{2\pi}^1$

$$\lim_{h\to 0+} \int_0^{2\pi} s(u)[h^{-\alpha}\Delta_h^\alpha f(u) - g(u)]\, du = 0,$$

i.e., $D^\alpha f = g$ exists in this weak-sense,*

(v) $I^{n-\alpha} f \in AC_{2\pi}^{n-1}$ and $(d/dx)^n I^{n-\alpha} f \in L_{2\pi}^\infty$, with $n = [\alpha] + 1$.

3.3 The point-wise Liouville-Grünwald fractional derivative

The conventional way to define a fractional derivative is via point-wise convergence. The foregoing norm-convergence approach is especially reasonable in view of the remark to part b) (iii) of Theorem 3.4. The connection of Definition (3.9) with a certain point-wise version of it, at least for $\alpha \ge 1$, is given by

Theorem 3.6. *a) If $f \in L^1_{2\pi}$ and $\alpha > 1$, then for almost all x*

$$\lim_{h \to 0+} J_{h,\alpha} f(x) = f(x).$$

b) If $f \in L^p_{2\pi}$, $1 \le p < \infty$, $\alpha \ge 1$, then $D^\alpha f \in L^p_{2\pi}$ iff there exists a $g \in L^1_{2\pi}$ such that $\Delta^\alpha_h f(x) = (h^\alpha/2\pi) \int_0^{2\pi} \chi_\alpha(x-u;h) g(u)\, du$ for all $x \in \mathbb{R}$ and $h \ge 0$, and the pointwise derivative $D^\alpha f$ belongs to $L^p_{2\pi}$.

The Grünwald-Liouville approach via fractional differences in the non-periodic case was considered by Samko [151]; see Samko et al [153, pp. 382–85].

3.4 Extensions and applications of the Liouville-Grünwald calculus

The foregoing rigorous and modern approach to the Liouville-Grünwald fractional calculus was further developed by Butzer, Dyckhoff, Görlich and Stens [13], Wilmes [178], [179], Taberski [164,165], Diaz and Osler [39], V.G Ponomarenko [138], D.P. Driavnov [44]. In particular, Butzer et al [13] defined moduli of continuity of fractional order, defined generalized Lipschitz spaces via these moduli and applied the matter to the theory of best approximation by *trigonometric* polynomials.

Basic in this respect is a *Bernstein-type inequality* for trigonometric polynomials in the fractional case. If $t_n(x)$ is a trigonometric polynomial of degree $n \in \mathbb{N}$, then it reads that

$$\|D^\alpha t_n\|_{X_{2\pi}} \le B(\alpha) n^\alpha \|t_n\|_{X_{2\pi}} \quad (\alpha > 0). \tag{3.18}$$

Needed in the proofs is further a *Jackson-type inequality* (cf. [24]) in the fractional case. It states: If $D^\alpha f \in X_{2\pi}$, then

$$E_n(f; X_{2\pi}) := \inf_{t_n} \|f - t_n\|_{X_{2\pi}} \le J(\alpha) n^{-\alpha} \|D^\alpha f\|_{X_{2\pi}}.$$

A certain generalization of inequality (3.18), is a fractional version of the so-called *M. Riesz inequality.* It states that (see Wilmes [178], [179])

$$\|D^\alpha t_n\|_{X_{2\pi}} \le (n/2)^\alpha \|\Delta^\alpha_{\pi/n} t_n\|_{X_{2\pi}}.$$

Above, $B(\alpha)$ and $J(\alpha)$ are constants depending on $\alpha > 0$.

Left-handed fractional order differences (with step 1) in the complex plane were considered by Diaz and Osler [39] in the form

$$\Delta^\alpha_1 f(z) = \sum_{j=0}^{\infty} (-1)^j \binom{\alpha}{j} f(z + \alpha - j), \quad z \in \mathbb{C},$$

with the main result

$$\Delta_1^\alpha f(z) = \frac{\Gamma(\alpha+1)}{2\pi i} \int_C \frac{f(u)\Gamma(u-z-\alpha)}{\Gamma(u-z+1)} du,$$

where the contour C envelopes the ray $L = \{u; u = z + \alpha - \xi,\ \xi \geq 0\}$ in the positive sense. Here f is assumed to be analytic in a domain containing L such that $|f(z)| \leq M|(-z)^{\alpha-p}|$ for some positive constants M and p. They also obtained a Leibniz formula for these differences.

The idea of fractional differentiation by Liouville-Grünwald allows other wide reaching generalizations as Samko et al [153, p. 446] report. Replacing the translation operator $(\tau_h f)(x) = f(x-h)$ in the definition (3.2), which can be rewritten as $\Delta_h^\alpha f(x) = (I - \tau_h)^\alpha f(x)$, by other generalized translation operators, one can obtain various forms of fractional differentiation. This idea was carried out by Butzer and Stens [26,27] for the so-called Chebyshev translation operator $\overline{\tau}_h$, defined by

$$\overline{\tau}_h f(x) = \frac{1}{2}[f(xh + \sqrt{(1-x^2)(1-h^2)}) + f(xh - \sqrt{(1-x^2)(1-h^2)})]$$

for $x, h \in [-1,1]$ with the associated fractional derivative

$$D^{(\alpha)} f = \lim_{h \to 1-} \frac{\overline{\Delta}_h^\alpha f(x)}{(1-h)^\alpha}, \quad \overline{\Delta}_h^\alpha f(x) := (-1)^{[\alpha]} \sum_{j=0}^{\infty} (-1)^j \binom{\alpha}{j} \overline{\tau}_h^j f(x).$$

The limit here is taken in the norm of the space $X = C[-1,1]$ or $X = L_w^p$ with norm $\|f\|_p = \{(1/\pi) \int_{-1}^1 |f(u)|^p w(u) du\}^{1/p}$, $w(x) = (1-x^2)^{-1/2}$. For a pointwise interpretation of $D^{(1)}$ in terms of ordinary derivatives, $D^{(1)} f(x) = (1-x^2) f''(x) - x f'(x)$; further $D^{(1/2)} f(x) = -\sqrt{1-x^2}(d/dx)(Hf)(x)$, the Hilbert transform of (3.36) now being defined by

$$(Hf)(x) = \lim_{r \to 1-} 2 \sum_{k=1}^{\infty} r^k f^\wedge(k) \sin(k \arccos x),$$

where $f^\wedge(k) = (1/\pi) \int_{-1}^1 f(u) \cos(k \arccos u) w(u) du$ is the Chebyshev transform of $f \in X$. The Chebyshev integral $I^{(\alpha)} f$ of order $\alpha > 0$ is defined by the convolution product $(I^{(\alpha)} f)(x) = (1/\pi) \int_{-1}^1 (\overline{\tau}_x f)(u) \psi_\alpha(u) w(u) du$, where $\psi_\alpha^\wedge(k) = (-1)^{[\alpha]} k^{-2\alpha}$, $k \in \mathbb{N}$. The basic theorem here is: *If* $f \in X$, *then* (i) $D^{(\alpha)} f \in X$ *iff* (ii) $\exists g_0 \in X$ *with* $(-1)^{[\alpha]} k^{2\alpha} f^\wedge(k) = g_0^\wedge(k)$, $k \in \mathbb{N}$, *iff* (iii) $\exists g_1 \in X$ *with* $f(x) = \left(I^{(\alpha)} g_1\right)(x) +$ const. (a.e.). *Then* $D^{(\alpha)} f(x) = g_i(x) - g_i^\wedge(0)$,

$i = 0, 1$. The fundamental theorem for fractional Chebyshev derivatives and integrals then reads: $D^{(\alpha)}(I^{(\alpha)}f) = f - f^\wedge(0) = I^{(\alpha)}(D^{(\alpha)}f)$, the latter equality holding if $D^{(\alpha)}f \in X$. This calculus enables one to present a simple and unified approach to the theory of best approximation by *algebraic* polynomials, which also covers the delicate behaviour on the boundary of the interval $[-1, 1]$.

Discrete fractional differences $\Delta^\alpha u_k := \sum_{j=0}^\infty (-1)^j \binom{\alpha}{j} u_{k+j}$ for sequences $\{u_k\}_{k\geq 0}$ of real numbers, introduced by A.F. Andersen (1928), the discrete fractional bounded variation spaces

$$bv_{\alpha+1} = \{\{u_k\} \in l^\infty; \quad \sum_{j=0}^{\infty} \binom{j+\alpha}{j} |\Delta^{\alpha+1} u_j| + \lim_{j\to\infty} |u_j| < \infty\}$$

and their continuous fractional counterparts $BV_\alpha(\mathbb{R}^+)$, introduced by W. Trebels [166], play important roles in various investigations of fractional differentiation in connection with Fourier- multiplier problems, of the Hankel, Jacobi-types, etc. In W. Trebels [166], G. Gasper-W. Trebels [57], [58] and H.J. Mertens-R.J. Nessel-G. Wilmes [117] also the Cossar derivative is applied.

D. Elliott [48] and D. Delbourgo-D. Elliott [38] treat algorithms for the numerical evaluation of certain Hadamard finite-part integrals (which are indeed fractional derivatives) by making use of the Grünwald fractional differences. The algorithms in question are based on the use of Bernstein polynomials; convergence rates are included.

3.5 The Marchaud fractional derivative

Marchaud's idea (1927) [111] was to try to define a fractional derivative of order α directly via ${}_{-\infty}I_x^\alpha f(x)$ of (1.8), replacing α by $-\alpha$. This suggests defining

$$I^{-\alpha} f(x) = \frac{1}{\Gamma(-\alpha)} \int_0^\infty u^{-\alpha-1} f(x-u)\, du \quad (\alpha > 0). \tag{3.19}$$

However, no matter how smooth f might be, the latter integral diverges due to the singularity of $u^{-\alpha-1}$ at the origin. Hence, following J. Hadamard (1892), first considering $0 < \alpha < 1$, one takes the "finite part" of (3.19) in the sense that one subtracts from (3.19) that part which makes it diverge, namely the term $\int_\epsilon^\infty u^{-\alpha-1} f(x)\, du = \alpha^{-1}\epsilon^{-\alpha} f(x)$ with $\epsilon \to 0+$. This yields the correct interpretation of (3.19), namely

$$\lim_{\epsilon\to 0+}\frac{1}{\Gamma(-\alpha)}\int_{\epsilon}^{\infty}u^{-\alpha-1}[f(x-u)-f(x)]\,du. \tag{3.20}$$

It will include the definition given by (1.21). Indeed, by partial integration one has (formally), noting $\Gamma(1-\alpha)=-\alpha\Gamma(-\alpha)$, that the limit expression (3.20) equals

$$\lim_{\epsilon\to 0+}\frac{1}{\Gamma(1-\alpha)}\int_{\epsilon}^{\infty}u^{-\alpha}f'(x-u)\,du={}_{-\infty}I_x^{1-\alpha}f'(x).$$

In order to extend equation (3.20) to the case $\alpha>1$, one tries to replace the first difference there by a difference of higher order l of f, with step u, i. e. ,

$$\Delta_u^l f(x):=\sum_{j=0}^{l}(-1)^j\binom{l}{j}f(x-ju). \tag{3.21}$$

In fact, the *Marchaud fractional derivative* of order α, $0<\alpha<l$, (Samko et al [153, p. 116 ff.]), reads

$$D^\alpha f(x)=\frac{1}{C_{\alpha,l}}\int_0^\infty\frac{\Delta_u^l f(x)}{u^{1+\alpha}}\,du \tag{3.22}$$

provided f is sufficiently smooth, where

$$C_{\alpha,l}=\int_0^\infty\frac{(1-e^{-u})^l}{u^{1+\alpha}}\,du \tag{3.23}$$

having the representation

$$C_{\alpha,l}:=\begin{cases}\Gamma(-\alpha)\sum_{j=1}^{l}(-1)^j\binom{l}{j}j^\alpha & (0<\alpha<l,\ l\notin\mathbb{N})\\ \frac{(-1)^{\alpha+1}}{\alpha!}\sum_{j=1}^{l}(-1)^j\binom{l}{j}j^\alpha\log j & (\alpha=1,2,\ldots,l-1).\end{cases} \tag{3.24}$$

To see that (3.22) is independent of l whenever $l > \alpha$, consider only the case $0 < \alpha < 1$. Then for any $l \in \mathbb{N}$,

$$\int_{\epsilon}^{\infty} \frac{\Delta_u^l f(x)}{u^{1+\alpha}} du = \frac{f(x)}{\alpha \epsilon^{\alpha}} + \sum_{j=1}^{l} (-1)^j \binom{l}{j} j^{\alpha} \int_{j\epsilon}^{\infty} \frac{f(x-u)}{u^{1+\alpha}} du \tag{3.25}$$

$$= \sum_{j=1}^{l} (-1)^j \binom{l}{j} j^{\alpha} \int_{j\epsilon}^{\infty} \frac{f(x-u) - f(x)}{u^{1+\alpha}} du$$

which yields the result as $\epsilon \to 0+$, noting (3.24). It should be observed that the functions $C_{\alpha,l}$ of (3.23) are practically the Stirling functions $S(\alpha, l)$ of Section 4. 3. In fact, $S(\alpha, l) = (-1)^l C_{\alpha,l}[l!\Gamma(-\alpha)]^{-1}$.

Usually, definition (3.22) can be interpreted in a more general sense in order to include larger classes of functions f for which a Marchaud derivative is defined. In fact, introducing the truncated integrals

$$D_{\varepsilon}^{\alpha,l} f(x) = \frac{1}{C_{\alpha,l}} \int_{\varepsilon}^{\infty} \frac{\Delta_u^l f(x)}{u^{1+\alpha}} du \quad (\varepsilon > 0), \tag{3.26}$$

their limit as $\varepsilon \to 0+$ may be regarded with respect to different types of convergence depending on the problems under consideration. We shall treat the case where convergence takes place in the norm of L^p, $1 \le p < \infty$, for functions f defined on the whole real line or on the half-axis $\mathbb{R}_+$. Though the latter class of functions may and will be regarded as a subclass of the former by extending f to be zero on the negative real axis, nicer and more well-rounded results may be obtained for functions defined on $\mathbb{R}_+$. Moreover, the Laplace transform is an effective tool in this case, which shall be considered briefly first. The treatment below is based on [6,7] and [173, I].

If $f \in L_{\text{loc}}(\mathbb{R}_+)$, its Laplace transform is defined by

$$\mathcal{L}[f](s) = \int_0^{\infty} e^{-su} f(u) du$$

for each complex number s for which the integral exists (see Section 5.3). Usually it suffices to confine oneself to real s. The following lemma shows how the Laplace transform acts on finite differences (3.21) and truncated integrals (3.26).

Lemma 3.7. *Let $f \in L_{\text{loc}}(\mathbb{R}_+)$ such that the Laplace transform of f is absolutely convergent for each $s > 0$. Then for $0 < \alpha < l$ and $s, t, \varepsilon > 0$,*

(i) $\mathcal{L}[t^{-\alpha}\Delta_t^l f](s) = s^\alpha \mathcal{L}[f](s) \cdot \mathcal{L}[p_{\alpha,l}](ts)$,

(ii) $\mathcal{L}[D_\varepsilon^{\alpha,l} f](s) = s^\alpha \mathcal{L}[f](s) \cdot \mathcal{L}[q_{\alpha,l}](\varepsilon s)$,

where the functions $p_{\alpha,l}$ and $q_{\alpha,l}$ are defined by

$$p_{\alpha,l}(u) = \Delta_1^l \frac{1}{\Gamma(\alpha)} u_+^{\alpha-1} = \frac{1}{\Gamma(\alpha)} \sum_{j=0}^{l} (-1)^j \binom{l}{j} (u-j)_+^{\alpha-1}$$

for $0 < \alpha \le l$ and

$$q_{\alpha,l}(u) = D_1^{\alpha,l} \frac{1}{\Gamma(\alpha)} u_+^{\alpha-1} = \frac{1}{C_{\alpha,l}\Gamma(\alpha+1)u} \sum_{j=0}^{l} (-1)^j \binom{l}{j} (u-j)_+^{\alpha}$$

for $0 < \alpha < l$. Both functions belong to $L^1(\mathbb{R}_+)$ and satisfy

$$\int_0^\infty p_{\alpha,l}(u)du = \begin{cases} 0 & \text{if} \quad 0 < \alpha < l \\ 1 & \quad \alpha = l \end{cases} \quad \text{and} \quad \int_0^\infty q_{\alpha,l}(u)du = 1.$$

Their Laplace transforms are given by

$$\mathcal{L}[p_{\alpha,l}](s) = s^{-\alpha}(1-e^{-s})^l \quad (s > 0)$$

and, respectively,

$$\mathcal{L}[q_{\alpha,l}](s) = \frac{s^{-\alpha}}{C_{\alpha,l}} \int_1^\infty u^{-\alpha-1}(1-e^{-su})^l du \quad (s > 0).$$

Part (ii) of Lemma 3.7 implies the following characterization of the Marchaud fractional derivative (see [6], [7]).

Theorem 3.8. *Let $f \in L_{\mathrm{loc}}(\mathbb{R}_+)$ such that the Laplace transform of f is absolutely convergent for each $s > 0$, and let $g \in L^p(\mathbb{R}_+)$, $1 \le p < \infty$. Then the following are equivalent for $0 < \alpha < l$:*

(i) $s^\alpha \mathcal{L}[f](s) = \mathcal{L}[g](s) \quad (s > 0)$,

(ii) ${}_0I_x^\alpha g = f$,

(iii) $D^\alpha f = g$, *i.e.,*
for each $\varepsilon > 0$, $D_\varepsilon^{\alpha,l} f \in L^p(\mathbb{R}_+)$ and $\lim_{\varepsilon\to 0+} \|D_\varepsilon^{\alpha,l} f - g\|_{L^p(\mathbb{R}_+)} = 0$.

Let us remark that the equivalence of (ii) and (iii) states that ${}_0I_x^\alpha$ and D^α are inverse operations. The implication (ii) $\Longrightarrow$ (iii) may be interpreted in the sense that if $g \in L^p(\mathbb{R}_+)$, then the integral equation ${}_0I_x^\alpha g = f$ has a solution which can be represented by the Marchaud fractional derivative $D^\alpha f$. As for the analogue of this result for functions defined on the whole real axis, note that $g \in L^p(\mathbb{R})$ implies the existence of the fractional integral ${}_{-\infty}I_x^\alpha g$, in general, only if $1 \leq p < 1/\alpha$. Otherwise one has to assume it expressly; cf. Theorem 3.10 below.

Also note that in the above theorem the function f itself need not belong to $L^p(\mathbb{R}_+)$. If, however, one is interested in a full theory for Marchaud fractional derivatives including, for example, the monotonicity and additivity laws, then it is appropriate to take a fixed L^p-space as a basis which implies that fractional derivatives are explained a priori only for functions belonging to the underlying space. Thus, considering now functions on the whole real line, we assume that the operator D^α has the domain

$$D(D^\alpha) := \{f \in L^p(\mathbb{R});\ \exists g \in L^p(\mathbb{R}) \text{ such that } \lim_{\varepsilon \to 0+} \|D_\varepsilon^{\alpha,l} f - g\|_{L^p(\mathbb{R})} = 0\}, \tag{3.27}$$

and for g as in (3.27) we set $D^\alpha f := g$.

Then, as a counterpart to Lemma 3.7, the following formulas may be regarded as representative for the whole theory: *For each* $f \in L^p(\mathbb{R})$ *and* $0 < \alpha < l$ *the convolution integrals*

$$(f * \frac{1}{t} p_{\alpha,l}(\frac{\cdot}{t}))(x) := \int_0^\infty f(x-u) p_{\alpha,l}(\frac{u}{t}) \frac{du}{t} \qquad (t > 0)$$

and

$$(f * \frac{1}{\varepsilon} q_{\alpha,l}(\frac{\cdot}{t}))(x) := \int_0^\infty f(x-u) q_{\alpha,l}(\frac{u}{\varepsilon}) \frac{du}{\varepsilon} \qquad (\varepsilon > 0)$$

belong to $D(D^\alpha)$ *and satisfy*

$$\begin{aligned} t^{-\alpha} \Delta_t^l f &= D^\alpha (f * \frac{1}{t} p_{\alpha,l}(\frac{\cdot}{t})), \\ D_\varepsilon^{\alpha,l} f &= D^\alpha (f * \frac{1}{\varepsilon} q_{\alpha,l}(\frac{\cdot}{\varepsilon})). \end{aligned}$$

Moreover, if $f \in D(D^\alpha)$, *then*

$$t^{-\alpha} \Delta_t^l f = D^\alpha f * \frac{1}{t} p_{\alpha,l}(\frac{\cdot}{t}), \tag{3.28}$$

$$D_\varepsilon^{\alpha,l} f = D^\alpha f * \frac{1}{\varepsilon} q_{\alpha,l}(\frac{\cdot}{\varepsilon}). \tag{3.29}$$

From these relations and others of similar type the following theorem is deduced. It is an application of results obtained in the more general framework of semigroups of operators; cf. [173, I], [176] and Chapter III of this volume.

Theorem 3.9. *a) For each $\alpha > 0$, D^α is a closed, densely defined linear operator in $L^p(\mathbb{R})$, $1 \le p < \infty$. There holds for $f \in L^p(\mathbb{R})$, $\alpha, \beta > 0$:*

(i) (Monotonicity law) If $f \in D(D^\alpha)$, then $f \in D(D^\beta)$ for any $0 < \beta < \alpha$ and

$$D^\beta f = \frac{1}{C_{\beta,l}} \int_0^\infty \frac{\Delta_u^l f}{u^{1+\beta}} du \quad (0 < \beta < l).$$

(ii) (Additivity law) There holds $f \in D(D^{\alpha+\beta})$ if and only if $f \in D(D^\beta)$ and $D^\beta f \in D(D^\alpha)$. In this case,

$$D^{\alpha+\beta} f = D^\alpha D^\beta f.$$

b) If $f \in L^p(\mathbb{R})$, $1 < p < \infty$, then the following statements are equivalent for $\alpha > 0$:

(i) $f \in D(D^\alpha)$,

(ii) $\|D_\varepsilon^{\alpha,l} f\|_{L^p(\mathbb{R})} = \mathcal{O}(1) \quad (\varepsilon \to 0+)$, where $0 < \alpha < l$,

(iii) there exists $g \in L^p(\mathbb{R})$ such that

$$\lim_{\delta \to 0+} \left\| \frac{1}{\Gamma(\alpha)} \int_0^\infty e^{-\delta u} u^{\alpha-1} g(\cdot - u) du - f(\cdot) \right\|_{L^p(\mathbb{R})} = 0.$$

In this event, $D^\alpha f = g$.

In Part b) (iii) of Theorem 3.9 the limit in L^p-norm of the integral

$$\frac{1}{\Gamma(\alpha)} \int_0^\infty e^{-\delta u} u^{\alpha-1} g(x-u) du$$

as $\delta \to 0+$ may be considered as a generalization of the fractional integral ${}_{-\infty}I_x^\alpha g(x)$ which need not exist, if $g \in L^p(\mathbb{R})$, as was mentioned above. In this wider sense, the fractional integral is the inverse operator of the fractional derivative defined by (3.27). This is the interpretation of the equivalence of (i) and (iii) in Theorem 3.9 b) above.

On the other hand, if for $g \in L^p(\mathbb{R})$ it is assumed that the integral ${}_{-\infty}I_x^\alpha g(x)$ does exist in the sense of Lebesgue for almost all real x, then for $0 < \alpha < l$ and $t > 0$ one has

$$t^{-\alpha}\Delta_t^l \; {}_{-\infty}I_x^\alpha g(x) = \int_0^\infty g(x-u) p_{\alpha,l}(\frac{u}{t})\frac{du}{t}$$

which implies, by Fubini's theorem, for $\varepsilon > 0$,

$$D_\varepsilon^{\alpha,l} \; {}_{-\infty}I_x^\alpha g(x) = \int_0^\infty g(x-u) q_{\alpha,l}(\frac{u}{\varepsilon})\frac{du}{\varepsilon}. \tag{3.30}$$

Note the similarity of these equations with formulas (3.28) and (3.29) above, which hold under different assumptions. The integral on the right-hand side of (3.30) belongs to $L^p(\mathbb{R})$ and converges to g in norm if $\varepsilon \to 0+$. Thus, there holds the following theorem which has its roots in the paper of Marchaud [111], cf. Samko [153, p. 125/6] and Rubin [147, p. 163].

Theorem 3.10. *Let $g \in L^p(\mathbb{R})$, $1 \le p < \infty$, and assume that the fractional integral $f(x) = {}_{-\infty}I_x^\alpha g(x)$ $(\alpha > 0)$ exists in the sense of Lebesgue for almost all real x. Then for $0 < \alpha < l$ and each $\varepsilon > 0$, $D_\varepsilon^{\alpha,l} f \in L^p(\mathbb{R})$ and*

$$\lim_{\varepsilon \to 0+} \|D_\varepsilon^{\alpha,l} f - g\|_{L^p(\mathbb{R})} = 0.$$

3.6 Equivalence of the Weyl and Marchaud fractional derivatives

At first to the situation for $0 < \alpha < 1$. If $f \in L^1_{2\pi}$, the Weyl fractional derivative

$$D^\alpha f(x) := \frac{d}{dx} I^{1-\alpha} f(x) = \frac{d}{dx}\left(\frac{1}{2\pi}\int_0^{2\pi} f(x-u)\psi_{1-\alpha}(u)\,du\right) \tag{3.31}$$

may, under (formal) differentiation under the integral sign and partial integration, be written as

$$\frac{1}{2\pi}\int_0^{2\pi} [f(x-u) - f(x)]\,\psi'_{1-\alpha}(u)\,du; \tag{3.32}$$

Samko et al [153, p. 352, 109] then speak of the *Weyl–Marchaud* derivative.

Denoting the truncated version of (3.32) by $\overline{D_\epsilon^{\alpha,1}} f(x)$, i.e.,

$$\overline{D_\epsilon^{\alpha,1}} f(x) = \frac{1}{2\pi} \int_\epsilon^{2\pi} [f(x-u) - f(x)]\, \psi'_{1-\alpha}(u)\, du,$$

then it can be shown for $f \in L_{2\pi}^p$, $1 \le p < \infty$, that the two truncated fractional derivatives $D_\epsilon^{\alpha,1} f(x)$ (cf. (3.26)) and $\overline{D_\epsilon^{\alpha,1}} f(x)$ converge for almost all x, and in $L_{2\pi}^p$ simultaneously, and

$$D^\alpha f(x) = \lim_{\epsilon \to 0} D_\epsilon^{\alpha,1} f(x) = \lim_{\epsilon \to 0} \overline{D_\epsilon^{\alpha,1}} f(x) =: \overline{D^\alpha} f(x), \qquad (3.33)$$

so that the Marchaud derivative for periodic functions coincides with that of Weyl (-Marchaud). Moreover the $L_{2\pi}^p$-convergence of the truncated Marchaud derivative, namely that $\exists g \in L_{2\pi}^p$ such that $\|D_\epsilon^{\alpha,1} f - g\|_{L_{2\pi}^p} \to 0$ for $\epsilon \to 0$, is equivalent to the representability of $f(x)$ as a Weyl fractional integral of some $g \in L_{2\pi}^p$, namely $f(x) = f^\wedge(0) + I^\alpha g(x)$. If $1 < p < \infty$, then this is also equivalent to $\|D_\epsilon^{\alpha,1} f\|_{L_{2\pi}^p} = \mathcal{O}(1)$, $\mathcal{O}$ being independent of ϵ.

Recalling Theorem 3.4, it is also clear that the strong Liouville- Grünwald derivative and the strong Marchaud derivative coincide for periodic functions in $L_{2\pi}^p$, $1 \le p < \infty$. The foregoing results may be extended from $0 < \alpha < 1$ to any $\alpha > 0$.

Marchaud's approach has proved to be very useful in building up integral representations of numerous operators in analysis and mathematical physics, in particular in finding explicit and approximate solutions to various integral equations, arising in mechanics, electrostatics, diffraction theory, and especially in constructing explicit inverses to various fractional integrals and operators of potential type. This matter is the main topic of B. Rubin's specialized volume [147]; see also the books by V.I. Fabrikant [53,54].

3.7 Riesz derivatives on $\mathbb{R}$

Let us here consider a modified type of fractional integrals which were introduced by M. Riesz (1949) [140]. If $0 < \alpha < 1$, the integral

$$R^\alpha f(x) = \frac{1}{2\Gamma(\alpha)\cos(\alpha\pi)/2} \int_{-\infty}^{\infty} \frac{f(u)}{|x-u|^{1-\alpha}} du \qquad (3.34)$$

is called the *Riesz potential* of f of order α. With (3.34) there is associated the so-called *conjugate Riesz potential* of order α defined by

$$\widetilde{R}^{\alpha} f(x) = \frac{1}{2\Gamma(\alpha)\sin(\alpha\pi)/2} \int_{-\infty}^{\infty} \frac{\operatorname{sgn}(x-u)}{|x-u|^{1-\alpha}} f(u)du. \tag{3.35}$$

Under appropriate assumptions on f, say, for instance, $f \in L^1(\mathbb{R}) \cap L^2(\mathbb{R})$, the integrals $R^{\alpha} f(x)$ and $\widetilde{R}^{\alpha} f(x)$ exist almost everywhere and belong to $L^1_{loc}(\mathbb{R})$. They are related to each other via the Hilbert transform H, namely

$$\widetilde{R}^{\alpha} f = R^{\alpha} H f,$$

where H is defined by the Cauchy principal value

$$(Hf)(x) \equiv f^{\sim}(x) = \lim_{\delta \to 0+} \frac{1}{\pi} \int_{|x-u| \geq \delta} \frac{f(u)}{x-u} du \tag{3.36}$$

at every point x for which this limit exists.

Moreover, provided f behaves well enough, then the following additivity laws are satisfied: If $\alpha > 0$, $\beta > 0$ such that $\alpha + \beta < 1$, then

$$\begin{aligned} R^{\alpha} R^{\beta} f &= R^{\alpha+\beta} f, \\ \widetilde{R}^{\alpha} \widetilde{R}^{\beta} f &= -R^{\alpha+\beta} f. \end{aligned}$$

The Riesz potential as well as its conjugate counterpart can be extended to all complex α with $Re\ \alpha \geq 0$ by distributional methods. See Samko et al [153, §25.2], where the matter is treated even in the multidimensional case.

To find an appropriate Riesz derivative of fractional order, let us look very formally at the Fourier transforms of $R^{\alpha} f$ and $\widetilde{R}^{\alpha} f$. (For notation and some basic properties of the Fourier transform, compare Section 5.1.) Since for $0 < \alpha < 1$

$$\int_{-\infty}^{\infty} \frac{e^{ivu}}{|u|^{1-\alpha}} du = 2|v|^{-\alpha} \int_{0}^{\infty} \frac{\cos u}{u^{1-\alpha}} du = 2|v|^{-\alpha}\Gamma(\alpha)\cos\frac{\pi\alpha}{2},$$

we have for $v \neq 0$, by the Fourier transform of convolutions,

$$[R^{\alpha} f]^{\wedge}(v) = |v|^{-\alpha} f^{\wedge}(v), \quad [\widetilde{R}^{\alpha} f]^{\wedge}(v) = (-i\ \operatorname{sgn} v)|v|^{-\alpha} f^{\wedge}(v),$$

and by the Fourier transform of derivatives,

$$\left[\frac{d}{dx} R^{1-\alpha} f(x)\right]^{\wedge}(v) = (iv)|v|^{\alpha-1} f^{\wedge}(v) = (i\ \operatorname{sgn} v)|v|^{\alpha} f^{\wedge}(v),$$

$$\left[\frac{d}{dx}\widetilde{R}^{1-\alpha}f(x)\right]^{\wedge}(v) = (iv)(-i\ \operatorname{sgn} v)|v|^{\alpha-1}f^{\wedge}(v) = |v|^{\alpha}f^{\wedge}(v). \tag{3.37}$$

Thus, comparing the last four formulas, one is led to take $\frac{d}{dx}\widetilde{R}^{1-\alpha}f(x)$ as a candidate for the α-th Riesz derivative, as this expression inverts, at least formally, the Riesz potential $R^{\alpha}f(x)$, while $\frac{d}{dx}R^{1-\alpha}f(x)$ plays the corresponding role with respect to the conjugate Riesz potential $\widetilde{R}^{\alpha}f(x)$. We want to follow up only the first of these two cases; the other may be treated in an analogous manner. Writing $\frac{d}{dx}\widetilde{R}^{1-\alpha}f(x)$ as the limit

$$\lim_{h\to 0}\frac{\widetilde{R}^{1-\alpha}f(x+h)-\widetilde{R}^{1-\alpha}f(x)}{h},$$

we can represent the numerator of the above difference quotient as the convolution integral

$$(f*n_{h,1-\alpha})(x) = \frac{1}{\sqrt{2\pi}}\int_{-\infty}^{\infty} f(x-u)n_{h,1-\alpha}(u)du, \tag{3.38}$$

where the kernel

$$n_{h,1-\alpha}(x) = \frac{\sqrt{2\pi}}{2\Gamma(1-\alpha)\cos(\alpha\pi)/2}\left\{\frac{\operatorname{sgn}(x+h)}{|x+h|^{\alpha}}-\frac{\operatorname{sgn} x}{|x|^{\alpha}}\right\}$$

belongs to $L^1(\mathbb{R})$ and has the Fourier transform

$$n^{\wedge}_{h,1-\alpha}(v) = (-i\ \operatorname{sgn} v)(e^{ihv}-1)|v|^{\alpha-1} \quad (v\neq 0).$$

Thus, for every $1\le p<\infty$, the integral (3.38) exists as a function in $L^p(\mathbb{R})$, whenever $f\in L^p(\mathbb{R})$. This gives rise to the following definition of a fractional Riesz derivative in the sense of the norm topology of L^p-function spaces:
Let $f\in L^p(\mathbb{R})$, $1\le p<\infty$. If $0<\alpha<1$ we define the α-th *strong Riesz derivative* $D^{\{\alpha\}}f$ of f by

$$\lim_{h\to 0}\|h^{-1}(f*n_{h,1-\alpha})-D^{\{\alpha\}}f\|_p = 0 \tag{3.39}$$

whenever this limit exists. If $\alpha=1$, (3.39) is replaced by

$$\lim_{h\to 0}\|h^{-1}[f^{\sim}(\cdot+h)-f^{\sim}(\cdot)]-D^{\{1\}}f\|_p = 0, \tag{3.40}$$

while for $\alpha > 1$ we proceed inductively. If $l < \alpha \leq l+1$, l being a positive integer, then we set

$$D^{\{\alpha\}}f := D^{\{\alpha-l\}}D^{\{l\}}f. \tag{3.41}$$

With this definition of an α-th strong Riesz derivative, formula (3.37) receives the following correct interpretation:
A function $f \in L^p(\mathbb{R})$, $1 \leq p \leq 2$, has a *strong Riesz derivative of order* $\alpha > 0$ if and only if the function $v \mapsto |v|^\alpha f^\wedge(v)$ is the Fourier transform of some function $g \in L^p(\mathbb{R})$. If so, g equals $D^{\{\alpha\}}f$.

Our principal aim in this section is, however, to characterize the strong Riesz derivative by an integral of Marchaud type. Indeed, making an attempt to define a Riesz potential of negative order one is led to use the method of Hadamard's finite part; for $0 < \alpha < 2$ it means to consider the integral

$$D_\varepsilon^{\{\alpha\}}f(x) = \frac{1}{K_{\alpha,2}} \int\limits_{|u|\geq\varepsilon} \frac{f(x-u)-f(x)}{|u|^{1+\alpha}}du \tag{3.42}$$

as $\varepsilon \to 0+$, where

$$K_{\alpha,2} = \begin{cases} 2\Gamma(-\alpha)\cos\frac{\pi\alpha}{2} & \text{if} \quad 0<\alpha<2, \quad \alpha \neq 1 \\ -\pi & \phantom{\text{if}} \quad \alpha = 1. \end{cases}$$

By a substitution, (3.42) may be rewritten as

$$D_\varepsilon^{\{\alpha\}}f(x) = \frac{1}{K_{\alpha,2}} \int\limits_\varepsilon^\infty \frac{f(x+u)-2f(x)+f(x-u)}{u^{1+\alpha}}du$$

with a central difference of f in the numerator of the integrand. This suggests for arbitrary $\alpha > 0$ the following regularizations

$$D_\varepsilon^{\{\alpha\}}f(x) = \frac{1}{K_{\alpha,2j}} \int\limits_\varepsilon^\infty \frac{\overline{\Delta}_u^{2j}f(x)}{u^{1+\alpha}}du \quad (0<\alpha<2j)$$

(more precisely $D_\varepsilon^{\{\alpha,2j\}}f$) as $\varepsilon \to 0+$, where the central difference of f of even order $2j$ is given by

$$\overline{\Delta}_u^{2j}f(x) = \sum_{k=0}^{2j}(-1)^k\binom{2j}{k}f(x+(j-k)u)$$

and

$$K_{\alpha,2j} = (-1)^j 2^{2j-\alpha} \int_0^\infty \frac{\sin^{2j} u}{u^{1+\alpha}} du.$$

The characterization of strong Riesz derivatives given in the next theorem is valid in L^p-spaces for all real $p \geq 1$; see Butzer-Trebels [28] and Butzer-Nessel [25], Sec. 11 (for $1 \leq p \leq 2$) and Sec. 13.2.4, 13.2.5 (for $p > 2$).

Theorem 3.11. *Let $\alpha > 0$. For $f \in L^p(\mathbb{R})$, $1 \leq p < \infty$, the following assertions are equivalent:*

(i) f has an α-th strong Riesz derivative,

(ii) there exists $g \in L^p(\mathbb{R})$ such that

$$\lim_{\varepsilon \to o+} \left\| \frac{1}{K_{\alpha,2j}} \int_\varepsilon^\infty \frac{\overline{\Delta}_u^{2j} f}{u^{1+\alpha}} du - g \right\|_p = 0, \tag{3.43}$$

where j is a positive integer chosen so that $0 < \alpha < 2j$.

In this case, $D^{\{\alpha\}} f = g$.

The above equivalence may be established by Fourier transform methods. In case of L^p-spaces, $1 \leq p \leq 2$, one may argue directly via the characterization of the strong Riesz derivative in terms of the Fourier transform which was mentioned subsequent to definitions (3.39) - (3.41), while for $p > 2$ dual methods are appropriate.

To give an idea how the Fourier transform acts on the Marchaud approximants $D_\varepsilon^{\{\alpha\}} f$, let us consider the case $0 < \alpha < 2$ more precisely. If $f \in L^p(\mathbb{R})$, $1 \leq p \leq 2$, and $0 < \alpha < 2$, then we have for almost all $v \in \mathbb{R}$

$$\left[D_\varepsilon^{\{\alpha\}} f\right]^\wedge (v) = \frac{2}{K_{\alpha,2}} |v|^\alpha f^\wedge(v) \int_{|\varepsilon v|}^\infty \frac{\cos u - 1}{u^{1+\alpha}} du,$$

which implies

$$\lim_{\varepsilon \to 0+} [D_\varepsilon^{\{\alpha\}} f]^\wedge(v) = |v|^\alpha f^\wedge(v). \tag{3.44}$$

On the other hand, there exists a function $k_\alpha \in L^1(\mathbb{R})$ satisfying $\int_{-\infty}^\infty k_\alpha(u) du = \sqrt{2\pi}$, such that

$$\frac{2}{K_{\alpha,2}} \int_{|v|}^\infty \frac{\cos u - 1}{u^{1+\alpha}} du = k_\alpha^\wedge(v) \qquad (v \in \mathbb{R})$$

(see Sunouchi [158] and [173, II]). Hence,

$$\left[D_\varepsilon^{\{\alpha\}} f\right]^\wedge (v) = |v|^\alpha f^\wedge(v) k_\alpha^\wedge(\varepsilon v). \tag{3.45}$$

Now, formulas (3.44) and (3.45) imply, by standard Fourier transform arguments, the equivalence of the statements (ii) and (iii) in the next theorem (for the case $0 < \alpha < 2$). For a different method of proof generalizing one due to Salem-Zygmund we refer to Butzer-Trebels [28], see also [25, p. 414]. Theorem 3.12 summarizes all results of this section for the spaces $L^p(\mathbb{R})$, $1 \le p \le 2$.

Theorem 3.12. *Let $\alpha > 0$. For $f \in L^p(\mathbb{R})$, $1 \le p \le 2$, the following assertions are equivalent:*

(i) f has an α-th strong Riesz derivative,

(ii) there exists $g \in L^p(\mathbb{R})$ such that $|v|^\alpha f^\wedge(v) = g^\wedge(v)$,

(iii) there exists $g \in L^p(\mathbb{R})$ such that eq. (3.43) holds for $0 < \alpha < 2j$,

(iv) if $1 < p \le 2$,

$$\left\| \frac{1}{K_{\alpha,2j}} \int_\varepsilon^\infty \frac{\overline{\Delta}_u^{2j} f}{u^{1+\alpha}} du \right\|_p = \mathcal{O}(1) \quad (\varepsilon \to 0+),$$

where $0 < \alpha < 2j$.

For the counterpart of (iv) in the space $L^1(\mathbb{R})$ we have the following characterization.

Proposition 3.13. *Let $\alpha > 0$. For $f \in L^1(\mathbb{R})$ the following assertions are equivalent:*

(i) There exists a function μ of bounded variation over $\mathbb{R}$ such that its Fourier-Stieltjes transform

$$\mu^\vee(v) = \frac{1}{\sqrt{2\pi}} \int_{-\infty}^{\infty} e^{-ivu} d\mu(u)$$

satisfies for almost all $v \in \mathbb{R}$

$$\mu^\vee(v) = |v|^\alpha f^\wedge(v).$$

(ii) If j is a positive integer such that $0 < \alpha < 2j$, then

$$\left\| \frac{1}{K_{\alpha,2j}} \int_\varepsilon^\infty \frac{\overline{\Delta}_u^{2j} f}{u^{1+\alpha}} du \right\|_1 = \mathcal{O}(1) \quad (\varepsilon \to 0+).$$

4 Various Applications

4.1 Integral representations of special functions

Fractional calculus may be used to construct non-trivial integral representations of special functions. One example is (2.16). Another is Poisson's integral representation of the Bessel function

$$J_\nu(z) = (\frac{z}{2})^\nu \sum_{k=0}^{\infty} \frac{(-1)^k}{k!\,\Gamma(\nu+k+1)} (\frac{z}{2})^{2k},$$

which is a solution of Bessel's equation $z^2D^2w + zDw + (z^2 - \nu^2)w = 0$, namely (see [121, p. 302, 114, 99, 89])

$$J_\nu(z) = \frac{2}{\Gamma(1/2)\Gamma(\nu+1/2)} (\frac{z}{2})^\nu \int_0^1 (1-u^2)^{\nu-1/2} \cos zu\, du \quad (\Re e\, \nu > -1/2).$$

As a third example, the Legendre *function* $P_\alpha(x)$ of the first kind and degree α may be expressed as a fractional derivative. It is given in terms of hypergeometric functions as

$$P_\alpha(x) = {}_2F_1(\alpha+1, -\alpha, 1; \frac{1}{2}(1-x)), \quad |1-x| < 2. \tag{4.1}$$

If α is a nonnegative integer n, then $P_\alpha(x)$ becomes the Legendre polynomial $P_n(x)$ which may be represented by the classical Rodrigues' formula

$$P_n(x) = \frac{1}{2^n n!} D_x^n (x^2-1)^n.$$

To obtain a similar expression for arbitrary α note that

$$x^\alpha(1-x)^\alpha = \frac{1}{\Gamma(-\alpha)} \sum_{k=0}^{\infty} \frac{\Gamma(k-\alpha)}{k!} x^{k+\alpha} \quad |x| < 1.$$

Differentiating termwise and using (recall (1.30))

$${}_0D_x^\alpha x^{k+\alpha} = \frac{\Gamma(\alpha+1+k)}{\Gamma(k+1)} x^k,$$

we obtain

$$\begin{aligned} {}_0D_x^\alpha x^\alpha (1-x)^\alpha &= \frac{1}{\Gamma(-\alpha)} \sum_{k=0}^{\infty} \frac{\Gamma(k-\alpha)}{k!} \frac{\Gamma(\alpha+1+k)}{\Gamma(k+1)} x^k \\ &= \Gamma(\alpha+1)\, {}_2F_1(\alpha+1, -\alpha, 1; x). \end{aligned}$$

Thus recalling (4.1),

$$P_\alpha(1-2x) = \frac{1}{\Gamma(\alpha+1)} {}_0D_x^\alpha[x^\alpha(1-x)^\alpha].$$

By the change of variable $t = 1-2x$, this becomes

$$P_\alpha(t) = \frac{1}{2^\alpha\Gamma(\alpha+1)} \, {}_tD_1^\alpha(1-t^2)^\alpha.$$

For further applications of fractional calculus to special functions see e.g. L.M.B.C. Campos [31], R.N. Kalia [87], Lavoie-Tremblay-Osler [102], V. Kiryakova [91].

4.2 Stirling functions of the first kind

The classical Stirling numbers, introduced by James Stirling in his "Methodus Differentialis" in 1730, which are said to be "as important as Bernoulli's, or even more so" (see Jordan ([85, 1959, p. 143])), play a major role in a variety of branches of mathematics, such as combinatorial theory, finite difference calculus, numerical analysis, interpolation theory and number theory. Those of the first kind, $s(n,k)$, can be defined via their exponential generating function

$$\frac{(\log(1+z))^k}{k!} = \sum_{n=k}^{\infty} \frac{s(n,k)}{n!} z^n \qquad (|z|<1,\ k \in \mathbb{N}_0) \tag{4.2}$$

or via their (horizontal) generating function

$$[z]_n = \sum_{k=0}^{n} s(n,k) z^k \qquad (z \in \mathbb{C},\ n \in \mathbb{N}_0) \tag{4.3}$$

where $[z]_n = z(z-1)\dots(z-n+1)$, and with the convention $s(n,0) = \delta_{n,0}$ (Kronecker's delta); see Riordan [141], Comtet [35].

The latter gives a natural possibility to define "Stirling numbers of fractional order" $s(\alpha,k)$ with $\alpha \in \mathbb{C}$ and $k \in \mathbb{N}_0$. In fact, these "Stirling functions", as one may call them, which were introduced by Butzer, Hauss and Schmidt [22], may be defined via the generating function

$$[z]_\alpha := \frac{\Gamma(z+1)}{\Gamma(z-\alpha+1)} = \sum_{k=0}^{\infty} s(\alpha,k) z^k \qquad (|z|<1, \alpha \in \mathbb{C}). \tag{4.4}$$

Theorem 4.1. *For $\alpha \in \mathbb{C}$ and $k > \Re e\ \alpha > -\infty$, $k \in \mathbb{N}$ there holds the integral representation*

$$s(\alpha, k) = \frac{1}{\Gamma(-\alpha)k!} \int_{0+}^{1-} \frac{(\log u)^k}{(1-u)^{\alpha+1}}\, du. \tag{4.5}$$

For a proof see [67, pp. 117-121].

This integral calls to mind the Riemann–Liouville derivative of order α with $\Re e\ \alpha > 0$ of the function $f(x) = (\log x)^k/k!$ taken at $x = 1$. In fact, equation (4.5) in this respect reads for $k > \Re e\ \alpha$, $k \in \mathbb{N}_0$,

$$s(\alpha, k) = \frac{1}{k!}\, {}_0\mathcal{D}_x^\alpha (\log x)^k \Big|_{x=1}. \tag{4.6}$$

The proof of (4.6) for $\Re e\ \alpha < 0$ is immediate from (4.5) and Definition (1.21). However, for $k > \Re e\ \alpha > 0$ it is rather long and technical (see [67, pp. 123-128]).

Formula (4.6) is the fractional counterpart of, which follows from (4.2),

$$s(n, k) = \frac{1}{k!} \left(\frac{d}{dz}\right)^n (\log(1+z))^k|_{z=0}. \tag{4.7}$$

Observe that the results (1.30) and (1.35) enable one to evaluate certain Stirling functions. Thus $s(\alpha, 0) = {}_0\mathcal{D}_x^\alpha 1|_{x=1} = x^{-\alpha}/\Gamma(1-\alpha)|_{x=1} = 1/\Gamma(1-\alpha)$, $s(1/2, 1) = \log 4/\sqrt{\pi}$, $s(-1/2, 1) = 2(\log 4 - 2)/\sqrt{\pi}$ (cf. Hauss [67, p. 107f]),

$$s(\alpha, 1) = \frac{\psi(1) - \psi(1-\alpha)}{\Gamma(1-\alpha)} = \frac{\alpha}{\Gamma(1-\alpha)} \sum_{j=1}^{\infty} \frac{1}{j(j-\alpha)} \quad (\alpha \notin \mathbb{N}), \tag{4.8}$$

$$\begin{aligned} s(\alpha, 2) &= \frac{1}{2\Gamma(1-\alpha)} \{[\psi(1) - \psi(1-\alpha)]^2 + \psi'(1) - \psi'(1-\alpha)\} \\ &= \frac{-\alpha}{\Gamma(1-\alpha)} \sum_{k=2}^{\infty} \left(\sum_{j=1}^{k-1} \frac{1}{j}\right) \frac{1}{k(k-\alpha)} \quad (\alpha \in \mathbb{R}\backslash\mathbb{N}). \end{aligned}$$

Observe that the representation (4.5) may also be associated with the Weyl-fractional derivative. In fact, under the substitution $u = (1+t)^{-1}$ the integral (4.5) turns into

$$s(\alpha, k) = \frac{(-1)^k}{\Gamma(-\alpha)k!} \int_0^\infty [\log(1+u)]^k u^{-1-\alpha}(1+u)^{\alpha-1}\, du$$

for $k > \Re e\ \alpha$, $k \in \mathbb{N}$. For $\Re e\ \alpha < 0$ it follows immediately that

$$s(\alpha, k) = {}_x\mathcal{D}^\alpha_\infty\{(1+x)^{\alpha-1}[\log(1+x)]^k(-1)^k/k!\}\Big|_{x=0}.$$

However a proof for such a representation for $k > \Re e\ \alpha > 0$ seems to be quite difficult.

Whether it is possible to evaluate the fractional order derivatives of the $s(\alpha, k)$ is an open question. In any case, the $s(\alpha, k)$ are arbitrarily often continuously differentiable in α and one has for $\alpha \in \mathbb{R}\backslash\mathbb{N}$ and $k \in \mathbb{N}_0$,

$$(d/d\alpha)s(\alpha, k) = \sum_{j=0}^{k} \frac{\psi^{(j)}(1-\alpha)}{j!} s(\alpha, k-j).$$

The $s(\alpha, k)$ are connected with the fundamental *Riemann zeta function* $\zeta(s) = \sum_{j=1}^\infty j^{-s}$, $\Re e\ s > 1$, in a striking fashion (see [19] and the literature cited there). Thus, for $m \in \mathbb{N}$,

$$\zeta(m+1) = \lim_{\alpha\to 0} \Gamma(-\alpha)(-1)^m s(\alpha, m) = \lim_{\alpha\to 0}(-1)^m\Gamma(-\alpha)\ {}_0I_x^{-\alpha}\ \frac{(\log x)^m}{m!}\Big|_{x=1},$$

as well as

$$\zeta(m+1) = \lim_{\alpha\to 0}(-1)^{m+1}(d/d\alpha)s(\alpha, m).$$

4.3 Stirling functions of the second kind

The Stirling numbers $S(n, k)$ of the second kind could be defined via their exponential generating function

$$\frac{(e^x-1)^k}{k!} = \sum_{n=k}^\infty S(n,k)\frac{x^n}{n!} \qquad (x \in \mathbb{R}; k \in \mathbb{N}_0) \tag{4.9}$$

or, equivalently, by

$$S(n,k) = \left(\frac{d}{dx}\right)^n \frac{(e^x-1)^k}{k!}\Big|_{x=0} \qquad (k, n \in \mathbb{N}_0), \tag{4.10}$$

or by the horizontal generating function

$$x^n = \sum_{k=0}^{n} S(n,k)[x]_k$$

with the associated

$$S(n,k) = \frac{1}{k!}\Delta^k x^n\Big|_{x=0} \qquad (k,n \in \mathbb{N}_0),$$

Δ being the forward (left-handed) difference operator, namely $\Delta f(x) := f(x+1) - f(x)$, $\Delta^{j+1} f(x) = \Delta(\Delta^j f)(x)$.

In analogy with the third definition, the *Stirling functions* $S(\alpha, k)$ of *second kind*, of fractional order $\alpha \in \mathbb{R}_+$, introduced in Butzer and Hauss [20], are defined by

$$S(\alpha,k) := \frac{1}{k!}\Delta^k x^\alpha\Big|_{x=0} \qquad (\alpha \in \mathbb{R}_+; k \in \mathbb{N}_0). \tag{4.11}$$

They may also be expressed by a finite sum as $S(\alpha,k) = \frac{1}{k!}\sum_{j=0}^{k}(-1)^{k+j}\binom{k}{j}j^\alpha$. There exists a real integral representation for $S(\alpha,k)$, namely there holds for $0 < \alpha < k$, $k \in \mathbb{N}$, (see [67, p. 71])

$$S(\alpha,k) = \frac{1}{k!\Gamma(-\alpha)}\int_0^\infty u^{-1-\alpha}(e^{-u}-1)^k\,du. \tag{4.12}$$

Now this integral looks formally like the *Weyl fractional derivative* of order $\alpha(> 0!)$ of the function $f(x) = (e^{-x}-1)^k/k!$, taken at $x = 0$. As a matter of fact,

Theorem 4.2. *For $k \in \mathbb{N}$, any $\alpha > 0$ there holds*

$$S(\alpha,k) = \frac{1}{k!}\,{}_xD_\infty^\alpha\left\{(e^{-x}-1)^k - (-1)^k\right\}\Big|_{x=0}. \tag{4.13}$$

As to the proof, recalling (1.38), for $p = j$,

$$\begin{aligned} {}_xD_\infty^\alpha\left\{(e^{-x}-1)^k - (-1)^k\right\}\Big|_{x=0} &= {}_xD_\infty^\alpha\left(\sum_{j=1}^{k}\binom{k}{j}(-1)^{k-j}e^{-jx}\right)\Big|_{x=0} \\ &= \sum_{j=1}^{k}\binom{k}{j}(-1)^{k-j}j^\alpha = k!S(\alpha,k). \end{aligned} \tag{4.14}$$

Observe that (4.13) is actually the counterpart of (4.10) in the fractional instance, since by (4.13), for $n \in \mathbb{N}$,

$$\begin{aligned} S(n,k) &= \frac{1}{k!}\, {}_xD^n_\infty \left\{(e^{-x}-1)^k - (-1)^k\right\}\Big|_{x=0} \\ &= \frac{(-1)^n}{k!}\left(\frac{d}{dx}\right)^n \left\{(e^{-x}-1)^k\right\}\Big|_{x=0} \end{aligned} \tag{4.15}$$

provided x in (4.10) is replaced by $-x$ (due to the fact that the integral (1.37) converges only for $p > 0$).

Let us finally observe that fractional order Stirling numbers $S(\alpha, k)$ have been considered indirectly, at least, in Westphal [173, p. 76], in her theory of fractional powers of infinitesimal generators of semigroup operators. They are related to the normalizing factor $C_{\alpha,k}$ –recall Section 3.5 – of the Marchaud fractional order derivative (3.22), (3.23), first treated by Marchaud. Also the proof of (4.12) in [20] is modelled upon [173].

4.4 Euler functions

The classical *Euler polynomials* $E_n(x)$ can be defined in terms of their exponential generating function by

$$\frac{2e^{xw}}{e^w+1} = \sum_{n=0}^{\infty} E_n(x)\frac{w^n}{n!} \qquad (x \in \mathbb{R}; |w| < \pi) \tag{4.16}$$

or, equivalently, by

$$E_n(x) = \left(\frac{d}{dw}\right)^n \left(\frac{2e^{xw}}{e^w+1}\right)\Big|_{w=0}. \tag{4.17}$$

These polynomials were extended to *Euler functions* $E_\alpha(z)$ with complex indices $\alpha \in \mathbb{C}$ and $\Re e\, z > 0$ in [14] by

$$E_\alpha(z) := \frac{\Gamma(\alpha+1)}{\pi i} \int_{\mathcal{C}} \frac{e^{\zeta z}}{1+e^\zeta}\zeta^{-\alpha-1}\, d\zeta, \tag{4.18}$$

$\mathcal{C}$ being a positively oriented loop around the negative real axis, composed of a circle C_2, of radius $0 < c < 2\pi$ around the origin together with the lower and upper edges $\mathcal{C}_1$ and $\mathcal{C}_3$ of the "cut" in the complex plane along $\mathbb{R}^-$. In fact, $\mathcal{C} = \mathcal{C}(-\infty, 0+)$ of (1.41).

These Euler functions have removable discontinuities at $\alpha \in \mathbb{Z}^-$ since the gamma function has simple poles at these points, and the integral vanishes by

Cauchy's theorem, since then the integrand is analytic. This yields (see [2]) that $E_\alpha(z)$ is an analytic function of $\alpha \in \mathbb{C}$ for $\Re e\ z > 0$. These functions can even be defined for all $z \in \mathbb{C}\backslash\mathbb{R}_0^-$; the $E_\alpha(z)$ are analytic there. There exists a Weyl-type integral representation, namely

$$E_\alpha(z) = \frac{1}{\Gamma(-\alpha)} \int_{-\infty}^{0} \frac{2e^{zu}}{1+e^u}(-u)^{-\alpha-1}\, du \tag{4.19}$$

for $\Re e\ \alpha < 0$ and $\Re e\ z > 0$. This integral can be regarded as the fractional Weyl *derivative* of order α of the generating function $f_z(y) := 2e^{zy}/(1+e^y)$ at $y = 0$. Thus $E_\alpha(z) = {}_{-\infty}D_y^\alpha f_z(y)|_{y=0} = (\frac{d}{dy})^m\ {}_{-\infty}W_y^{m-\alpha} f_z(y)|_{y=0}$ holds for $m - 1 \leq \Re e\ \alpha < m$, which is the counterpart of formula (4.17). In Section 1.3 we presented three examples dealing with the evaluation of the functions x^b, $\log x$ and e^{-px} according to the Riemann-Liouville-Weyl definitions. Let us now apply the Liouville-Grünwald definition, and take the Euler functions as the example. In order to apply it we shall take the 1–antiperiodic continuation of $E_\alpha(x)$ for $x \in (0,1]$ and denote it by $\mathcal{E}_\alpha(x)$. It has the Fourier series representation

$$\mathcal{E}_\alpha(x) = 2\Gamma(\alpha+1) \sum_{k=-\infty}^{\infty} \frac{e^{(2k+1)\pi i x}}{[(2k+1)\pi i]^{\alpha+1}}$$

for $\alpha \in \mathbb{C}$, $\Re e\ \alpha > -1$. Whereas $(d/dx)E_n(x) = nE_{n-1}(x)$ for $n \in \mathbb{N}$, $(d/dx)E_\alpha(x) = \alpha E_{\alpha-1}(x)$ for $x \in \mathbb{R}_0^-$, $\Re e\ \alpha > 0$. As to the fractional order differential operator D^β for $\beta > 0$ of Definition (3.9), taken in the norm of $L^1_{2\pi}$–space, we have

$$D^\beta \mathcal{E}_\alpha(x) = \frac{\Gamma(\alpha+1)}{\Gamma(\alpha-\beta+1)} \mathcal{E}_{\alpha-\beta}(x) \quad (x \in \mathbb{R})$$

provided $\alpha > \beta > -\infty$. As to the proof, one has for all $\nu \in \mathbb{Z}$,

$$\begin{aligned}(i\nu)^\beta [\mathcal{E}_\alpha(\cdot)]^\wedge(\nu) &= \begin{cases} \frac{2\Gamma(\alpha+1)}{[(2k+1)\pi i]^{\alpha-\beta+1}} & ,\nu = (2k+1)n \text{ for one } k \in \mathbb{Z}, \\ 0 & ,\nu \neq (2k+1)n \text{ for all } k \in \mathbb{Z} \end{cases} \\ &= \frac{\Gamma(\alpha+1)}{\Gamma(\alpha-\beta+1)} [\mathcal{E}_{\alpha-\beta}(\cdot)]^\wedge(\nu).\end{aligned}$$

An application of Theorem 3.4 then completes the proof. For details see [14].

4.5 Eulerian numbers $E(\alpha, k)$ for $\alpha \in \mathbb{R}$

The *Eulerian numbers* are well known, especially in combinatorics and discrete mathematics, but also in geometry, statistical applications and spline theory. These numbers, $E(n, k)$, defined by

$$E(n,k) := \sum_{j=0}^{k} (-1)^j \binom{n+1}{j} (k+1-j)^n \quad (k, n \in \mathbb{N}_0), \tag{4.20}$$

satisfy the recurrence formula (in n)

$$E(n,k) = (k+1)E(n-1,k) + (n-k)E(n-1,k-1) \quad (k, n \in \mathbb{N}). \tag{4.21}$$

They are also quite useful since they connect the monomials x^n with the consecutive binomial coefficients, namely

$$x^n = \sum_{k=0}^{n-1} E(n,k) \binom{x+k}{n} \qquad (x \in \mathbb{R}; n \in \mathbb{N}_0). \tag{4.22}$$

There is a natural extension to *Eulerian functions* $E(\alpha, k)$, where $n \in \mathbb{N}$ is replaced by $\alpha \in \mathbb{R}$, defined for $k \in \mathbb{N}_0$ by

$$E(\alpha,k) := \sum_{j=0}^{k} (-1)^j \binom{\alpha+1}{j} (k+1-j)^\alpha. \tag{4.23}$$

This matter was first carried out in [21]. The first four Eulerian functions are given for $\alpha \in \mathbb{R}$ by

$$\begin{aligned}
E(\alpha,0) &= 1, \\
E(\alpha,1) &= 2^\alpha - (\alpha+1), \\
E(\alpha,2) &= 3^\alpha - (\alpha+1)2^\alpha + \binom{\alpha+1}{2}, \\
E(\alpha,3) &= 4^\alpha - (\alpha+1)3^\alpha + \binom{\alpha+1}{2}2^\alpha - \binom{\alpha+1}{3}.
\end{aligned}$$

There exists a horizontal generating function for the $E(\alpha, k)$, namely

$$\sum_{k=0}^{\infty} E(\alpha,k)x^k = \frac{A_\alpha(x)}{x} \qquad (\alpha \in \mathbb{R}; |x| < 1), \tag{4.24}$$

where $A_\alpha(x)$ are the generalized Eulerian "polynomials", i. e.

$$A_\alpha(x) := (1-x)^{\alpha+1} \sum_{k=1}^{\infty} k^\alpha x^k; \tag{4.25}$$

the radius of convergence of the latter series is 1. This yields a representation of these functions in terms of derivatives, namely

$$E(\alpha, k) = \frac{1}{k!} \left(\frac{d}{dx}\right)^k \left\{ \frac{A_\alpha(x)}{x} \right\} \Bigg|_{x=0} \qquad (\alpha \in \mathbb{R}; k \in \mathbb{N}_0). \tag{4.26}$$

Another such representation is that due to Worpitzky (1883) in the classical instance, namely

$$E(n, k) = \left(\frac{d}{dx}\right)^n \left(\sum_{j=0}^{k} (-1)^j \binom{n+1}{j} e^{(k-j+1)x} \right) \Bigg|_{x=0}. \tag{4.27}$$

The counterpart for the $E(\alpha, k)$ for arbitrary $\alpha \in \mathbb{R}$ can be expressed in terms of a *Weyl fractional* derivative. Indeed,

Theorem 4.3. *a) For any $\alpha < 0$, $k \in \mathbb{N}_0$ there holds*

$$E(\alpha, k) = {}_xW_\infty^{-\alpha} \left(\sum_{j=0}^{k} (-1)^j \binom{\alpha+1}{j} e^{-(k-j+1)x} \right) \Bigg|_{x=0}. \tag{4.28}$$

b) For $\alpha > 0$, $k \in \mathbb{N}_0$ one has

$$E(\alpha, k) = {}_xD_\infty^{\alpha} \left(\sum_{j=0}^{k} (-1)^j \binom{\alpha+1}{j} e^{-(k-j+1)x} \right) \Bigg|_{x=0}. \tag{4.29}$$

The proof follows basically from (1.37), (1.38) with $p = k - j + 1$. For details see [21].

The $E(\alpha, k)$ have interesting properties, asymptotic relations, and are connected with the Stirling functions of Sec. 4.3. In particular,

$$\sum_{k=1}^{m+1} k^\alpha = \sum_{k=0}^{m} \binom{\alpha+m-k+1}{\alpha+1} E(\alpha,k) \qquad (\alpha \in \mathbb{R}; m \in \mathbb{N}_0);$$

$$(m+1)^\alpha = \sum_{k=0}^{m} \binom{\alpha+m-k}{\alpha} E(\alpha,k);$$

$$\lim_{\alpha\to\infty} \frac{E(\alpha,k)}{(k+1)^\alpha} = 1, \quad \lim_{\alpha\to\infty} \frac{E(\alpha+1,k)}{E(\alpha,k)} = k+1;$$

$$\lim_{\alpha\to-\infty} \frac{E(\alpha,k)}{(-\alpha)^k} = \frac{1}{k!} \qquad (k \in \mathbb{N});$$

$$E(\alpha,k) = (-1)^{k+1} \sum_{j=1}^{k+1} (-1)^j \binom{\alpha-j}{k-j+1} j! S(\alpha,j) \qquad (\alpha > 0; k \in \mathbb{N}_0);$$

$$E(\alpha,k) = \Gamma(\alpha+1) \int_k^{k+1} p_\alpha(u)du \qquad (\alpha > 1; k \in \mathbb{N}_0);$$

$$\sum_{k=0}^{\infty} E(\alpha,k) = \Gamma(\alpha+1) \qquad (\alpha > 1).$$

The second to last result connects the Eulerian numbers with the basic function $p_\alpha(x)$ of Proposition 3.1; it yields the last one since $\int_0^\infty p_\alpha(u)du = 1$. Further, $p_\alpha(k+1) = [\Gamma(\alpha)]^{-1}E(\alpha-1,k)$ for $\alpha > 0$. The implications may be worthwhile to consider.

In any case, the foregoing connection suggests it seems to be possible to generalize the Eulerian functions $E(\alpha,k)$ to the instance when also k is allowed to be fractional in terms of the $p_\alpha(x)$ function, namely

$$E(\alpha,\beta) = \Gamma(\alpha+1)p_{\alpha+1}(\beta+1)$$

where β is real with $\beta > -1$. For the role of Eulerian numbers in mathematics see the excellent account in Hilton et al [83, pp. 217–248] and its literature.

4.6 The Bernoulli functions $B_\alpha(x)$ for $\alpha \in \mathbb{R}$

The classical Bernoulli polynomials $B_n(x)$ can be defined via their exponential generating function by

$$\frac{we^{wx}}{e^w-1} = \sum_{n=0}^{\infty} B_n(x)\frac{w^n}{n!} \quad (x \in \mathbb{R}; |w| < 2\pi), \tag{4.30}$$

or equivalently by

$$B_n(x) = \left(\frac{d}{dw}\right)^n \left(\frac{we^{wx}}{e^w - 1}\right)\bigg|_{w=0}. \tag{4.31}$$

But they can also be defined in terms of their Fourier series, at first for $0 \le x < 1$ by $B_n(x) = \mathcal{B}_n(x)$, where $\mathcal{B}_n(x)$ is the one-periodic function

$$\mathcal{B}_n(x) := -2n! \sum_{k=1}^{\infty} \frac{\cos(2\pi kx - n\pi/2)}{(2\pi k)^n}, \quad (x \in \mathbb{R}, n \ge 2). \tag{4.32}$$

This led Butzer et al [17] to define the *Bernoulli functions* $B_\alpha(x)$ for $\alpha \in \mathbb{C}$, firstly for $\Re e\ \alpha > 1$ by $B_\alpha(x) = \mathcal{B}_\alpha(x)$ on $0 \le x < 1$, where

$$\mathcal{B}_\alpha(x) := -2\Gamma(\alpha+1) \sum_{k=1}^{\infty} \frac{\cos(2\pi kx - \alpha\pi/2)}{(2\pi k)^\alpha}, \quad (x \in \mathbb{R}, \Re e\ \alpha > 1), \tag{4.33}$$

have period 1. The $B_\alpha(x)$ are then extended in a suitable way to $x \in \mathbb{R}_0^+$ and $\mathbb{R}^-$. They can also be extended (noting (4.31)) beyond the line $\Re e\ \alpha = 1$ to the whole α-plane $\mathbb{C}$ as well as to complex $x = z$, via the contour integral

$$B_\alpha(z) := \frac{\Gamma(\alpha+1)}{2\pi i} \int_C \frac{e^{wz}}{e^w - 1} w^{-\alpha}\, dw \quad (\alpha \in \mathbb{C}) \tag{4.34}$$

for $z \in \mathbb{C}\backslash R_0^-$, at first for $\Re e\ z > 0$, where C denotes the positively oriented loop following Defintion (4.18), and then for any $z \in \mathbb{C}\backslash\mathbb{R}_0^-$ by

$$B_\alpha(z - m) := B_\alpha(z) - \alpha \sum_{k=0}^{m-1} (z - k - 1)^{\alpha-1}, \tag{4.35}$$

where $\Re e\ z \in (0,1)$, $z \notin \mathbb{R}$, $m \in \mathbb{N}$. The $B_\alpha(z)$ are now holomorphic functions of $\alpha \in \mathbb{C}$ for $z \in \mathbb{C}\backslash\mathbb{R}_0^-$.

$B_\alpha(z)$ can also be represented as a Weyl integral

$$B_\alpha(z) = \frac{1}{-\Gamma(-\alpha)} \int_{-\infty}^{0} \frac{e^{zu}}{e^u - 1} (-u)^{-\alpha}\, du \tag{4.36}$$

for $\Re e\ \alpha < 0$ and $\Re e\ z > 0$. This integral suggests that the fractional Weyl derivative of order α of the generating function $f_z(y) = (ye^{zy})/(e^y - 1)$ at $y = 0$, equals

$$B_\alpha(z) = {}_{-\infty}D_y^\alpha f_z(y)|_{y=0} = \left(\frac{d}{dy}\right)^m {}_{-\infty}W_y^{m-\alpha} f_z(y)|_{y=0} \tag{4.37}$$

for $m-1 \leq \Re e\alpha < m$. It is the counterpart of formula (4.31) in the fractional instance.

The foregoing Bernoulli functions can be connected in an interesting way with Liouville-Grünwald derivatives. Indeed, given such a derivative $D^\alpha f \in L^p_{2\pi}$, $1 \leq p < \infty$ of order α, then the $B_\alpha(x)$ may be used to recapture $f(x)$ itself from

$$f(x) = -\frac{(2\pi)^{\alpha-1}}{\Gamma(\alpha+1)} \int_0^{2\pi} D^\alpha f(x-u) B_\alpha(u/2\pi)\, du + \frac{1}{2\pi} \int_0^{2\pi} f(u)\, du. \tag{4.38}$$

It is due to the fact that the function $\psi_\alpha(x)$ of (3.13), which is associated with the Weyl-fractional integral, can be represented in an interesting fashion in terms of $B_\alpha(x)$ i. e. ,

$$\psi_\alpha(x) = \frac{-(2\pi)^\alpha}{\Gamma(\alpha+1)} B_\alpha\left(\frac{x}{2\pi}\right), \quad 0 < x < 2\pi. \tag{4.39}$$

The connection (4.39) was noted in the particular case $\alpha = n \in \mathbb{N}$ by Samko et al [153]. The result (4.38) follows from Theorem 3.4.

4.7 Ordinary and partial differential equations and other applications

There exist a variety of papers dealing with applications of fractional calculus to (ordinary) differential equations of fractional order, the first discussion of which goes back at least to L.O' Shaughnessy (1918) and E.L. Post (1919). See especially the monographs by Miller-Ross [121], Samko et al [153, pp. 795–872], and the work of M. Fujiwae (1933), E. Pitcher and W.E. Sewell (1938), G.H. Hardy (1945), J.M. Barret (1954), M.A. Al-Bassam (1966), M.M. Dzherbashyan and A.B. Nersesyan (1968), L.M.B.C. Campos (1990), D. Delbosco and L. Rodino (1996). See e.g. the recent N. Hayek et al [69] and the literature cited there. An algorithm for the numerical solution of nonlinear differential equations of fractional order was developed by K. Diethelm and A.D. Freed [40,41].

As to the partial differential equations there also exist a large number of publications. For some earlier material see Oldham-Spanier [133, pp. 197–218] and for more recent results Rusev et al [150], and especially the succeeding chapters of this volume.

Thus H. Berens and U. Westphal already in 1968 [7] considered the Cauchy problem for the generalized wave equation

$$\frac{d}{dx} w(x,t) + {}_0D_t^\gamma w(x,t) = 0 \quad (x,t > 0; 0 < \gamma < 1)$$

where ${}_0D_t^\gamma$ is the Riemann-Liouville fractional differentiation operator applied with respect to t and suitably restricted to functions of $L^p(0,\infty)$. The solution has the form $w(x,t) = [W_\gamma(x)f](t)$, where $\{W_\gamma(x); x \geq 0\}$ is a (C_0)-semigroup in $L^p(0,\infty)$.

In the particular case $\gamma = 1/2$ it can be given explicitly as $[W_{1/2}(x)f](t) = (x/\sqrt{4\pi}) \cdot \int_0^t f(t-u)u^{-3/2}exp(-x^2/4u)du$ with $f \in L^p(0,\infty)$. The proofs are carried out by precise applications of semigroup theory and Laplace transform analysis.

This investigation of the fractional wave equation, which is discussed briefly in Samko et al [153, p. 865], seems to be unknown among mathematical physicists and physicists. Of the literature they cite, it is generally restricted to the standard works of Oldham-Spanier [133] and Ross [144]; sometimes Ross [146], McBride-Roach [115] and Nishimoto [124, Vols. I and II] are also added (see e.g. H. Beyer and S. Kempfle [8]).

Explicit solutions of fractional wave and diffusion equations were given by Schneider and Wyss [157], Hilfer [77], B.J. West et al [171] and R. Metzler and Nonnenmacher [119], in terms of Fox functions (see A.M. Mathai and R.K. Saxena [112]). F. Mainardi in a series of papers studied especially the fractional diffusion equation, based on Laplace transforms and special functions of E.M. Wright-type (cf. Kiryakova [91]; see [110], and El-Sayed [49] for a semigroup approach. Schneider [156] and Hilfer [80,81] established the relation between fractional diffusion and stochastic processes, master equations and continuous time random walks.

A review of the early contributors of the applications of fractional calculus to the theory of *viscoelasticity* has been given by R.L. Bagley and P.J. Torvik [5]. In later papers these authors and R.C. Koeller [94] show the connection between fractional calculus and Abel's integral for materials with memory and that the fractional calculus constitutive equation allows for a continuous transition from the solid state to the fluid state when the memory parameter varies from zero to one. In particular, Koeller continued the Bagley-Torvik work and Yu. N. Rabotnov's (1980) theory of hereditary solid mechanics using the method of integral equations (instead of Laplace transforms); it leads to results expressed in terms of the Mittag-Leffler function which depends on the fractional derivative parameter in question. See especially R. Gorenflo and F. Mainardi ([64]) in this respect. Chapters VII and VIII in this volume give a review of more recent work on viscoelasticity and rheology. As to statics and dynamics of polymers, Douglas [43] has emphasized the appearance of fractional integral equations for surface interacting polymers and other systems. See also Chapter VI of this volume.

Of recent increasing interest in the area of fractals and nonlinear dynamics are fractional dynamical systems. Such systems can be discussed in the frame of abstract ergodic theory as flows or semiflows on measure spaces involving fractional time derivatives. They are related to problems of time irreversibility and ergodicity breaking. For literature in this respect see e.g. R. Hilfer ([78,79]), H. Hayakawa [69].

A new class of phase transitions, called anequilibrium transitions, has recently been introduced by R. Hilfer [73]-[75]. It characterizes each phase transition by its generalized noninteger order and a slowly varying function (in the sense of E. Seneta, see e.g. Jansche [84]) Thermodynamically this characterization arises from generalizing the classification of P. Ehrenfest (1933) to noninteger orders $\lambda \geq 1$ but also $\lambda < 1$.

In the setting of Hamiltonian chaotic analysis anomalous kinetics of particles has been studied by G.M. Zaslavsky in real systems. Such systems have fractal sets of islands or tori (cylinders) in the phase space, and these sets are responsible for the anomalous kinetics. In this respect a new fractional Fokker-Planck-Kolmogorov equation describes the particle kinetics; it can be considered as a new universal scheme for those processes where chaotic dynamics meet strongly correlated ballistic dynamics similar to the Lévy flights. See [180] and its literature as well as Chapters IV and V of this volume for more information on this subject.

5 Integral Transforms and Fractional Calculus

5.1 Fourier transforms

The Fourier transform of a function $f : \mathbb{R} \to \mathbb{C}$, defined by

$$\mathcal{F}[f](v) = f^\wedge(v) = \frac{1}{\sqrt{2\pi}} \int_{\mathbb{R}} f(u) e^{-ivu} du, \quad v \in \mathbb{R}, \tag{5.1}$$

is a powerful tool in the analysis of operators commuting with the translation operator (see e.g. [25]). Its inverse is given by

$$f(x) = \mathcal{F}^{-1}[f^\wedge(v)](x) = \frac{1}{\sqrt{2\pi}} \int_{\mathbb{R}} f^\wedge(v) e^{ixv} dv$$

for almost all $x \in \mathbb{R}$, if f and $f^\wedge$ belong to $L^1(\mathbb{R})$. Two of the basic properties of the Fourier transform are

$$\mathcal{F}[f^{(n)}](v) = (iv)^n f^\wedge(v), \quad v \in \mathbb{R} \tag{5.2}$$

$$[\mathcal{F}f]^{(n)}(v) = \mathcal{F}[(-ix)^n f(x)](v) \tag{5.3}$$

valid for sufficiently smooth functions f; (5.2) holds if, for example, $f \in L^1(\mathbb{R}) \cap AC_{\text{loc}}^{n-1}(\mathbb{R})$ with $f^{(n)} \in L^1(\mathbb{R})$, while for (5.3) it is sufficient that f as well as $x^n f(x)$ belong to $L^1(\mathbb{R})$. The convolution theorem

$$\mathcal{F}[f * g](v) = \mathcal{F}[f](v)\mathcal{F}[g](v) \tag{5.4}$$

holds, for example, for $f, g \in L^1(\mathbb{R})$, where

$$(f * g)(x) = \frac{1}{\sqrt{2\pi}} \int_{\mathbb{R}} f(x-u)g(u)du. \tag{5.5}$$

The fractional integral ${}_{-\infty}I_x^\alpha f$ of (1.12) is such a commuting operator. The counterpart of (5.2) reads for such operators (see e.g. [147, p. 32]

Lemma 5.1. *If* $f \in L^1(\mathbb{R})$ *and* $0 < \Re e\, \alpha < 1$, *then*

$$\mathcal{F}[{}_{-\infty}I_x^\alpha f](v) = (iv)^{-\alpha} f^\wedge(v). \tag{5.6}$$

Indeed, changing the order of integration,

$$\int_{-N}^{N} [{}_{-\infty}I_x^\alpha f(x)] e^{-ivx} dx = \frac{1}{\Gamma(\alpha)} \int_{-\infty}^{-N} e^{-itv} f(t)dt \int_{-N-t}^{N-t} e^{-iyv} y^{\alpha-1} dy$$

$$+ \frac{1}{\Gamma(\alpha)} \int_{-N}^{N} e^{-itv} f(t)dt \int_{0}^{N-t} e^{-iyv} y^{\alpha-1} dy.$$

But for any fixed $v \neq 0$, $0 < \Re e\, \alpha < 1$,

$$\int_0^\infty e^{-iyv} y^{\alpha-1} dy = (iv)^{-\alpha}\Gamma(\alpha). \tag{5.7}$$

Here the Fourier integral on the left is to be understood in the sense of a principal value. Hence, taking the limit above for $N \to \infty$, there results (5.6). Thus the action of fractional integration under Fourier transforms is reduced to dividing the Fourier transform by $(iv)^\alpha$.

Equation (5.6) cannot be extended to values $\Re e\, \alpha \geq 1$ immediately since the left-hand side of (5.6) may then not exist even for very smooth functions, e.g. $f \in C_0^\infty$, the space of all infinitely differentiable functions with compact support in $\mathbb{R}$. Indeed, for $\alpha = 1$, ${}_{-\infty}I_x^1 f(x) = \int_{-\infty}^x f(u)du$, so that ${}_{-\infty}I_x^1 f(x) \to$ const. as $x \to \infty$, implying that $\mathcal{F}[{}_{-\infty}I_x^1 f]$ does not exist in the usual sense (see e.g. [153, p. 139]).

If $\alpha > 1$ take a non-negative function $f \in C_0^\infty$ to be positive on some interval $[a, b]$. Then for $x > b$

$$ {}_{-\infty}I_x^\alpha f(x) \geq \frac{1}{\Gamma(\alpha)} \int_a^b (x-u)^{\alpha-1} f(u)du \geq \min_{a \leq u \leq b} f(u) \frac{(x-a)^\alpha - (x-b)^\alpha}{\Gamma(\alpha+1)}, $$

so that ${}_{-\infty}I_x^\alpha f(x) \sim Cx^{\alpha-1}/\Gamma(\alpha)$ as $x \to \infty$, implying again that $\mathcal{F}[{}_{-\infty}I_x^\alpha f]$ does not exist in the usual sense. Thus one can only expect to extend (5.6) to all α with $\Re e\, \alpha > 0$ for a special class of functions f.

For this purpose, let $S = S(\mathbb{R})$ be the space of Schwartz test functions, i.e., those $f(x)$, $x \in \mathbb{R}$ which are infinitely differentiable and rapidly decreasing together with their derivatives as $|x| \to \infty$, thus

$$ \lim_{|x| \to \infty} (1+x^2)^m f^{(j)}(x) = 0, \qquad \text{all } m \geq 0, j \geq 0. $$

$S(\mathbb{R})$ is a topological vector space, the topology being given by the family of seminorms $\|f\|_{j,m} = \sup_{x \in \mathbb{R}} |(1+x^2)^m f^{(j)}(x)|$; it is metrizable and is complete (thus a Frechet space). Now let $\Psi = \Psi(\mathbb{R})$ be the space of functions $f \in S$ which are equal to zero at the point $x = 0$ together with their derivatives, i.e., $\Psi := \{f \in S; \quad f^{(k)}(0) = 0, \quad k \in \mathbb{N}_0.\}$ Then the Lizorkin space Φ is the space of functions in S the Fourier transforms of which belong to Ψ, thus $\Phi = \{f; \quad f \in S, \quad f^\wedge \in \Psi\}$. In fact, Φ may be characterized as the space of $f \in S$ which are orthogonal to all polynomials:

$$ \int_{\mathbb{R}} u^k f(u)du = 0, \quad k \in \mathbb{N}_0, \tag{5.8} $$

since $\int_{\mathbb{R}} e^{-iou} u^k f(u)du = i^k \sqrt{2\pi} (f^\wedge)^{(k)}(0) = 0$.

Now if f belongs to Φ, then equation (5.6) is indeed valid for all $\Re e\, \alpha \geq 0$. Let us establish it for $1 \leq \Re e\, \alpha < 2$. Indeed,

$$ \Gamma(\alpha)\mathcal{F}[{}_{-\infty}I_x^\alpha f](v) = \lim_{N \to \infty} \int_{-\infty}^N e^{-itv} dt \int_{-\infty}^t (t-u)^{\alpha-1} f(u)du $$

$$ = \lim_{N \to \infty} \int_{-\infty}^N e^{-ivu} f(u)du \int_0^{N-u} y^{\alpha-1} e^{-ivy} dy. $$

If $\Re e\,\alpha \neq 1$, integration by parts yields

$$(iv)\Gamma(\alpha)\mathcal{F}[_{-\infty}I_x^\alpha f](v) = \lim_{N\to\infty}\{-N^{\alpha-1}e^{-ivN}\cdot\int_{-\infty}^{N}(1-\frac{u}{N})^{\alpha-1}f(u)du+$$

$$(\alpha-1)\int_{-\infty}^{N} f(u)e^{-ivu}du\cdot\int_{0}^{N-u} y^{\alpha-2}e^{-ivy}dy\}.$$

The first term on the right tends to zero by (5.8) (noting L'Hospital's rule), while in the second one can pass to the limit directly since $\Re e\,\alpha < 2$. Noting (5.7), this yields $\Gamma(\alpha)\mathcal{F}[_{-\infty}I^\alpha f](v) = (\alpha-1)\Gamma(\alpha-1)(iv)^{-\alpha}f^\wedge(v)$, as required. The case $\alpha = 1$ is simple, since, by (5.8),

$$\mathcal{F}[_{-\infty}I_x^1 f](v) = \lim_{N\to\infty}\int_{-\infty}^{N}\frac{e^{-iuv}-e^{-ivN}}{iv}f(u)du = \frac{1}{iv}f^\wedge(v).$$

For the case $\Re e\ \alpha = 1$, $\alpha \neq 1$, i.e., $\alpha = 1 + i\vartheta$, $\vartheta \neq 0$, which is quite technical (see [153, p. 149]).

As to the fractional counterpart of (5.2), we have, by Lemma 5.1 and (5.2) for sufficiently smooth functions f, for example, $f \in S$ $(n = [\Re e\ \alpha] + 1)$,

$$\mathcal{F}[_{-\infty}D_x^\alpha f](v) = \mathcal{F}[_{-\infty}I_x^{n-\alpha}f^{(n)}(x)](v) = (iv)^{-(n-\alpha)}(f^{(n)})^\wedge(v)$$

$$= (iv)^{-(n-\alpha)}(iv)^n f^\wedge(v) = (iv)^\alpha f^\wedge(v).$$

Let us finally remark that there exists a *fractional* Fourier transform $\mathcal{F}_\alpha$ of order α, introduced by N. Wiener (1929). It is a unitary integral transform on $L^2(\mathbb{R})$, the eigenvalues and eigenfunctions of which are, respectively, $e^{-in\alpha}$ and the normalized Hermite functions h_n $(n \in \mathbb{N}_0)$. It is defined for all $\alpha \in \mathbb{R}$ by the limit in $L^2(\mathbb{R})$-norm

$$\mathcal{F}_\alpha f(\cdot) = l.i.m._{N\to\infty}\int_{-N}^{N} K_\alpha(\cdot,u)f(u)du,$$

where

$$K_\alpha(v,u) = \begin{cases} A_\alpha \exp\left(\frac{iv^2}{2}\cot\alpha\right)\exp\left(-\frac{ivu}{\sin\alpha}+\frac{iu^2}{2}\cot\alpha\right) & \alpha \notin \{l\pi; l\in\mathbb{Z}\} \\ \delta(u-v) & \alpha\in\{2\pi l; l\in\mathbb{Z}\} \\ \delta(u+v) & \alpha\in\{2\pi l+\pi; l\in\mathbb{Z}\} \end{cases}$$

and $A_\alpha = \exp[-i\left(\frac{\pi}{4}\text{sgn }\alpha - \frac{\alpha}{2}\right)]/\sqrt{2\pi|\sin\alpha|}$.

For $\alpha = \pi/2$ we obtain the classical Fourier transform $\mathcal{F}$ and for $\alpha = -\pi/2$ its inverse transform $\mathcal{F}^{-1}$. For $\alpha = n\pi/2$, $n \in \mathbb{Z}$, we have $\mathcal{F}^n$, the nth power of $\mathcal{F}$. $\mathcal{F}_\alpha$ satisfies the index law $\mathcal{F}_\alpha \mathcal{F}_\beta f = \mathcal{F}_{\alpha+\beta} f$ forall $f \in L^2(\mathbb{R})$. The inverse of $\mathcal{F}_\alpha$ is $\mathcal{F}_{-\alpha}$. This transform is of interest in sampling theory of signal analysis in case the samples are not equally spaced apart but are irregular. It is of primary interest in the case of complex valued functions. See e.g. B. Bittner [9] and especially the literature cited there. For sampling theory itself one may consult Higgins [71] and Higgins-Stens [72].

5.2 Mellin transforms

The Mellin transform of $f : \mathbb{R}^+ \to \mathbb{C}$, defined by

$$\mathcal{M}[f(u)](s) = f^\wedge_\mathcal{M}(s) = \int_0^\infty u^{s-1} f(u)du \quad (s = c + it) \tag{5.9}$$

for $f \in X_c := \{f : \mathbb{R}^+ \to \mathbb{C},\ \|f\|_{X_c} = \int_0^\infty |f(u)|u^{c-1}du < \infty\}$, for some $c \in \mathbb{R}$, has its inverse given by

$$f(x) = \mathcal{M}^{-1}[f^\wedge_\mathcal{M}(s)](x) = \frac{1}{2\pi i}\int_{c-i\infty}^{c+i\infty} f^\wedge_\mathcal{M}(s)x^{-s}ds, \tag{5.10}$$

provided $f^\wedge_\mathcal{M} \in L^1(\{c\} \times i\mathbb{R})$. The Mellin convolution of f and g is given by

$$(f \circ g)(x) := \int_0^\infty f\Big(\frac{x}{u}\Big)g(u)\frac{du}{u} \tag{5.11}$$

provided it exists, and the associated convolution theorem reads for $f, g \in X_c$,

$$[f \circ g]^\wedge_\mathcal{M}(s) = [f]^\wedge_\mathcal{M}(s)[g]^\wedge_\mathcal{M}(s). \tag{5.12}$$

For the Mellin differential operator Θ_c, defined for $c \in \mathbb{R}$ by

$$\Theta_c f(x) := xf'(x) + cf(x), \quad x \in \mathbb{R}_+ \tag{5.13}$$

in case it exists, and that of order $n \in \mathbb{N}$ iteratively by $\Theta_c^1 f := \Theta_c f$, $\Theta_c^n f = \Theta_c(\Theta_c^{n-1} f)$, one has for $s = c + it$, $t \in \mathbb{R}$,

$$[\Theta_c^n f]^\wedge_\mathcal{M}(s) = (-it)^n f^\wedge_\mathcal{M}(s) \tag{5.14}$$

provided $f \in AC_{\rm loc}^{n-1}(\mathbb{R}_+) \cap X_c$ and $\Theta_c^n f \in X_c$. See especially [23] and the literature cited there for the foregoing approach to Mellin transforms.

Whereas this is the natural operator of differentiation in the Mellin setting, for the classical nth order derivative $D_x^n f(x)$ we have

$$\mathcal{M}[D_x^n f](s) = \frac{(-1)^n \Gamma(s)}{\Gamma(s-n)} f_{\mathcal{M}}^{\wedge}(s-n), \tag{5.15}$$

provided $f \in AC_{\rm loc}^{n-1}(\mathbb{R}_+)$ and $D_x^k f \in X_{c+k-n}$ for $k = 0, \dots, n$. Now for the classical fractional integral on $\mathbb{R}_+$, given by

$${}_0I_x^\alpha f(x) = \frac{1}{\Gamma(\alpha)} \int_0^x (x-u)^{\alpha-1} f(u) du \tag{5.16}$$

one has

$$\mathcal{M}[{}_0I_x^\alpha f(x)](s) = \frac{\Gamma(1-\alpha-s)}{\Gamma(1-s)} f_{\mathcal{M}}^{\wedge}(s+\alpha) \tag{5.17}$$

if $f \in X_{c+\Re e\, \alpha}$ and $c + \Re e\, \alpha < 1$. In fact, formally

$$\mathcal{M}\Big[\frac{x^\alpha}{\Gamma(\alpha)} \int_1^\infty \Big(1-\frac{1}{y}\Big)^{\alpha-1} f(\frac{x}{y}) \frac{dy}{y^2}\Big](s) = \mathcal{M}[x^\alpha \cdot (g_\alpha \circ f)(x)](s),$$

where $g_\alpha(x) = \Gamma(\alpha)^{-1}(x-1)_+^{\alpha-1} x^{-\alpha}$ (see [147, p. 43]) with $\mathcal{M}[g_\alpha(x)](s) = \Gamma(1-s)/\Gamma(1-s+\alpha)$, $\Re e\, s < 1$. Now $\mathcal{M}[x^\alpha f(x)](s) = f_{\mathcal{M}}^{\wedge}(s+\alpha)$ so that (5.17) follows since

$$\mathcal{M}[x^\alpha (g_\alpha \circ f)(x)](s) = [g_\alpha \circ f]_{\mathcal{M}}^{\wedge}(s+\alpha) = \frac{\Gamma(1-s-\alpha)}{\Gamma(1-(s+\alpha)+\alpha)} \cdot f_{\mathcal{M}}^{\wedge}(s+\alpha).$$

As to the fractional derivative on $\mathbb{R}^+$, one has by (5.15) under suitable conditions

$$\mathcal{M}[{}_0D_x^\alpha f](s) = \frac{(-1)^n \Gamma(s)}{\Gamma(s-n)} \mathcal{M}[{}_0I_x^{n-\alpha} f](s-n) \quad (n-1 \le \Re e\, \alpha < n). \tag{5.18}$$

Hence by (5.17)

$$\mathcal{M}[{}_0D_x^\alpha f](s) = \frac{(-1)^n \Gamma(s)}{\Gamma(s-n)} \frac{\Gamma(1-(s-\alpha))}{\Gamma(1-s+n)} f_{\mathcal{M}}^{\wedge}(s-\alpha).$$

By applying the functional equation

$$\Gamma(1-(s-\alpha)) = \frac{\pi}{\Gamma(s-\alpha)\sin\pi(s-\alpha)}$$

as well as ditto for $\alpha = n$, then

$$\begin{aligned}\mathcal{M}[{}_0D_x^\alpha f](s) &= (-1)^n\Gamma(s)\frac{\pi}{\Gamma(s-\alpha)\sin\pi(s-\alpha)}\frac{\sin\pi(s-n)}{\pi}f_{\mathcal{M}}^{\wedge}(s-\alpha) \\ &= \frac{\sin\pi s}{\sin\pi(s-\alpha)}\frac{\Gamma(s)}{\Gamma(s-\alpha)}\cdot f_{\mathcal{M}}^{\wedge}(s-\alpha).\end{aligned} \tag{5.19}$$

Sufficient conditions for the validity of (5.19) in case $n-1 \leq \Re e\ \alpha < n$ are, for instance, that $c+n-\Re e\ \alpha < 1$, $f \in AC_{\text{loc}}^{n-1}(\mathbb{R}_+)$ and $D_x^k f \in X_{c+k-\Re e\ \alpha}$ for $k = 0, \ldots, n$. Note that (5.19) reduces to (5.15) for $\alpha = n$. Thus the factor $(\sin\pi s)/\sin\pi(s-\alpha)$ is a replacement for $(-1)^n$ in the fractional case.

Observe that the fractional integral defined by (5.16) is not the natural one connected with a possible generalization of the differential operator given via (5.14) to the fractional case in the sense that one is the inverse to the other (in the sense of Theorem 11 of [23]). What would be needed is a fractional integral which is the true counterpart of the new differential operator (5.13), at first in the particular case $\alpha = n$. In this respect see also the theory of fractional powers of some classes of operators, including Riemann-Liouville and Weyl fractional integrals, via a Mellin transform approach due to McBride [114]. Rooney [143] presented an approach via Mellin multipliers.

The fractional integral $\Gamma(\alpha)\ {}_{-\infty}I_x^\alpha f(x) = \int_0^\infty u^{\alpha-1}f(x-u)du$, $\Re e\ \alpha > 0$ is the Mellin transform of $f(x-u)$. Applying the inverse Mellin transform $\mathcal{M}^{-1}$, we deduce the following integral representation of the function $f(x)$ in terms of its fractional integral, namely

$$f(x-u) = \frac{1}{2\pi i}\int_{\Re e\ \alpha-i\infty}^{\Re e\ \alpha+i\infty}\Gamma(\alpha)\ {}_{-\infty}I_x^\alpha f(x)u^{-\alpha}d\alpha \tag{5.20}$$

provided, for fixed $x \in \mathbb{R}$ and some $c > 0$, the function $\Gamma(\alpha)_{-\infty}I_x^\alpha f(x)$ belongs to $L^1(\{c\} \times i\mathbb{R})$ with respect to α.

This formula may be regarded as another integral counterpart of the Taylor series expansion of $f(x-u)$. Further, taking $x = 0$, we have

$$f(-u) = \frac{1}{2\pi i}\int_{\Re e\ \alpha-i\infty}^{\Re e\ \alpha+i\infty}\Gamma(\alpha)\ {}_{-\infty}I_x^\alpha\ f(x)|_{x=0}\,u^{-\alpha}d\alpha \quad (u > 0).$$

This implies that $f(x)$ can be recovered from the values of its fractional integrals ${}_{-\infty}I_x^\alpha f(x)$ taken at only the one point $x = 0$ if these are known for all α on the vertical line $\Re e\, \alpha = \text{ const } > 0$.

5.3 Laplace transforms and characterizations of fractional derivatives

The Laplace transform of a locally integrable function $f : \mathbb{R}_+ \to \mathbb{C}$, defined by

$$\mathcal{L}[f](s) = f_{\mathcal{L}}^{\wedge}(s) = \int_0^\infty e^{-su} f(u) du \tag{5.21}$$

is another powerful tool in solving differential and integral equations of fractional analysis. In this framework it is appropriate to assume that the integral in (5.21) converges absolutely in the complex half-plane $\Re e\, s > 0$. If, in addition, f is of bounded variation in a neighbourhood of some $x > 0$, then there holds the inversion formula

$$\mathcal{L}^{-1}[f_{\mathcal{L}}^{\wedge}](x) = PV \frac{1}{2\pi i} \int_{\sigma - i\infty}^{\sigma + i\infty} e^{xs} f_{\mathcal{L}}^{\wedge}(s) ds = \frac{f(x+0) + f(x-0)}{2} \quad (x > 0) \tag{5.22}$$

for each fixed $\sigma > 0$. The associated convolution relation is given by

$$(f \otimes g)(x) = \int_0^x f(x-u) g(u) du \tag{5.23}$$

the transform of which is $\mathcal{L}[f \otimes g](s) = \mathcal{L}[f](s)\mathcal{L}[g](s)$. Thus the transform of the fractional integral ${}_0I_x^\alpha f$, $\Re e\, \alpha > 0$, namely the Laplace convolution

$${}_0I_x^\alpha f(x) = \left(f \otimes \frac{x_+^{\alpha-1}}{\Gamma(\alpha)} \right)(x),$$

is given by, noting $\mathcal{L}[x_+^{\alpha-1}/\Gamma(\alpha)](s) = s^{-\alpha}$ for $\Re e\, \alpha > 0$,

$$\mathcal{L}[{}_0I_x^\alpha f](s) = s^{-\alpha}\mathcal{L}[f](s), \tag{5.24}$$

valid for sufficiently good functions.

As to fractional differentiation, first recall that for $n \in \mathbb{N}$

$$\mathcal{L}[\frac{d^n}{dx^n} f](s) = s^n \mathcal{L}[f](s) - \sum_{k=0}^{n-1} s^k f^{(n-1-k)}(0+), \tag{5.25}$$

so that

$$\begin{aligned} \mathcal{L}[{}_0D_x^\alpha f](s) &= \mathcal{L}[\frac{d^n}{dx^n}\, {}_0I_x^{n-\alpha} f](s) \\ &= s^n \mathcal{L}[{}_0I_x^{n-\alpha} f](s) - \sum_{k=0}^{n-1} s^k \frac{d^{n-1-k}}{dx^{n-1-k}}\, {}_0I_x^{n-\alpha} f(0+) \\ &= s^\alpha \mathcal{L}[f](s) - \sum_{k=0}^{n-1} s^k\, {}_0D_x^{\alpha-1-k} f(0+) \quad (n-1 < \alpha < n). \end{aligned} \tag{5.26}$$

If, however, one would work with definition (1.26) of fractional differentiation, i.e. $\overline{{}_0D_x^\alpha} f :={}_0 I_x^{n-\alpha} f^{(n)}(x)$, then

$$\mathcal{L}[{}_0\overline{D}_x^\alpha f](s) = s^\alpha \mathcal{L}[f](s) - \sum_{k=0}^{n-1} s^{\alpha-1-k} f^{(k)}(0+) \quad (n-1 < \alpha < n). \tag{5.27}$$

The essential difference between the expansions (5.26) and (5.27) is that the former involves the fractional derivatives ${}_0D_x^{\alpha-1-k} f(0+)$ $(k = 0, 1, .., n-1)$, whereas the latter involves only the integer derivatives $f^{(k)}(0+)$.

G. Doetsch [42, p. 163ff.] defines the fractional derivative as the solution g of the integral equation

$${}_0I_x^\alpha g(x) = \frac{1}{\Gamma(\alpha)} \int_0^x (x-u)^{\alpha-1} g(u)du = f(x)$$

in case it exists. Let us denote it by ${}_0^*D_x^\alpha f$. Then there holds the following connection between the three definitions: *Let $f, g \in L_{\text{loc}}(\mathbb{R}_+)$ and let their Laplace transforms converge absolutely for each s with $\Re e\ s > 0$. Then the following three assertions (i) - (iii) are equivalent for $n-1 \leq \alpha < n$:*

(i) $s^\alpha f_{\mathcal{L}}^\wedge(s) = g_{\mathcal{L}}^\wedge(s) \quad (\Re e\ s > 0)$,

(ii) $g = {}_0^*D_x^\alpha f$,

(iii) $\frac{d^k}{dx^k}[{}_0I_x^{n-\alpha}]f$ *are locally absolutely continuous on* $[0, \infty)$ *and equal to zero at* $x = 0$ *for* $k = 0, 1, \ldots, n-1$, *together with* ${}_0D_x^\alpha f = g$.

Each of these assertions is implied by the following sufficient condition:

(iv) $f^{(k)}(x)$ *are locally absolutely continuous on* $[0, \infty)$ *and equal to zero at* $x = 0$ *for* $k = 0, 1, \ldots, n-1$*, together with* ${}_0\overline{D}_x^\alpha f = g$.

For functions belonging to $L^p(\mathbb{R}_+)$, $1 \leq p < \infty$, these concepts can be related to the Marchaud-type derivative considered in Section 3.5; cf. Theorems 3.8–3.10.

References

1. M. Aigner, "Kombinatorik", I. Springer-Verlag, Berlin, 1975.
2. T.M. Apostol, "Introduction to Analytic Number Theory, Springer, New York 1976.
3. M.A. Al-Bassam, "On fractional analysis and its applications." In: Modern Analysis and its Applications (H.L. Manocha, ed.), Prentice Hall of India Ltd., New Delhi 1986, pp. 269-307.
4. S. Bagdasarov, "Chebyshev Splines and Kolmogorov Inequalities." Birkhäuser, Basel 1998.
5. R.L. Bagley and P.J. Torvik, "A theoretical basis for the application of fractional calculus to viscoelasticity." J. of Rheology 27 (3) (1983), 201–210.
6. H. Berens and U. Westphal, "Zur Charakterisierung von Ableitung nichtganzer Ordnung im Rahmen der Laplace-Transformation." Math. Nachr. 38 (1968), 115–129.
7. H. Berens and U. Westphal, "A Cauchy problem for a generalized wave equation." Acta Sci. Math. (Szeged), 29 (1968), 93–106.
8. H. Beyer and S. Kempfle, "Definition of physically consistent damping laws with fractional derivatives." ZAMM 75 (1995), 623–635.
9. B. Bittner, "Irregular sampling based on the fractional Fourier transform." In: P.J.S.G. Ferreira [56], pp. 431–436.
10. B. Bonilla, J. Trujillo and M. Rivero, "On fractional order continuity, integrability and derivability of real functions." In: Rusev-Dimovski-Kiryakova [150], pp. 48–55.
11. P.L. Butzer, "Beziehungen zwischen den Riemannschen, Taylorschen und gewöhnlichen Ableitungen reellwertiger Funktionen." Math. Ann. 144 (1961), 275-298.
12. P.L. Butzer and H. Berens, "Semi-Groups of Operators and Approximation." Grundlehren Math. Wiss. 145, Springer, Berlin 1967.
13. P.L. Butzer, H. Dyckhoff, E. Görlich and R.L. Stens, "Best trigonometric approximation, fractional order derivatives and Lipschitz classes."

Canad. J. Math. 29 (1977), 781–793.

14. P.L. Butzer, S. Flocke and M. Hauss, "Euler functions $E_\alpha(z)$ with complex α and applications." In: ***Approximation, Probability and Related Fields.*** Proc. of Int. Conference, Santa Barbara, May 1993, (G.A. Anastassiou and S.T. Rachev, eds.), Plenum Press, New York, 1994, pp. 127–150.
15. P.L. Butzer and A. Gessinger, "Ergodic theorems for semigroups and cosine operator functions at zero and infinity with rates; Applications to partial differential equations. A survey." Contemporary Mathematics, Vol. 190 (1995), 67–93.
16. P.L. Butzer and A. Gessinger, "The mean ergodic theorem for cosine operator functions with optimal and non-optimal rates." Acta Math. Hungar. 68(4) (1995), 319–351.
17. P.L. Butzer, M. Hauss and M. Leclerc, "Bernoulli numbers and polynomials of arbitrary complex indices." Appl. Math. Lett. 5, No. 6 (1992), 83–88.
18. P.L. Butzer and M. Hauss, "Stirling functions of first and second kind; some new applications. In: Approximation, Interpolation and Summability." (Proc. Conf. in honour of Prof. Jakimovski, Tel Aviv, June 1990), Israel Meth. Conf. Proceedings, Vol 4 (1991), Weizmann Press, Israel, pp. 89-108.
19. P.L. Butzer and M. Hauss, "Applications of sampling theory to combinatorial analysis, Stirling numbers, special functions and the Riemann zeta function." In: J.R. Higgins and R.L. Stens, eds. [72].
20. P.L. Butzer and M. Hauss, "On Stirling functions of the second kind." Stud. Appl. Math. 84 (1991), 71-79.
21. P.L. Butzer and M. Hauss, "Eulerian numbers with fractional order parameters." Aequationes Math. 46 (1993), 119-142.
22. P.L. Butzer, M. Hauss and M. Schmidt, "Factorial functions and Stirling numbers of fractional order." Resultate Math. 16 (1989), 16-48.
23. P.L. Butzer, S. Jansche, "A direct approach to the Mellin transform." J. Fourier Anal. Appl. 3 (1997), 325–376.
24. P.L. Butzer, J. Junggeburth, "On Jackson-type inequalities in approximation theory." In: General Inequalities I (Proc. Conf. Oberwolfach 1976, E.F. Beckenbach, Ed.) ISNM 41, Birkhäuser, Basel 1978, pp. 85–114.
25. P.L. Butzer, R.J. Nessel, "Fourier Analysis and Approximation." Vol. I, One Dimensional Theory. Birkhäuser Basel, and Academic Press, New York 1971.
26. P.L. Butzer and R. Stens, "The operational properties of the Chebyshev

transform. II: Fractional Derivatives." In: Theory of Approximation of Functions (Russian). (Proc. Int. Conf. Theory of Approximation of Functions; Kaluga, July 1975; Eds. S.B. Steckin-S.A. Teljakovskii) Moscow 1977, pp. 49–62.

27. P.L. Butzer and R. Stens, "Chebyshev transform methods in the solution of the fundamental theorem of best algebraic approximation in the fractional case." In: Fourier Analysis and Approximation Theory (Colloquia Math. Soc. János Bolyai, 19; Proc. Conf. Budapest, August 1976; Eds. G. Alexits - P. Turán) Amsterdam 1978, pp. 191–212.
28. P.L. Butzer and W. Trebels, "Hilberttransformation, gebrochene Integration und Differentiation". Westdeutscher Verlag, Köln and Opladen 1968, 81 pp.
29. P.L. Butzer and U. Westphal, "The mean ergodic theorem and saturation." Indiana Univ. Math. J. 20 (1971), 1163–1174.
30. P.L. Butzer and U. Westphal, "An access to fractional differentiation via fractional difference quotients." In: Fractional Calculus and its Applications, Lecture Notes in Math., Springer, Berlin 1975, pp. 116-145
31. L.M.B.C. Campos, "On a concept of derivative of complex order with applications to special functions." Portug. Math. 43 (1985), 347–376.
32. M. Caputo, "Elasticità e Dissipazione." Zanichelli, Bologna 1969.
33. A. Carpinteri and F. Mainardi (Eds.), "Fractals and Fractional Calculus in Continuum Mechanics." Springer, Vienna and New York 1997.
34. Y.W. Chen, "Entire solutions of a class of differential equations of mixed type." Comm. Pure and Appl. Math. 14 (1961), 229–255.
35. L. Comtet, "Advanced Combinatorics." D. Reidel Publishing Company, Dordrecht 1974 (Revised Edition).
36. J. Cossar, "A theorem on Cesàro summability." J. London Math. Soc. 16 (1941), 56–68.
37. H.T. Davis, "The Theory of Linear Operators." Principia Press, Bloomington, Indiana 1936.
38. D. Delbourgo and D. Elliott, "On the approximate evaluation of Hadamard finite-part integrals." IMA J. Numer. Anal. 14 (1994), 485–500.
39. J.B. Diaz and T.J. Osler, "Differences of fractional order." Math. Comput. 28 (1974), 185–202.
40. K. Diethelm and A.D. Freed, "On the solution of nonlinear fractional differential equations used in the modeling of viscoplasticity." In: Scientific Computing and Chemical Engineering I: Computational Fluid Dynamics, Reaction Engineering, and Molecular Properties (F. Keil, W. Mackens, H. Voß and J. Werther, Eds.), Springer, Berlin 1999, pp. 217–224

41. K. Diethelm and A.D. Freed, "The FracPECE subroutine for the numerical solution of differential equations of fractional order". In: Forschung und wissenschaftliches Rechnen 1998 (T. Plesser and S. Heinzel, Eds.), Gesellschaft für wissenschaftliche Datenverarbeitung, Göttingen 1999.
42. G. Doetsch, "Handbuch der Laplace Transform, Band III." Birkhäuser, Basel 1956.
43. J.F. Douglas, "Surface interacting polymers: An integral equation and fractional calculus approach." Macromolecules 22 (1989), 1786–1797
44. D.P. Drianov, "Equivalence between fractional average modulus of smoothness and fractional K-functional." C.R. Acad. Bulg. Sci. 38 (1985), 1609–1612.
45. M.M. Dzherbashyan, "Integral Transforms and Representations of Functions in the Complex Domain." (Russian), Nauka, Moscow 1966.
46. M.M. Dzherbashyan, "A generalized Riemann-Liouville operator." (Russian), Dokl. Akad. Nauk SSSR, 177(4) (1967), 767–770.
47. M.M. Dzherbashyan, "The criterion of the expansion of the functions to the Dirichlet series." (Russian), Izv. Akad. Nauk Armyan. SSR Ser. Fiz.-Mat. Nauk 11(5) (1958), 85–108.
48. D. Elliott, "An asymptotic analysis of two algorithms for certain Hadamard finite-part integrals." IMA J. Numer. Anal. 13 (1993), 445–462.
49. A.M.A. El-Sayed, "Fractional order diffusion-wave equation." Int. J. Theor. Phys. 35 (1996), 311-322.
50. H. Engler, "Similarity solutions for a class of hyperbolic integrodifferential equations." Differential Integral Eqns. (1997) (to appear).
51. A. Erdélyi, "On fractional integration and its applications to the theory of Hankel transforms." Quart. J. Math. (Oxford), 11 (1940), 293–303.
52. A. Erdélyi, W. Magnus, F. Oberhettinger and F.G. Tricomi, "Tables of Integral Transforms". Vol. 2, pp. 185–200, McGraw-Hill, New York, 1954.
53. V.I. Fabrikant, "Applications of Potential Theory in Mechanics: A Selection of New Results." Kluwer Academic Publ., Dordrecht 1989.
54. V.I. Fabrikant, "Mixed Boundary Value Problems of Potential Theory and their Applications in Engineering." Kluwer, Dordrecht 1991.
55. S. Fenyö and H.W. Stolle, "Theorie und Praxis der linearen Integralgleichungen." Deutscher Verlag d. Wiss., Berlin 1963.
56. P.J.S.G. Ferreira (Ed.), "SAMPTA '97 Int. Workshop on Sampling Theory and Applications." Universidade de Aveiro, Proc. Conf. June 1997, Aveiro 1997.
57. G. Gasper and W. Trebels, "Multiplier criteria of Hörmander type for

Jacobi series and application to Jacobi series and Hankel transforms." Math. Ann. 242 (1979), 225–240.
58. G. Gasper and W. Trebels, "Necessary conditions for Hankel multipliers." Indiana Univ. Math. J. 31 (1982), 403–414.
59. M.C. Gaer – L.A. Rubel, "The fractional derivative and entire functions." In: B. Ross, ed. [144, pp. 171–206].
60. I. Genebashvili, A. Gogatishvili, V. Kokilashvili and M. Krbec "Weight Theory of Integral Transforms on Spaces of Homogeneous Type." Pitman Monographs and Surveys in Pure and Applied Mathematics 92, Longman, Harlow 1998.
61. A. Gessinger, "Connections between the approximation and ergodic behaviour of cosine operators and semigroups." Proc. 3rd Int. Conf. Functional Analysis and Approximation Theory, Maratea, Sept. 1996 (ed. by F. Altomare). In: Serie II, Suppl. Rend. Circ. Mat. Palermo 52 (1998), 475–489.
62. A. Gessinger, "Der Zusammenhang zwischen dem approximations- und ergodentheoretischen Verhalten von Halbgruppen und Kosinusoperatorfunktionen; neue Anwendungen der Kosinusoperatortheorie." Shaker Verlag, Aachen 1997, 141 pp.
63. R. Gorenflo and F. Mainardi, "Fractional calculus: integral and differential equations of fractional order." In: Carpinteri et al [33].
64. R. Gorenflo and F. Mainardi, "Fractional oscillations and Mittag-Leffler functions." In: Recent Advances in Applied Mathematics (RAAM '96; Proc. Int. Workshop, Kuwait, May 1996), Kuwait 1996, pp. 193–208.
65. R. Gorenflo and S. Vessella, "Abel Integral Equations: Analysis and Applications." Springer, Berlin 1991.
66. A.K. Grünwald, "Über "begrenzte" Derivationen und deren Anwendung." Zeit. für Mathematik und Physik 12 (1867), 441–480.
67. M. Hauss, "Über die Theorie der fraktionierten Stirling-Zahlen und deren Anwendung." Diplomarbeit, Lehrstuhl A für Mathematik RWTH Aachen 1990, 161 pp.
68. H. Hayakawa, "Fractional dynamics in phase ordering processes." Fractals 1 (1993), 947–953.
69. N. Hayek, J. Trujillo, M. Rivero, B. Bonilla and J.C. Moreno, "An extension of Picard-Lindelöf theorem to fractional differential eqations." Appl. Anal. 70(1999), 347–361.
70. H.P. Heinig,"Weighted inequalities in Fourier analysis." Nonlinear Analysis, Function Spaces and Appl. 4, Proc. Spring School 1990, Teubner-Verlag, Leipzig 1990, pp. 42–85.
71. J.R. Higgins, "Sampling Theory in Fourier and Signal Analysis. Foun-

dations." Clarendon Press, Oxford 1996.
72. J.R. Higgins and R.L. Stens (Eds.), "Sampling Theory in Fourier and Signal Analysis. Advanced Topics." Clarendon Press, Oxford 1999. (in print)
73. R. Hilfer, "Thermodynamic scaling derived via analytic continuation from the classification of Ehrenfest", Physica Scripta 44 (1991),321–322
74. R. Hilfer, "Multiscaling and the classification of continuous phase transitions", Phys. Rev. Lett. 68,(1992), 190–192
75. R. Hilfer, "Classification theory for anequilibrium phase transitions", Phys.Rev. E 48 (1993), 2466–2475
76. R. Hilfer, "On a new class of phase transitions." In: Random magnetism and high-temperature superconductivity. W. Beyermann, N. Huang-Liu and D. MacLaughlin, Eds., World Scientific Publ. Co., Singapore 1994, pp. 85–99.
77. R. Hilfer, "Exact solutions for a class of fractal time random walks", Fractals 3 (1995), 211–216
78. R. Hilfer, "Foundations of fractional dynamics", Fractals 3 (1995), 549–556
79. R. Hilfer, "An extension of the dynamical foundation for the statistical equilibrium concept", Physica A 221 (1995), 89–96 Fractals 3 (1995), 549–556
80. R. Hilfer, "On fractional diffusion and its relation with continuous time random walks." In: Anomalous Diffusion: From Basis to Applications. R. Kutner, A. Pekalski and K. Sznajd-Weron, Eds., Springer Verlag, Berlin 1999, pp. 77–82.
81. R. Hilfer and L. Anton, "Fractional master equations and fractal time random walks", Phys.Rev. E 51 (1995), R848–R851
82. E. Hille, "Generalizations of Landau's inequality to linear operators." In: Linear Operators and Approximation (Proc. Conf. Oberwolfach, August 1971, P.L. Butzer, J.P. Kahane and B. Sz. Nagy, Eds.), ISNM, Vol. 20, Birkhäuser, Basel 1972, pp. 20–32.
83. P. Hilton, D. Holton and J. Pedersen, "Mathematical Reflections; in a Room with many Mirrors." Springer, New York 1996.
84. S. Jansche, "$\mathcal{O}$-regularly varying functions in approximation theory." J. Inequal. & Appl. 1 (1997), 253–274.
85. C. Jordan, "Calculus of Finite Differences." Chelsea Publishing, New York 1965.
86. J. Junggeburth, K. Scherer and W. Trebels, "Zur besten Approximation auf Banachräumen mit Anwendungen auf ganze Funktionen." Forschungsberichte des Landes Nordrhein-Westfalen Nr. 2311, West-

deutscher Verlag Opladen 1973.
87. R.N. Kalia (Ed.), Recent Advances in Fractional Calculus. Global Publ. Co., Sauk Rapids, Minnesota, 1993.
88. R.N. Kalia "Generalized derivatives and special functions." In: R.N. Kalia [87], pp. 292–307.
89. R.N. Kalia "Fractional calculus; Its brief history and recent advances". In: R.N. Kalia [87], pp. 1–30.
90. L. Kalla and R.K. Saxena "Integral operators involving generalized hypergeometric functions." Math. Z. 108 (1969), 231–234.
91. V. Kiryakova, "Generalized Fractional Calculus and Applications." Pitman Research Notes in Mathematics No. 301. Longman Scientific & Technical, Harlow 1994.
92. V. Kiryakova (Chief-Ed.), "Fractional Calculus & Applied Analysis." Inst. of Math. & Informatics, Bulg. Acad. Sci., Sofia (a journal that first appeared 1998).
93. H. Kober, "On fractional integrals and derivatives." Quart. J. Math. (Oxford), 11 (1940), 193–211.
94. R.C. Koeller, "Applications of fractional calculus to the theory of viscoelasticity." J. Appl. Mech. 51 (1984), 299–307.
95. V. Kokilashvili and M. Krbec, "Weighted inequalities in Lorentz and Orlicz Spaces." World Scientific. Singapore, etc., 1991.
96. V. Kokilashvili and A. Meskhi, "Boundedness and compactness criteria for fractional integral transforms." (in print)
97. V. Kokilashvili and A. Meskhi, "Criteria for the boundness and compactness of operators with power-logarithmic kernels." Analysis Mathematica (to appear)
98. H. Komatsu, "Fractional powers of operators IV: Potential operators." J. Math. Soc. Japan 21 (1969), 221–228.
99. M.K. Kwong and A. Zettl, "Norm Inequalities for Derivatives and Differences." Lecture Notes in Mathematics, No. 1536, Springer, Berlin 1992.
100. S.F. Lacroix, "Traité du Calcul Differential et du Calcul Integral." Vol. 3, pp. 409–410, Courcier, Paris ²1819.
101. J.L. Lavoie, T.J. Osler and R. Tremblay, "Fundamental properties of fractional derivatives via Pochhammer integrals." In B. Ross, ed. [144, pp. 323–356].
102. J.L. Lavoie, T.J. Osler and R. Tremblay, "Fractional derivatives and special functions." SIAM Rev. 18 (1976), 240–268.
103. J. Liouville, "Memoire sur quelques quéstions de géometrie et de mécanique, et sur un noveau genre pour réspondre ces quéstions." Jour.

École Polytech. 13 (1832), 1–69.
104. P.I. Lizorkin, "Bounds for trigonometrical integrals and the Bernstein inequality for fractional derivatives (Russian)." Izv. Akad. Nauk SSSR, Ser. Mat. 29(1) (1965), 109–126.
105. E.R. Love, "Two index laws for fractional integrals and derivatives." J. Australian Math. Soc. 14 (1972), 385–410.
106. E.R. Love and Y.C. Young, "On fractional integration by parts." Proc. Lond. Math. Soc. 44 (1938), 1–35.
107. J. Lützen, "Joseph Liouville 1809–1882. Master of Pure and Applied Mathematics." Springer, New York, Berlin, Heidelberg, 1990.
108. W. Magnus, F. Oberhettinger and R.P. Soni, "Formulas and Theorems for the Special Functions of Mathematical Physics." Springer, New York 1966.
109. F. Mainardi, "Fractional relaxation-oscillation and fractional diffusion - wave phenomena." Chaos, Solitons & Fractals 7 (1996), 1461–1477.
110. F. Mainardi, "Fractional calculus: Some basic problems in continuum and statistical mechanics." In: Carpinteri et al [33].
111. A. Marchaud, "Sur les dérivées et sur les différences des fonctions des variables réelles." J. Math. Pures Appl. (9) 6 (1927), 337–425.
112. A.M. Mathai and R.K. Saxena, "The H-Function with Applications to Statistics and Other Disciplines." Wiley, New Delhi 1978.
113. A.C. McBride, "Fractional Calculus and Integral Transforms of Generalized Functions." Pitman Research Notes in Mathematics, ♯ 31, Pitman, London 1979.
114. A.C. McBride, "A Mellin transform approach to fractional calculus on $(0, \infty)$." In: McBride - Roach [115], pp. 99–139.
115. A.C. McBride - G.F. Roach, "Fractional Calculus". Pitman, Research Notes in Math. 138, 1985.
116. H.J. Mertens and R.J. Nessel, "A sufficient condition for α-fold quasiconvex multipliers of strong convergence." In: Approximation and Function Spaces (Proc. Conf. Gdańsk, 27.-31.8.1979, Ed. Z. Ciesielski). Polish Sci. Publ., Warszawa 1981, and North-Holland, Amsterdam 1981, pp. 436–447.
117. H.J. Mertens, R.J. Nessel and G. Wilmes, "Über Multiplikatoren zwischen verschiedenen Banach-Räumen im Zusammenhang mit diskreten Orthogonalentwicklungen." Forsch.-Bericht d. Landes Nordrhein-Westfalen Nr. 2599, Westdeutscher Verlag, Opladen 1976, 1–55 pp.
118. A. Meskhi, "Solution of some weight problems for the Riemann-Liouville and Weyl operators." Georgian Math. J. 5(1998), 565–574.
119. R. Metzler and T.F. Nonnenmacher, "Fractional diffusion: exact repre-

sentations of spectral functions." J. Phys. A: Math. Gen. 30 (1997), 1089–1093.

120. M. Mikolás, "On the recent trends in the development, theory and applications of fractional calculus, in B. Ross, ed. [144], pp. 357– 375.
121. K.S. Miller and B. Ross, "An Introduction to Fractional Calculus and Fractional Differential Equations". Wiley, New York 1993.
122. R.J. Nessel, "Towards a survey of Paul Butzer's contributions to approximation theory." In: Mathematical Analysis, Wavelets and Signal Processing. Contemp. Math. 190, Amer. Math. Soc., Providence, RI, 1995, pp. 31–65.
123. R.J. Nessel and W. Trebels, "Gebrochene Differentiation und Integration und Charakterisierungen von Favard–Klassen." In: Constructive Theory of Functions (Proc. Conf. Budapest, August 1969, Eds. G. Alexits, S.B. Steckin). Akad. Kiadò, Budapest 1972, pp. 331–341.
124. K. Nishimoto, "Fractional Calculus." Vols. I-IV, Descartes Press, Koriyama 1984–1991.
125. K. Nishimoto (Ed.), "Fractional Calculus and its Applications." College of Engineering, Nihon University, Japan 1990.
126. K. Nishimoto (Ed.), "Journal of Fractional Calculus." Descartes Pres, Koriyama, Japan 963 (Volume 8 appeared in November 1995).
127. K. Nishimoto and Shih-Tong Tu, "Complementary functions in Nishimoto's calculus." J. Frac. Calc. 3 (1993), 39–48.
128. T.F. Nonnenmacher, "Fractional integral and differential equations for a class of Lévy-type probability densities." J. Phys. A: Math. Gen. 23 (1990), L697–L700.
129. T.F. Nonnenmacher and W.G. Glöckle, "A fractional model for mechanical stress relaxation." Phil. Mag. Lett. 64 (2) (1991), 89–93.
130. T.F. Nonnenmacher and R. Metzler, "On the Riemann-Liouville fractional calculus and some recent applications." Fractals 3 (1995), 557–566.
131. G.O. Okikiolu, "Fractional integrals of the H-type." Quart. J. Math. (Oxford), 18 (1967), 33–42.
132. G.O. Okikiolu, "Aspects of the Theory of Bounded Integral Operators in L_p-spaces." Academic Press, London 1971.
133. K.B. Oldham and J. Spanier, "The Fractional Calculus." Academic Press, New York 1974.
134. T.J. Osler, "Leibniz rule for fractional derivatives, generalized and an application to infinite series." SIAM J. Appl. Math. 18 (1970), 658–674.
135. T.J. Osler, "An integral analogue of Taylor's series and its use in computing Fourier transforms." Math. Comput. 26 (1972), 449–460.
136. T.J. Osler, "The integral anlog of the Leibniz rule." Math. Comput. 26

(1972), 903–915.
137. T.J. Osler, "A further extension of the Leibniz rule to fractional derivatives and its relation to Parseval's formula." SIAM J. Math. Anal. 3 (1972), 1–16.
138. V.G. Ponomarenko, "Modulus of smoothness of fractional order and the best approximation in L_p, $1 < p < \infty$." (Russian). In: Constructive Function Theory '81, (Proc. Conf. Varna, June 1981). Publ. House Bulg. Acad. Sci., Sofia 1983, pp. 129–133.
139. B. Riemann, "Versuch einer allgemeinen Auffassung der Integration und Differentiation." Gesammelte Mathematische Werke und Wissenschaftlicher Nachlass. Teubner, Leipzig 1876 (Dover, New York, 1953), pp. 331–344.
140. M. Riesz, "L'intégrale de Riemann-Liouville et le problème de Cauchy." Acta Math. 81 (1949), 1–223.
141. J. Riordan, "An Introduction to Combinatorial Analysis." John Wiley & Sons, New York, 1958.
142. J. Riordan, "Combinatorial Identities." Wiley 1968. Corrected ed. , R. E. Krieger Publishing Co. , Huntington, New York 1979.
143. P.G.Rooney, "A survey of Mellin multipliers." In: McBride and Roach [115], pp. 176–187.
144. B. Ross (Ed.), "Fractional Calculus and its Applications." Lecture Notes in Mathematics, ♯ 457, Springer, Berlin 1975.
145. B. Ross, "The development of fractional calculus 1695–1900." Historia Math. 4 (1977), 75–89.
146. B. Ross, "Fractional calculus." Math. Magazine 50 (1977), 115–122.
147. B. Rubin, "Fractional Integrals and Potentials." Pitman Monographs and Surveys in Pure and Applied Mathematics no. 82, Longman 1996.
148. Th. Runst and W. Sickel, "Sobolev Spaces of Fractional Orders." Nemytskij Operators, and Nonlinear Partial Differential Equations." De Gruyter, Berlin 1996.
149. S. Ruscheweyh, "New criteria for univalent functions." Proc. Amer. Math. Soc. 49 (1975), 109–115.
150. P. Rusev, I. Dimovski and V. Kiryakova (Eds.), "Transform Methods & Special Functions." (Proc. Sec. Int. Workshop, August 1996, Varna). Institute of Mathematics & Informatics, Bulg. Acad. Sciences, Sofia 1998.
151. S.G. Samko, "The coincidence of Grünwald–Letnikov differentiation with other forms of fractional differentiation. The periodic and non–periodic cases (Russian)." In: Reports of the extended Session of the Seminar of the I.N. Vekua Inst. Appl. Math., Tbilisi 1985. Tbiliss. Gos. Univ. 1,

no. 1, 183–186.

152. S.G. Samko, "Hypersingular Integrals and their Applications." (Russian). Izdat. Rostov Univ, Rostov-on-Don 1984.
153. S.G. Samko, A. A. Kilbas and O. I. Marichev, "Fractional Integrals and Derivatives. Theory and Applications." Gordon and Breach, Amsterdam 1993.
154. R.K. Saxena. "On fractional integration operators." Math. Z. 96 (1967), 288–291.
155. H. Schiessel and A. Blumen, "Fractal aspects in polymer science." Fractals 3 (1995), 483–490.
156. W.R. Schneider, "Grey noise." In: 'Ideas and Methods in Mathematical Analysis. Stochastics, and Applications, Vol. 1, S. Albeverio, J.E. Fenstad, H. Holden and T. Lindstrøm, eds., Cambridge University Press, 1992, pp. 261–282
157. W.R. Schneider and W. Wyss, "Fractional diffusion and wave equations." J. Math. Phys. 30 (1989), 134–144.
158. G. Sunouchi. "Direct theorems in the theory of approximation." Acta Math. Acad. Sci. Hungar. 20 (1969), 409–420.
159. I.N. Sneddon, "Mixed Boundary Value Problems in Potential Theory." North- Holland Publ., Amsterdam 1966.
160. H.M. Srivastava and S. Owa (Eds.), "Univalent Functions, Fractional Calculus, and their Applications." Halsted (Horwood), Chichester; Wiley, New York 1989.
161. H.M. Srivastava, "Some applications of fractional calculus involving the gamma and related functions." In R.N Kalia [87], pp. 179–202.
162. H.M. Srivastava and H.L. Manocha, "A Treatise on Generating Functions." Cambridge University Press, Cambridge 1966.
163. R. Taberski, "Differences, moduli and derivatives of fractional orders." Ann. Soc. Math. Polon. Comm. Prace Mat. 19 (2) (1977), 389–400.
164. R. Taberski, "Estimates for entire functions of exponential type." Funct. Approx. Comment. Math. 13 (1982), 129–147.
165. R. Taberski, "Contribution to fractional calculus and exponential approximation." Funct. Approx. Comment. Math. 15 (1986), 81–106.
166. W. Trebels, "Multipliers for (C, α)-Bounded Fourier Expansions in Banach Spaces and Approximation Theory." Lecture Notes in Math. No. 329, Springer, Berlin 1973.
167. W. Trebels and U. Westphal, "On the Landau-Kallman-Rota-Hille inequality." In: Linear Operators and Approximation (Proc. Conf. Oberwolfach, August 1971, P.L. Butzer, J.P. Kahane and B. Sz. Nagy, Eds.), ISNM, Vol. 20, Birkhäuser, Basel 1972, pp. 115–119.

168. J. Trujillo, M. Rivero and B. Bonilla, "On a Riemann-Liouville generalized Taylor's formula." J. Math. Anal. Appl. 231 (1999), 255-265.
169. Vu Kim Tuan and R. Gorenflo, "The Grünwald-Letnikov difference operator and regularization of the Weyl fractional differentiation." Z. Anal. Anwendungen 13 (1994), 537–545.
170. Y. Watanabe, "Notes on the generalized derivative of Riemann-Liouville and its application to Leibniz's formula." I. and II. Tôhoku Math. J. 34 (1931), 8-27 and 28-41.
171. B.J. West, P. Grigolini, R. Metzler and T.F. Nonnenmacher, "Fractional diffusion and Lévy stable processes." Physical Review E, 55 (1997), 99–106.
172. U. Westphal, "An approach to fractional powers of operators via fractional differences." Proc. London Math. Soc. (3) 29 (1974), 557–576.
173. U. Westphal, "Ein Kalkül für gebrochene Potenzen infinitesimaler Erzeuger von Halbgruppen und Gruppen von Operatoren; Teil I: Halbgruppenerzeuger; Teil II: Gruppenerzeuger." Compositio Mathematica 22 (1970), 67–103; 104–136.
174. U. Westphal, "Gebrochene Potenzen abgeschlossener Operatoren, definiert mit Hilfe gebrochener Differenzen." In: 'Linear Operators and Approximation II' (Proc. Oberwolfach Conf.; Ed. by P.L. Butzer and B. Sz.-Nagy), ISNM Vol. 25, Birkhäuser, Basel, 1974, pp. 23–27.
175. U. Westphal, "Fractional powers of infinitesimal generators." This Volume.
176. U. Westphal, "A generalized version of the Abelian mean ergodic theorem with rates for semigroup operators and fractional powers of infinitesimal generators." Result. Math. 34 (1998), 381-394.
177. H. Weyl, "Bemerkungen zum Begriff des Differentialquotienten gebrochener Ordnung." Vierteljschr. Naturforsch. Ges. Zürich 62 (1917), 296–302.
178. G. Wilmes, "On Riesz-type inequalities and K-functionals related to Riesz potentials in $\mathbb{R}^n$." Numer. Funct. Anal. Optim. 1 (1979), 57-77.
179. G. Wilmes, "Some inequalities for Riesz potentials of trigonometric polynomials of several variables." Proc. Sympos. Pure Math. Harmon. Anal. Euclidean Spaces, Amer. Math. Soc. (Williamstown, 1978) Providence, RI, 1979, pp. 175–182.
180. G.M. Zaslavsky, "Anomalous transport in systems with Hamiltonian chaos." In: Transport, Chaos and Plasma Physics (Conference Marseille 1993; S. Benkadda, F. Doveil and V. Eiskens, Eds.). World Scientific Publishing, Singapore 1994.

181. A. Zygmund, "Trigonometric Series." Vol. I, II. Cambridge Univ. Press, Cambridge 1968 (2. Edition, reprinted).

CHAPTER II

FRACTIONAL TIME EVOLUTION

R. HILFER
ICA-1, Universität Stuttgart, 70569 Stuttgart, Germany
Institut für Physik, Universität Mainz, 55099 Mainz, Germany

Contents

1 Introduction

A large number of problems in theoretical physics, including Schrödingers, Maxwells and Newtons equations, can be formulated as initial value problems for dynamical evolution equations of the form

$$\frac{\mathrm{d}}{\mathrm{d}t} f(t) = \mathrm{B} f(t) \tag{1}$$

where $t \in \mathbb{R}$ denotes time and B is an operator on a Banach space. Depending on the initial data $f(0) = f_0$ describing the state or observable of the system at time $t = 0$ the problem is to find the state or observable $f(t)$ of the system at later times $t > 0$. [a]

Many authors, mostly driven by the needs of applied problems, have considered generalizations of equation (1) of the form

$$\frac{\text{"}\mathrm{d}^\alpha\text{"}}{\mathrm{d}t^\alpha} f(t) = \mathrm{B} f(t) \tag{2}$$

in which the first order time derivative $\mathrm{d}/\mathrm{d}t$ is replaced with a certain fractional time derivative "$\mathrm{d}^\alpha/\mathrm{d}t^\alpha$" of order $\alpha > 0$ (see e.g. [1]–[24] and the Chapters IV –VIII in this book). A number of fundamental questions are raised by such a replacement. In order to appreciate these it is useful to recall that the appearance of $\mathrm{d}/\mathrm{d}t$ in eq. (1) reflects not only a basic symmetry of nature but also the basic principle of locality. Of course, the symmetry in question is time translation invariance. Remember that

$$\frac{\mathrm{d}}{\mathrm{d}s} f(s) = \lim_{t \to 0} \frac{f(s) - f(s-t)}{t} = -\lim_{t \to 0} \frac{\mathcal{T}(t) f(s) - f(s)}{t} \tag{3}$$

[a] In classical mechanics the states are points in phase space, the observables are functions on phase space, and the operator B is specified by a vector field and Poisson brackets. In quantum mechanics (with finitely many degrees of freedom) the states correspond to rays in a Hilbert space, the observables to operators on this space, and the operator B to the Hamiltonian. In field theories the states are normalized positive functionals on an algebra of operators or observables, and then B becomes a derivation on the algebra of observables. The equations (1) need not be first order in time. An example is the initial-value problem for the wave equation for $g(t, x)$

$$\frac{\partial^2 g}{\partial t^2} = c^2 \frac{\partial^2 g}{\partial x^2}$$

in one dimension. It can be recast into the form of eq. (1) by introducing a second variable h and defining

$$f = \begin{pmatrix} g \\ h \end{pmatrix} \quad \text{and} \quad \mathrm{B} = \begin{pmatrix} 0 & 1 \\ 1 & 0 \end{pmatrix} c \frac{\partial}{\partial x}$$

identifies $-\mathrm{d}/\mathrm{d}t$ as the infinitesimal generator of time translations[b] defined by

$$\mathcal{T}(t)f(s) = f(s-t). \tag{4}$$

Equation (2) abandons $\mathcal{T}(t)$ as the general time evolution, and this raises the question what replaces eq. (4), and how a fractional derivative can arise as the generator of a physical time evolution. Most workers in fractional calculus have avoided these questions, and my purpose in this chapter is to review and discuss an answer provided recently in [6,7,8,9,10,11].

Derivatives of fractional order $0 < \alpha \leq 1$ were found to emerge quite generally as the infinitesimal generators of coarse grained macroscopic time evolutions given by [6,7,8,9,10,11]

$$\mathrm{T}_\alpha(t)f(t_0) = \int_0^\infty \mathcal{T}(s)f(t_0)h_\alpha\left(\frac{s}{t}\right)\frac{\mathrm{d}s}{t} \tag{5}$$

where $t \geq 0$ and $0 < \alpha \leq 1$. Explicit expressions for the kernels $h_\alpha(x)$ for all $0 < \alpha \leq 1$ are given in eq. (69) below. It is the main objective of this chapter to show that (in a certain sense) all macroscopic time evolutions have the form of eq. (5), and that fractional time derivatives are their infinitesimal generators.

Given the great difference between $\mathrm{T}_\alpha(t)$ in eq. (5) and $\mathrm{T}_1(t) = \mathcal{T}(t)$ in eq. (4) it becomes clear that basic issues, such as irreversibility, translation symmetry, or the meaning of stationarity are inevitably involved when proposing fractional dynamics. Let me therefore advance the basic postulate that all time evolutions of physical systems are irreversible. Obviously this *law of irreversibility* must be considered to be an empirical law of nature equal in rank to the law of energy conservation. Reversible behaviour is an idealization. Its validity or applicability in physical experiments depends on the degree to which the system can be isolated (or decoupled) from its past history and its environment. According to this view the irreversible flow of time is more fundamental than the time reversal symmetry of Newtons or other equations. My starting point is therefore that for a general time evolution operator $\mathrm{T}(t)$ the evolution parameter t is not a time instant (which could be positive or negative), but a duration, which cannot be negative.

An immediate consequence of the postulated law of irreversibility is that the classical irreversibility problem of theoretical physics becomes reversed.

[b] A simple translation with unit "speed" reflects the idea of time "flowing" uniformly with constant velocity. This idea is embodied in measuring time by comparison with periodic processes (clocks). A competing idea, related to the flow of time represented by eq. (5), is to measure time by comparison with nonperiodic clocks such as decay or aging processes.

Now the theoretical task is not to explain how irreversibility arises from reversible evolution equations, but how seemingly reversible equations arise as idealizations from an underlying irreversible time evolution. A possible explanation is provided by the present theory based on eq. (5). It turns out that the case $\alpha = 1$ in eq. (5) is of predominant mathematical and physical importance, because it is in a quantifiable sense a strong universal attractor. In this case the kernel $h_1(x)$ becomes

$$h_1(x) = \lim_{\alpha \to 1^-} h_\alpha(x) = \delta(x-1), \tag{6}$$

and the time evolution $\mathrm{T}_1(t) = \mathcal{T}(t)$ in (5) reduces to a simple translation as in eq.(4). $T_1(t)$ with $t \geq 0$ is a representation of the time semigroup $(\mathbb{R}_+, +)$. It can be extended to one of the full group $(\mathbb{R}, +)$. This is not possible for T_α with $0 < \alpha < 1$. The physical interpretation of α is seen from $\operatorname{supp} h_\alpha = \mathbb{R}_+$ for $\alpha \neq 1$ and $\operatorname{supp} h_1 = \{1\}$ for $\alpha = 1$. Hence the parameter α classifies and quantifies the influence of the past history. Small values of α correspond to a strong influence of the past history. For $\alpha = 1$ the influence of the past history is minimal in the sense that it enters only through the present state.

The basic result in eq. (5) was given in [6] and subsequently rationalized within ergodic theory by investigating the recurrence properties of induced automorphisms on subsets of measure zero [9,10,11]. In these investigations the existence of a recurrent subset of measure zero had to be assumed. Such an assumption becomes plausible from observations in low dimensional chaotic systems (see e.g. [25,26] and Chapter V). A rigorous proof for any given dynamical system, however, appears difficult, and it is therefore of interest to rederive the emergence of $\mathrm{T}_\alpha(t)$ from a different, and more general, approach.

2 Foundations

2.1 Basic Desiderata for Time Evolutions

The following basic requirements define a time evolution in this chapter.

1. Semigroup
   A time evolution is a pair $(\{\mathrm{T}_\tau(t) : 0 \leq t < \infty\}, (B_\tau, \|\cdot\|))$ where $\mathrm{T}_\tau(t) = \mathrm{T}(t\tau)$ is a semigroup of operators $\{\mathrm{T}(t) : 0 \leq t < \infty\}$ mapping the Banach space $(B_\tau(\mathbb{R}), \|\cdot\|)$ of functions $f_\tau(s) = f(s\tau)$ on $\mathbb{R}$ to itself. The argument $t \geq 0$ of $\mathrm{T}_\tau(t)$ represents a time duration, the argument $s \in \mathbb{R}$ of $f_\tau(s)$ a time instant. The index $\tau > 0$ indicates the units (or scale) of time. Below, τ will again be frequently suppressed to simplify the notation. The elements $f_\tau(s) = f(s\tau) \in B_\tau$ represent observables or

the state of a physical system as function of the time coordinate $s \in \mathbb{R}$. The semigroup conditions require

$$\mathrm{T}_\tau(t_1)\mathrm{T}_\tau(t_2)f_\tau(t_0) = \mathrm{T}_\tau(t_1+t_2)f_\tau(t_0) \tag{7}$$

$$\mathrm{T}_\tau(0)f_\tau(t_0) = f_\tau(t_0) \tag{8}$$

for $t_1, t_2 > 0$, $t_0 \in \mathbb{R}$ and $f_\tau \in B_\tau$. The first condition may be viewed as representing the unlimited divisibility of time.

2. Continuity
   The time evolution is assumed to be strongly continuous in t by demanding

$$\lim_{t\to 0} \|\mathrm{T}(t)f - f\| = 0 \tag{9}$$

   for all $f \in B$.

3. Homogeneity
   The homogeneity of the time coordinate requires commutativity with translations

$$\mathcal{T}(t_1)\mathrm{T}(t_2)f(t_0) = \mathrm{T}(t_2)\mathcal{T}(t_1)f(t_0) \tag{10}$$

   for all $t_2 > 0$ and $t_0, t_1 \in \mathbb{R}$. This postulate allows to shift the origin of time and it reflects the basic symmetry of time translation invariance.

4. Causality
   The time evolution operator should be causal in the sense that the function $g(t_0) = (\mathrm{T}(t)f)(t_0)$ should depend only on values of $f(s)$ for $s < t_0$.

5. Coarse Graining
   A time evolution operator $\mathrm{T}(t)$ should be obtainable from a coarse graining procedure. A precise definition of coarse graining is given in Definition 2.3 below. The idea is to combine a time average $\frac{1}{t}\int_{s-t}^{s} f(t')\,\mathrm{d}t'$ in the limit $t, s \to \infty$ with a rescaling of s and t.

While the first four requirements are conventional the fifth requires comment. Averages over long intervals may themselves be timedependent on much longer time scales. An example would be the position of an atom in a glass. On short time scales the position fluctuates rapidly around a well defined average position. On long time scales the structural relaxation processes in the glass can change this average position. The purpose of any coarse graining procedure is to connect microscopic to macroscopic scales. Of course, what is microscopic depends on the physical situation. Any microscopic time evolution may itself be viewed as macroscopic from the perspective of an underlying more microscopic theory. Therefore it seems physically necessary and natural to demand that a time evolution should generally be obtainable from a coarse graining procedure.

2.2 Evolutions, Convolutions and Averages

There is a close connection and mathematical similarity between the simplest time evolution $\mathrm{T}(t) = \mathcal{T}(t)$ and the operator $\mathrm{M}(t)$ of time averaging defined as the mathematical mean

$$\mathrm{M}(t)f(s) = \frac{1}{t}\int_{s-t}^{s} f(y)\,\mathrm{d}y, \tag{11}$$

where $t > 0$ is the length of the averaging interval. Rewriting this formally as

$$\mathrm{M}(t)f(s) = \frac{1}{t}\int_{0}^{t} f(s-y)\,\mathrm{d}y = \frac{1}{t}\int_{0}^{t} \mathcal{T}(y)f(s)\,\mathrm{d}y \tag{12}$$

exhibits the relation between $\mathrm{M}(t)$ and $\mathcal{T}(t)$. It shows also that $\mathrm{M}(t)$ commutes with translations (see eq. (10)).

A second even more suggestive relationship between $\mathrm{M}(t)$ and $\mathcal{T}(t)$ arises because both operators can be written as convolutions. The operator $\mathrm{M}(t)$ may be written as

$$\begin{aligned}\mathrm{M}(t)f(s) &= \frac{1}{t}\int_{0}^{t} f(s-y)\,\mathrm{d}y = \int_{-\infty}^{\infty} f(s-y)\frac{1}{t}\chi_{[0,1]}\left(\frac{y}{t}\right)\,\mathrm{d}y\\ &= \int_{0}^{s} f(s-y)\frac{1}{t}\chi_{[0,1]}\left(\frac{y}{t}\right)\,\mathrm{d}y,\end{aligned} \tag{13}$$

where the kernel

$$\chi_{[0,1]}(x) = \begin{cases} 1 & \text{for } x \in [0,1] \\ 0 & \text{for } x \notin [0,1] \end{cases} \tag{14}$$

is the characteristic function of the unit interval. The Laplace convolution in the last line requires $t < s$. The translations $\mathcal{T}(t)$ on the other hand may be written as

$$\mathcal{T}(t)f(s) = f(s-t) = \int_{-\infty}^{\infty} f(s-y)\frac{1}{t}\delta\left(\frac{y}{t}-1\right)\,\mathrm{d}y$$

$$= \int_{0}^{s} f(s-y)\frac{1}{t}\delta\left(\frac{y}{t}-1\right)\,\mathrm{d}y \tag{15}$$

where again $0 < t < s$ is required for the Laplace convolution in the last equation. The similarity between eqs. (15) and (13) suggests to view the time translations $\mathcal{T}(t)$ as a degenerate form of averaging f over a single point. The operators $\mathrm{M}(t)$ and $\mathcal{T}(t)$ are both convolution operators. By Lebesgues theorem $\lim_{t\to 0}\mathrm{M}(t)f(s) = f(s)$ so that $\mathrm{M}(0)f(t) = f(t)$ in analogy with eq. (8) which holds for $\mathcal{T}(t)$. However, while the translations $\mathcal{T}(t)$ fulfill eq. (7) and form a convolution semigroup whose kernel is the Dirac measure at 1, the averaging operators $\mathrm{M}(t)$ do not form a semigroup as will be seen below.

The appearance of convolutions and convolution semigroups is not accidental. Convolution operators arise quite generally from the symmetry requirement of eq. (10) above. Let $L^p(\mathbb{R}^n)$ denote the Lebesgue spaces of p-th power integrable functions, and let $\mathcal{S}$ denote the Schwartz space of test functions for tempered distributions [27]. It is well established that all bounded linear operators on $L^p(\mathbb{R}^n)$ commuting with translations (i.e. fulfilling eq. (10)) are of convolution type [27].

Theorem 2.1. *Suppose the operator* $\mathrm{B} : L^p(\mathbb{R}^n) \to L^q(\mathbb{R}^n)$, $1 \leq p, q, \leq \infty$ *is linear, bounded and commutes with translations. Then there exists a unique tempered distribution* g *such that* $\mathrm{B}h = g * h$ *for all* $h \in \mathcal{S}$.

For $p = q = 1$ the tempered distributions in this theorem are finite Borel measures. If the measure is bounded and positive this means that the operator B can be viewed as a weighted averaging operator. In the following the case $n = 1$ will be of interest. A positive bounded measure μ on $\mathbb{R}$ is uniquely determined by its distribution function $\widetilde{\mu} : \mathbb{R} \to [0, 1]$ defined by

$$\widetilde{\mu}(x) = \frac{\mu(]-\infty, x[)}{\mu(\mathbb{R})}. \tag{16}$$

The tilde will again be omitted to simplify the notation. Physically a weighted average $\mathrm{M}(t;\mu)f(s)$ represents the measurement of a signal $f(s)$ using an apparatus with response characterized by μ and resolution $t > 0$. Note that the resolution (length of averaging interval) is a duration and cannot be negative.

Definition 2.1 (Averaging). *Let μ be a (probability) distribution function on $\mathbb{R}$, and $t > 0$. The weighted (time) average of a function f on $\mathbb{R}$ is defined as the convolution*

$$\mathrm{M}(t;\mu)f(s) = (f * \mu(\cdot/t))(s) = \int_{-\infty}^{\infty} f(s-s')\,\mathrm{d}\mu(s'/t) = \int_{-\infty}^{\infty} \mathcal{T}(s')f(s)\,\mathrm{d}\mu(s'/t) \tag{17}$$

whenever it exists. The average is called causal if the support of μ is in $\mathbb{R}_+$. It is called degenerate if the support of μ consists of a single point.

The weight function or kernel $m(x)$ corresponding to a distribution $\mu(x)$ is defined as $m(x) = \mathrm{d}\mu/\mathrm{d}x$ whenever it exists.

The averaging operator $\mathrm{M}(t)$ in eq. (11) corresponds to a measure with distribution function

$$\mu_{\chi}(x) = \begin{cases} 0 & \text{for } x \leq 0 \\ x & \text{for } 0 \leq x \leq 1 \\ 1 & \text{for } x \geq 1 \end{cases} \tag{18}$$

while the time translation $\mathcal{T}(t)$ corresponds to the (Dirac) measure $\delta(x-1)$ concentrated at 1 with distribution function

$$\mu_{\delta}(x) = \begin{cases} 0 & \text{for } x < 1 \\ 1 & \text{for } x \geq 1. \end{cases} \tag{19}$$

Both averages are causal, and the latter is degenerate.

Repeated averaging leads to convolutions. The convolution κ of two distributions μ, ν on $\mathbb{R}$ is defined through

$$\kappa(x) = (\mu * \nu)(x) = \int_{-\infty}^{\infty} \mu(x-y)\mathrm{d}\nu(y) = \int_{-\infty}^{\infty} \nu(x-y)\mathrm{d}\mu(y). \tag{20}$$

The Fourier transform of a distribution is defined by

$$\mathcal{F}\{\mu(t)\}(\omega) = \widehat{\mu}(\omega) = \int_{-\infty}^{\infty} e^{i\omega t}\mathrm{d}\mu(t) = \int_{-\infty}^{\infty} e^{i\omega t}m(t)\,\mathrm{d}t \tag{21}$$

where the last equation holds when the distribution admits a weight function. A sequence $\mu_n(x)$ of distributions is said to converge weakly to a limit $\mu(x)$, written as

$$\lim_{n\to\infty} \mu_n = \mu \tag{22}$$

if

$$\lim_{n\to\infty} \int_{-\infty}^{\infty} f(x)\mathrm{d}\mu_n(x) = \int_{-\infty}^{\infty} f(x)\mathrm{d}\mu(x) \tag{23}$$

holds for all bounded continuous functions f.

The operators $\mathrm{M}(t)$ and $\mathcal{T}(t)$ above have positive kernels, and preserve positivity in the sense that $f \geq 0$ implies $\mathrm{M}(t)f \geq 0$. For such operators one has

Theorem 2.2. *Let* T *be a bounded operator on* $L^p(\mathbb{R})$, $1 \leq p < \infty$ *that is translation invariant in the sense that*

$$\mathrm{T}\mathcal{T}(t)f = \mathcal{T}(t)\mathrm{T}f \tag{24}$$

for all $t \in \mathbb{R}$ *and* $f \in L^p(\mathbb{R})$, *and such that* $f \in L^p(\mathbb{R})$ *and* $0 \leq f \leq 1$ *almost everywhere implies* $0 \leq \mathrm{T}f \leq 1$ *almost everywhere. Then there exists a uniquely determined bounded measure* μ *on* $\mathbb{R}$ *with mass* $\mu(\mathbb{R}) \leq 1$ *such that*

$$\mathrm{T}f(t) = (\mu * f)(t) = \int_{-\infty}^{\infty} f(t-s)\mathrm{d}\mu(s) \tag{25}$$

Proof. For the proof see [28]. □

The preceding theorem suggests to represent those time evolutions that fulfill the requirements 1.– 4. of the last section in terms of convolution semigroups of measures.

Definition 2.2 (Convolution semigroup). *A family* $\{\mu_t : t > 0\}$ *of positive bounded measures on* $\mathbb{R}$ *with the properties that*

$$\mu_t(\mathbb{R}) \leq 1 \quad \text{for } t > 0, \tag{26}$$

$$\mu_{t+s} = \mu_t * \mu_s \quad \text{for } t, s > 0, \tag{27}$$

$$\delta = \lim_{t\to 0} \mu_t \tag{28}$$

is called a convolution semigroup of measures on $\mathbb{R}$.

Here δ is the Dirac measure at 0 and the limit is the weak limit. The desired characterization of time evolutions now becomes

Corollary 2.1. *Let* $\mathrm{T}(t)$ *be a strongly continuous time evolution fulfilling the conditions of homogeneity and causality, and being such that* $f \in L^p(\mathbb{R})$ *and* $0 \leq f \leq 1$ *almost everywhere implies* $0 \leq \mathrm{T}f \leq 1$ *almost everywhere. Then* $\mathrm{T}(t)$ *corresponds uniquely to a convolution semigroup of measures* μ_t *through*

$$\mathrm{T}(t)f(s) = (\mu_t * f)(s) = \int_{-\infty}^{\infty} f(s-s')\mathrm{d}\mu_t(s') \tag{29}$$

with $\operatorname{supp}\mu_t \subset \mathbb{R}_+$ *for all* $t \geq 0$.

Proof. Follows from Theorem 2.2 and the observation that $\operatorname{supp}\mu_t \cap \mathbb{R}_- \neq \emptyset$ would violate the causality condition. □

Equation (29) establishes the basic convolution structure of the assertion in eq. (5). It remains to investigate the requirement that $\mathrm{T}(t)$ should arise from a coarse graining procedure, and to establish the nature of the kernel in eq. (5).

2.3 Time Averaging and Coarse Graining

The purpose of this section is to motivate the definition of coarse graining. A first possible candidate for a coarse grained macroscopic time evolution could be obtained by simply rescaling the time in a microscopic time evolution as

$$\mathrm{T}_\infty(\bar{t})f(s) = \lim_{\tau\to\infty} \mathrm{T}_\tau(\bar{t})f(s) = \lim_{\tau\to\infty} \mathrm{T}(\tau\bar{t})f(s) = \lim_{\tau\to\infty} f(s-\tau\bar{t}) \tag{30}$$

where $0 < \bar{t} < \infty$ would be macroscopic times. However, apart from special cases, the limit will in general not exist. Consider for example a sinusoidal $f(t)$ oscillating around a constant. Also, the infinite translation T_∞ is not an average, and this conflicts with the requirement above, that coarse graining should be a smoothing operation.

A second highly popular candidate for coarse graining is therefore the averaging operator $\mathrm{M}(t)$. If the limit $t \to \infty$ exists and $f(t)$ is integrable in the finite interval $[s_1, s_2]$ then the average

$$\overline{f} = \lim_{t\to\infty} \mathrm{M}(t)f(s_1) = \lim_{t\to\infty} \mathrm{M}(t)f(s_2) \tag{31}$$

is a number independent of the instant s_i. Thus, if one wants to study the macroscopic time dependence of $\overline{f}$, it is necessary to consider a scaling limit in which also $s \to \infty$. If the scaling limit $s, t \to \infty$ is performed such that $s/t = \overline{s}$ is constant then

$$\lim_{\substack{t,s\to\infty \\ s=t\overline{s}}} \mathrm{M}(t)f(s) = \int_{\overline{s}-1}^{\overline{s}} f_\infty(z)\, \mathrm{d}z = \mathrm{M}(1)f_\infty(\overline{s}) \tag{32}$$

becomes again an averaging operator over the infinitely rescaled observable. Now M(1) still does not qualify as a coarse grained time evolution because $\mathrm{M}(1)\mathrm{M}(1) \neq \mathrm{M}(2)$ as will be shown next.

Consider again the operator $\mathrm{M}(t)$ defined in eq. (11). It follows that

$$\mathrm{M}^2(t)f(s) = \left(\frac{1}{t}\chi_{[0,1]}\left(\frac{\cdot}{t}\right) * \frac{1}{t}\chi_{[0,1]}\left(\frac{\cdot}{t}\right) * f\right)(s) \tag{33}$$

and

$$\frac{1}{t^2}\int_0^x \chi_{[0,1]}\left(\frac{x-y}{t}\right)\chi_{[0,1]}\left(\frac{y}{t}\right)\, \mathrm{d}y = \begin{cases} 0 & \text{for } x \leq 0 \\ \dfrac{x}{t^2} & \text{for } 0 \leq x \leq t \\ \dfrac{2}{t} - \dfrac{x}{t^2} & \text{for } t \leq x \leq 2t \\ 0 & \text{for } x \geq 2t \end{cases} \tag{34}$$

Thus twofold averaging may be written as

$$\mathrm{M}^2(t)f(s) = \int_0^s f(s-y)\frac{1}{t}\chi^{(2)}\left(\frac{y}{t}\right)\, \mathrm{d}y \tag{35}$$

where

$$\chi^{(2)}(x) = \begin{cases} x & \text{for } 0 \leq x \leq 1 \\ 2-x & \text{for } 1 \leq x \leq 2 \\ 0 & \text{otherwise} \end{cases} \tag{36}$$

is the new kernel. It follows that $\mathrm{M}^2(t) \neq \mathrm{M}(2t)$, and hence the averaging operators $\mathrm{M}(t)$ do not form a semigroup.

Although $\mathrm{M}^2(t) \neq \mathrm{M}(2t)$ the iterated average is again a convolution operator with support $[0, 2t]$ compared to $[0, t]$ for $\mathrm{M}(t)$. Similarly $\mathrm{M}^3(t)$ has support $[0, 3t]$. This suggests to investigate the iterated average $\mathrm{M}^n(t)f(s)$ in a scaling limit $n, s \to \infty$. The limit $n \to \infty$ smoothes the function by enlarging the averaging window to $[0, nt]$, and the limit $s \to \infty$ shifts the origin to infinity. The result may be viewed as a coarse grained time evolution in the sense of a time evolution on time scales "longer than infinitely long". [c] It is therefore necessary to rescale s. If the rescaling factor is called $\sigma_n > 0$ one is interested in the limit $n, s \to \infty$ with $\overline{s} = s/\sigma_n$ fixed, and $\sigma_n \to \infty$ with $n \to \infty$ and fixed $t > 0$

$$\lim_{\substack{n,s\to\infty \\ s=\sigma_n\overline{s}}} (\mathrm{M}(t)^n f)(s) = \lim_{n\to\infty} (\mathrm{M}(t)^n f)(\sigma_n \overline{s}) \tag{37}$$

whenever this limit exists. Here $\overline{s} > 1$ denotes the macroscopic time.

To evaluate the limit note first that eq. (11) implies

$$\mathrm{M}(t) f(\sigma_n \overline{s}) = \int_0^{\overline{s}} f_{\sigma_n}(\overline{s} - z) \frac{\sigma_n}{t} \chi_{[0,1]}\left(\frac{\sigma_n z}{t}\right) \mathrm{d}z \tag{38}$$

where $f_\tau(t) = f(t\tau)$ denotes the rescaled observable with a rescaling factor τ. The n-th iterated average may now be calculated by Laplace transformation with respect to $\overline{s}$. Note that

$$\mathcal{L}\left\{\frac{1}{c}\chi_{[0,1]}\left(\frac{x}{c}\right)\right\}(u) = \frac{1 - e^{-cu}}{cu} = E_{1,2}(-cu) \tag{39}$$

for all $c \in \mathbb{R}$, where $E_{1,2}(x)$ is the generalized Mittag-Leffler function defined as

$$E_{a,b}(x) = \sum_{k=0}^{\infty} \frac{x^k}{\Gamma(ak + b)} \tag{40}$$

for all $a > 0$ and $b \in \mathbb{C}$. Using the general relation

$$E_{a,b}(x) = \frac{1}{\Gamma(b)} + x E_{a,a+b}(x) \tag{41}$$

gives with eqs. (37) and (38)

$$\mathcal{L}\left\{\mathrm{M}(t)^n f(\sigma_n \overline{s})\right\}(\overline{u}) = \left(1 - \frac{t\overline{u}}{\sigma_n} E_{1,3}\left(-\frac{t\overline{u}}{\sigma_n}\right)\right)^n \frac{1}{\sigma_n} \mathcal{L}\left\{f(s)\right\}\left(\frac{\overline{u}}{\sigma_n}\right) \tag{42}$$

where $f(\overline{u})$ is the Laplace transform of $f(\overline{s})$. Noting that $E_{1,3}(0) = 1/2$ it becomes apparent that a limit $n \to \infty$ will exist if the rescaling factors are

[c] The scaling limit was called "ultralong time limit" in [10]

chosen as $\sigma_n \sim n$. With the choice $\sigma_n = \sigma n/2$ and $\sigma > 0$ one finds for the first factor

$$\lim_{n\to\infty}\left(1-\frac{2t\overline{u}}{n\sigma}E_{1,3}\left(-\frac{2t\overline{u}}{n\sigma}\right)\right)^n = e^{-t\overline{u}/\sigma}. \tag{43}$$

Concerning the second factor assume that for each $\overline{u}$ the limit

$$\lim_{n\to\infty}\frac{2}{n}\mathcal{L}\{f(s)\}\left(\frac{2\overline{u}}{n}\right) = \overline{f}(\overline{u}) \tag{44}$$

exists and defines a function $\overline{f}(\overline{u})$. Then

$$\lim_{n\to\infty}\frac{1}{\sigma_n}\mathcal{L}\{f(\overline{s})\}\left(\frac{\overline{u}}{\sigma_n}\right) = \frac{1}{\sigma}\overline{f}\left(\frac{\overline{u}}{\sigma}\right), \tag{45}$$

and it follows that

$$\lim_{n\to\infty}\mathcal{L}\{\mathrm{M}(t)^n f(\sigma_n\overline{s})\}(\overline{u}) = e^{-t\overline{u}/\sigma}\frac{1}{\sigma}\overline{f}\left(\frac{\overline{u}}{\sigma}\right). \tag{46}$$

With $\overline{t} = t/\sigma$ Laplace inversion yields

$$\lim_{\substack{n,s\to\infty\\ s=\sigma_n\overline{s}}}(\mathrm{M}(t)^n f)(s) = \int_0^{\overline{s}}\overline{f}(\sigma\overline{s}-\sigma\overline{y})\delta(\overline{y}-\overline{t})\,\mathrm{d}\overline{y} = \overline{f}_\sigma(\overline{s}-\overline{t}). \tag{47}$$

Using eq. (12) the result (47) may be expressed symbolically as

$$\lim_{\substack{n,s\to\infty\\ s/n=\sigma\overline{s}/2}}\left(\frac{1}{t}\int_0^t \mathrm{T}(y)\,\mathrm{d}y\right)^n f(s) = \overline{f}_\sigma(\overline{s}-\overline{t}) = \overline{\mathrm{T}}(\overline{t})\,\overline{f}_\sigma(\overline{s}) \tag{48}$$

with $\overline{t} = t/\sigma$. This expresses the macroscopic or coarse grained time evolution $\overline{\mathrm{T}}(\overline{t})$ as the scaling limit of a microscopic time evolution $\mathrm{T}(t)$. Note that there is some freedom in the choice of the rescaling factors σ_n expressed by the prefactor σ. This freedom reflects the freedom to choose the time units for the coarse grained time evolution.

The coarse grained time evolution $\overline{\mathrm{T}}(\overline{t})$ is again a translation. The coarse grained observable $\overline{f}(\overline{s})$ corresponds to a microscopic average by virtue of the following result [29].

Proposition 2.1. *If $f(x)$ is bounded from below and one of the limits*

$$\lim_{y\to\infty} \frac{1}{y} \int_0^y f(x)\,\mathrm{d}x$$

or

$$\lim_{z\to 0} z \int_0^\infty f(x) e^{-zx}\,\mathrm{d}x$$

exists then the other limit exists and

$$\lim_{y\to\infty} \frac{1}{y} \int_0^y f(x)\,\mathrm{d}x = \lim_{z\to 0} z\mathcal{L}\left\{f(x)\right\}(z)\,. \tag{49}$$

Comparison of the last relation with eq. (44) shows that $\overline{f}(\overline{s})$ is a microscopic average of $f(s)$. While s is a microscopic time coordinate, the time coordinate $\overline{s}$ of $\overline{f}$ is macroscopic.

The preceding considerations justify to view the time evolution $\overline{\mathcal{T}}(\overline{t})$ as a coarse grained time evolution. Every observation or measurement of a physical quantity $f(s)$ requires a minimum duration t determined by the temporal resolution of the measurement apparatus. The value $f(s)$ at the time instant s is always an average over this minimum time interval. The averaging operator $\mathrm{M}(t)$ with kernel $\chi_{[0,1]}$ defined in equation (11) represents an idealized averaging apparatus that can be switched on and off instantaneously, and does not otherwise influence the measurement. In practice one is usually confronted with finite startup and shutdown times and a nonideal response of the apparatus. These imperfections are taken into account by using a weighted average with a weight function or kernel that differs from $\chi_{[0,1]}$. The weight function reflects conditions of the measurement, as well as properties of the apparatus and its interaction with the system. It is therefore of interest to consider causal averaging operators $\mathrm{M}(t;\mu)$ defined in eq. (17) with general weight functions. A general coarse graining procedure is then obtained from iterating these weighted averages.

Definition 2.3 (Coarse Graining). *Let μ be a probability distribution on $\mathbb{R}$, and $\sigma_n > 0$, $n \in \mathbb{N}$ a sequence of rescaling factors. A coarse graining limit is defined as*

$$\lim_{\substack{n,s\to\infty\\ s=\sigma_n\overline{s}}} (\mathrm{M}(t;\mu)^n f)(s) \tag{50}$$

whenever the limit exists. The coarse graining limit is called causal if $\mathrm{M}(t;\mu)$ *is causal, i.e. if* $\operatorname{supp}\mu \subset \mathbb{R}_+$.

2.4 Coarse Graining Limits and Stable Averages

The purpose of this section is to investigate the coarse graining procedure introduced in Definition 2.3. Because the coarse graining procedure is defined as a limit it is useful to recall the following well known result for limits of distribution functions [30]. For the convenience of the reader its proof is reproduced in the appendix.

Proposition 2.2. *Let* $\mu_n(s)$ *be a weakly convergent sequence of distribution functions. If* $\lim_{n\to\infty}\mu_n(s) = \mu(s)$, *where* $\mu(s)$ *is nondegenerate then for any choice of* $a_n > 0$ *and* b_n *there exist* $a > 0$ *and* b *such that*

$$\lim_{n\to\infty}\mu_n(a_n x + b_n) = \mu(ax+b). \tag{51}$$

The basic result for coarse graining limits can now be formulated.

Theorem 2.3 (Coarse Graining Limit). *Let* $f(s)$ *be such that the limit* $\lim_{a\to 0} a\widehat{f}(a\omega) = \widehat{\overline{f}}(\omega)$ *defines the Fourier transform of a function* $\overline{f}(s)$. *Then the coarse graining limit exists and defines a convolution operator*

$$\lim_{\substack{n,s\to\infty \\ s=\sigma_n\overline{s}}}(\mathrm{M}(t;\mu)^n f)(s) = \int_{-\infty}^{\infty}\overline{f}(\overline{s}-\overline{s}')\,\mathrm{d}\nu(\overline{s}'/t;\mu) \tag{52}$$

if and only if for any $a_1, a_2 > 0$ *there are constants* $a > 0$ *and* b *such that the distribution function* $\nu(x) = \nu(x;\mu)$ *obeys the relation*

$$\nu(a_1 x) * \nu(a_2 x) = \nu(ax+b)\ . \tag{53}$$

Proof. In the previous section the coarse graining limit was evaluated for the distribution μ_χ from eq. (18) and the corresponding ν was found in eq. (47) to be degenerate. A degenerate distribution ν trivially obeys eq. (53). Assume therefore from now on that neither μ nor ν are degenerate.

Employing equation (17) in the form

$$\mathrm{M}(t;\mu)f(\sigma_n\overline{s}) = \int_{-\infty}^{\infty} f(\sigma_n\overline{s}-\sigma_n y)\mathrm{d}\mu(\sigma_n y/t) \tag{54}$$

one computes the Fourier transformation of $\mathrm{M}(t;\mu)^n f$ with respect to $\overline{s}$

$$\mathcal{F}\{\mathrm{M}(t;\mu)^n f(\sigma_n \overline{s})\}(\overline{\omega}) = \left[\widehat{\mu}\left(\frac{t\overline{\omega}}{\sigma_n}\right)\right]^n \frac{1}{\sigma_n}\widehat{f}\left(\frac{\overline{\omega}}{\sigma_n}\right). \tag{55}$$

By assumption $\widehat{f}(\overline{\omega}/\sigma_n)/\sigma_n$ has a limit whenever $\sigma_n \to \infty$ with $n \to \infty$. Thus the coarse graining limit exists and is a convolution operator whenever $[\widehat{\mu}(t\overline{\omega}/\sigma_n)]^n$ converges to $\widehat{\nu}(\overline{\omega})$ as $n \to \infty$. Following [30] it will be shown that this is true if and only if the characterization (53) and $\sigma_n \to \infty$ with $n \to \infty$ apply. To see that

$$\lim_{n\to\infty} \sigma_n = \infty \tag{56}$$

holds, assume the contrary. Then there is a subsequence σ_{n_k} converging to a finite limit. Thus

$$|\widehat{\mu}(t\omega/\sigma_{n_k})|^{n_k} = |\widehat{\nu}(\omega)|(1 + o(1)) \tag{57}$$

so that

$$|\widehat{\mu}(\omega)| = |\widehat{\nu}(\omega\sigma_{n_k}/t)|^{1/n_k}(1 + o(1)) \tag{58}$$

for all ω. As $n_k \to \infty$ this leads to $|\widehat{\mu}(\omega)| = 1$ for all ω and hence μ must be degenerate contrary to assumption.

Next, it will be shown that

$$\lim_{n\to\infty} \frac{\sigma_{n+1}}{\sigma_n} = 1. \tag{59}$$

From eq. (56) it follows that $\lim_{n\to\infty} |\widehat{\mu}(\omega/\sigma_n)| = 1$ and therefore

$$|\widehat{\mu}(t\omega/\sigma_n)|^n = |\widehat{\nu}(\omega)|(1 + o(1)) \tag{60}$$

and

$$|\widehat{\mu}(t\omega/\sigma_{n+1})|^{n+1} = |\widehat{\nu}(\omega)|(1 + o(1)). \tag{61}$$

Substituting ω by $\sigma_n\omega/\sigma_{n+1}$ in eq. (60) and by $\sigma_{n+1}\omega/\sigma_n$ in eq. (61) shows that

$$\lim_{n\to\infty}\left|\frac{\widehat{\nu}(\sigma_{n+1}\omega/\sigma_n)}{\widehat{\nu}(\omega)}\right| = \lim_{n\to\infty}\left|\frac{\widehat{\nu}(\sigma_n\omega/\sigma_{n+1})}{\widehat{\nu}(\omega)}\right| = 1\,. \tag{62}$$

If $\lim_{n\to\infty} \sigma_{n+1}/\sigma_n \neq 1$ then there exists a subsequence of either (σ_{n+1}/σ_n) or (σ_n/σ_{n+1}) converging to a constant $A < 1$. Therefore eq. (62) implies $\widehat{\nu}(\omega) = \widehat{\nu}(A\omega)$ which upon iteration yields

$$|\widehat{\nu}(\omega)| = |\widehat{\nu}(A^n\omega)|. \tag{63}$$

Taking the limit $n \to \infty$ then gives $|\widehat{\nu}(0)| = 1$ implying that ν is degenerate contrary to assumption.

Now let $0 < a_1 < a_2$ be two constants. Because of (56) and (59) it is possible to choose for each $\varepsilon > 0$ and sufficiently large $n > n_0(\varepsilon)$ an index $m(n)$ such that

$$0 \leq \frac{\sigma_m}{\sigma_n} - \frac{a_2}{a_1} < \varepsilon. \tag{64}$$

Consider the identity

$$\left[\widehat{\mu}\left(\frac{a_1 t\overline{\omega}}{\sigma_n}\right)\right]^{n+m} = \left[\widehat{\mu}\left(\frac{a_1 t\overline{\omega}}{\sigma_n}\right)\right]^{n} \left[\widehat{\mu}\left(\frac{\sigma_m}{\sigma_n}\frac{a_1 t\overline{\omega}}{\sigma_m}\right)\right]^{m} \tag{65}$$

By hypothesis the distribution functions corresponding to $[\widehat{\mu}(t\overline{\omega}/\sigma_n)]^n$ converge to $\nu(\overline{s})$ as $n \to \infty$. Hence each factor on the right hand side converges and their product converges to $\nu(a_1\overline{s}) * \nu(a_2\overline{s})$. It follows that the distribution function on the left hand side must also converge. By Proposition 2.2 there must exist $a > 0$ and b such that the left hand side differs from $\nu(\overline{s})$ only as $\nu(a\overline{s} + b)$.

Finally the converse direction that the coarse graining limit exists for $\mu = \nu$ is seen to follow from eq. (53). This concludes the proof of the theorem. □

The theorem shows that the coarse graining limit, if it exists, is again a macroscopic weighted average $\mathrm{M}(t;\nu)$. The condition (53) says that this macroscopic average has a kernel that is stable under convolutions, and this motivates the

Definition 2.4 (Stable Averages). *A weighted averaging operator* $\mathrm{M}(t;\mu)$ *is called stable if for any* $a_1, a_2 > 0$ *there are constants* $a > 0$ *and* $b \in \mathbb{R}$ *such that*

$$\mu(a_1 x) * \mu(a_2 x) = \mu(ax + b) \tag{66}$$

holds.

This nomenclature emphasizes the close relation with the limit theorems of probability theory [30,31]. The next theorem provides the explicit form for distribution functions satisfying eq. (66). The proof uses Bernsteins theorem and hence requires the concept of complete monotonicity.

Definition 2.5. *A C^∞-function $f :]0,\infty[\to \mathbb{R}$ is called completely monotone if*

$$(-1)^n \frac{\mathrm{d}^n f}{\mathrm{d}x^n} \geq 0 \tag{67}$$

for all integers $n \geq 0$.

Bernsteins theorem [31, p.439] states that a function is completely monotone if and only if it is the the Laplace transform ($u > 0$)

$$\mu(u) = \mathfrak{L}\left\{\mu(x)\right\}(u) = \int_0^\infty e^{-ux}\mathrm{d}\mu(x) = \int_0^\infty e^{-ux} m(x)\,\mathrm{d}x \tag{68}$$

of a distribution μ or of a density $m = \mathrm{d}\mu/\mathrm{d}x$.

In the next theorem the explicit form of stable averaging kernels is found to be a special case of the general H-function. Because the H-function will reappear in other results its general definition and properties are presented separately in Section 4.

Theorem 2.4. *A causal average is stable if and only if its weight function is of the form*

$$h_\alpha(x;b,c) = \frac{1}{b^{1/\alpha}} h_\alpha\left(\frac{x-c}{b^{1/\alpha}}\right) = \frac{1}{\alpha(x-c)} H^{10}_{11}\left(\frac{b^{1/\alpha}}{x-c}\,\middle|\, \begin{matrix}(0,1)\\(0,1/\alpha)\end{matrix}\right) \tag{69}$$

where $0 < \alpha \leq 1$, $b > 0$ and $c \in \mathbb{R}$ are constants and $h_\alpha(x) = h_\alpha(s;1,0)$.

Proof. Let $c = 0$ without loss of generality. The condition (66) together with $\operatorname{supp}\mu \subset [0,\infty[$ defines one sided stable distribution functions [31]. To derive the form (69) it suffices to consider condition (66) with $b = 0$. Assume thence that for any $a_1, a_2 > 0$ there exists $a > 0$ such that

$$\mu(a_1 x) * \mu(a_2 x) = \mu(ax) \tag{70}$$

where the convolution is now a Laplace convolution because of the condition $\operatorname{supp} \subset [0,\infty[$. Laplace tranformation yields

$$\mu(u/a_1)\mu(u/a_2) = \mu(u/a). \tag{71}$$

Iterating this equation (with $a_1 = a_2 = 1$) shows that there is an n-dependent constant $a(n)$ such that

$$\mu(u)^n = \mu(u/a(n)) \tag{72}$$

and hence

$$\mu\left(\frac{u}{a(nm)}\right) = \mu(u)^{nm} = \mu\left(\frac{u}{a(n)}\right)^m = \mu\left(\frac{u}{(a(n)a(m)}\right). \tag{73}$$

Thus $a(n)$ satisfies the functional equation

$$a(nm) = a(n)a(m) \tag{74}$$

whose solution is $a(n) = n^{1/\gamma}$ with some real constant written as $1/\gamma$ with hindsight. Inserting $a(n)$ into eq.(72) and substituting the function $g(x) = \log \mu(x)$ gives

$$ng(u) = g(un^{-1/\gamma}). \tag{75}$$

Taking logarithms and substituting $f(x) = \log g(e^x)$ this becomes

$$\log n + f(\log u) = f\left(\log u - \frac{\log n}{\gamma}\right). \tag{76}$$

The solution to this functional equation is $f(x) = -\gamma x$. Substituting back one finds $g(x) = x^{-\gamma}$ and therefore $\mu(u)$ is of the general form $\mu(u) = \exp(u^{-\gamma})$ with $\gamma \in \mathbb{R}$. Now μ is also a distribution function. Its normalization requires $\mu(u = 0) = 1$ and this restricts γ to $\gamma < 0$. Moreover, by Bernsteins theorem $\mu(u)$ must be completely monotone. A completely monotone function is positive, decreasing and convex. Therefore the power in the exponent must have a negative prefactor, and the exponent is restricted to the range $-1 \leq \gamma < 0$. Summarizing, the Laplace transform $\mu(u)$ of a distribution satisfying (70) is of the form

$$\mu(u) = h_\alpha(u; b, 0) = e^{-bu^\alpha} \tag{77}$$

with $0 < \alpha \leq 1$ and $b > 0$. Checking that $h_\alpha(u; b, 0)$ does indeed satisfy eq. (70) yields $a^{-\alpha} = a_1^{-\alpha} + a_2^{-\alpha}$ as the relation between the constants. For the proof of the general case of eq. (66) see Refs. [30,31].

To invert the Laplace transform it is convenient to use the relation

$$\mathcal{M}\{m(x)\}(s) = \frac{\mathcal{M}\{\mathcal{L}\{m(x)\}(u)\}(1-s)}{\Gamma(1-s)} \tag{78}$$

between the Laplace transform and the Mellin transform

$$\mathcal{M}\{m(x)\}(s) = \int\limits_0^\infty x^{s-1} m(t)\, \mathrm{d}x \tag{79}$$

of a function $m(x)$. Using the Mellin transform [32]

$$\mathcal{M}\left\{e^{-bx^\alpha}\right\}(s) = \frac{\Gamma(s/\alpha)}{\alpha b^{s/\alpha}} \tag{80}$$

valid for $\alpha > 0$ and $\operatorname{Re} s > 0$ it follows that

$$\mathcal{M}\left\{h_\alpha(x;b,0)\right\}(s) = \frac{1}{\alpha b^{(1-s)/\alpha}} \frac{\Gamma((1-s)/\alpha)}{\Gamma(1-s)}. \tag{81}$$

The general relation $\mathcal{M}\left\{x^{-1}f(x^{-1})\right\}(s) = \mathcal{M}\left\{f(x)\right\}(1-s)$ then implies

$$\mathcal{M}\left\{x^{-1}h_\alpha(x^{-1};b,0)\right\}(s) = \frac{1}{\alpha b^{s/\alpha}} \frac{\Gamma(s/\alpha)}{\Gamma(s)} \tag{82}$$

which leads to

$$x^{-1}h_\alpha(x^{-1};b,0) = \frac{1}{\alpha} H^{10}_{11}\left(b^{1/\alpha}x \,\middle|\, \begin{matrix}(0,1)\\(0,1/\alpha)\end{matrix}\right) \tag{83}$$

by identification with eq. (153) below. Restoring a shift $c \neq 0$ yields the result of eq. (69). □

Note that $h_\alpha(x) = h_\alpha(s;1,0)$ is the standardized form used in eq. (5). It remains to investigate the sequence of rescaling factors σ_n. For these one finds

Corollary 2.2. *If the coarse graining limit exists and is nondegenerate then the sequence σ_n of rescaling factors has the form*

$$\sigma_n = n^{1/\alpha}\Lambda(n) \tag{84}$$

where $0 < \alpha \leq 1$ and $\Lambda(n)$ is slowly varying, i.e. $\lim_{n\to\infty}\Lambda(bn)/\Lambda(n) = 1$ for all $b > 0$ (see Chapter IX , Section 2.3).

Proof. [33] Let $\widehat{\mu}_n(\omega) = \widehat{\mu}(\omega)^n$. Then for all ω and any fixed k

$$|\widehat{\mu}_n(\omega/\sigma_n)| = e^{-b|\omega|^\alpha}(1+o(1)) = |\widehat{\mu}_{kn}(\omega/\sigma_{kn})|. \tag{85}$$

On the other hand

$$|\widehat{\mu}_{kn}(\omega/\sigma_{kn})| = |\widehat{\mu}_n((\omega\sigma_n/\sigma_{kn})/\sigma_n)|^k = e^{-b|\omega|^\alpha}(1+o(1)) \tag{86}$$

where the remainder tends uniformly to zero on every finite interval. Suppose that the sequence σ_n/σ_{kn} is unbounded so that there is a subsequence with $\sigma_{kn_j}/\sigma_{n_j} \to 0$. Setting $\omega = \sigma_{kn_j}/\sigma_{n_j}$ in eq. (86) and using eq. (85) gives

$\exp(-bk) = 1$ which cannot be satisfied because $b, k > 0$. Hence σ_n/σ_{kn} is bounded. Now the limit $n \to \infty$ in eqs. (85) and (86) gives

$$e^{-b|\omega|^\alpha} = e^{-bk|\omega|^\alpha(\sigma_n/\sigma_{kn})^\alpha}(1 + o(1)). \tag{87}$$

This requires that

$$\lim_{n\to\infty} \frac{\sigma_{kn}}{\sigma_n} = k^{1/\alpha} \tag{88}$$

implying eq. (84) by virtue of the Characterization Theorem 2.2 in Chapter IX . (For more information on slow and regular variation see Chapter IX and references therein). □

2.5 Macroscopic Time Evolutions

The preceding results show that a coarse graining limit is characterized by the quantities (α, b, c, Λ). These quantities are determined by the coarsening weight μ. The following result, whose proof can be found in [33, p. 85], gives their relation with the coarsening weight.

Theorem 2.5 (Universality Classes of Time Evolutions). *In order that a causal coarse graining limit based on* $\mathrm{M}(t;\mu)$ *gives rise to a macroscopic average with* $h_\alpha(x;b,c)$ *it is necessary and sufficient that* $\widehat{\mu}(\omega)$ *behaves as*

$$\log\widehat{\mu}(\omega) = ic\omega - b|\omega|^\alpha\Lambda(\omega) \tag{89}$$

in a neighbourhood of $\omega = 0$, *and that* $\Lambda(\omega)$ *is slowly varying for* $\omega \to 0$. *In case* $0 < \alpha \le 1$ *the rescaling factors can be chosen as*

$$\sigma_n^{-1} = \inf\{\omega > 0 : |\omega|^\alpha\Lambda(\omega) = b/n\} \tag{90}$$

while the case $\alpha > 1$ *reduces to the degenerate case* $\alpha = 1$.

The preceding theorem characterizes the domain of attraction of a universality class of time evolutions. Summarizing the results gives a characterization of macroscopic time evolutions arising from coarse graining limits.

Theorem 2.6 (Macroscopic Time Evolution). *Let* $f(s)$ *be such that the limit* $\lim_{a\to 0} a\widehat{f}(a\omega) = \widehat{\overline{f}}(\omega)$ *defines the Fourier transform of a function* $\overline{f}(s)$. *If* $\mathrm{M}(t;\mu)$ *is a causal average whose coarse graining limit exists with* α, b, c *as* *in the preceding theorem then*

$$\lim_{\substack{n,s\to\infty \\ s=\sigma_n\overline{s}}} (\mathrm{M}(t;\mu)^n f)(s) = \int_{\overline{c}}^{\infty} \overline{f}(\overline{s}-y) h_\alpha\left(\frac{y}{\overline{t}}\right)\frac{\mathrm{d}y}{\overline{t}} = \int_{\overline{c}}^{\infty} \overline{\mathcal{T}}_y \overline{f}(\overline{s}) h_\alpha\left(\frac{y}{\overline{t}}\right)\frac{\mathrm{d}y}{\overline{t}}$$

$$= \mathrm{M}(\overline{t};h_\alpha)\overline{f}(\overline{s}-\overline{c}) = \overline{\mathrm{T}}_\alpha(\overline{t})\overline{f}(\overline{s}-\overline{c}) \tag{91}$$

defines a family of one parameter semigroups $\overline{\mathrm{T}}_\alpha(\overline{t})$ *with parameter* $\overline{t} = t^\alpha b$ *indexed by* α. *Here* $\overline{\mathcal{T}}_{\overline{t}}\overline{f}(\overline{s}) = \overline{f}(\overline{s}-\overline{t})$ *denotes the translation semigroup, and* $\overline{c} = c/(tb)^{1/\alpha}$ *is a constant.*

Proof. Noting that $\operatorname{supp} h_\alpha(x) \subset \mathbb{R}_+$ and combining Theorems 2.3 and 2.4 gives

$$\lim_{\substack{n,s\to\infty \\ s=\sigma_n\overline{s}}} (\mathrm{M}(t;\mu)^n f)(s) = \int_{c}^{\infty} \overline{f}(\overline{s}-\overline{s}')\frac{1}{tb^{1/\alpha}} h_\alpha\left(\frac{\overline{s}'-c}{tb^{1/\alpha}}\right)\mathrm{d}\overline{s}' = \overline{\mathrm{T}}_\alpha(\overline{t})\overline{f}(\overline{s}-\overline{c}) \tag{92}$$

where $0 < \alpha \leq 1$, $b > 0$ and $c \in \mathbb{R}$ are the constants from theorem 2.4 and the last equality defines the operators $\overline{\mathrm{T}}_\alpha(\overline{t})$ with $\overline{t} = t^\alpha b$ and $\overline{c} = c/(tb)^{1/\alpha}$. Fourier transformation then yields

$$\mathcal{F}\left\{(\overline{\mathrm{T}}_\alpha(\overline{t})\overline{f})(\overline{s}-\overline{c})\right\}(\overline{\omega}) = e^{-ic\overline{\omega}-\overline{t}(i\overline{\omega})^\alpha}, \tag{93}$$

and the semigroup property (7) follows from

$$\mathcal{F}\left\{(\overline{\mathrm{T}}_\alpha(\overline{t}_1)\overline{\mathrm{T}}_\alpha(\overline{t}_2)\overline{f})(\overline{s}-\overline{c})\right\}(\overline{\omega}) = e^{-ic\overline{\omega}-\overline{t}_1(i\overline{\omega})^\alpha-\overline{t}_2(i\overline{\omega})^\alpha}$$

$$= \mathcal{F}\left\{(\overline{\mathrm{T}}_\alpha(\overline{t}_1+\overline{t}_2)\overline{f})(\overline{s}-\overline{c})\right\}(\overline{\omega}) \tag{94}$$

by Fourier inversion. Condition (8) is checked similarly. □

The family of semigroups $\overline{\mathrm{T}}_\alpha(\overline{t})$ indexed by α that can arise from coarse graining limits are called *macroscopic time evolutions.* These semigroups are also holomorphic, strongly continuous and equibounded (see Chapter III).

From a physical point of view this result emphasizes the different role played by $\overline{s}$ and $\overline{t}$. While $\overline{s}$ is the macroscopic time coordinate whose values are $\overline{s} \in \mathbb{R}$, the duration $\overline{t} > 0$ is positive. If the dimension of a microscopic time duration t is [s], then the dimension of the macroscopic time duration $\overline{t}$ is [s$^\alpha$].

2.6 Infinitesimal Generators

The importance of the semigroups $\overline{\mathrm{T}}_\alpha(\bar{t})$ for theoretical physics as universal attractors of coarse grained macroscopic time evolutions seems not to have been noticed thus far. This is the more surprising as their mathematical importance for harmonic analysis and probability theory has long been recognized [31,34,35,28]. The infinitesimal generators are known to be fractional derivatives [31,35,36,37]. The infinitesimal generators are defined as

$$\mathrm{A}_\alpha \bar{f}(\bar{s}) = \lim_{\bar{t}\to 0} \frac{\overline{\mathrm{T}}_\alpha(\bar{t})\bar{f}(\bar{s}) - \bar{f}(\bar{s})}{\bar{t}}. \tag{95}$$

For more details on semigroups and their infinitesimal generators see Chapter III .

Formally one calculates A_α by applying direct and inverse Laplace transformation with $\bar{c} = 0$ in eq. (91) and using eq. (77)

$$\begin{aligned}
\mathrm{A}_\alpha \bar{f}(\bar{s}) &= \lim_{\bar{t}\to 0} \frac{1}{2\pi i} \int\limits_{\eta-i\infty}^{\eta+i\infty} e^{\bar{s}\bar{u}} \left(\frac{e^{-\bar{t}\bar{u}^\alpha} - 1}{\bar{t}} \right) \bar{f}(\bar{u})\, \mathrm{d}\bar{u} \\
&= \frac{1}{2\pi i} \int\limits_{\eta-i\infty}^{\eta+i\infty} e^{\bar{s}\bar{u}} \lim_{\bar{t}\to 0} \left(\frac{e^{-\bar{t}\bar{u}^\alpha} - 1}{\bar{t}} \right) \bar{f}(\bar{u})\, \mathrm{d}\bar{u} \\
&= -\frac{1}{2\pi i} \int\limits_{\eta-i\infty}^{\eta+i\infty} e^{\bar{s}\bar{u}} \bar{u}^\alpha \bar{f}(\bar{u})\, \mathrm{d}\bar{u}.
\end{aligned} \tag{96}$$

The result can indeed be made rigorous and one has

Theorem 2.7. *The infinitesimal generator* A_α *of the macroscopic time evolutions* $\overline{\mathrm{T}}_\alpha(\bar{t})$ *is related to the infinitesimal generator* $\mathrm{A} = -\mathrm{d}/\mathrm{d}\bar{t}$ *of* $\overline{\mathcal{T}}_{\bar{t}}$ *through*

$$\begin{aligned}
\mathrm{A}_\alpha \bar{f}(\bar{s}) = -(-\mathrm{A})^\alpha \bar{f}(\bar{s}) = -\mathrm{D}^\alpha \bar{f}(\bar{s}) &= -\frac{1}{\Gamma(-\alpha)} \int\limits_0^\infty \frac{\bar{f}(\bar{s} - y) - \bar{f}(\bar{s})}{y^{\alpha+1}}\, \mathrm{d}y \\
&= -\frac{1}{\Gamma(-\alpha)} \int\limits_0^\infty y^{-\alpha-1} (\overline{\mathcal{T}}_y - 1) \bar{f}(\bar{s})\, \mathrm{d}y\,.
\end{aligned} \tag{97}$$

Proof. See Chapter III . □

The theorem shows that fractional derivatives of Marchaud type arise as the infinitesimal generators of coarse grained time evolutions in physics. The order α of the derivative lies between zero and unity, and it is determined by the decay of the averaging kernel. The order α gives a quantitative measure for the decay of the averaging kernel. The case $\alpha \neq 1$ indicates that memory effects and history dependence may become important.

3 Applications

3.1 Fractional Invariance and Stationarity

To simplify the notation $\overline{\mathrm{T}}_\alpha(\bar{t})$ will be denoted as $\mathrm{T}_\alpha(t)$ in the following. A first application of fractional time evolutions $T_\alpha(t)$ concerns the important notion of stationarity. This amounts to setting the left and right hand sides in eq. (2) to zero. Surprisingly, the importance of the condition "$\mathrm{d}^\alpha f/\mathrm{d}t^\alpha$"$= 0$ for the infinitesimal generators of fractional dynamics has rarely been noticed. Stationary states $f(s)$ may be defined more generally as states that are invariant under the time evolution after a sufficient amount of time has elapsed during which all the transients have had time to decay.

Definition 3.1. *An observable or state $f(t)$ is called stationary or asymptotically invariant under the time evolution $\mathrm{T}_\alpha(t)$ if*

$$\mathrm{T}_\alpha(t)f(s) = f(s) \tag{98}$$

holds for $s/t \to \infty$. It is called stationary in the strict sense, or strictly invariant under $\mathrm{T}_\alpha(t)$, if condition (98) holds for all $t \geq 0$ and $s \in \mathbb{R}$.

The function $f(s) = f_0$ where f_0 is a constant is asymptotically and strictly stationary under the fractional time evolutions $\mathrm{T}_\alpha(t)$. This follows readily by insertion into the definition, and by noting that $h_\alpha(x)$ is a probability density.

In addition to the conventional constants there exists a second class of stationary states given by

$$f(s) = \begin{cases} f_0 s^{\gamma-1} & \text{for } s > 0 \\ 0 & \text{for } s \leq 0 \end{cases} \tag{99}$$

where f_0 and γ are constants. To see this one evaluates

$$\mathrm{T}_\alpha(t)f(s) = \int\limits_0^\infty f(s-x)\frac{1}{t}h_\alpha\left(\frac{x}{t}\right)\,dx$$

$$= f_0 \int\limits_0^s (s-x)^{\gamma-1}\frac{1}{\alpha t}H_{11}^{01}\left(\frac{x}{t}\,\middle|\,\begin{matrix}(1-1/\alpha,1/\alpha)\\(0,1)\end{matrix}\right)\mathrm{d}x \tag{100}$$

where relations (170) and (172) were used to rewrite the H-function in eq. (69). Using the integral (178), the reduction formulae (167) and (169), and property (171) one finds

$$\mathrm{T}_\alpha(t)f(s) = f_0 s^{\gamma-1}\Gamma(\gamma)H_{11}^{01}\left(\left(\frac{s}{t}\right)^\alpha\,\middle|\,\begin{matrix}(1,1)\\(1-\gamma,\alpha)\end{matrix}\right). \tag{101}$$

An application of the series expansion (181) gives

$$\mathrm{T}_\alpha(t)f(s) = f_0 s^{\gamma-1}\Gamma(\gamma)\sum_{k=0}^\infty \frac{(-1)^k(t/s)^{k\alpha}}{k!\Gamma(\gamma-k\alpha)}. \tag{102}$$

For $s/t \to \infty$ only the $k=0$ term in the series contributes and this shows that $\mathrm{T}_\alpha(t)f(s) = f(s)$ in the limit. These considerations show that fractional time evolutions have the usual constants as strict stationary states, but admit also algebraic behaviour as a novel type of stationary states.

To elucidate the significance of the new type of stationary states it is useful to consider the infinitesimal form, $\mathrm{A}_\alpha f = 0$, of the stationarity condition. The nature of the limit $s/t \to \infty$ suggests that their appearance might be related to the initial conditions. To incorporate initial conditions into the infinitesimal generator it is necessary to consider a Riemann-Liouville representation of the fractional time derivative.

The Riemann-Liouville algorithm for fractional differentiation is based on integer order derivatives of fractional integrals.

Definition 3.2 (Riemann-Liouville fractional integral). *The right-sided Riemann-Liouville fractional integral of order $\alpha > 0, \alpha \in \mathbb{R}$ of a locally integrable function f is defined as*

$$(\mathrm{I}_{a+}^\alpha f)(x) = \frac{1}{\Gamma(\alpha)}\int\limits_a^x (x-y)^{\alpha-1}f(y)\,\mathrm{d}y \tag{103}$$

for $x > a$, the left-sided Riemann-Liouville fractional integral is defined as

$$(\mathrm{I}^{\alpha}_{a-}f)(x) = \frac{1}{\Gamma(\alpha)} \int_x^a (y-x)^{\alpha-1} f(y)\,\mathrm{d}y \tag{104}$$

for $x < a$.

The following generalized definition, based on differentiating fractional integrals, seems to be new.

Definition 3.3 (Fractional derivatives). *The (right-/left-sided) fractional derivative of order $0 < \alpha < 1$ and type $0 \leq \beta \leq 1$ with respect to x is defined by*

$$\mathrm{D}^{\alpha,\beta}_{a\pm} f(x) = \left(\pm \mathrm{I}^{\beta(1-\alpha)}_{a\pm} \frac{\mathrm{d}}{\mathrm{d}x} (\mathrm{I}^{(1-\beta)(1-\alpha)}_{a\pm} f) \right)(x) \tag{105}$$

for functions for which the expression on the right hand side exists.

The Riemann-Liouville fractional derivative $\mathrm{D}^{\alpha}_{a\pm} := \mathrm{D}^{\alpha,0}_{a\pm}$ corresponds to $a > -\infty$ and type $\beta = 0$. Fractional derivatives of type $\beta = 1$ are discussed in Chapter I and were employed in [4]. It seems however that fractional derivatives of general type $0 < \beta < 1$ have not been considered previously. A relation between fractional derivatives of the same order but different types is given in Chapter IX . For subsequent calculations it is useful to record the Laplace-Transformation

$$\mathfrak{L}\left\{ \mathrm{D}^{\alpha,\beta}_{a+} f(x) \right\}(u) = u^{\alpha} \mathfrak{L}\left\{ f(x) \right\}(u) - u^{\beta(\alpha-1)} (\mathrm{D}^{(1-\beta)(\alpha-1),0}_{a+} f)(0+) \tag{106}$$

where the inital value $(\mathrm{D}^{(1-\beta)(\alpha-1),0}_{a+} f)(0+)$ is the Riemann-Liouville derivative for $t \to 0+$. Note that fractional derivatives of type 1 involve nonfractional initial values.

It is now possible to discuss the infinitesimal form of fractional stationarity where the generator A_{α} for initial conditions of type $0 \leq \beta \leq 1$ is represented by $\mathrm{D}^{\alpha,\beta}_{0+}$. The fractional differential equation

$$\mathrm{D}^{\alpha,\beta}_{0+} f(t) = 0 \tag{107}$$

for f with initial condition

$$\mathrm{I}^{(1-\beta)(1-\alpha)}_{0+} f(0+) = f_0 \tag{108}$$

defines fractional stationarity of order α and type β. Of course, for $\alpha = 1$ this definition reduces to the conventional definition of stationarity. Equation (107) is solved by

$$f(t) = \frac{f_0\, t^{(1-\beta)(\alpha-1)}}{\Gamma((1-\beta)(\alpha-1)+1)}. \tag{109}$$

This may be seen by inserting $f(t)$ into the definition

$$\mathrm{D}_{0+}^{\alpha,\beta} f(x) = \left(\mathrm{I}_{0+}^{\beta(1-\alpha)} \frac{\mathrm{d}}{\mathrm{d}x} (\mathrm{I}_{0+}^{(1-\beta)(1-\alpha)} f) \right)(x) \tag{110}$$

and using the basic fractional integral

$$\mathrm{I}_{a+}^{\alpha} (x-a)^{\beta} = \frac{\Gamma(\beta+1)}{\Gamma(\alpha+\beta+1)} (x-a)^{\alpha+\beta} \tag{111}$$

(derived in eq. (1.30) in Chapter I). Note that the fractional integral

$$\mathrm{I}_{0+}^{(1-\beta)(1-\alpha)} f(t) = f_0 \tag{112}$$

remains conserved and constant for all t while the function itself varies. In particular $\lim_{t\to 0} f(t) = \infty$ and $\lim_{t\to\infty} f(t) = 0$. For $\beta = 1$ and for $\alpha = 1$ one recovers $f(t) = f_0$ as usual.

The new types of stationary states for which a fractional integral rather than the function itself is constant were first discussed in [6,9]. It seems to me that the lack of knowledge about fractional stationarity is partially responsible for the difficulty of deciding which type of fractional derivative should be used when generalizing traditional equations of motion.

Another simple instance of a fractional differential equation is the equation

$$\mathrm{D}_{0+}^{\alpha,\beta} f(t) = C \tag{113}$$

with $C \in \mathbb{R}$ a constant, and with initial condition

$$\mathrm{I}_{0+}^{(1-\beta)(1-\alpha)} f(0+) = f_0 \tag{114}$$

as before. Laplace transformation using eq. (106) gives

$$f(u) = \frac{C}{u^{\alpha+1}} + \frac{f_0}{u^{\alpha+\beta(1-\alpha)}} \tag{115}$$

and thence

$$f(t) = \frac{C\, t^{\alpha}}{\Gamma(\alpha+1)} + \frac{f_0\, t^{(1-\beta)(\alpha-1)}}{\Gamma((1-\beta)(1-\alpha)+1)}. \tag{116}$$

For $\beta = 1$ this reduces to

$$f(t) = \frac{C\, t^\alpha}{\Gamma(\alpha+1)} + f_0. \tag{117}$$

3.2 Generalized Fractional Relaxation

Consider the fractional Cauchy problem

$$\mathrm{D}_{0+}^{\alpha,\beta} f(t) = -C\, f(t) \tag{118}$$

for f with initial condition

$$\mathrm{I}_{0+}^{(1-\beta)(1-\alpha)} f(0+) = f_0 \tag{119}$$

where C is a ("fractional relaxation") constant. Laplace Transformation gives

$$f(u) = \frac{u^{\beta(\alpha-1)}\, f_0}{C + u^\alpha}. \tag{120}$$

To invert the Laplace transform rewrite this equation as

$$f(u) = \frac{u^{\alpha-\gamma}}{C+u^\alpha} = u^{-\gamma}\frac{1}{Cu^{-\alpha}+1} = \sum_{k=0}^{\infty}(-C)^k u^{-\alpha k-\gamma} \tag{121}$$

with

$$\gamma = \alpha + \beta(1-\alpha). \tag{122}$$

Inverting the series term by term using $\mathcal{L}\left\{x^{\alpha-1}/\Gamma(\alpha)\right\} = u^{-\alpha}$ yields the result

$$f(t) = x^{\gamma-1}\sum_{k=0}^{\infty}\frac{(-Cx^\alpha)^k}{\Gamma(\alpha k+\gamma)}\ . \tag{123}$$

The solution may be written as

$$f(t) = f_0\, t^{(1-\beta)(\alpha-1)} E_{\alpha,\alpha+\beta(1-\alpha)}(-Ct^\alpha) \tag{124}$$

using the generalized Mittag-Leffler function defined by

$$E_{a,b}(x) = \sum_{k=0}^{\infty}\frac{x^k}{\Gamma(ak+b)} \tag{125}$$

for all $a > 0, b \in \mathbb{C}$. This function is an entire function of order $1/a$ [38]. Moreover it is completely monotone if and only if $0 < a \leq 1$ and $b \geq a$ [39].

For $C = 0$ the result reduces to eq. (109) because $E_{a,b}(0) = 1/\Gamma(b)$. Of special interest is again the case $\beta = 1$. It has the well known solution

$$f(t) = f_0 \; E_\alpha(-Ct^\alpha) \tag{126}$$

where $E_\alpha(x) = E_{\alpha,1}(x)$ denotes the ordinary Mittag-Leffler function.

3.3 Generalized Fractional Diffusion

Consider the fractional partial differential equation for $f : \mathbb{R}^d \times \mathbb{R}_+ \to \mathbb{R}$

$$\mathrm{D}_{0+}^{\alpha,\beta} f(\boldsymbol{r}, t) = C \; \Delta f(\boldsymbol{r}, t) \tag{127}$$

with Laplacian Δ and fractional "diffusion" constant C. The function $f(\boldsymbol{r}, t)$ is assumed to obey the initial condition

$$\mathrm{I}_{0+}^{(1-\beta)(1-\alpha)} f(\boldsymbol{r}, 0+) = f_{0\boldsymbol{r}} = f_0 \delta(\boldsymbol{r}) \tag{128}$$

where $\delta(\boldsymbol{r})$ is the Dirac measure at the origin. Fourier Transformation, defined as

$$\mathcal{F}\{f(\boldsymbol{r})\}(\boldsymbol{q}) = \int_{\mathbb{R}^d} e^{i\boldsymbol{q}\cdot\boldsymbol{r}} f(\boldsymbol{r}) \mathrm{d}\boldsymbol{r}, \tag{129}$$

and Laplace transformation of eq. (127) now yields

$$f(\boldsymbol{q}, u) = \frac{u^{\beta(\alpha-1)} \; f_0}{C\boldsymbol{q}^2 + u^\alpha}. \tag{130}$$

Using the result (124) for the inverse Laplace transform of (120) gives

$$f(\boldsymbol{q}, t) = f_0 \; t^{(1-\beta)(\alpha-1)} E_{\alpha,\alpha+\beta(1-\alpha)}(-C\boldsymbol{q}^2 t^\alpha). \tag{131}$$

Setting $\boldsymbol{q} = 0$ shows that the solution of (127) cannot be a probability density except for $\beta = 1$. For $\beta \neq 1$ the spatial integral is time dependent, and f would need to be divided by $t^{(1-\beta)(\alpha-1)}$ to admit a probabilistic interpretation.

To invert eq. (130) completely it seems advantageous to first invert the Fourier transform and then the Laplace transform. The Fourier transform may be inverted by noting the formula [40]

$$(2\pi)^{-d/2} \int e^{i\boldsymbol{q}\cdot\boldsymbol{r}} \left(\frac{|\boldsymbol{r}|}{m}\right)^{1-(d/2)} K_{(d-2)/2}(m|\boldsymbol{r}|) \; \mathrm{d}\boldsymbol{r} = \frac{1}{\boldsymbol{q}^2 + m^2} \tag{132}$$

which leads to

$$f(\boldsymbol{r},u) = f_0(2\pi C)^{-d/2}\left(\frac{r}{\sqrt{C}}\right)^{1-(d/2)} u^{\beta(\alpha-1)+\alpha(d-2)/4} K_{(d-2)/2}\left(\frac{r u^{\alpha/2}}{\sqrt{C}}\right) \tag{133}$$

with $r = |\boldsymbol{r}|$. To invert the Laplace transform one uses again the relation (78) with the Mellin transform defined in eq. (79). Setting $A = r/\sqrt{C}$, $\lambda = \alpha/2$, $\nu = (d-2)/2$ and $\mu = \beta(\alpha-1) + \alpha(d-2)/4$ and using the general relation

$$\mathcal{M}\left\{x^q g(bx^p)\right\}(s) = \frac{1}{p} b^{-(s+q)/p} g\left(\frac{s+q}{p}\right) \qquad (b,p>0) \tag{134}$$

leads to

$$\mathcal{M}\left\{f(r,u)\right\}(s) = \frac{f_0}{\lambda}(2\pi C)^{-d/2} A^{1-(d/2)} A^{-(s+\mu)/\lambda} \mathcal{M}\left\{K_\nu(u)\right\}((s+\mu)/\lambda)\,. \tag{135}$$

The Mellin transform of the Bessel function reads [32]

$$\mathcal{M}\left\{K_\nu(x)\right\}(s) = 2^{s-2}\Gamma\left(\frac{s+\nu}{2}\right)\Gamma\left(\frac{s-\nu}{2}\right). \tag{136}$$

Inserting this, using eq.(78), and restoring the original variables then yields

$$\mathcal{M}\left\{f(r,t)\right\}(s) = \frac{f_0}{\alpha(r^2\pi)^{d/2}}\left(\frac{r}{2\sqrt{C}}\right)^{2(1-\beta)(1-(1/\alpha))}\left(\frac{r}{2\sqrt{C}}\right)^{2s/\alpha}$$
$$\frac{\Gamma\left(\frac{d}{2} + (\beta-1)(1-\frac{1}{\alpha}) - \frac{s}{\alpha}\right)\Gamma\left(1 + (\beta-1)(1-\frac{1}{\alpha}) - \frac{s}{\alpha}\right)}{\Gamma(1-s)} \tag{137}$$

for the Mellin transform of f. Comparing this with the Mellin transform of the H-function in eq. (175) allows to identify the H-function parameters as $m = 0$,$n = 2$,$p = 2$,$q = 1$, $A_1 = A_2 = 1/\alpha$, $a_1 = 1 - (d/2) - (\beta-1)(1-(1/\alpha))$, $a_2 = (1-\beta)(1-(1/\alpha))$, $b_1 = 0$ and $B_1 = 1$ if $(\alpha d/2) + (\beta-1)(\alpha-1) > 0$. Then the result becomes

$$f(r,t) = \frac{f_0}{\alpha(r^2\pi)^{d/2}}\left(\frac{r}{2\sqrt{C}}\right)^{2(1-\beta)(1-(1/\alpha))}$$
$$H^{02}_{21}\left(\left(\frac{2\sqrt{C}}{r}\right)^{2/\alpha} t \,\middle|\, \begin{array}{l} (1-\frac{d}{2}+(1-\beta)(1-\frac{1}{\alpha}),\frac{1}{\alpha}),((1-\beta)(1-\frac{1}{\alpha}),\frac{1}{\alpha}) \\ (0,1) \end{array}\right). \tag{138}$$

This may be simplified using eqs.(170), (171) and (172) to become finally

$$f(r,t) = \frac{f_0\, t^{(1-\beta)(\alpha-1)}}{(r^2\pi)^{d/2}} H_{12}^{20}\left(\frac{r^2}{4Ct^\alpha}\,\middle|\, \begin{matrix}(1+(1-\beta)(\alpha-1),\alpha)\\ (d/2,1),(1,1)\end{matrix}\right). \tag{139}$$

This result reduces to the known case of type $\beta = 1$ in which case $f(\boldsymbol{r},t)$ is also a probability density. For $\beta \neq 1$ the function $f(\boldsymbol{r},t)$ does not have a probabilistic interpretation because its normalization decays as $t^{(1-\beta)(\alpha-1)}$.

3.4 Relation with Continuous Time Random Walk

The fractional diffusion eq. (127) of type $\beta = 1$ has a probabilistic interpretation as noted after eq. (131). $f(\boldsymbol{r},t)$ may be viewed as the probability density for a random walker or diffusing object to be at position $\boldsymbol{r}$ at time t under the condition that it started from the origin $\boldsymbol{r} = \boldsymbol{0}$ at time $t = 0$. This probabilistic interpretation is very helpful for understanding the meaning of the fractional time derivative appearing in eq. (127). Rewriting equation (127) in integral form it becomes

$$f(\boldsymbol{r},t) = \delta_{\boldsymbol{r}\boldsymbol{0}} + \frac{C}{\Gamma(\alpha)} \int_0^t (t-s)^{\alpha-1} \Delta f(\boldsymbol{r},t)\,\mathrm{d}s \tag{140}$$

where the initial condition has been incorporated. This integral equation is very reminiscent of the integral equation for continuous time random walks [41,42].

In a continuous time random walk one imagines a random walker that starts at $\boldsymbol{r} = \boldsymbol{0}$ at time $t = 0$ and proceeds by successive random jumps [43,44,45,46,47,48]. The probability density for a time interval of length t between two consecutive jumps is denoted $\psi(t)$ and the probability density of a displacement by a vector $\boldsymbol{r}$ in a single jump is denoted $p(\boldsymbol{r})$. Then the integral equation of continuous time random walk theory reads

$$f(\boldsymbol{r},t) = \delta_{\boldsymbol{r}\boldsymbol{0}}\Phi(t) + \int_0^t \psi(t-s) \int_{\mathbb{R}^d} p(\boldsymbol{r}-\boldsymbol{r}') f(\boldsymbol{r},t)\,\mathrm{d}\boldsymbol{r}'\,\mathrm{d}s \tag{141}$$

where $\Phi(t)$ is the probability that the walker survives at the origin for a time of length t. Here the walker is assumed to be prepared in its initial position from which it develops according to $\psi(t)$. In general the first step needs special consideration [49,49,45]. The survival probablity $\Phi(t)$ is related to the waiting time density through

$$\Phi(t) = 1 - \int_0^t \psi(t')\, dt' \tag{142}$$

The formal similarity between eqs. (141) and (140) suggests that there exists a relation between them. To establish the relation note that eq. (130) for $\beta = 1$ gives the solution of eq. (127) in Fourier-Laplace space as

$$f(\boldsymbol{q}, u) = \frac{u^{\alpha-1}}{C\boldsymbol{q}^2 + u^\alpha}. \tag{143}$$

The Fourier-Laplace solution of eq.(141) is [44,50,51,46]

$$f(\boldsymbol{q}, u) = \frac{1}{u}\frac{1-\psi(u)}{1-\psi(u)p(\boldsymbol{q})}. \tag{144}$$

Equating these two equations yields

$$\frac{1-p(\boldsymbol{q})}{C\boldsymbol{q}^2} = \frac{1-\psi(u)}{u^\alpha \psi(u)} \quad . \tag{145}$$

Because the left hand side does not depend on u and the right hand side is independent of $\boldsymbol{q}$ they must both equal a common constant τ_0^α. It follows that

$$p(\boldsymbol{q}) = 1 - C\tau_0^\alpha \boldsymbol{q}^2 \tag{146}$$

identifying the constant $C\tau_0^\alpha$ as the mean square displacement of a single jump. For the waiting time density one finds

$$\psi(u) = \frac{1}{1 + \tau_0^\alpha u^\alpha} \tag{147}$$

which may be inverted in the same way as eq. (120) to give

$$\psi(t; \alpha, \tau_0) = \frac{1}{\tau_0}\left(\frac{t}{\tau_0}\right)^{\alpha-1} E_{\alpha,\alpha}\left(-\frac{t^\alpha}{\tau_0^\alpha}\right) \tag{148}$$

where $E_{a,b}(x)$ is again the Mittag-Leffler function defined in eq. (40).

For $\alpha = 1$ the waiting time density becomes exponential

$$\psi(t; 1, \tau_0) = \frac{1}{\tau_0} e^{-t/\tau_0}. \tag{149}$$

For $0 < \alpha < 1$ chracteristic differences arise from the asymptotic behaviour for $t \to 0$ and $t \to \infty$. The asymptotic behaviour of $\psi(t)$ for $t \to 0$ is obtained by noting that $E_{\alpha,\alpha}(0) = 1$, and hence

$$\psi(t) \sim t^{\alpha-1} \tag{150}$$

for $t \to 0$. For $\alpha < 1$ the waiting time density is singular at the origin implying a statistical abundance of short intervals between jumps compared to the exponential case $\alpha = 1$. For large $t \to \infty$ recall the asymptotic series expansion [52]

$$E_{a,b}(z) = -\sum_{n=1}^{N} \frac{z^{-n}}{\Gamma(b - an)} + O(|z|^N) \tag{151}$$

valid for $|\arg(-z)| < (1 - (a/2))\pi$ and $z \to \infty$. It follows that $E_{a,a}(-x) \sim x^{-2}$ for $x \to \infty$ and hence

$$\psi(t) \sim t^{-1-\alpha} \tag{152}$$

for $t \to \infty$. This shows that fractional diffusion is equivalent to a continuous time random walk whose waiting time density is a generalized Mittag-Leffler function. The waiting time density has a long time tail of the form usually assumed in the general theory [53,49,54,46] and exhibits a power law divergence at the origin. The exponent of both power laws is given by the order of the fractional derivative.

4 H-Functions

4.1 Definition

The H-function of order $(m, n, p, q) \in \mathbb{N}^4$ and with parameters $A_i \in \mathbb{R}_+ (i = 1, \ldots, p)$, $B_i \in \mathbb{R}_+ (i = 1, \ldots, q)$, $a_i \in \mathbb{C}(i = 1, \ldots, p)$, and $b_i \in \mathbb{C}(i = 1, \ldots, q)$ is defined for $z \in \mathbb{C}, z \neq 0$ by the contour integral [55,56,57,58,59]

$$H^{m,n}_{p,q}\left(z \,\middle|\, \begin{matrix} (a_1, A_1), \ldots, (a_p, A_p) \\ (b_1, B_1), \ldots, (b_q, B_q) \end{matrix}\right) = \frac{1}{2\pi i} \int_{\mathcal{L}} \eta(s) z^{-s} \, \mathrm{d}s \tag{153}$$

where the integrand is

$$\eta(s) = \frac{\prod_{i=1}^{m} \Gamma(b_i + B_i s) \prod_{i=1}^{n} \Gamma(1 - a_i - A_i s)}{\prod_{i=n+1}^{p} \Gamma(a_i + A_i s) \prod_{i=m+1}^{q} \Gamma(1 - b_i - B_i s)}. \tag{154}$$

In (153) $z^{-s} = \exp\{-s\log|z| - i\arg z\}$ and $\arg z$ is not necessarily the principal value. The integers m, n, p, q must satisfy

$$0 \le m \le q, \qquad 0 \le n \le p \tag{155}$$

and empty products are interpreted as being unity. The parameters are restricted by the condition

$$\mathbb{P}_a \cap \mathbb{P}_b = \emptyset \tag{156}$$

where

$$\begin{aligned} \mathbb{P}_a &= \{\text{poles of } \Gamma(1 - a_i - A_i s)\} = \left\{\frac{1 - a_i + k}{A_i} \in \mathbb{C} : i = 1, \ldots, n; k \in \mathbb{N}_0\right\} \\ \mathbb{P}_b &= \{\text{poles of } \Gamma(b_i + B_i s)\} = \left\{\frac{-b_i - k}{B_i} \in \mathbb{C} : i = 1, \ldots, m; k \in \mathbb{N}_0\right\} \end{aligned} \tag{157}$$

are the poles of the numerator in (154). The integral converges if one of the following conditions holds [59]

$$\mathcal{L} = \mathcal{L}(c - i\infty, c + i\infty; \mathbb{P}_a, \mathbb{P}_b); \quad |\arg z| < C\pi/2; \quad C > 0 \tag{158a}$$

$$\mathcal{L} = \mathcal{L}(c - i\infty, c + i\infty; \mathbb{P}_a, \mathbb{P}_b); \quad |\arg z| = C\pi/2; \quad C \ge 0; \quad cD < -\operatorname{Re} F \tag{158b}$$

$$\mathcal{L} = \mathcal{L}(-\infty + i\gamma_1, -\infty + i\gamma_2; \mathbb{P}_a, \mathbb{P}_b); \quad D > 0; \quad 0 < |z| < \infty \tag{159a}$$

$$\mathcal{L} = \mathcal{L}(-\infty + i\gamma_1, -\infty + i\gamma_2; \mathbb{P}_a, \mathbb{P}_b); \quad D = 0; \quad 0 < |z| < E^{-1} \tag{159b}$$

$$\mathcal{L} = \mathcal{L}(-\infty + i\gamma_1, -\infty + i\gamma_2; \mathbb{P}_a, \mathbb{P}_b); \quad D = 0; \quad |z| = E^{-1}; C \ge 0; \operatorname{Re} F < 0 \tag{159c}$$

$$\mathcal{L} = \mathcal{L}(\infty + i\gamma_1, \infty + i\gamma_2; \mathbb{P}_a, \mathbb{P}_b); \quad D < 0; \quad 0 < |z| < \infty \tag{160a}$$

$$\mathcal{L} = \mathcal{L}(\infty + i\gamma_1, \infty + i\gamma_2; \mathbb{P}_a, \mathbb{P}_b); \quad D = 0; \quad |z| > E^{-1} \tag{160b}$$

$$\mathcal{L} = \mathcal{L}(\infty + i\gamma_1, \infty + i\gamma_2; \mathbb{P}_a, \mathbb{P}_b); \quad D = 0; \quad |z| = E^{-1}; C \ge 0; \operatorname{Re} F < 0 \tag{160c}$$

where $\gamma_1 < \gamma_2$. Here $\mathcal{L}(z_1, z_2; \mathbb{G}_1, \mathbb{G}_2)$ denotes a contour in the complex plane starting at z_1 and ending at z_2 and separating the points in $\mathbb{G}_1$ from those in

$\mathbb{G}_2$, and the notation

$$C = \sum_{i=1}^{n} A_i - \sum_{i=n+1}^{p} A_i + \sum_{i=1}^{m} B_i - \sum_{i=m+1}^{q} B_i \tag{161}$$

$$D = \sum_{i=1}^{q} B_i - \sum_{i=1}^{p} A_i \tag{162}$$

$$E = \prod_{i=1}^{p} A_i^{A_i} \prod_{i=1}^{q} B_i^{-B_i} \tag{163}$$

$$F = \sum_{i=1}^{q} b_i - \sum_{i=1}^{p} a_j + (p-q)/2 + 1 \tag{164}$$

was employed. The H-functions are analytic for $z \neq 0$ and multivalued (single valued on the Riemann surface of $\log z$).

4.2 Basic Properties

From the definition of the H-functions follow some basic properties. Let $S_n (n \geq 1)$ denote the symmetric group of n elements, and let π_n denote a permutation in S_n. Then the product structure of (154) implies that for all $\pi_n \in S_n, \pi_m \in S_m, \pi_{p-n} \in S_{p-n}$ and $\pi_{q-m} \in S_{q-m}$

$$H^{m,n}_{p,q}\left(z \,\middle|\, \begin{matrix}(a_1, A_1), \ldots, (a_p, A_p)\\ (b_1, B_1), \ldots, (b_q, B_q)\end{matrix}\right) = H^{m,n}_{p,q}\left(z \,\middle|\, \begin{matrix}P_n, P_{p-n}\\ P_m, P_{q-m}\end{matrix}\right) \tag{165}$$

where the parameter permutations

$$\begin{aligned}
P_n &= (a_{\pi_n(1)}, A_{\pi_n(1)}), \ldots, (a_{\pi_n(n)}, A_{\pi_n(n)})\\
P_{p-n} &= (a_{\pi_{p-n}(n+1)}, A_{\pi_{p-n}(n+1)}), \ldots, (a_{\pi_{p-n}(p)}, A_{\pi_{p-n}(p)})\\
P_m &= (b_{\pi_m(1)}, B_{\pi_m(1)}), \ldots, (b_{\pi_m(m)}, B_{\pi_m(m)})\\
P_{q-m} &= (b_{\pi_{q-m}(m+1)}, B_{\pi_{q-m}(m+1)}), \ldots, (b_{\pi_{q-m}(q)}, B_{\pi_{q-m}(q)})
\end{aligned} \tag{166}$$

have to be inserted on the right hand side. If any of $n, m, p-n$ or $q-m$ vanishes the corresponding permutation is absent.

The order reduction formula

$$\begin{aligned}
&H^{m,n}_{p,q}\left(z \,\middle|\, \begin{matrix}(a_1, A_1), (a_2, A_2) \ldots, (a_p, A_p)\\ (b_1, B_1), (b_2, B_2) \ldots, (b_{q-1}, B_{q-1})(a_1, A_1)\end{matrix}\right)\\
&= H^{m,n-1}_{p-1,q-1}\left(z \,\middle|\, \begin{matrix}(a_2, A_2), \ldots, (a_p, A_p)\\ (b_1, B_1), \ldots, (b_{q-1}, B_{q-1})\end{matrix}\right)
\end{aligned} \tag{167}$$

holds for $n \geq 1$ and $q > m$, and similarly

$$H^{m,n}_{p,q}\left(z \middle| \begin{matrix} (a_1, A_1), (a_2, A_2) \dots, (a_{p-1}, A_{p-1})(b_1, B_1) \\ (b_1, B_1), (b_2, B_2) \dots, (b_q, B_q) \end{matrix}\right)$$

$$= H^{m-1,n}_{p-1,q-1}\left(z \middle| \begin{matrix} (a_1, A_1), \dots, (a_{p-1}, A_{p-1}) \\ (b_2, B_2), \dots, (b_q, B_q) \end{matrix}\right) \tag{168}$$

for $m \geq 1$ and $p > n$. The formula

$$H^{m,n}_{p,q}\left(z \middle| \begin{matrix} (a, 0), (a_2, A_2) \dots, (a_p, A_p) \\ (b_1, B_1), \dots, (b_q, B_q) \end{matrix}\right)$$

$$= \Gamma(1-a) H^{m,n-1}_{p-1,q}\left(z \middle| \begin{matrix} (a_2, A_2), \dots, (a_p, A_p) \\ (b_1, B_1), \dots, (b_q, B_q) \end{matrix}\right) \tag{169}$$

holds for $n \geq 1$ and $\mathrm{Re}(1-a) > 0$. Analogous formulae are readily found if a parameter pair $(a, 0)$ or $(b, 0)$ appears in one of the other groups.

A change of variables in (153) shows

$$H^{m,n}_{p,q}\left(z \middle| \begin{matrix} (a_1, A_1), \dots, (a_p, A_p) \\ (b_1, B_1), \dots, (b_q, B_q) \end{matrix}\right)$$

$$= H^{n,m}_{q,p}\left(\frac{1}{z} \middle| \begin{matrix} (1-b_1, B_1), \dots, (1-b_q, B_q) \\ (1-a_1, A_1), \dots, (1-a_p, A_p) \end{matrix}\right) \tag{170}$$

which allows to transform an H-function with $D > 0$ and $\arg z$ to one with $D < 0$ and $\arg(1/z)$. For $\gamma > 0$

$$\frac{1}{\gamma} H^{m,n}_{p,q}\left(z \middle| \begin{matrix} (a_1, A_1), \dots, (a_p, A_p) \\ (b_1, B_1), \dots, (b_q, B_q) \end{matrix}\right)$$

$$= H^{m,n}_{p,q}\left(z^\gamma \middle| \begin{matrix} (a_1, \gamma A_1), \dots, (a_p, \gamma A_p) \\ (b_1, \gamma B_1), \dots, (b_q, \gamma B_q) \end{matrix}\right) \tag{171}$$

while for $\gamma \in \mathbb{R}$

$$z^\gamma H^{m,n}_{p,q}\left(z \middle| \begin{matrix} (a_1, A_1), \dots, (a_p, A_p) \\ (b_1, B_1), \dots, (b_q, B_q) \end{matrix}\right)$$

$$= H^{m,n}_{p,q}\left(z \middle| \begin{matrix} (a_1 + \gamma A_1, A_1), \dots, (a_p + \gamma A_p, A_p) \\ (b_1 + \gamma B_1, B_1), \dots, (b_q + \gamma B_q, B_q) \end{matrix}\right) \tag{172}$$

holds.

For $m = 0$ with conditions (159) the integrand is analytic and thus

$$H_{p,q}^{0,n}\left(z\,\middle|\,\begin{matrix}(a_1,A_1),\dots,(a_p,A_p)\\(b_1,B_1),\dots,(b_q,B_q)\end{matrix}\right) = 0. \tag{173}$$

4.3 Integral Transformations

The definition of an H-function in eq. (153) becomes an inverse Mellin transform if $\mathfrak{L}$ is chosen parallel to the imaginary axis inside the strip

$$\max_{1\le i\le m} \operatorname{Re}\frac{-b_i}{B_i} < s < \min_{1\le i\le m} \operatorname{Re}\frac{1-a_i}{A_i} \tag{174}$$

by the Mellin inversion theorem [60]. Therefore

$$\mathcal{M}\left\{H_{p,q}^{m,n}(z)\right\}(s) = \eta(s) = \frac{\prod\limits_{i=1}^{m}\Gamma(b_i+B_i s)\prod\limits_{i=1}^{n}\Gamma(1-a_i-A_i s)}{\prod\limits_{i=n+1}^{p}\Gamma(a_i+A_i s)\prod\limits_{i=m+1}^{q}\Gamma(1-b_i-B_i s)} \tag{175}$$

whenever the inequality

$$\max_{1\le i\le m} \operatorname{Re}\frac{-b_i}{B_i} < \min_{1\le i\le m} \operatorname{Re}\frac{1-a_i}{A_i} \tag{176}$$

is fulfilled.

The Laplace transform of an H-function is obtained from eq. (175) by using eq. (78). One finds

$$\begin{aligned}\mathfrak{L}\left\{H_{p,q}^{m,n}(z)\right\}(u) &= \int_0^\infty e^{-ux}H_{p,q}^{m,n}\left(x\,\middle|\,\begin{matrix}(a_1,A_1),\dots,(a_p,A_p)\\(b_1,B_1),\dots,(b_q,B_q)\end{matrix}\right)\mathrm{d}x\\ &= H_{q,p+1}^{n+1,m}\left(u\,\middle|\,\begin{matrix}(1-b_1-B_1,B_1),\dots,(1-b_q-B_q,B_q)\\(0,1)(1-a_1-A_1,A_1),\dots,(1-a_p-A_p,A_p)\end{matrix}\right)\\ &= \frac{1}{u}H_{p+1,q}^{m,n+1}\left(\frac{1}{u}\,\middle|\,\begin{matrix}(0,1)(a_1,A_1),\dots,(a_p,A_p)\\(b_1,B_1),\dots,(b_q,B_q)\end{matrix}\right)\end{aligned} \tag{177}$$

valid for $\operatorname{Re}s > 0$, $C > 0$, $|\arg z| < \frac{1}{2}C\pi$ and $\min_{1\le j\le m}\operatorname{Re}(b_j/B_j) > -1$.

The definite integral found in [59, 2.25.2.2.]

$$\int_0^y x^{\beta-1}(y-x)^{\gamma-1} H^{m,n}_{p,q}\left(Cx^{\delta}(y-x)^{\eta} \,\middle|\, \begin{matrix}(a_1,A_1),\ldots,(a_p,A_p)\\(b_1,B_1),\ldots,(b_q,B_q)\end{matrix}\right)\mathrm{d}x$$

$$= y^{\beta+\gamma-1} H^{m,n+2}_{p+2,q+1}\left(Cy^{\delta+\eta} \,\middle|\, \begin{matrix}(1-\beta,\delta),(1-\gamma,\eta),(a_1,A_1),\ldots,(a_p,A_p)\\(b_1,B_1),\ldots,(b_q,B_q),(1-\beta-\gamma,\delta+\eta)\end{matrix}\right) \tag{178}$$

contains as a special case the fractional Riemann-Liouville integral [58, (2.7.13)]

$$\mathrm{I}^{\alpha}_{0+} H^{m,n}_{p,q}(y) = \frac{1}{\Gamma(\alpha)}\int_0^y (y-x)^{\alpha-1} H^{m,n}_{p,q}\left(x \,\middle|\, \begin{matrix}(a_1,A_1),\ldots,(a_p,A_p)\\(b_1,B_1),\ldots,(b_q,B_q)\end{matrix}\right)\mathrm{d}x$$

$$= y^{\alpha} H^{m,n+1}_{p+1,q+1}\left(y \,\middle|\, \begin{matrix}(0,1),(a_1,A_1),\ldots,(a_p,A_p)\\(b_1,B_1),\ldots,(b_q,B_q),(-\alpha,1)\end{matrix}\right) \tag{179}$$

valid if $\min_{1\leq j\leq m} \mathrm{Re}(b_j/B_j) > 0$. The fractional Riemann-Liouville derivative is obtained from this formula by analytic continuation to $\alpha < 0$.

4.4 Series Expansions

The H-functions may be represented as the series [56,57,58,59]

$$H^{m,n}_{p,q}\left(z \,\middle|\, \begin{matrix}(a_1,A_1),\ldots,(a_p,A_p)\\(b_1,B_1),\ldots,(b_q,B_q)\end{matrix}\right) = \sum_{i=1}^{m}\sum_{k=0}^{\infty} c_{ik}\frac{(-1)^k}{k!B_i} z^{(b_i+k)/B_i} \tag{180a}$$

where

$$c_{ik} = \frac{\prod\limits_{\substack{j=1\\j\neq i}}^{m}\Gamma(b_j-(b_i+k)B_j/B_i)\prod\limits_{j=1}^{n}\Gamma(1-a_j+(b_i+k)A_j/B_i)}{\prod\limits_{j=m+1}^{q}\Gamma(1-b_j+(b_i+k)B_j/B_i)\prod\limits_{j=n+1}^{p}\Gamma(a_j-(b_i+k)A_j/B_i)} \tag{180b}$$

whenever $D \geq 0$, $\mathfrak{L}$ is as in (158) or (159) and the poles in $\mathbb{P}_b$ are simple. Similarly

$$H^{m,n}_{p,q}\left(z \,\middle|\, \begin{matrix}(a_1,A_1),\ldots,(a_p,A_p)\\(b_1,B_1),\ldots,(b_q,B_q)\end{matrix}\right) = \sum_{i=1}^{n}\sum_{k=0}^{\infty} c_{ik}\frac{(-1)^k}{k!A_i} z^{-(1-a_i+k)/A_i} \tag{181a}$$

where

$$c_{ik} = \frac{\prod_{\substack{j=1\\j\neq i}}^{n} \Gamma(1-a_j-(1-a_i+k)A_j/A_i) \prod_{j=1}^{m} \Gamma(b_j+(1-a_i+k)B_j/A_i)}{\prod_{j=n+1}^{p} \Gamma(a_j+(1-a_i+k)A_j/A_i) \prod_{j=m+1}^{q} \Gamma(1-b_j-(1-a_i+k)B_j/A_i)} \tag{181b}$$

whenever $D \leq 0$, $\mathfrak{L}$ is as in (158) or (160) and the poles in $\mathbb{P}_a$ are simple.

5 Appendix: Proof of Proposition 2.2

The proof given below follows Ref. [30]. Suppose $\lim_{n\to\infty} \mu_n(s) = \mu(s)$ and $\lim_{n\to\infty} \mu_n(a_n s + b_n) = \nu(s)$ with $\mu(s)$ and $\nu(s)$ both nondegenerate. Then it must be shown that there exist $a > 0$ and b such that

$$\mu(s) = \nu(as + b) \tag{182}$$

Pick a sequence of integers $n_1 < n_2 < \ldots < n_k < \ldots$ such that $\lim_{k\to\infty} a_{n_k} = a$ and $\lim_{k\to\infty} b_{n_k} = b$ exists with $0 \leq a \leq \infty, -\infty \leq b \leq \infty$. Consider this sequence of indices from now on as fixed. Then, to simplify the notation, suppose without loss of generality that $\lim_{k\to\infty} a_k = a$ and $\lim_{k\to\infty} b_k = b$.

First it will be shown that $0 < a < \infty$. Suppose $a = \infty$. Let

$$u = \sup\{x : \limsup_{n\to\infty}(a_n x + b_n) < \infty\}. \tag{183}$$

Then for $v < x < u$

$$\limsup_{n\to\infty}(a_n v + b_n) \leq \limsup_{n\to\infty}(v - x)a_n + \limsup_{n\to\infty}(a_n x + b_n), \tag{184}$$

and hence for every $v < u$, it follows that $\nu(v) = 0$ because $(a_n v + b_n) \to -\infty$ with $n \to \infty$. For $v > u$ on the other hand, $\limsup(a_n v + b_n) = \infty$ and hence $\nu(v) = 1$ for $v > u$. Thus the assumption $a = \infty$ contradicts to $\nu(s)$ being nondegenerate.

It follows that also b must be finite. In fact if $\lim_{n\to\infty}(a_n x + b_n) = \infty$ then $\nu(x) = 1$ while for $\lim_{n\to\infty}(a_n x + b_n) = -\infty$ follows $\nu(x) = 0$.

Suppose now that $a = 0$. Then for every x and $\varepsilon > 0$

$$b - \varepsilon \leq a_n x + b_n \leq b + \varepsilon \tag{185}$$

if n is chosen sufficiently large. By monotonicity of μ_n it follows that

$$\mu_n(b-\varepsilon) \le \mu_n(a_n x + b_n) \le \mu_n(b+\varepsilon). \tag{186}$$

If ε is chosen so that $\mu(x)$ is continuous at the points $b-\varepsilon$ and $b+\varepsilon$, then

$$\mu(b-\varepsilon) \le \nu(x) \le \mu(b+\varepsilon). \tag{187}$$

Because x was arbitrary it follows that $\mu(b-\varepsilon) = 0$ and $\mu(b+\varepsilon) = 1$. Hence $\mu(x)$ is degenerate, contrary to the conditions above.

Finally, let x be such that $\mu(x)$ is continuous at the point $ax+b$, and that $\nu(x)$ is continuous at x. Then

$$\lim_{n\to\infty} \mu_n(a_n x + b_n) = \nu(x). \tag{188}$$

On the other hand because $\lim_{n\to\infty}(a_n x + b_n) = ax + b$ one has for sufficiently large n that

$$ax + b - \varepsilon \le a_n x + b_n \le ax + b + \varepsilon \tag{189}$$

where $\varepsilon > 0$ is chosen such that the distribution function μ is continuous at the points $ax+b-\varepsilon$ and $ax+b+\varepsilon$. Hence by monotonicity

$$\mu_n(ax+b-\varepsilon) \le \mu_n(a_n x + b_n) \le \mu_n(ax+b+\varepsilon) \tag{190}$$

and for $n \to \infty$

$$\mu(ax+b-\varepsilon) \le \liminf_{n\to\infty} \mu_n(a_n x + b_n) \le \limsup_{n\to\infty} \mu_n(a_n x + b_n) \le \mu(ax+b+\varepsilon). \tag{191}$$

Because $ax + b$ is a point of continuity for $\mu(x)$ and ε is arbitrary it follows that

$$\lim_{n\to\infty} \mu_n(a_n x + b_n) = \mu(ax+b) \tag{192}$$

and hence $\nu(x) = \mu(ax+b)$ proving the assertion.

References

1. G.W. Scott-Blair and J.E. Caffyn. An application of the theory of quasi-properties to the treatment of anomalous stress-strain relations. *Phil. Mag.*, 40:80, 1949.
2. H. Berens and U. Westphal. A Cauchy problem for a generalized wave equation. *Acta Sci. Math. (Szeged)*, 29:93, 1968.

3. K.B. Oldham and J.S. Spanier. The replacement of Fick's law by a formulation involving semidifferentiation. *J. Electroanal. Chem. Interfacial Electrochem.*, 26:331, 1970.
4. M. Caputo and F. Mainardi. Linear models of dissipation in anelastic solids. *Riv.Nuovo Cim.*, 1:161, 1971.
5. K.B. Oldham and J.S. Spanier. *The Fractional Calculus.* Academic Press, New York, 1974.
6. R. Hilfer. Classification theory for anequilibrium phase transitions. *Phys. Rev. E*, 48:2466, 1993.
7. R. Hilfer. On a new class of phase transitions. In W.P. Beyermann, N.L. Huang-Liu, and D.E. MacLaughlin, editors, *Random Magnetism and High-Temperature Superconductivity*, page 85, Singapore,, 1994. World Scientific Publ. Co.
8. R. Hilfer. Exact solutions for a class of fractal time random walks. *Fractals*, 3(1):211, 1995.
9. R. Hilfer. Fractional dynamics, irreversibility and ergodicity breaking. *Chaos, Solitons & Fractals*, 5:1475, 1995.
10. R. Hilfer. Foundations of fractional dynamics. *Fractals*, 3:549, 1995.
11. R. Hilfer. An extension of the dynamical foundation for the statistical equilibrium concept. *Physica A*, 221:89, 1995.
12. R.L. Bagley and P.J. Torvik. A theoretical basis for the application of fractional calculus to viscoelasticity. *J. Rheology*, 27:201, 1983.
13. R.L. Bagley and P.J. Torvik. On the fractional calculus model of viscoelastic behaviour. *J. Rheology*, 30:133, 1986.
14. W. Wyss. The fractional diffusion equation. *J. Math. Phys.*, 27:2782, 1986.
15. W.R. Schneider and W. Wyss. Fractional diffusion and wave equations. *J. Math. Phys.*, 30:134, 1989.
16. A.M.A. El-Sayed. Fractional-order diffusion-wave equation. *Int. J. Theor. Phys.*, 35:311, 1996.
17. A. Compte. Stochastic foundations of fractional dynamics. *Phys.Rev. E*, 55:4191, 1996.
18. T.F. Nonnenmacher and W.G. Glöckle. A fractional model for mechanical stress relaxation. *Phil. Mag. Lett.*, 64:89, 1991.
19. L. Gaul, P. Klein, and S. Kempfle. Damping description involving fractional operators. *Mechanical Systems and Signal Processing*, 5:81, 1991.
20. H. Beyer and S. Kempfle. Definition of physically consistent damping laws with fractional derivatives. *A. angew. Math. Mech.*, 75:623, 1995.
21. R.R. Nigmatullin. The realization of the generalized transfer equation in a medium with fractal geometry. *phys. stat. sol. b*, 133:425, 1986.
22. S. Westlund. Dead matter has memory ! *Physica Scripta*, 43:174, 1991.
23. G. Jumarie. A Fokker-Planck equation of fractional order with respect to time. *J. Math. Phys.*, 33:3536, 1992.

24. H. Schiessel and A. Blumen. Fractal aspects in polymer science. *Fractals*, 3:483, 1995.
25. G. Zaslavsky. Fractional kinetic equation for Hamiltonian chaos. *Physica D*, 76:110, 1994.
26. G. Zaslavsky. From Hamiltonian chaos to Maxwell's demon. *Chaos*, 5:653, 1995.
27. E. Stein and G. Weiss. *Introduction to Fourier Analysis on Euclidean Spaces.* Princeton University Press, Princeton, 1971.
28. C. Berg and G. Forst. *Potential Theory on Locally Compact Abelian Groups.* Springer, Berlin, 1975.
29. A.P. Prudnikov, Yu.A. Brychkov, and O.I. Marichev. *Integrals and Series*, volume 4. Gordon and Breach, New York, 1992.
30. B.V. Gnedenko and A.N. Kolmogorov. *Limit Distributions for Sums of Independent Random Variables.* Addison-Wesley, Cambridge, 1954.
31. W. Feller. *An Introduction to Probability Theory and Its Applications*, volume II. Wiley, New York, 1971.
32. F. Oberhettinger. *Tables of Mellin Transforms.* Springer Verlag, Berlin, 1974.
33. I.A. Ibragimov and Yu.V. Linnik. *Independent and Stationary Sequences of Random Variables.* Wolters-Nordhoff Publishing, Groningen, 1971.
34. S. Bochner. *Harmonic Analysis and the Theory of Probability.* University of California Press, Berkeley, 1955.
35. K. Yosida. *Functional Analysis.* Springer, Berlin, 1965.
36. A.V. Balakrishnan. Fractional powers of closed operators and the semigroups generated by them. *Pacific J. Math.*, 10:419, 1960.
37. U. Westphal. Ein Kalkül für gebrochene Potenzen infinitesimaler Erzeuger von Halbgruppen und Gruppen von Operatoren. *Compositio Math.*, 22:67, 1970.
38. L. Bieberbach. *Lehrbuch der Funktionentheorie*, volume II. Teubner, Leipzig, 1931.
39. W.R. Schneider. Completely monotone generalized Mittag-Leffler functions. *Expo. Math.*, 14:3, 1996.
40. P.L. Butzer and R. J. Nessel. *Fourier Analysis and Approximation*, volume 1. Birkhäuser Verlag, Basel, 1971.
41. R. Hilfer and L. Anton. Fractional master equations and fractal time random walks. *Phys.Rev.E, Rapid Commun.*, 51:848, 1995.
42. R. Hilfer. On fractional diffusion and its relation with continuous time random walks. In A. Pekalski R. Kutner and K. Sznajd-Weron, editors, *Anomalous Diffusion: From Basis to Applications*, page 77. Springer, 1999.

43. E.W. Montroll and G.H. Weiss. Random walks on lattices II. *J. Math. Phys.*, 6:167, 1965.
44. M.N. Barber and B.W. Ninham. *Random and Restricted Walks.* Gordon and Breach Science Publ., New York, 1970.
45. J.W. Haus and K. Kehr. Diffusion in regular and disordered lattices. *Phys.Rep.*, 150:263, 1987.
46. J. Klafter, A. Blumen, and M.F. Shlesinger. Stochastic pathway to anomalous diffusion. *Phys. Rev. A*, 35:3081, 1987.
47. B.D. Hughes. *Random Walks and Random Environments*, volume 1. Clarendon Press, Oxford, 1995.
48. B.D. Hughes. *Random Walks and Random Environments*, volume 2. Clarendon Press, Oxford, 1996.
49. J.K.E. Tunaley. Some properties of the asymptotic solutions of the Montroll-Weiss equation. *J. Stat. Phys.*, 12:1, 1975.
50. E.W. Montroll and B.J. West. On an enriched collection of stochastic processes. In E.W. Montroll and J.L Lebowitz, editors, *Fluctuation Phenomena*, page 61, Amsterdam, 1979. North Holland Publ. Co.
51. G.H. Weiss and R.J. Rubin. Random walks: Theory and selected applications. *Adv. Chem. Phys.*, 52:363, 1983.
52. A. Erdelyi (et al.). *Higher Transcendental Functions*, volume III. Mc Graw Hill Book Co., New York, 1955.
53. M.F. Shlesinger. Asymptotic solutions of continuous time random walks. *J. Stat. Phys.*, 10:421, 1974.
54. M.F. Shlesinger, J. Klafter, and Y.M. Wong. Random walks with infinite spatial and temporal moments. *J. Stat. Phys.*, 27:499, 1982.
55. C. Fox. The G and H functions as symmetrical Fourier kernels. *Trans. Am. Math. Soc.*, 98:395, 1961.
56. B.L.J. Braaksma. Asymptotic expansions and anlytic continuations for a class of Barnes-integrals. *Compos.Math.*, 15:239, 1964.
57. A.M. Mathai and R.K. Saxena. *The H-function with Applications in Statistics and Other Disciplines.* Wiley, New Delhi, 1978.
58. H.M. Srivastava, K.C. Gupta, and S.P. Goyal. *The H-functions of One and Two Variables with Applications.* South Asian Publishers, New Delhi, 1982.
59. A.P. Prudnikov, Yu.A. Brychkov, and O.I. Marichev. *Integrals and Series*, volume 3. Gordon and Breach, New York, 1990.
60. I.N. Sneddon. *The Use of Integral Transforms.* Mc Graw Hill, New York, 1972.

CHAPTER III

FRACTIONAL POWERS OF INFINITESIMAL GENERATORS OF SEMIGROUPS

U. WESTPHAL
Institut für Mathematik, Universität Hannover, Welfengarten 1, D-30167 Hannover, Germany

Contents

1 Introduction

Many results on fractional integrals and derivatives are valid with respect to a norm of a suitable function space as, for instance, the Lebesgue spaces $L^p(0,\infty)$ or $L^p(-\infty,\infty)$, $1 \leq p < \infty$. Hence, it is natural and useful to look at these operations from a more functional analytic point of view by interpreting them as powers of arbitrary order of the operator of differentiation. A first step in this more abstract direction is to consider a one-parameter family of bounded linear operators $T(t)$ on a Banach space X to itself with the property that

$$T(t)T(s) = T(t+s)$$

for all values t, s in the interval $\mathbb{R}_+$ of nonnegative real numbers. We shall assume that the operator valued function $t \mapsto T(t)$ is continuous with respect to the norm topology of X, in particular $T(t)x$ converges to $T(0)x = x$ as $t \to 0+$ for all elements x of the space X. We refer to such a family as a (C_0)-semigroup of bounded linear operators. Mostly, we shall assume in addition that the semigroup is equibounded which means that the norms of the operators $T(t)$ are bounded above by a constant $M \geq 1$ independent of $t \in \mathbb{R}_+$.

The right-hand derivative of the operator function $t \mapsto T(t)$ at $t = 0$ taken in the norm topology of X is called the infinitesimal generator A of the semigroup. In analogy to the numerically valued exponential function $t \mapsto e^{ta}$, where a denotes a fixed real or complex number, the semigroup operators $T(t)$ are sometimes rewritten as e^{tA}.

Some of the simplest and most important examples of semigroups of bounded linear operators arise from the operation of left translation $[T(t)f](u) = f(u+t)$ and right translation $[T(t)f](u) = f(u-t)$, respectively, in spaces of continuous functions f as well as in Lebesgue spaces. Under appropriate assumptions, the infinitesimal generator in these cases is the operator of differentiation, $A = \frac{d}{du}$ and $A = -\frac{d}{du}$, respectively.

One of the most fruitful characterizations of the fractional derivative $D^\alpha f$ of a function f, say in $L^p(\mathbb{R})$, is the representation via a Marchaud integral, that is, in case $0 < \alpha < 1$,

$$D^\alpha f(u) = \lim_{\varepsilon \to 0+} \frac{1}{-\Gamma(-\alpha)} \int_\varepsilon^\infty t^{-\alpha-1}[f(u) - f(u-t)]\, dt.$$

Using the semigroup of right translations this may be rewritten as

$$(-A)^{\alpha} f = \lim_{\varepsilon \to 0+} \frac{1}{-\Gamma(-\alpha)} \int_{\varepsilon}^{\infty} t^{-\alpha-1}[f - T(t)f]\, dt, \tag{1.1}$$

where the limit is taken with respect to the norm of $L^p(\mathbb{R})$. This suggests in case of an arbitrary equibounded (C_0)-semigroup $\{T(t);\, t \in \mathbb{R}_+\}$ with generator A the definition

$$(-A)^{\alpha} x = \lim_{\varepsilon \to 0+} \frac{1}{C_{\alpha,m}} \int_{\varepsilon}^{\infty} t^{-\alpha-1}[I - T(t)]^m x\, dt \tag{1.2}$$

for each $x \in X$ for which the limit exists in the norm topology of X. Here α is a positive real number, m an integer strictly larger than α and $C_{\alpha,m}$ a suitable normalizing constant. (The case $m = 1$ with values of α between zero and one is the direct analogue of (1.1).)

Definition (1.2) is also motivated by a result of Lions and Peetre [22] who proved a representation of the form (1.2) for integral powers of $(-A)$ which are customarily defined by iteration.

A systematic treatment of fractional powers starting from definition (1.2) has been given in Westphal [45]. Alternatively, one may proceed from the following definition considered in [46],

$$(-A)^{\alpha} x = \lim_{t \to 0+} t^{-\alpha}[I - T(t)]^{\alpha} x, \tag{1.3}$$

which is motivated by the characterization of fractional derivatives due to Liouville, Grünwald and Letnikov. The methods used in both of these papers heavily depend on the Laplace transform which is a very natural tool when dealing with families of operators that satisfy the functional equation of the exponential function. Indeed, Hille [13] developed an operational calculus for infinitesimal generators of a wide class of semigroups that is based on an algebra of numerically valued functions which are representable as Laplace transforms of Borel measures on $\mathbb{R}_+$. This calculus was extended by Phillips in a series of papers [32,33,34]; see Chapter XV of the monograph [13] of Hille and Phillips.

The principal feature of the Hille-Phillips calculus is the following: Given the Laplace transform $F(z)$ of a Borel measure μ on $\mathbb{R}_+$,

$$F(z) = \int_0^{\infty} e^{-zt}\, d\mu(t),$$

then with the formal correspondence $e^{-zt} \sim e^{tA}$ in mind, an operator valued Function $F(-A)$ is defined by

$$F(-A)x = \int_0^\infty T(t)x \, d\mu(t), \qquad x \in X, \tag{1.4}$$

where the complex number z and the measure μ are suitably restricted for the convergence of the integrals involved. Note that the Laplace transform $F(z)$ is holomorphic in a right half plane of the complex plane while the spectrum of the infinitesimal generator A is contained in a left half plane. This explains the appearance of the minus sign in $F(-A)$. To avoid this, some authors define the negative of A as the infinitesimal generator of the semigroup. We prefer, however, to maintain the original meaning and accept that the notation is sometimes a bit clumsy.

The operators $F(-A)$ provided by the Hille-Phillips operational calculus are bounded linear operators on X to itself and hence, in general, cannot include power functions of $(-A)$, since these are expected to be closed operators but not bounded, in general, as it is the case for integral powers. Nevertheless, this calculus is successful in a more indirect manner. Phillips [33] defined the fractional powers $-(-A)^\alpha$ as infinitesimal generators of 'new' semigroups which were obtained from the originally given semigroup by the operational calculus. This method generalizes one which was proposed by Bochner [6] and Feller [11] in connection with their investigations of stochastic processes. For more recent studies in this direction we refer to Berg, Boyadzhiev and deLaubenfels [5] and Schilling [39].

In his thesis of 1954 Balakrishnan [1] extended the Hille-Phillips calculus by admitting also functions that are multipliers with respect to the Laplace transform. The corresponding operators $F(-A)$ are closed linear with dense domain. The construction of such an $F(-A)$ occurs in two steps. The first step provides a concrete integral representation of the form (1.4) for elements x of a certain subset of the expected domain of $F(-A)$. Then, in a second step, the restricted operator obtained in this way is shown to be closable; thus its smallest closed extension is uniquely determined, and this is taken as the very definition of $F(-A)$. As for a complex number α with positive real part, the function $F(z) = z^\alpha$ is a multiplier with respect to the Laplace transform, fractional powers $(-A)^\alpha$ are covered by Balakrishnan's extension of the operational calculus.

About the same time, Nelson [30] developed an operational calculus for semigroup generators which is based on an algebra of Laplace transforms of distributions. His construction of the operators $F(-A)$ is in some respect similar to that of Balakrishnan, at least in the sense that at first a restriction of $F(-A)$ on a subset of its domain is defined and then the full operator is obtained by a process of closed extension. The class of distributions considered by Nelson is essentially that which was introduced by L. Schwartz [40] in the framework of a much more general representation theory of semigroups.

Following Schwartz, a distribution U on the set of real numbers $\mathbb{R}$ is said to be $\mathbb{R}_+$-summable if it has support in $\mathbb{R}_+$ and if it is a finite sum of derivatives of finite Borel measures, say,

$$U = \sum_{j=1}^{m} D^j \mu_j. \tag{1.5}$$

Nelson heavily argues with this representation, while in the calculus introduced by Schwartz also an equivalent characterization of (1.5) plays an essential role, namely the fact that the convolution of U with every infinitely differentiable test function φ of compact support contained in $\mathbb{R}_+$, $U * \varphi$, is a function of $L^1(\mathbb{R}_+)$. Hence, if the underlying (C_0)-semigroup is equibounded one obtains a family of bounded linear operators defined by the integrals

$$\int_0^\infty (U * \varphi)(t)\, T(t)x\, dt, \qquad x \in X. \tag{1.6}$$

The operator $F(-A)$ that corresponds to the Laplace transform $F(z)$ of the distribution U is defined by Schwartz as the limit of the bounded operators (1.6), that is,

$$F(-A)x = \lim_{\varphi} \int_0^\infty (U * \varphi)(t)\, T(t)x\, dt,$$

where φ approaches the Dirac measure δ within a certain class of test functions. If U is chosen as the α-th fractional derivative of δ, that is the distribution δ^α, one obtains a definition of the fractional power $(-A)^\alpha$ via a limit process. This application of Schwartz' functional calculus is mentioned, for instance, by Faraut [9].

In the late eighties, the subject was taken up by Lanford and Robinson [24]. They analyzed fractional powers of infinitesimal generators starting from the definition according to the functional calculus due to Schwartz. Instead of the Laplace transform they used the theory of Fourier transforms of tempered distributions. Moreover, their results include also other topologies than the norm topology of the Banach space X.

An approach to fractional powers of infinitesimal generators via Schwartz' operational calculus is, in fact, very satisfactory from a theoretical point of view. It allows us to interpret the complex function $F(z) = z^\alpha$ as a Laplace transform, not of a function, but of a distribution, namely δ^α. We want to emphasize, however, that, in principle, the calculus can be reduced to dealing with Laplace transforms of L^1-functions; cf. [45] and [46]. This is not surprising if one has in mind that an $\mathbb{R}_+$-summable distribution convolved with a suitable test function has to belong to $L^1(\mathbb{R}_+)$.

It is the main purpose of this chapter to describe a systematic approach to fractional powers of infinitesimal generators that is based on the Laplace transform and starts with the definition proposed by the operational calculus of Schwartz. The treatment presented here did not appear in this form before. In Section 2 we introduce some basic facts of the operational calculus due to Schwartz and fix the definition of powers of an infinitesimal generator A of fractional order $\alpha > 0$. In Section 3 we establish the equivalence of the definition with the representations indicated in (1.2) and (1.3). Section 4 is devoted to basic properties of fractional powers, as the monotonicity law and the power rules $(-A)^\alpha(-A)^\beta = (-A)^{\alpha+\beta}$ for $\alpha, \beta > 0$ and $((-A)^\alpha)^\beta = (-A)^{\alpha\beta}$ for $0 < \alpha < 1$ and $\beta > 0$. In Section 5 it is shown how powers with negative exponents can be included in the framework of the present approach. The results on abstract Riesz potentials considered in Section 6 generalize the investigations of Bochner and Riesz on fractional powers of the Laplace operator on the one hand and fractional potentials of negative order on the other hand, which stood at the very beginning of the development of fractional powers of infinitesimal generators. Finally, in Section 7 we outline some extensions to more general classes of operators than infinitesimal generators.

2 Operational Calculus for an Infinitesimal Generator

Let X be a Banach space with elements $x, y, \ldots$ having norms $||x||, ||y||, \ldots$ and let $\mathcal{E}(X)$ be the Banach algebra of bounded linear operators from X into itself.

Let $\{T(t);\, t \in \mathbb{R}_+\}$ be an equibounded (C_0)-semigroup of linear operators in $\mathcal{E}(X)$, that is,

$$T(t)T(s) = T(t+s) \text{ for } t, s > 0,\ T(0) = I \text{ (identity operator)},$$
$$\lim_{t\to 0+} ||T(t)x - x|| = 0 \text{ for each } x \in X,$$
$$||T(t)|| \leq M \text{ for all } t \in \mathbb{R}_+,\ M \text{ being a constant.}$$

The infinitesimal generator of $\{T(t);\, t \in \mathbb{R}_+\}$ is defined by

$$Ax = \lim_{t\to 0+} t^{-1}[T(t)x - x],$$

the domain $D(A)$ being the set of all $x \in X$ for which the limit exists in the norm topology of X. In general, A is not a bounded operator; it is, however, closed and its domain is a dense subspace of X. See e.g. [7] and [13].

Let μ be a finite complex Borel measure on $\mathbb{R}_+$. Then we define a linear operator $G(\mu)$ in $\mathcal{E}(X)$ by the formula

$$G(\mu)x := \int_0^\infty T(t)x\, d\mu(t), \qquad x \in X.$$

If, in particular, μ has a density with respect to the Lebesgue measure, say f, then we set

$$G(f)x := \int_0^\infty f(t)T(t)x\, dt, \qquad x \in X.$$

Let us give two simple examples.

a) If δ_t is the Dirac measure at $t \in \mathbb{R}_+$, then

$$G(\delta_t) = T(t).$$

b) For each complex number z whose real part, $\operatorname{Re} z$, is nonnegative, let e_z be the complex-valued function defined by

$$e_z(t) := \begin{cases} e^{-zt} & \text{if} \quad t \geq 0, \\ 0 & \phantom{\text{if}} \quad t < 0. \end{cases}$$

Then the operators $G(e_z)$ are just the resolvent operators of the infinitesimal generator A:

$$G(e_z)x = \int_0^\infty e^{-zt}T(t)x\, dt = (zI - A)^{-1}x, \qquad \operatorname{Re} z > 0,\ x \in X.$$

The mapping G which assigns to each finite Borel measure μ on $\mathbb{R}_+$ the operator $G(\mu)$ in $\mathcal{E}(X)$ shall be extended now to a class of distributions.

For the general theory of distributions we refer to the standard books of Schwartz [41], Gelfand-Shilov [12], and Zemanian [50], for instance. The latter is recommended, in particular, with respect to results on convolutions and Laplace transforms of distributions having supports that are bounded below.

Let us just introduce some notation. In the following $\mathcal{D}(\mathbb{R})$ denotes the usual space of test functions on $\mathbb{R}$, that is the space of infinitely differentiable functions with compact support contained in $\mathbb{R}$. $\mathcal{D}'(\mathbb{R})$ will be the space of distributions acting on $\mathcal{D}(\mathbb{R})$. If $\varphi \in \mathcal{D}(\mathbb{R})$ and $U \in \mathcal{D}'(\mathbb{R})$ we denote the value of U at φ by the symbol $\langle \varphi, U \rangle$. The abbreviation supp φ (supp U) stands for the support of φ (U). The convolution $U * \varphi$ of $U \in \mathcal{D}'(\mathbb{R})$ and $\varphi \in \mathcal{D}(\mathbb{R})$ is defined as the infinitely differentiable function

$$(U * \varphi)(t) := \langle \varphi(t-u), U(u) \rangle,$$

where the notation $U(u)$ means that the distribution U acts on the function $\varphi(t-u)$ when the latter is regarded as a function of the variable u. The space of Lebesgue integrable functions on $\mathbb{R}$ is denoted by $L^1(\mathbb{R})$.

Following Schwartz [40] let us introduce

Definition 2.1. *A distribution $U \in \mathcal{D}'(\mathbb{R})$ is said to be $\mathbb{R}_+$-summable, if (i) supp $U \subset \mathbb{R}_+$ and (ii) U is a finite sum of derivatives of finite Borel measures on $\mathbb{R}$, or, equivalently, (ii') $U * \varphi \in L^1(\mathbb{R})$ for each $\varphi \in \mathcal{D}(\mathbb{R})$. The space of $\mathbb{R}_+$-summable distributions will be denoted by $\mathcal{D}'_{L^1}(\mathbb{R}_+)$.*

Each $U \in \mathcal{D}'_{L^1}(\mathbb{R}_+)$ is, in fact, a tempered distribution. Thus the Laplace transform of U

$$\mathcal{L}(U)(z) := \langle e_z, U \rangle$$

is well defined for all complex numbers z with $\operatorname{Re} z > 0$. Moreover, if $U, V \in \mathcal{D}'_{L^1}(\mathbb{R}_+)$, then the convolution $U * V$ defined by

$$\langle \varphi, U * V \rangle = \langle \langle \varphi(t+\tau), U(t) \rangle,\ V(\tau) \rangle, \qquad \varphi \in \mathcal{D}(\mathbb{R}),$$

also belongs to $\mathcal{D}'_{L^1}(\mathbb{R}_+)$ and satisfies the *product theorem* for the Laplace transformation

$$\mathcal{L}(U * V)(z) = \mathcal{L}(U)(z) \cdot \mathcal{L}(V)(z), \qquad \operatorname{Re} z > 0.$$

Throughout this chapter, mostly one of the distributions that are to be convolved will be determined by a function, say f, that belongs to $L^1(\mathbb{R}_+)$. As usual, such a regular distribution will be denoted by the same symbol as its generating function f. Recall that the Laplace transform of f is given by the integral

$$\mathcal{L}(f)(z) = \int_0^\infty e^{-zt} f(t)\, dt, \qquad \operatorname{Re} z \geq 0.$$

If $U, V \in \mathcal{D}'_{L^1}(\mathbb{R}_+)$ such that

$$\mathcal{L}(U)(z) = \mathcal{L}(V)(z), \qquad \operatorname{Re} z > 0,$$

then, by the *uniqueness theorem* for the Laplace transformation, $U = V$.

To define the operator $G(U)$ for a distribution U that is $\mathbb{R}_+$-summable, U will be approximated by regularizations $U * \varphi$ with test functions φ tending to the Dirac measure δ. (Recall that $\langle \psi, \delta \rangle = \psi(0)$ for all $\psi \in \mathcal{D}(\mathbb{R})$.) To be more precise, let $\mathcal{F}$ be the filter having for a base the sets F_ε, $\varepsilon > 0$, which are defined by

$$F_\varepsilon := \left\{ \varphi \in \mathcal{D}(\mathbb{R});\ \operatorname{supp} \varphi \subset [0, \varepsilon),\ \varphi \geq 0,\ \left| \int_0^\infty \varphi(t)\, dt - 1 \right| \leq \varepsilon \right\}.$$

This filter converges to δ with respect to the topology of $\mathcal{D}'(\mathbb{R})$. Moreover, we have for each $x \in X$

$$\lim_{\mathcal{F}} G(\varphi)x = x. \tag{2.1}$$

Here and in the following $\lim_{\mathcal{F}}$ denotes the limit with respect to the norm topology of X, as φ approaches δ following the filter $\mathcal{F}$.

The next definition provides an operational calculus for the infinitesimal generator A which is wide enough to cover also power functions of fractional order.

Definition 2.2. *Let $U \in \mathcal{D}'_{L^1}(\mathbb{R}_+)$. The operator $G(U)$ with domain*

$$D(G(U)) := \{x \in X;\ \lim_{\mathcal{F}} G(U * \varphi)x \text{ exists}\}$$

is defined by the formula

$$G(U)x = \lim_{\mathcal{F}} G(U * \varphi)x, \qquad x \in D(G(U)).$$

In general, $G(U)$ is no longer a bounded operator. We shall see, however, that it is closed and its domain is a dense subspace of X.

It is clear from Definition 2.2. that the mapping $U \mapsto G(U)$ is linear. We now want to consider its behaviour under convolution which is crucial for the operational calculus. As the original literature due to Schwartz [40] is sometimes difficult of access, we shall sketch the proof of the general convolution theorem. It is convenient to begin with a special case.

Lemma 2.3. *Let $U \in \mathcal{D}'_{L^1}(\mathbb{R}_+)$ and $\psi \in L^1(\mathbb{R}_+)$. If $x \in D(G(U))$, then $x \in D(G(\psi * U))$ and*

$$G(\psi * U)x = G(\psi)G(U)x.$$

Proof. Let $\varphi \in \mathcal{D}(\mathbb{R})$ such that supp $\varphi \subset \mathbb{R}_+$. As ψ and $U * \varphi$ are functions from $L^1(\mathbb{R}_+)$, their convolution also belongs to $L^1(\mathbb{R}_+)$. Thus, $G(\psi)$, $G(U * \varphi)$ and $G(\psi * (U * \varphi))$ are bounded linear operators from X into itself. Due to the semigroup property of the family $\{T(t);\, t \in \mathbb{R}_+\}$ one obtains for every $x \in X$

$$G(\psi * (U * \varphi))x = G(\psi)G(U * \varphi)x.$$

In this equation we may replace the operator on the left-hand side by $G((\psi * U) * \varphi)$, since convolution is associative. Thus we have

$$G((\psi * U) * \varphi)x = G(\psi)G(U * \varphi)x, \qquad x \in X. \tag{2.2}$$

Now, if $x \in D(G(U))$, then, since $G(\psi)$ is bounded, the limit $\lim_{\mathcal{F}} G(\psi)G(U * \varphi)x$ exists and equals $G(\psi)G(U)x$. But then also the left-hand side of (2.2) is convergent as $\varphi \to \delta$ following the filter $\mathcal{F}$. Hence, $x \in D(G(\psi * U))$ and

$$G(\psi * U)x = G(\psi)G(U)x.$$

From Lemma 2.3 we deduce the general convolution theorem.

Theorem 2.4. *Let $U, V \in \mathcal{D}'_{L^1}(\mathbb{R}_+)$ and $x \in D(G(V))$. Then $x \in D(G(U * V))$ if and only if $G(V)x \in D(G(U))$. If so, then*

$$G(U * V)x = G(U)G(V)x.$$

Proof. Let $\varphi \in \mathcal{D}(\mathbb{R})$ such that supp $\varphi \subset \mathbb{R}_+$ and apply the previous lemma to $\psi := U * \varphi$. Then, if $x \in D(G(V))$, we obtain

$$G((U * V) * \varphi)x = G((U * \varphi) * V)x = G(U * \varphi)G(V)x.$$

The result now follows, as φ tends to δ.

As a consequence of Theorem 2.4 we obtain the following properties of an operator $G(U)$.

Theorem 2.5. *Let $U \in \mathcal{D}'_{L^1}(\mathbb{R}_+)$.*

(i) For each $\varphi \in \mathcal{D}(\mathbb{R})$ with supp $\varphi \subset \mathbb{R}_+$ and each $x \in X$, the element $G(\varphi)x$ belongs to $D(G(U))$ and

$$G(U)G(\varphi)x = G(U * \varphi)x.$$

If $x \in D(G(U))$, then

$$G(\varphi)G(U)x = G(U * \varphi)x.$$

(ii) The operator $G(U)$ is closed and its domain $D(G(U))$ is dense in X.

(iii) For each positive integer n, we have

$$G(\delta^n) = (-A)^n,$$

where δ^n denotes the n-th derivative of δ.

Proof. Statement (i) is clear from Theorem 2.4. To show that $D(G(U))$ is dense in X we observe that, due to (2.1), each element $x \in X$ may be approximated by elements $G(\varphi)x$ ($\varphi \to \delta$), which belong to $D(G(U))$ according to (i). To see that $G(U)$ is closed, suppose $x_k \in D(G(U))$, $k = 1, 2, \ldots$ such that $x_k \to x$ and $G(U)x_k \to y$ as $k \to \infty$. Then by (i), we have for each $\varphi \in \mathcal{D}(\mathbb{R})$ with supp $\varphi \subset \mathbb{R}_+$ that

$$G(\varphi)G(U)x_k = G(U * \varphi)x_k,$$

and as $k \to \infty$

$$G(\varphi)y = G(U * \varphi)x.$$

If $\varphi \to \delta$ following the filter $\mathcal{F}$, the left-hand side of the last equation tends to y. Hence, the limit of the right-hand side exists, too, and equals y, which means that $x \in D(G(U))$ and $G(U)x = y$. This proves statement (ii). As for (iii), let n be a positive integer. Then for each function $\varphi \in \mathcal{D}(\mathbb{R})$, whose support is contained in $\mathbb{R}_+$, one can show by induction on n that

$$G(\delta^n * \varphi)x = G(\varphi^{(n)})x = \begin{cases} (-A)^n G(\varphi)x, & x \in X \\ G(\varphi)(-A)^n x, & x \in D(A^n). \end{cases}$$

If we take the limit in these equations as $\varphi \to \delta$ following the filter $\mathcal{F}$, then the conclusion follows by (2.1) and the fact that $(-A)^n$ is closed.

Part (iii) of Theorem 2.5 suggests to define the fractional power $(-A)^\alpha$ for real numbers $\alpha > 0$ by $G(\delta^\alpha)$, where δ^α means the derivative of order α of the Dirac measure δ. The distribution δ^α is, in fact, defined for all complex numbers α. See for instance, [12; I, §3], [15; Sec. 3.2], [50; Sec. 2.5]. If $\operatorname{Re}\alpha < 0$, δ^α is the regular distribution generated by the locally integrable function

$$\frac{1}{\Gamma(-\alpha)} u_+^{-\alpha-1} := \begin{cases} \frac{1}{\Gamma(-\alpha)} u^{-\alpha-1} & \text{if} \quad u > 0, \\ 0 & \phantom{\text{if}} \quad u \le 0, \end{cases}$$

where Γ denotes the Gamma function. This distribution can be extended to all complex numbers α by analytic continuation or, equivalently, by the method of Hadamard's finite part. Thus, if m is a nonnegative integer and α a complex number such that $\operatorname{Re}\alpha < m$, then δ^α may be represented by

$$\langle \varphi, \delta^\alpha \rangle = \frac{(-1)^m}{\Gamma(m-\alpha)} \int_0^\infty u^{m-\alpha-1} \varphi^{(m)}(u)\, du, \qquad \varphi \in \mathcal{D}(\mathbb{R}).$$

In the strip $m - 1 < \operatorname{Re}\alpha < m$ (m a positive integer), this formula can be transformed to

$$\langle \varphi, \delta^\alpha \rangle = \frac{1}{\Gamma(-\alpha)} \int_0^\infty u^{-\alpha-1} \Big[\varphi(u) - \sum_{j=0}^{m-1} \frac{u^j}{j!} \varphi^{(j)}(0) \Big]\, du, \qquad \varphi \in \mathcal{D}(\mathbb{R}).$$

We are interested in the distribution δ^α only for real positive α. In this case, it is easily seen that δ^α is $\mathbb{R}_+$-summable and its Laplace transform is given by

$$\mathcal{L}(\delta^\alpha)(z) = z^\alpha, \qquad \alpha > 0, \operatorname{Re} z > 0.$$

Note that the principal branch of z^α is understood, that is, if z is a nonzero complex number, then z^α means $\exp(\alpha \log z)$, where the imaginary part of $\log z$ is chosen to be in $(-\pi, \pi]$.

Definition 2.6. *For $\alpha > 0$ we define*

$$(-A)^\alpha = G(\delta^\alpha).$$

By Theorem 2.4, $(-A)^\alpha$ is a closed linear operator with domain $D((-A)^\alpha)$ dense in X. It is one of our purposes to verify the power rules of additivity

$$(-A)^\alpha (-A)^\beta = (-A)^{\alpha+\beta}$$

and multiplicativity

$$((-A)^\alpha)^\beta = (-A)^{\alpha\beta}.$$

Concerning additivity, we have $\delta^\alpha * \delta^\beta = \delta^{\alpha+\beta}$, if $\alpha, \beta > 0$. Thus Theorem 2.4 gives

$$(-A)^\alpha(-A)^\beta x = (-A)^{\alpha+\beta} x \tag{2.3}$$

under the restrictive hypothesis that $x \in D((-A)^\beta)$: In this case, $x \in D((-A)^{\alpha+\beta})$ if and only if $(-A)^\beta x \in D((-A)^\alpha)$, and if so, (2.3) is true. Of course, this statement can be improved in the sense that whenever one side of (2.3) is meaningful, then the other is so and both sides coincide. We shall come back to this extended version in Section 4 after having established the inclusion $D((-A)^{\alpha+\beta}) \subset D((-A)^\beta)$.

3 Representations of Fractional Powers of an Infinitesimal Generator

Since the definition of the fractional power $(-A)^\alpha$ is based on a rather abstract limit process which involves filter convergence, one may ask for more convenient characterizations. We want to give two representations of $(-A)^\alpha$, which, though also by limit processes, yield more concrete approximants. They are modelled after the approaches to fractional derivatives due to Marchaud and Liouville-Grünwald (compare Chapter I). For a whole class of integral representations of fractional powers of infinitesimal generators we refer to Stafney [42].

3.1 Marchaud type representation

Given a positive integer m and a function $\varphi \in \mathcal{D}(\mathbb{R})$, the m-th right-handed difference of φ at $u \in \mathbb{R}$ with increment $t \in \mathbb{R}$ is defined by

$$\Delta_t^m \varphi(u) = \sum_{j=0}^{m} (-1)^j \binom{m}{j} \varphi(u - jt).$$

It is well-known (see for instance [36], Section 10.2) that the distribution δ^α $(0 < \alpha < m)$ can be expressed by

$$\langle \varphi, \delta^\alpha \rangle = \frac{1}{C_{\alpha,m}} \int_0^\infty t^{-\alpha-1} \, \Delta_{-t}^m \varphi(0) \, dt, \qquad \varphi \in \mathcal{D}(\mathbb{R}),$$

where

$$C_{\alpha,m} = \int_0^\infty t^{-\alpha-1}(1-e^{-t})^m \, dt.$$

Note that

$$(\delta^\alpha * \varphi)(u) = \frac{1}{C_{\alpha,m}} \int_0^\infty t^{-\alpha-1} \Delta_t^m \varphi(u) \, dt, \qquad \varphi \in \mathcal{D}(\mathbb{R}),$$

is just the Marchaud fractional derivative of φ of order α (compare Chapter I , Section 3.5 of this volume).

We now approximate δ^α by the family $\{\mu_\varepsilon^{\alpha,m};\ \varepsilon > 0\}$ of finite Borel measures on $\mathbb{R}_+$ defined by the following truncated integrals

$$\langle \varphi, \mu_\varepsilon^{\alpha,m} \rangle = \frac{1}{C_{\alpha,m}} \int_\varepsilon^\infty t^{-\alpha-1} \Delta_{-t}^m \varphi(0) \, dt, \qquad \varphi \in \mathcal{D}(\mathbb{R}).$$

These measures generate the bounded linear operators $G(\mu_\varepsilon^{\alpha,m})$ given by

$$G(\mu_\varepsilon^{\alpha,m})x = \frac{1}{C_{\alpha,m}} \int_\varepsilon^\infty t^{-\alpha-1}[I - T(t)]^m x \, dt, \qquad x \in X.$$

To show that the operators $G(\mu_\varepsilon^{\alpha,m})$ approximate the fractional power $(-A)^\alpha$ as $\varepsilon \to 0+$, we make use of a relationship between the measures $\mu_\varepsilon^{\alpha,m}$ and the distribution δ^α which may be described most nicely by their Laplace transforms.

Lemma 3.1. *For $0 < \alpha < m$, $m = 1, 2, \ldots,$ and $\varepsilon > 0$ the Laplace transform*

$$\mathcal{L}(\mu_\varepsilon^{\alpha,m})(z) = \frac{1}{C_{\alpha,m}} \int_\varepsilon^\infty t^{-\alpha-1}(1-e^{-zt})^m \, dt$$

of the measures $\mu_\varepsilon^{\alpha,m}$ may be represented by

$$\mathcal{L}(\mu_\varepsilon^{\alpha,m})(z) = z^\alpha \cdot \mathcal{L}\left(\frac{1}{\varepsilon} q_{\alpha,m}\left(\frac{\cdot}{\varepsilon}\right)\right)(z), \qquad Re\, z > 0, \tag{3.1}$$

where $q_{\alpha,m}$ is a function from $L^1(\mathbb{R}_+)$ defined by

$$q_{\alpha,m}(u) = \frac{1}{C_{\alpha,m}\,\Gamma(1+\alpha)\,u} \sum_{j=0}^{m} (-1)^j \binom{m}{j} (u-j)_+^\alpha$$

and satisfying

$$\int_0^\infty q_{\alpha,m}(u)\,du = 1.$$

For a proof of this lemma we refer to [45,I].

By the product and the uniqueness theorems for the Laplace transformation we obtain from (3.1)

$$\mu_\varepsilon^{\alpha,m} = \delta^\alpha * \frac{1}{\varepsilon} q_{\alpha,m}\left(\frac{\cdot}{\varepsilon}\right). \tag{3.2}$$

Thus, the regularizations $\delta^\alpha * \varphi$ with test functions φ running through the sets of the filter $\mathfrak{F}$, have been replaced by convolutions of the distribution δ^α with specific $L^1(\mathbb{R}_+)$-functions. Though these new approximants of δ^α are no longer functions from $L^1(\mathbb{R}_+)$ they are still finite measures on $\mathbb{R}_+$ and therefore lead to bounded linear operators when the operational calculus is applied to them.

Part (ii) of the next theorem is the Marchaud type representation of fractional powers $(-A)^\alpha$. Compare [4] and [45,I].

Theorem 3.2. *(i) For each $x \in X$ the integral $\int_0^\infty q_{\alpha,m}\left(\frac{t}{\varepsilon}\right) T(t)x \frac{dt}{\varepsilon}$ belongs to $D((-A)^\alpha)$ and*

$$(-A)^\alpha \int_0^\infty q_{\alpha,m}\left(\frac{t}{\varepsilon}\right) T(t)x \frac{dt}{\varepsilon} = \frac{1}{C_{\alpha,m}} \int_\varepsilon^\infty t^{-\alpha-1}[I - T(t)]^m x\,dt.$$

If $x \in D((-A)^\alpha)$, the operator $(-A)^\alpha$ may be drawn under the integral sign.

(ii) The fractional power $(-A)^\alpha$ is characterized by

$$(-A)^\alpha x = \lim_{\varepsilon\to 0+} \frac{1}{C_{\alpha,m}} \int_\varepsilon^\infty t^{-\alpha-1}[I - T(t)]^m x\,dt,$$

where the limits exists if and only if $x \in D((-A)^\alpha)$.

Part (i) of Theorem 3.2. is an immediate consequence of the convolution theorem 2.4 applied to formula (3.2). Part (ii) is then deduced from (i) when the limit is taken there as $\varepsilon \to 0+$ noting that $(-A)^\alpha$ is a closed operator and

$$\lim_{\varepsilon \to 0+} \int_0^\infty q_{\alpha,m}\left(\frac{t}{\varepsilon}\right) T(t)x \,\frac{dt}{\varepsilon} = x$$

for each $x \in X$.

3.2 Liouville-Grünwald type representation

The fractional derivative of order $\alpha > 0$ of a function φ which, for convenience, is assumed to belong to $\mathcal{D}(\mathbb{R})$, may be characterized by the limit

$$\lim_{t \to 0} \frac{\Delta_t^\alpha \varphi(u)}{t^\alpha},$$

where

$$\Delta_t^\alpha \varphi(u) = \sum_{j=0}^{\infty} (-1)^j \binom{\alpha}{j} \varphi(u - jt)$$

is the right-handed difference of φ of fractional order $\alpha > 0$ (compare Chapter I , Section 3.1 of this volume). This representation which goes back to Liouville, Grünwald and Letnikov suggests to consider the family $\{\nu_t^\alpha;\, t > 0\}$ of finite Borel measures on $\mathbb{R}_+$

$$\nu_t^\alpha := \frac{1}{t^\alpha} \sum_{j=0}^{\infty} (-1)^j \binom{\alpha}{j} \delta_{jt}$$

as approximants of the distribution δ^α.

The operational calculus then yields the bounded linear operators

$$G(\nu_t^\alpha)x = \frac{1}{t^\alpha} \sum_{j=0}^{\infty} (-1)^j \binom{\alpha}{j} T(jt)x, \qquad x \in X.$$

The right-hand side of this formula may be regarded as a difference quotient of fractional order α

$$\frac{[I - T(t)]^\alpha}{t^\alpha} x,$$

where the fractional power $[I - T(t)]^\alpha$ is to be understood in the sense of Definition 2.6. Indeed, if $t > 0$ is fixed, $T(t) - I$ generates an equibounded, uniformly continuous semigroup $\{S_t(\tau); \tau \in \mathbb{R}_+\}$ of operators in $\mathcal{E}(X)$ which may be represented by the exponential power series expansion

$$S_t(\tau) := e^{-\tau} \sum_{k=0}^{\infty} \frac{\tau^k}{k!} T(kt).$$

Since $\|T(kt)\| \le M$ for all $k = 0, 1, \ldots$, one has $\|S_t(\tau)\| \le M$ for all $\tau \in \mathbb{R}_+$. For $\varphi \in \mathcal{D}(\mathbb{R})$, supp $\varphi \subset \mathbb{R}_+$, one can show that

$$\int_0^\infty (\delta^\alpha * \varphi)(\tau) S_t(\tau) x \, d\tau = \sum_{j=0}^{\infty} (-1)^j \binom{\alpha}{j} T(jt) \int_0^\infty \varphi(\tau) S_t(\tau) x \, d\tau, \qquad x \in X,$$

which implies

$$[I - T(t)]^\alpha x = \sum_{j=0}^{\infty} (-1)^j \binom{\alpha}{j} T(jt) x, \qquad x \in X,$$

as φ tends to δ following the filter $\mathcal{F}$.

Now, in principal, the procedure is the same as in the Marchaud case. However, the proofs of some details are more involved. For the following lemma we refer to [46].

Lemma 3.3. *For $\alpha > 0$ and $t > 0$ the Laplace transform*

$$\mathcal{L}(\nu_t^\alpha)(z) = \left(\frac{1 - e^{-zt}}{t} \right)^\alpha, \qquad \operatorname{Re} z > 0,$$

of the measure ν_t^α may be represented by

$$\mathcal{L}(\nu_t^\alpha)(z) = z^\alpha \, \mathcal{L}\Big(\frac{1}{t} p_\alpha \left(\frac{\cdot}{t} \right) \Big)(z), \qquad \operatorname{Re} z > 0, \tag{3.3}$$

where the function p_α is defined by

$$p_\alpha(u) = \frac{1}{\Gamma(\alpha)} \sum_{j=0}^{\infty} (-1)^j \binom{\alpha}{j} (u - j)_+^{\alpha - 1}.$$

p_α belongs to $L^1(\mathbb{R}_+)$ and is normalized, that is

$$\int_0^\infty p_\alpha(u) \, du = 1.$$

Let us mention that the most delicate step in the proof of the previous lemma is to show that p_α belongs to $L^1(\mathbb{R}_+)$. In [46] this result is deduced from an asymptotic formula for power functions of fractional order due to Ingham [17]. For an alternative proof using Fourier transform methods we refer to [24] and [37].

Again by the product and uniqueness theorems for the Laplace transformation one obtains from (3.3)

$$\nu_t^\alpha = \delta^\alpha * \frac{1}{t} p_\alpha \left(\frac{\cdot}{t}\right).$$

The functional calculus applied to this formula then leads to the representation of fractional powers in the sense of Liouville-Grünwald given in part (ii) of the next theorem.

Theorem 3.4. *(i) For each $x \in X$ the integral $\int_0^\infty p_\alpha(\frac{u}{t})T(u)x \frac{du}{t}$ belongs to $D((-A)^\alpha)$ and*

$$(-A)^\alpha \int_0^\infty p_\alpha \left(\frac{u}{t}\right) T(u)x \frac{du}{t} = \frac{[I - T(t)]^\alpha}{t^\alpha} x.$$

If $x \in D((-A)^\alpha)$, then the operator $(-A)^\alpha$ may be drawn under the integral sign.
(ii) The fractional power $(-A)^\alpha$ is characterized by

$$(-A)^\alpha x = \lim_{t \to 0+} t^{-\alpha} [I - T(t)]^\alpha x,$$

where the limit exists if and only if $x \in D((-A)^\alpha)$.

4 Power Rules

At first we want to show the monotonicity law which says that $x \in D((-A)^\alpha)$ implies that $x \in D((-A)^\beta)$ for each $\beta < \alpha$. For the sake of simplicity, we may assume that $0 < \alpha - \beta < 1$. We have for all complex numbers z with $\operatorname{Re} z > 0$ that

$$\left(\frac{1 - e^{-zt}}{t}\right)^\beta - z^\beta = t^{\alpha-\beta} z^\alpha \left[\frac{(1 - e^{-zt})^\beta}{(tz)^\alpha} - \frac{1}{(tz)^{\alpha-\beta}}\right]. \qquad (4.1)$$

The expression in brackets on the right-hand side of (4.1) may be written as Laplace transform $\mathcal{L}(\frac{1}{t}r_{\alpha,\beta}(\frac{\cdot}{t}))(z)$, where $r_{\alpha,\beta}$ is an element of $L^1(\mathbb{R}_+)$ defined by

$$r_{\alpha,\beta}(u) = \frac{1}{\Gamma(\alpha)} \sum_{j=0}^{\infty} (-1)^j \binom{\beta}{j} (u-j)_+^{\alpha-1} - \frac{1}{\Gamma(\alpha-\beta)} u_+^{\alpha-\beta-1}.$$

It follows from (4.1) that for $0 < \beta < \alpha < \beta+1$ and $t > 0$

$$\nu_t^\beta - \delta^\beta = t^{\alpha-\beta}\delta^\alpha * \frac{1}{t} r_{\alpha,\beta}\left(\frac{\cdot}{t}\right).$$

If we apply the operational calculus to this formula, we obtain a Taylor-type expansion for the semigroup operators $T(t)$, $t \in \mathbb{R}_+$, which implies the desired monotonicity law. See [46] and also [47] for a generalization of the Taylor expansion.

Theorem 4.1. *If $x \in D((-A)^\alpha)$ for some $\alpha > 0$, then $x \in D((-A)^\beta)$ for each positive $\beta < \alpha$, and for $0 < \alpha - \beta < 1$ we have the formula*

$$[I - T(t)]^\beta x - t^\beta(-A)^\beta x = t^\alpha \int_0^\infty r_{\alpha,\beta}\left(\frac{u}{t}\right) T(u)(-A)^\alpha x \,\frac{du}{t}.$$

We may now give the additivity law for fractional powers in full generality.

Theorem 4.2. *Let $\alpha, \beta > 0$. Then an element $x \in X$ belongs to $D((-A)^{\alpha+\beta})$ if and only if $x \in D((-A)^\beta)$ and $(-A)^\beta x \in D((-A)^\alpha)$. If so, then*

$$(-A)^{\alpha+\beta}x = (-A)^\alpha(-A)^\beta x.$$

The next step is to prove the multiplicativity law

$$((-A)^\alpha)^\beta = (-A)^{\alpha\beta}$$

which makes sense in the present setting only if $-(-A)^\alpha$ generates an equibounded semigroup of operators of class (C_0). This is true if $0 < \alpha < 1$. Thus, in the following, α is restricted to these values.

Let us consider the family $\{f_{\alpha,t};\, t > 0\}$ of stable densities on $\mathbb{R}_+$ in the sense of P. Lévy which are defined by their Laplace transforms

$$e^{-tz^\alpha} = \int_0^\infty e^{-zu} f_{\alpha,t}(u)\,du, \qquad t > 0,\ \operatorname{Re} z \geq 0,\ 0 < \alpha < 1.$$

The functions $f_{\alpha,t}$ are nonnegative, belong to $L^1(\mathbb{R}_+)$ with

$$\int_0^\infty f_{\alpha,t}(u)\,du = 1, \qquad t > 0,$$

and satisfy the semigroup property with respect to convolution

$$f_{\alpha,t} * f_{\alpha,s} = f_{\alpha,t+s}, \qquad t, s > 0.$$

Moreover,

$$f_{\alpha,t}(u) = t^{-1/\alpha} f_{\alpha,1}(t^{-1/\alpha}u), \qquad t > 0.$$

Now, the bounded linear operators $T_\alpha(t)$ defined by $T_\alpha(0) = I$ and

$$T_\alpha(t)x := G(f_{\alpha,t})x = \int_0^\infty f_{\alpha,t}(u)T(u)x\,du, \qquad t > 0,$$

form an equibounded holomorphic semigroup of class (C_0).

For these and further properties we refer to Yosida [49; Sec. IX,11]. The semigroup $\{T_\alpha(t);\, t \in \mathbb{R}_+\}$ is an example for a method of constructing new semigroups out of old ones which goes back to N.P. Romanoff [35] and in connection with stochastic processes to S. Bochner [6] and was developed further by R.S. Phillips [33], A.V. Balakrishnan [1] and J. Faraut [9]; the latter, in particular, revealed the connection with the operational calculus for distributions due to Schwartz. Moreover, see Berg, Boyadzhiev and deLaubenfels [5] and Schilling [39].

To show that the infinitesimal generator A_α of $\{T_\alpha(t);\, t \in \mathbb{R}_+\}$ equals $-(-A)^\alpha$ let us note that

$$\mathcal{L}\left(\frac{f_{\alpha,t} - \delta}{t}\right)(z) = \frac{e^{-tz^\alpha} - 1}{t} = -z^\alpha \mathcal{L}(g_{\alpha,t})(z), \qquad \operatorname{Re} z > 0, \tag{4.2}$$

where $g_{\alpha,t}$ is the nonnegative function

$$g_{\alpha,t}(u) = \frac{1}{t}\int_0^t f_{\alpha,s}(u)\,ds, \qquad t, u > 0,$$

which belongs to $L^1(\mathbb{R}_+)$ and satisfies

$$\int_0^\infty g_{\alpha,t}(u)\,du = 1, \qquad t > 0, \tag{4.3}$$

as well as

$$g_{\alpha,t}(u) = t^{-1/\alpha} g_{\alpha,1}(t^{-1/\alpha}u), \qquad t, u > 0. \tag{4.4}$$

Thus, we have the representation

$$t^{-1}(f_{\alpha,t} - \delta) = -\delta^{\alpha} * g_{\alpha,t}, \qquad t > 0$$

which implies that for each $x \in X$ the integral $\int_0^\infty g_{\alpha,t}(u)T(u)x\,du$ belongs to $D((-A)^\alpha)$ and satisfies

$$t^{-1}[T_\alpha(t)x - x] = -(-A)^\alpha \int_0^\infty g_{\alpha,t}(u)T(u)x\,du; \tag{4.5}$$

if $x \in D((-A)^\alpha)$, then the operator $(-A)^\alpha$ may be drawn under the integral sign. By (4.3) and (4.4) one has that

$$\lim_{t\to 0+} \int_0^\infty g_{\alpha,t}(u)T(u)x\,du = x, \qquad x \in X.$$

Thus, as t tends to zero in the equation (4.5) as well as in its counterpart for $x \in D((-A)^\alpha)$, we obtain that $x \in D(A_\alpha)$ if and only if $x \in D((-A)^\alpha)$, and if so, then

$$A_\alpha x = -(-A)^\alpha x.$$

Theorem 4.3. *Let* $0 < \alpha < 1$ *and* $\beta > 0$. *Then*

$$((-A)^\alpha)^\beta = (-A)^{\alpha\beta}.$$

Proof. Let us denote by G_α the mapping that is associated with the operational calculus for the infinitesimal generator $A_\alpha = -(-A)^\alpha$. Then by definition,

$$((-A)^\alpha)^\beta x = \lim_{\mathcal{F}} G_\alpha(\delta^\beta * \varphi)x,$$

whenever this limit exists as $\varphi \to \delta$ following the filter $\mathcal{F}$. Note that, if h is any function of $L^1(\mathbb{R}_+)$, then we obtain by interchanging the order of integration that

$$G_\alpha(h)x = G(h_\alpha)x, \qquad x \in X, \tag{4.6}$$

where h_α denotes the function of $L^1(\mathbb{R}_+)$ defined by

$$h_\alpha(u) = \int_0^\infty h(t) f_{\alpha,t}(u)\, dt.$$

We now show that given $\varphi \in \mathcal{D}(\mathbb{R})$ such that supp $\varphi \subset \mathbb{R}_+$, then for each $x \in X$ the element $G_\alpha(\varphi)x$ belongs to $D((-A)^{\alpha\beta})$ and

$$G_\alpha(\delta^\beta * \varphi)x = (-A)^{\alpha\beta} G_\alpha(\varphi)x; \tag{4.7}$$

if $x \in D((-A)^{\alpha\beta})$, then

$$G_\alpha(\delta^\beta * \varphi)x = G_\alpha(\varphi)(-A)^{\alpha\beta}x. \tag{4.8}$$

Indeed, the Laplace transform of $(\delta^\beta * \varphi)_\alpha$ may be evaluated as

$$\mathcal{L}((\delta^\beta * \varphi)_\alpha)(z) = \mathcal{L}(\delta^\beta * \varphi)(z^\alpha) = z^{\alpha\beta}\mathcal{L}(\varphi)(z^\alpha) = z^{\alpha\beta}\mathcal{L}(\varphi_\alpha)(z), \quad \operatorname{Re} z > 0.$$

Thus,

$$(\delta^\beta * \varphi)_\alpha = \delta^{\alpha\beta} * \varphi_\alpha$$

which, by the operational calculus, implies

$$G((\delta^\beta * \varphi)_\alpha)x = \begin{cases} (-A)^{\alpha\beta} G(\varphi_\alpha)x, & x \in X \\ G(\varphi_\alpha)(-A)^{\alpha\beta}x, & x \in D((-A)^{\alpha\beta}). \end{cases}$$

Noting the relation (4.6) between G_α and G we obtain the formulas (4.7) and (4.8). Now, in these formulas let φ tend to δ following the filter $\mathcal{F}$. Then, since for each $x \in X$

$$\lim_{\mathcal{F}} G_\alpha(\varphi)x = x,$$

the desired multiplicativity law follows.

5 Negative Powers

Since the regular distribution $\delta^{-\alpha}$ ($\alpha > 0$) does not belong to $\mathcal{D}'_{L^1}(\mathbb{R}_+)$, we cannot use the operational calculus directly to define negative powers of an infinitesimal generator. We may, however, try to multiply the distribution $\delta^{-\alpha}$

by an exponential convergence factor in order to obtain approximants that are covered by the operational calculus. In fact, for positive α and λ the functions

$$r_\lambda^\alpha(u) = \frac{1}{\Gamma(\alpha)} u_+^{\alpha-1} e_\lambda(u)$$

belong to $L^1(\mathbb{R}_+)$ and thus generate the operators $R_\lambda^\alpha \in \mathcal{E}(X)$ defined by

$$R_\lambda^\alpha x := G(r_\lambda^\alpha)x = \frac{1}{\Gamma(\alpha)} \int_0^\infty u^{\alpha-1} e^{-\lambda u} T(u)x\, du, \qquad x \in X.$$

Recall that for $\alpha = 1$ the operator $R_\lambda := R_\lambda^1$ is just the resolvent $(\lambda I - A)^{-1}$ of the infinitesimal generator A. If α is a positive integer m, then

$$R_\lambda^m = G(r_\lambda^m) = (R_\lambda)^m.$$

Thus, the operators R_λ^α extend the integral powers of the resolvent R_λ to arbitrary $\alpha > 0$. For convenience, we will interpret R_λ^0 as the identity operator I.

Our purpose is to determine the behaviour of R_λ^α as λ approaches zero. At first note that the operators $\lambda^\alpha R_\lambda^\alpha$ are uniformly bounded with respect to positive λ and α, that is

$$\|\lambda^\alpha R_\lambda^\alpha\| \le M \qquad \text{for all } \lambda, \alpha > 0.$$

Moreover, the additivity law

$$R_\lambda^\alpha R_\lambda^\beta = R_\lambda^{\alpha+\beta} \qquad \alpha, \beta > 0$$

is easily verified. To see how R_λ^α is related to $(-A)^\alpha$, let us consider

$$\mathcal{L}(\delta^\alpha * r_\lambda^\alpha)(z) = z^\alpha \frac{1}{(\lambda + z)^\alpha} = \mathcal{L}(\varrho_\lambda^\alpha)(z), \qquad \operatorname{Re} z > 0,$$

where ϱ_λ^α denotes the finite Borel measure on $\mathbb{R}_+$ which is given by

$$\varrho_\lambda^\alpha = \delta + \sum_{j=1}^\infty (-1)^j \binom{\alpha}{j} \lambda^j r_\lambda^j.$$

Then, $G(\varrho_\lambda^\alpha)$ is the bounded linear operator defined by

$$G(\varrho_\lambda^\alpha)x = \sum_{j=0}^\infty (-1)^j \binom{\alpha}{j} \lambda^j R_\lambda^j x, \qquad x \in X.$$

This formula may be rewritten as

$$G(r_\lambda^\alpha)x = [I - \lambda R_\lambda]^\alpha x, \qquad x \in X,$$

where the fractional power on the right-hand side is to be understood in the sense of Definition 2.6. This can be shown in the same manner as the corresponding assertion for $[I - T(t)]^\alpha$ in Section 3.2, since $\lambda R_\lambda - I$ generates an equibounded uniformly continuous semigroup of operators in $\mathcal{E}(X)$. Now, if Theorem 2.4 is applied to

$$\delta^\alpha * r_\lambda^\alpha = \varrho_\lambda^\alpha,$$

we obtain the following basic lemma; compare [48].

Lemma 5.1. *Let $\alpha > 0$ and $\lambda > 0$. For each $x \in X$, $R_\lambda^\alpha x$ belongs to $D((-A)^\alpha)$ and*

$$(-A)^\alpha R_\lambda^\alpha x = [I - \lambda R_\lambda]^\alpha x.$$

If $x \in D((-A)^\alpha)$, then

$$R_\lambda^\alpha (-A)^\alpha x = [I - \lambda R_\lambda]^\alpha x.$$

In particular, this lemma implies that for each $\alpha > 0$ the family of operators $\{(-A)^\alpha R_\lambda^\alpha;\ \lambda > 0\}$ is uniformly bounded with respect to $\lambda > 0$, namely,

$$\|(-A)^\alpha R_\lambda^\alpha\| \leq M \sum_{j=0}^{\infty} \left| \binom{\alpha}{j} \right|.$$

Concerning convergence in the norm topology of X as $\lambda \to 0+$, we can deduce the following theorem. (Note that $\overline{R((-A)^\alpha)}$ denotes the closure of the range of $(-A)^\alpha$).

Theorem 5.2. *For $\alpha > 0$ and $x \in X$ the following assertions are equivalent:*

(i) $x \in \overline{R((-A)^\alpha)}$,

(ii) $\lim_{\lambda \to 0+} \lambda^\alpha R_\lambda^\alpha x = 0$,

(iii) $x \in \overline{R(A)}$,

(iv) $\lim_{\lambda \to 0+} (-A)^\alpha R_\lambda^\alpha x = x$.

In particular, we have for each $\alpha > 0$,

$$\overline{R((-A)^\alpha)} = \overline{R(A)}.$$

Moreover, one realizes from Theorem 5.2 at a first glimpse that a necessary condition for $R_\lambda^\alpha x$ to be convergent for some $x \in X$ as $\lambda \to 0+$, is that x belongs to $\overline{R(A)}$. As $\overline{R(A)}$ is an invariant subspace of X for R_λ^α, that is $R_\lambda^\alpha(\overline{R(A)}) \subset \overline{R(A)}$, the limit of $R_\lambda^\alpha x$, if it exists, belongs to $\overline{R(A)}$, too. More precisely, we have the following corollary of Theorem 5.2.

Corollary 5.3. *For $\alpha > 0$ and $x \in X$ the limit $\lim_{\lambda\to 0+} R_\lambda^\alpha x$ exists in the norm topology of X if and only if there is an element $y \in D((-A)^\alpha) \cap \overline{R(A)}$ such that $x = (-A)^\alpha y$. If so, y is uniquely determined and $\lim_{\lambda\to 0+} R_\lambda^\alpha x = y$.*

Now, let A_0 denote the restriction of A to $\overline{R(A)}$, in sign

$$A_0 = A|_{\overline{R(A)}},$$

which means that $D(A_0) = D(A) \cap \overline{R(A)}$ is the domain of A_0 and

$$A_0 x = Ax \qquad \text{for each } x \in D(A_0).$$

Then, A_0 is an operator whose domain and range are contained in $\overline{R(A)}$. If we consider $\overline{R(A)}$ as a Banach space of its own, endowed with the norm which is induced by that of X, then the family of operators $T(t)$, $t \in \mathbb{R}_+$, restricted to $\overline{R(A)}$, forms an equibounded (C_0)-semigroup of linear operators in $\mathcal{E}(\overline{R(A)})$, and A_0 is just its infinitesimal generator. Therefore, $(-A_0)^\alpha$ is well defined for each $\alpha > 0$ in the sense of Definition 2.6. It is not difficult to see that

$$(-A_0)^\alpha = (-A)^\alpha|_{\overline{R(A)}}.$$

With this terminology, Corollary 5.3 may be rewritten in shortened form

$$\lim_{\lambda\to 0+} R_\lambda^\alpha(-A_0)^\alpha y = y \qquad \text{for each } y \in D((-A_0)^\alpha),$$

$$(-A_0)^\alpha \lim_{\lambda\to 0+} R_\lambda^\alpha x = x \qquad \text{whenever } \lim_{\lambda\to 0+} R_\lambda^\alpha x \text{ exists.}$$

This means that for each $\alpha > 0$, $(-A_0)^\alpha$ has an inverse $((-A_0)^\alpha)^{-1}$ which is given by the limit of R_λ^α as $\lambda \to 0+$. Hence we introduce the following

Definition 5.4. *Let $\alpha > 0$. The operator $(-A_0)^{-\alpha}$ with domain*

$$D((-A_0)^{-\alpha}) = \{x \in X;\ \lim_{\lambda\to 0+} R_\lambda^\alpha x \text{ exists}\}$$

is defined by the formula

$$(-A_0)^{-\alpha}x = \lim_{\lambda\to 0+} R_\lambda^\alpha x, \qquad x \in D((-A_0)^{-\alpha}).$$

By the discussion preceding Definition 5.4 we have for each $\alpha > 0$,

$$(-A_0)^{-\alpha} = ((-A_0)^\alpha)^{-1}.$$

We remark that if $\{T(t);\, t \in \mathbb{R}_+\}$ is an exponentially decaying semigroup, that is, for all $t \in \mathbb{R}_+$

$$\|T(t)\| \leq M\, e^{-\omega t}$$

for some $M \geq 1$ and $\omega > 0$, then the infinitesimal generator A has a bounded inverse in $\mathcal{E}(X)$. Thus in this case A_0 and A coincide. Furthermore, for each $\alpha > 0$, $(-A)^{-\alpha}$ also belongs to $\mathcal{E}(X)$ and is given by the integral

$$(-A)^{-\alpha}x = \frac{1}{\Gamma(\alpha)} \int_0^\infty u^{\alpha-1} T(u)x\, du, \qquad x \in X.$$

In the general case, $(-A_0)^{-\alpha}$ has the following properties.

Theorem 5.5.

(i) *For each* $\alpha > 0$, $(-A_0)^{-\alpha}$ *is a closed linear operator whose domain is dense in* $\overline{R(A)}$.

(ii) *If* $0 < \beta < \alpha$, *then* $D((-A_0)^{-\alpha}) \subset D((-A_0)^{-\beta})$ *and for each* $x \in D((-A_0)^{-\alpha})$,

$$(-A_0)^{-\beta}x = (-A_0)^{\alpha-\beta}(-A_0)^{-\alpha}x.$$

(iii) *For all* $\alpha, \beta > 0$,

$$(-A_0)^{-\alpha}(-A_0)^{-\beta} = (-A_0)^{-(\alpha+\beta)}.$$

Combining the additivity laws for positive and negative powers, we obtain the following corollary, where $(-A_0)^0$ is to be interpreted as the identity operator in $\overline{R(A)}$.

Corollary 5.6. *Let* $\alpha, \beta \in \mathbb{R}$. *If* $x \in D((-A_0)^\beta)$ *and* $(-A_0)^\beta x \in D((-A_0)^\alpha)$, *then* $x \in D((-A_0)^{\alpha+\beta}$ *and*

$$(-A_0)^\alpha(-A_0)^\beta x = (-A_0)^{\alpha+\beta}x.$$

6 Abstract Riesz Potentials

Let us assume that the given semigroup $\{T(t);\, t \in \mathbb{R}_+\}$ may be extended to an equibounded (C_0)-group $\{T(t);\, t \in \mathbb{R}\}$ of linear operators in $\mathcal{E}(X)$, that is,

$$T(t)T(s) = T(t+s) \text{ for } t, s \in \mathbb{R},\ T(0) = I,$$
$$\lim_{t\to 0} \|T(t)x - x\| = 0 \text{ for each } x \in X,$$
$$\|T(t)\| \le M \text{ for all } t \in \mathbb{R},\ M \text{ being a constant.}$$

The family $\{T(-t);\, t \in \mathbb{R}_+\}$ is then an equibounded (C_0)-semigroup with infinitesimal generator $(-A)$, if A generates the original semigroup $\{T(t);\, t \in \mathbb{R}_+\}$. We call the operator A the infinitesimal generator of the group; it is characterized by the two-sided limit

$$Ax = \lim_{t\to 0} t^{-1}[T(t)x - x], \qquad x \in D(A).$$

By the theorem of Hille-Yosida it may be shown that A^2 also generates an equibounded (C_0)-semigroup, say $\{W(t);\, t \in \mathbb{R}_+\}$, which may be represented by the integrals

$$W(t)x = \frac{1}{2\sqrt{\pi t}} \int_{-\infty}^{\infty} e^{-u^2/4t}\, T(u)x\, du, \qquad t > 0,\ x \in X, \tag{6.1}$$

with the kernel $2^{-1}(\pi t)^{-1/2} e^{-u^2/4t}$ of Gauss-Weierstrass.

Further, we know from Section 4 that for $0 < \alpha < 2$ the fractional power $-(-A^2)^{\alpha/2}$ is the infinitesimal generator of the equibounded (C_0)-semigroup $\{V_\alpha(t);\, t \in \mathbb{R}_+\}$ which is given by the integrals

$$V_\alpha(t) = \int_0^{\infty} f_{\alpha/2,t}(u) W(u)x\, du, \qquad t > 0,\ x \in X. \tag{6.2}$$

Substituting (6.1) into (6.2) we obtain for $0 < \alpha < 2$

$$V_\alpha(t) = \int_{-\infty}^{\infty} a_{\alpha,t}(u) T(u)x\, du, \qquad t > 0,\ x \in X, \tag{6.3}$$

where the stable densities $a_{\alpha,t}$ are nonnegative even functions on $\mathbb{R}$ that belong to $L(\mathbb{R})$ and have the Fourier transforms

$$\int_{-\infty}^{\infty} e^{-ivu} a_{\alpha,t}(u)\, du = \mathcal{L}(f_{\alpha/2,t})(v^2) = e^{-t|v|^\alpha}, \qquad v \in \mathbb{R}.$$

In the special case $\alpha = 1$, $a_{1,t}(u) = t\pi^{-1}(u^2 + t^2)^{-1}$ is the kernel of Cauchy-Poisson. We denote the corresponding semigroup by $\{P(t);\, t \in \mathbb{R}_+\}$ and its infinitesimal generator by A_P, thus

$$P(t)x = \frac{t}{\pi}\int_{-\infty}^{\infty} \frac{1}{u^2 + t^2} T(u)x\, du, \qquad t > 0,\, x \in X,$$

and $A_P = -(-A^2)^{1/2}$. By the multiplicativity law given in Theorem 4.3, we have for all $\alpha > 0$

$$((-A^2)^{1/2})^\alpha = (-A^2)^{\alpha/2};$$

that means that the fractional powers of the infinitesimal generators of the semigroups $\{P(t);\, t \in \mathbb{R}_+\}$ and $\{W(t);\, t \in \mathbb{R}_+\}$ are related by

$$(-A_P)^\alpha = (-A^2)^{\alpha/2}. \tag{6.4}$$

Concerning the historical background of these facts let us consider the group of translations on $L^p(\mathbb{R})$, $1 \le p < \infty$, defined by

$$[T(t)f](u) = f(u + t), \qquad f \in L^p(\mathbb{R}).$$

Its infinitesimal generator A is the differential operator

$$[Af](u) = \frac{d}{du} f(u) = f'(u)$$

with domain

$$D(A) = \{f \in L^p(\mathbb{R});\; f \in AC_{\text{loc}}(\mathbb{R}),\; f' \in L^p(\mathbb{R})\},$$

where $AC_{\text{loc}}(\mathbb{R})$ is the set of functions which are locally absolutely continuous on $\mathbb{R}$. A^2 is given by

$$[A^2 f](u) = \frac{d^2}{du^2} f(u) = f''(u)$$

and

$$D(A^2) = \{f \in L^p(\mathbb{R});\; f, f' \in AC_{\text{loc}}(\mathbb{R}),\; f'' \in L^p(\mathbb{R})\}.$$

The infinitesimal generator A_P of the semigroup $\{P(t);\, t \in \mathbb{R}_+\}$ associated with the group of translations is given by

$$A_P f = -\left(\frac{d}{du} H\right) f = -(Hf)'$$

with domain

$$D(A_P) = \{f \in L^p(\mathbb{R});\ Hf \in AC_{\text{loc}}(\mathbb{R}),\ (Hf)' \in L^p(\mathbb{R})\},$$

where H is the Hilbert transform defined by the Cauchy principal value

$$(Hf)(u) = \lim_{\delta\to 0+} \frac{1}{\pi} \int\limits_{|u-w|\geq\delta} \frac{f(w)}{u-w}\, dw.$$

For $\alpha > 0$, $\left(\frac{d}{du}H\right)^\alpha$ equals the α-th strong Riesz derivative $D^{\{\alpha\}}$ defined by (3.39)–(3.41) of Chapter I . On the other hand, we have by (6.4) for all $\alpha > 0$

$$\left(\frac{d}{du}H\right)^\alpha f = \left(-\frac{d^2}{du^2}\right)^{\alpha/2} f,$$

whenever one of these two terms exists for some $f \in L^p(\mathbb{R})$.

In case $0 < \alpha < 2$ it was Bochner [6] who defined $-\left(-\frac{d^2}{du^2}\right)^{\alpha/2}$ by the infinitesimal generator of the semigroup $\{V_\alpha(t);\ t \in \mathbb{R}_+\}$ associated with the group of translations while Feller [11] interpreted this generator as a Riesz potential of negative order $(-\alpha)$. We refer to Section 3.7 of Chapter I for further characterizations.

We take up here the aspect of a Riesz derivative to extend it to the general case of an arbitrary equibounded (C_0)-group. To this end, let us consider for $0 < \alpha < 2n$, $n = 1, 2, \dots$ the truncated integrals of Marchaud type

$$\frac{1}{K_{\alpha,2n}} \int\limits_\varepsilon^\infty t^{-\alpha-1} \left[T\left(\frac{t}{2}\right) - T\left(-\frac{t}{2}\right)\right]^{2n} x\, dt, \qquad x \in X, \tag{6.5}$$

with the central difference

$$\left[T\left(\frac{t}{2}\right) - T\left(-\frac{t}{2}\right)\right]^{2n} = \sum_{j=0}^{2n} (-1)^j \binom{2n}{j} T((n-j)t)$$

and the normalizing factor

$$K_{\alpha,2n} = (-1)^n\, 2^{2n-\alpha} \int\limits_0^\infty t^{-\alpha-1} \sin^{2n} t\, dt.$$

Then the operator defined by the limit

$$\lim_{\varepsilon\to 0+} \frac{1}{K_{\alpha,2n}} \int\limits_\varepsilon^\infty t^{-\alpha-1} \left[T\left(\frac{t}{2}\right) - T\left(-\frac{t}{2}\right)\right]^{2n} x\, dt \tag{6.6}$$

may be considered as the *abstract Riesz potential* of order $(-\alpha)$ associated with the group $\{T(t);\, t \in \mathbb{R}\}$. Compare Balakrishnan [3] and Westphal [45,II]. We want to show how this abstract Riesz potential is related to the fractional powers of the infinitesimal generators of the semigroups $\{W(t);\, t \in \mathbb{R}_+\}$ and $\{P(t);\, t \in \mathbb{R}_+\}$. This will generalize the results given in Section 3.7 of Chapter I for the special case of the translation group.

Let us remark first that the operators in $\mathcal{E}(X)$ defined by the integrals (6.5) can be interpreted in the framework of an operational calculus for group generators which could be developed analogously to that for semigroup generators introduced in Section 2. The role of the Laplace transform can then be taken over by the Fourier transform. For the purposes of this section it is, however, not necessary to carry out the details of such an approach in general.

As an analogue of Lemma 3.1 we have in the present case (see [45,II])

Lemma 6.1. *For $0 < \alpha < 2n$, $n = 1, 2, \ldots$, and $\varepsilon > 0$ the integral*

$$\frac{1}{K_{\alpha,2n}} \int_{\varepsilon}^{\infty} t^{-\alpha-1} \left[e^{-ivt/2} - e^{ivt/2}\right]^{2n} dt, \qquad v \in \mathbb{R},$$

is the Fourier transform of a finite Borel measure on $\mathbb{R}$ and can be represented by

$$|v|^{\alpha} \int_{-\infty}^{\infty} e^{-ivt}\, \overline{q}_{\alpha,2n}\left(\frac{t}{\varepsilon}\right) \frac{dt}{\varepsilon}, \qquad v \in \mathbb{R},$$

where $\overline{q}_{\alpha,2n}$ is an even function which belongs to $L^1(\mathbb{R})$ and satisfies

$$\int_{-\infty}^{\infty} \overline{q}_{\alpha,2n}(t)dt = 1.$$

Note that for each $x \in X$,

$$\lim_{\varepsilon \to 0+} \int_{-\infty}^{\infty} \overline{q}_{\alpha,2n}\left(\frac{t}{\varepsilon}\right) T(t)x \frac{dt}{\varepsilon} = x$$

in the norm topology of X.

Now, let $\varphi \in \mathcal{D}(\mathbb{R})$ such that supp $\varphi \subset \mathbb{R}_+$. Then by the group property of the family $\{T(t);\, t \in \mathbb{R}\}$ and using Lemma 6.1 we can prove for $0 < \alpha < 2n$

and each $x \in X$ the identity

$$\int_0^\infty (\delta^{\alpha/2} * \varphi)(u)\, W(u)\, du \int_{-\infty}^\infty \overline{q}_{\alpha,2n}\left(\frac{t}{\varepsilon}\right) T(t)x \,\frac{dt}{\varepsilon} =$$

$$= \frac{1}{K_{\alpha,2n}} \int_\varepsilon^\infty t^{-\alpha-1}\left[T\left(\frac{t}{2}\right) - T\left(-\frac{t}{2}\right)\right]^{2n} dt \int_0^\infty \varphi(u) W(u) x\, du.$$

If in this formula φ tends to δ following the filter $\mathfrak{F}$ we obtain the first part of the next theorem from which the second part immediately follows.

Theorem 6.2. *(i) For each $x \in X$ the integral $\int_{-\infty}^\infty \overline{q}_{\alpha,2n}\left(\frac{t}{\varepsilon}\right) T(t)x \,\frac{dt}{\varepsilon}$ belongs to $D((-A^2)^{\alpha/2})$ and*

$$(-A^2)^{\alpha/2} \int_{-\infty}^\infty \overline{q}_{\alpha,2n}\left(\frac{t}{\varepsilon}\right) T(t)x \,\frac{dt}{\varepsilon} =$$

$$= \frac{1}{K_{\alpha,2n}} \int_\varepsilon^\infty t^{-\alpha-1}\left[T\left(\frac{t}{2}\right) - T\left(-\frac{t}{2}\right)\right]^{2n} x\, dt.$$

If $x \in D((-A^2)^{\alpha/2})$, the operator $(-A^2)^{\alpha/2}$ may be drawn under the integral sign.

(ii) For an element $x \in X$ the limit

$$\lim_{\varepsilon\to 0+} \frac{1}{K_{\alpha,2n}} \int_\varepsilon^\infty t^{-\alpha-1}\left[T\left(\frac{t}{2}\right) - T\left(-\frac{t}{2}\right)\right]^{2n} x\, dt$$

exists if and only if $x \in D((-A^2)^{\alpha/2})$. If so, the limit equals $(-A^2)^{\alpha/2}x$.

Combining this theorem with equation (6.4) and Theorem 3.2 applied to the semigroup $\{W(t);\, t \in \mathbb{R}_+\}$ and $\{P(t);\, t \in \mathbb{R}_+\}$, respectively, we obtain the following characterization of abstract Riesz potentials (6.6).

Theorem 6.3. *Let $0 < \alpha < 2n$, $n = 1, 2, \ldots$ and consider for an element $x \in X$ the families of integrals*

(i) $\dfrac{1}{K_{\alpha,2n}} \int_\varepsilon^\infty t^{-1-\alpha}\left[T\left(\frac{t}{2}\right) - T\left(-\frac{t}{2}\right)\right]^{2n} x\, dt,$

(ii) $\dfrac{1}{C_{\alpha/2,n}} \int_\varepsilon^\infty t^{-1-\alpha/2}[I - W(t)]^n x\, dt,$

(iii) $\frac{1}{C_{\alpha,2n}} \int_{\varepsilon}^{\infty} t^{-1-\alpha}[I - P(t)]^{2n} x \, dt.$

If one of these families converges in norm as $\varepsilon \to 0+$, then also the other two families are convergent in the same sense. If so, the three limits coincide and are equal to $(-A^2)^{\alpha/2}x = (-A_P)^{\alpha}x$.

7 Miscellaneous Comments

In this section we shall briefly outline some extensions to fractional powers of more general operators than infinitesimal generators. One can find several surveys on this subject in books on differential equations (see, for instance, Krein [22], Tanabe [44], Fattorini [10]). Thus, our main purpose is to outline some newer developments which, however, require the knowledge of the older ones. In any case, these final remarks are by no means exhaustive.

If A is the infinitesimal generator of an equibounded (C_0)-semigroup, then if $x \in D(A)$ and $0 < \alpha < 1$ we have the representation

$$(-A)^{\alpha}x = \frac{1}{-\Gamma(-\alpha)} \int_0^{\infty} t^{-\alpha-1}[I - T(t)]x \, dt$$

which can be rewritten by means of the resolvent of A as

$$(-A)^{\alpha}x = \frac{\sin \alpha\pi}{\pi} \int_0^{\infty} \lambda^{\alpha-1} R_{\lambda}(-A)x \, d\lambda.$$

This formula motivates the definition of fractional powers which was introduced in 1960 by Balakrishnan [2] for the following class $\mathcal{K}$ of operators containing the infinitesimal generators as a proper subclass.

A closed linear operator A with domain and range in a Banach space X is said to belong to the class $\mathcal{K}$ if the interval $(0, \infty)$ is contained in its resolvent set $\varrho(A)$ and the resolvent $R_{\lambda} = (\lambda I - A)^{-1}$ satisfies

$$\|\lambda R_{\lambda}\| \leq M \qquad \text{for all } \lambda > 0,$$

M being a positive constant independent of λ. In the literature, the notation 'non-negative operator' has become customary for $(-A)$, if $A \in \mathcal{K}$. Following Balakrishnan we define a linear operator J^{α} for complex α with $\operatorname{Re}\alpha > 0$:

If $0 < \operatorname{Re}\alpha < 1$, then $D(J^{\alpha}) = D(A)$ and

$$J^{\alpha}x = \frac{\sin \alpha\pi}{\pi} \int_0^{\infty} \lambda^{\alpha-1} R_{\lambda}(-A)x \, d\lambda, \qquad x \in D(A). \tag{7.1}$$

J^α can be extended to other values of α as follows. If $x \in D(A^2)$, (7.1) can be transformed to

$$J^\alpha x = \frac{\sin \alpha\pi}{\pi} \int_0^\infty \lambda^{\alpha-1} \left[R_\lambda - \frac{\lambda}{1+\lambda^2} I\right] (-A)x \, d\lambda + \sin\frac{\alpha\pi}{2}(-A)x. \tag{7.2}$$

But the integral in (7.2) actually converges in $0 < \operatorname{Re}\alpha < 2$. Thus (7.2) is a holomorphic extension of $J^\alpha x$ to $0 < \operatorname{Re}\alpha < 2$. Moreover, if $1 < \operatorname{Re}\alpha < 2$ and $x \in D(A^2)$, then one easily verifies that

$$J^{\alpha-1}(-A)x = J^\alpha x.$$

This gives rise to the following iteration:

If $n - 1 < \operatorname{Re}\alpha < n$, $n = 1, 2, \ldots$, then J^α with domain $D(J^\alpha) = D(A^n)$ is defined by

$$J^\alpha x = J^{\alpha-n+1}(-A)^{n-1}x, \qquad x \in D(A^n). \tag{7.3}$$

If $\operatorname{Re}\alpha = n$, then we set $D(J^\alpha) = D(A^{n+1})$ and

$$J^\alpha x = J^{\alpha-n+1}(-A)^{n-1}x, \qquad x \in D(A^{n+1}),$$

where $J^{\alpha-n+1}$ is given by (7.2) in this case.

For each complex α with $\operatorname{Re}\alpha > 0$, J^α is a closable linear operator. Thus, we can consider the smallest closed extension of J^α, denoted by $\overline{J^\alpha}$. If $\alpha = n$, $n = 1, 2, \ldots$, then $\overline{J^n} = (-A_+)^n$, where A_+ denotes the restriction of A to the set $\{x \in D(A);\ Ax \in \overline{D(A)}\}$. If A is densely defined, then $A_+ = A$. Hence in this case the definition

$$(-A)^\alpha = \overline{J^\alpha}, \qquad \operatorname{Re}\alpha > 0, \tag{7.4}$$

is meaningful. Otherwise one has to replace, in the above formula, A by A_+.

Fractional powers of operators of the class $\mathcal{K}$ were intensively studied in a series of papers by Komatsu [20]. His definitions (he gives, in fact, two) are equivalent to that of Balakrishnan, but differ with respect to the domains and integral representations of the operators J^α. These modifications were made with the intention to point out essential connections with the interpolation spaces due to Lions-Peetre [25]. Komatsu was the first who proved the additivity law

$$(-A)^\alpha(-A)^\beta = (-A)^{\alpha+\beta}, \qquad \operatorname{Re}\alpha > 0, \operatorname{Re}\beta > 0,$$

for densely defined operators $A \in \mathcal{K}$ in full generality. He also considered negative powers, more precisely, powers with exponents whose real parts are negative. Moreover, he treated infinitesimal generators of equibounded (C_0)-semigroups as a special subclass of $\mathcal{K}$. Compare [20,II] for the Marchaud type characterization given in Section 3.1 of this chapter. Analogously, the following characterization of powers of densely defined operators $A \in \mathcal{K}$ is valid (cf. [20,II])

$$(-A)^\alpha x = \lim_{N\to\infty} \frac{\Gamma(m)}{\Gamma(\alpha)\Gamma(m-\alpha)} \int_0^N \lambda^{\alpha-1}[I - \lambda R_\lambda]^m x \, d\lambda, \tag{7.5}$$

where $0 < \operatorname{Re}\alpha < m$, and the limit exists if and only if $x \in D((-A)^\alpha)$.

In Hövel-Westphal [16], fractional powers of densely defined operators from $\mathcal{K}$ were studied on the basis of formula (7.5) as a definition. This treatment heavily depends on the use of the Stieltjes transform, comparable with the role of the Laplace transform in the case of semigroup generators. In particular, the following analogue of the formula given in Theorem 3.2 (i) is valid:

If $0 < \alpha < 1$, then for each $x \in X$,

$$\frac{1}{\Gamma(\alpha)\Gamma(1-\alpha)} \int_0^N \lambda^{\alpha-1}[I - \lambda R_\lambda]x \, d\lambda = (-A)^\alpha \int_0^\infty \tilde{p}_\alpha(\lambda/N) R_\lambda x \, d\lambda,$$

where $\tilde{p}_\alpha$ is an appropriate numerically valued function satisfying $\lambda^{-1}\tilde{p}_\alpha(\lambda) \in L^1(\mathbb{R}_+)$. Again $(-A)^\alpha$ can be drawn under the integral sign, if $x \in D((-A)^\alpha)$. Formulas of this type are useful in many respects when dealing with fractional powers. For further developments concerning an operational calculus based on the Stieltjes transform we refer to Hirsch [14].

If $0 < \alpha \leq 1/2$, then for a densely defined operator $A \in \mathcal{K}$ the fractional power $-(-A)^\alpha$ generates an equibounded (C_0)-semigroup, even if A itself is not an infinitesimal generator. This result which is very important for applications to differential equations was proved already by Balakrishnan [2].

Fractional powers of operators belonging to subclasses of $\mathcal{K}$ were independently introduced by Krasnosel'skii-Sobolevskii [21] and Kato [18,19] about the same time as Balakrishnan's paper [2] appeared. The equivalence of the various definitions for a densely defined A was proved by Nollau [31].

Let us consider somewhat more precisely the case that $A \in \mathcal{K}$ and $0 \in \varrho(A)$, that is, A has a bounded inverse A^{-1} defined on all of X. Then fractional powers of $(-A)$ of negative order can be defined by the Cauchy formula

$$(-A)^{-\alpha} = \frac{1}{2\pi i} \int_C (-\lambda)^{-\alpha} R_\lambda \, d\lambda, \qquad \operatorname{Re}\alpha > 0, \tag{7.6}$$

where C is an appropriately chosen contour contained in $\varrho(A)\backslash\mathbb{R}_+$ and 'surrounding' the half-line $\mathbb{R}_+$. (Note that the principal branch of $(-\lambda)^{-\alpha}$ is a holomorphic function of λ in the complex plane cut along $\mathbb{R}_+$). The operators $(-A)^{-\alpha}$ are bounded and one-to-one. Thus one defines $(-A)^{\alpha}$ for $\operatorname{Re}\alpha > 0$ as the inverse of $(-A)^{-\alpha}$,

$$(-A)^{\alpha} = ((-A)^{-\alpha})^{-1}. \tag{7.7}$$

The definitions described so far for operators of the class $\mathcal{K}$ lead to satisfactory theories of fractional powers in case A is densely defined or 0 belongs to the resolvent set of A. As there are, however, many examples of operators from $\mathcal{K}$ which do not fulfill these conditions, Martinez, Sanz and Marco looked for appropriate extensions of the operators $\overline{J^{\alpha}}$ and presented two generalizations of the previous definitions of fractional powers in [29] and [27]. The first definition given in [29] is based on the observation that if $A \in \mathcal{K}$, then for each $\varepsilon > 0$, $A - \varepsilon I \in \mathcal{K}$, too, and $0 \in \varrho(A - \varepsilon I)$. Thus, $(-A + \varepsilon I)^{\alpha}$ exists in the sense of definition (7.7) for $\operatorname{Re}\alpha > 0$, and $(-A)^{\alpha}$ may be defined by the formula

$$(-A)^{\alpha} x = \lim_{\varepsilon \to 0+} (-A + \varepsilon I)^{\alpha} x, \qquad \operatorname{Re}\alpha > 0, \tag{7.8}$$

for all $x \in X$ for which the limit exists in the norm topology of X. In [27], there is introduced an equivalent definition, namely,

$$(-A)^{\alpha} x = (I - A)\overline{J^{\alpha}}(I - A)^{-1} x, \qquad \operatorname{Re}\alpha > 0, \tag{7.9}$$

for all $x \in X$ for which the right-hand side exists. Formula (7.9) has the advantage that the properties of $(-A)^{\alpha}$ can be derived directly from those already known for $\overline{J^{\alpha}}$. In any case, the operators defined by (7.8) and (7.9), respectively, interpolate the integral powers $(-A)^n$, $n = 1, 2, \ldots$ and satisfy the properties that are expected for fractional powers. In [27], also the question of uniqueness of a definition of fractional powers is discussed from a more general point of view. In this context we also refer to [26] concerning n-th roots. In [28], fractional powers in Fréchet spaces are studied. For extensions of fractional powers of operators to Fréchet spaces see also Lamb [23] and Schiavone [38].

In [43], B. Straub investigated fractional powers of a closed densely defined linear operator A with a polynomially bounded resolvent, that is, there exist an angle $\varphi \in (0, \frac{\pi}{2})$ and a nonnegative integer k such that the sector $\{\lambda \in \mathbb{C};\ |\arg\lambda| \leq \varphi\} \cup \{0\}$ is contained in $\varrho(A)$ and the resolvent satisfies

$$||R_\lambda|| \leq M(1 + |\lambda|)^k$$

for all λ in this sector, M being a positive constant. In this case, a modification of the Cauchy formula (7.6) leads to the following definition:

For $\alpha \in \mathbb{C}$, $(-A)^\alpha$ is defined as the smallest closed extension of the operator defined on $D(A^{[\alpha]+k+2})$ by the integrals

$$\frac{1}{2\pi i}\int_C (-\lambda)^\alpha R_\lambda x\, d\lambda, \qquad \text{if } \operatorname{Re}\alpha < 0,$$

and

$$\frac{1}{2\pi i}\int_C (-\lambda)^{\alpha-[\alpha]-1} R_\lambda (-A)^{[\alpha]+1} x\, d\lambda, \qquad \text{if } \operatorname{Re}\alpha \geq 0,$$

where C is an appropriate contour in $\varrho(A)\backslash\mathbb{R}_+$. In particular, $(-A)^\alpha$ belongs to $\mathcal{E}(X)$, if $\operatorname{Re}\alpha < -(k+1)$. Moreover, it is shown that $-(-A)^\alpha$ generates a holomorphic semigroup of growth order $(k+1)/\alpha$, if $0 < \alpha \leq \frac{1}{2}$. This fact is important concerning applications to Cauchy problems. For extensions of the above investigations to a more general class of operators we refer to [8].

References

1. A.V. Balakrishnan, An operational calculus for infinitesimal generators of semigroups, *Trans. Amer. Math. Soc.* **91** (1959), 330–353.
2. A.V. Balakrishnan, Fractional powers of closed operators and the semigroups generated by them, *Pacific J. Math.* **10** (1960), 419–437.
3. A.V. Balakrishnan, Representation of abstract Riesz potentials of the elliptic type, *Bull. Amer. Math. Soc.* **64**, no. 5 (1958), 288–289.
4. H. Berens, P.L. Butzer and U. Westphal, Representation of fractional powers of infinitesimal generators of semigroups, *Bull. Amer. Math. Soc.* **74** (1968), 191–196.
5. C. Berg, K. Boyadzhiev and R. deLaubenfels, Generation of generators of holomorphic semigroups, *J. Austral. Math. Soc. Ser. A* **55** (1993), 246–269.
6. S. Bochner, Diffusion equation and stochastic processes, *Proc. Nat. Acad. Sci. USA* **35** (1949), 368–370.

7. P.L. Butzer and H. Berens, *Semi-Groups of Operators and Approximation*, Grundlehren Math. Wiss. 145, Springer, Berlin 1967.
8. R. deLaubenfels, F. Yao and S. Wang, Fractional powers of operators of regularized type, *J. Math. Anal. Appl.* **199** (1996), 910–933.
9. J. Faraut, Semi-groupes de mesures complexes et calcul symbolique sur les générateurs infinitesimaux de semi-groupes d'opérateurs, *Ann. Inst. Fourier (Grenoble)* **20** (1970), 235–301.
10. H.O. Fattorini, *The Cauchy Problem*, Encyclopedia of Mathematics and its Applications, Vol. 18, Addison-Wesley Publishing Company, Reading, Massachusetts 1983.
11. W. Feller, On a generalization of Marcel Riesz' potentials and the semi-groups generated by them, *Comm. Sém. Math. Univ. Lund. Tome Supplémentaire* (1952), 72–81.
12. I.M. Gelfand and G.E. Shilov, *Generalized Functions*, Vol. I, Academic Press Inc., New York 1964.
13. E. Hille and R.S. Phillips, *Functional Analysis and Semi-Groups*, Amer. Math. Soc. Colloq. Publ., Vol. 31, Providence, R.I. 1957.
14. F. Hirsch, Intégrales de résolvantes et calcul symbolique, *Ann. Inst. Fourier (Grenoble)* **22** (1972), 239–264.
15. L. Hörmander, *The Analysis of Linear Partial Differential Operators I*, Springer-Verlag, Berlin, Heidelberg, New York, Tokyo 1983.
16. H.W. Hövel and U. Westphal, Fractional powers of closed operators, Studia Math. **42** (1972), 177–194.
17. A.E. Ingham, The equivalence theorem for Cesàro and Riesz summability, Publ. Ramanujan Inst. **1** (1969), 107–113.
18. T. Kato, Note on fractional powers of linear operators, *Proc. Japan Acad.* **36** (1960), 94–96.
19. T. Kato, Fractional powers of dissipative operators, *J. Math. Soc. Japan* **13** (1961), 246–274, II, ibidem **14** (1962), 242–248.
20. H. Komatsu, Fractional powers of operators, *Pacific J. Math.* **19** (1966), 285–346, II: Interpolation spaces, *Pacific J. Math.* **21** (1967), 89–111, III: Negative powers, *J. Math. Soc. Japan* **21** (1969), 205–220, IV: Potential operators, *J. Math. Soc. Japan* **21** (1969), 221–228, V: Dual operators, *J. Fac. Sci. Univ. Tokyo*, Sec. I A, **17** (1970), 373–396, VI: Interpolation of non-negative operators and imbedding theorems, *J. Fac. Sci. Univ. Tokyo*, **19** (1972), 1–63.
21. M.A. Krasnosel'skii and P.E. Sobolevskii, Fractional powers of operators acting in Banach spaces, *Dokl. Akad. Nauk SSSR* **129** (1959), 499–502 (Russian).
22. S.G. Krein, *Linear Differential Equations in Banach Space*, Translations

of Mathematical Monographs, Vol. 29, Amer. Math. Soc., Providence, Rhode Island 1971.

23. W. Lamb, Fractional powers of operators defined on a Fréchet space, *Proc. Edinburgh Math. Soc.* (2) **27** (1984), 165–180.
24. O.E. Lanford and D.W. Robinson, Fractional powers of generators of equicontinuous semigroups and fractional derivatives, *J. Austral. Math. Soc. Ser. A* **46** (1989), 473–504.
25. J.L. Lions et J. Peetre, Sur une classe d'espaces d'interpolation, *Inst. Hautes Études Sci. Publ. Math.* **19** (1964), 5–68.
26. C. Martinez and M. Sanz, n-th roots of a non-negative operator, *Manuscripta Math.* **64** (1989), 403–417.
27. C. Martinez and M. Sanz, Fractional powers of non-densely defined operators, *Ann. Scuola Norm. Sup. Pisa Cl. Sci.* (4) **18** (1991), 443–454.
28. C. Martinez, M. Sanz and V. Calvo, Fractional powers of non-negative operators in Fréchet spaces, *Internat. J. Math. Math. Sci.* **12** (1989), 309–320.
29. C. Martinez, M. Sanz and L. Marco, Fractional powers of operators, *J. Math. Soc. Japan* **40** (1988), 331–347.
30. E. Nelson, A functional calculus using singular Laplace integrals, *Trans. Amer. Math. Soc.* **88** (1958), 400–413.
31. V. Nollau, Über Potenzen von linearen Operatoren in Banachschen Räumen, *Acta Sci. Math. (Szeged)* **28** (1967), 107–121.
32. R.S. Phillips, Spectral theory for semi-groups of linear operators. *Trans. Amer. Math. Soc.* **71** (1951), 393–415.
33. R.S. Phillips, On the generation of semi-groups of linear operators, *Pacific J. Math.* **2** (1952), 343–369.
34. R.S. Phillips, Semi-groups of operators, *Bull. Amer. Math. Soc.* **61** (1955), 16–33.
35. N.P. Romanoff, On one parameter groups of linear transformations I, *Ann. of Math.* **48** (1947), 216-233.
36. B. Rubin, *Fractional Integrals and Potentials*, Addison Wesley Longman Limited, Harlow 1996.
37. S.G. Samko, A.A. Kilbas and O.I. Marichev, *Fractional Integrals and Derivatives, Theory and Applications*, Gordon and Breach Science Publishers, Amsterdam 1993.
38. S.E. Schiavone, Fractional powers of operators and Riesz fractional integrals, *Proc. Roy. Soc. Edinburgh Sect. A* **112** (1989), 237–247.
39. R.L. Schilling, Subordination in the sense of Bochner and a related functional calculus, *J. Austral. Math. Soc. Ser. A* **64** (1998), 368–396.
40. L. Schwartz, *Lectures on Mixed Problems in Partial Differential Equa-*

tions and Representations of Semi-Groups, Tata Institute of Fundamental Research, Bombay 1958.
41. L. Schwartz, *Théorie des Distributions*, Hermann, Paris 1966.
42. J.D. Stafney, Integral representations of fractional powers of infinitesimal generators, *Illinois J. Math.* **20** (1976), 124–133.
43. B. Straub, Fractional powers of operators with polynomially bounded resolvent and the semigroups generated by them, *Hiroshima Math. J.* **24** (1994), 529–548.
44. H. Tanabe, *Equations of Evolution*, Pitman, London 1979.
45. U. Westphal, Ein Kalkül für gebrochene Potenzen infinitesimaler Erzeuger von Halbgruppen und Gruppen von Operatoren, Teil I: Halbgruppenerzeuger, *Compositio Math.* **22** (1970), 67–103, Teil II: Gruppenerzeuger, *Compositio Math.* **22** (1970), 104–136.
46. U. Westphal, An approach to fractional powers of operators via fractional differences, *Proc. London Math. Soc.* **(3)29** (1974), 557–576.
47. U. Westphal, Gebrochene Potenzen abgeschlossener Operatoren, definiert mit Hilfe gebrochener Differenzen, in: *Linear Operators and Approximation II* (Proc. Conf. Oberwolfach, March 1974, ed. by P.L. Butzer and B. Sz.-Nagy), Birkhäuser, Basel, Stuttgart 1974, pp. 23–27.
48. U. Westphal, A generalized version of the Abelian mean ergodic theorem with rates for semigroup operators and fractional powers of infinitesimal generators, *Results Math.*, to appear.
49. K. Yosida, *Functional Analysis*, Springer-Verlag, Berlin 1978.
50. A.H. Zemanian, *Distribution Theory and Transform Analysis, An Introduction to Generalized Functions, with Applications*, McGraw-Hill, New York 1965.

CHAPTER IV

FRACTIONAL DIFFERENCES, DERIVATIVES AND FRACTAL TIME SERIES

BRUCE J. WEST, PAOLO GRIGOLINI
Center for Nonlinear Science, University of North Texas, Denton, Texas 76203, USA

Contents

1 Introduction

It seems quite remarkable that nearly thirty years ago one of us (BJW), as a graduate student, sat in a seminar room at the University of Rochester and listened to Benoit Mandelbrot talk about why the night sky was not uniformly illuminated (Olber's paradox) and how income is distributed in western societies (Pareto's Law). It would be another ten years before he (Mandelbrot) coined the word fractal to take cognizance of the fact that there is a large class of natural and social phenomena that traditional statistical physics is not equipped to describe. In the past decade there has been a blossoming literature on fractal random processes, those with inverse power-law spectra characteristic of long-time memory, episodic processes with Lévy stable distribution functions, and their applications to the physical and life sciences [1,2,3,4,5]. Three separate approaches have been used to model such phenomena and it is impossible to list even a representative sample of that work here and so we confine our references to a few books and review articles that we have found useful.

One approach to modeling fractal time series is by means of low-dimensional, nonlinear, deterministic dynamical equations having intermittent chaotic solutions [6]. The spectra of such systems spread themselves into broad band, inverse power laws, indicative of fractal random time series. Of course such scaling processes are generated by colored noise as well, which leads us to the second method of modeling, that being stochastic differential equations. In particular, such equations are often generated using random walks with long-time correlations in the random fluctuations, yielding fractional diffusion equations to describe the evolution of the probability density for the random walk variable [7]. The statistics of a system's response to such fluctuations is often found to deviate strongly from that usually expected using the central limit theorem (CLT). For example, the second moment of the random walk variable may diverge. A generalized version of the CLT yields Lévy statistical distributions to describe the system random response to such correlated fluctuations, see for example Montroll and West [8]. This last work done at the end of the seventies was the harbinger of what was to be an avalanche of research into the nature of phenomena whose evolution cannot be described by differential equations. Subsequently, it was found that both the inverse power-law spectra and the Lévy statistical distribution are consequences of scaling and fractals [9] as was discussed in the second edition of ref.[[8]] and in its sequel [10].

The third method for generating fractal time series, and probably the least well known in the physical science literature, is by means of fractional differences in discrete stochastic equations [11,12,13]. This technique has until very recently had only relatively modest acceptance in the field of economics where it was first introduced, see ref.[[14]] for a historical review and some recent applications. This should come as no surprise, the work of Mandelbrot [15], first published in the economics literature thirty years ago, has only in the past few year begun to influence that community. But as we shall see, fractional differences and fractals serve our purposes in modeling physical and biophysical processes with long-time memory very well.

Each of the above approaches explains the erratic behavior in time series from a particular perspective. Just as a simple random walk is the discrete time analog of Brownian motion [8], a fractional-difference process driven by discrete white noise is the discrete time analog of fractional Brownian motion (*fBm)*, that is, a process with long-time memory and Gaussian statistics [11]. Herein we examine extensions of these latter arguments to non-Gaussian statistics, in particular to α-stable Lévy statistical processes. Before presenting this discussion, however, it is necessary to relate the continuum limits of our fractional differences to fractional derivatives, since it turns out that the evolution of Lévy distributions are described by a fractional partial differential equation as first noted by West and Seshadri [16] and later rediscovered by a number of authors [17,18,19,20].

In Section 2 we briefly review some of the difficulties in modeling stochastic processes with long-time memory. In particular those problems associated with representing *fBm* using stochastic differential equations and describing anomalous diffusion using either *fBm* or Lévy stable distributions. In Section 3 we describe fractional differences between discrete time steps and define the fractional derivatives in terms of the continuum limit of the step length. This discussion is extended to discrete stochastic processes in Section 4 where we relate the fractional difference to the long-term memory of the system's response to white noise. Here the variance of the system's response is found to be anomalous, that is, the variance increases as t^{2H} where t is the time, H is the Hurst coefficient and $H \neq 1/2$. A value of H=1/2 would correspond to normal diffusion, and also to the limit of the fractional difference becoming a normal difference. In Section 5 the fractional differences are shown to give rise to a fractional derivatives that can be used to construct a fractional diffusion equation. This last equation can describe the evolution of a probability density describing the statistics of the fractal time series generated from fractional-difference stochastic equations. We show that the solution to this fractional diffusion equation is the Lévy distribution in agreement with ref.[[16]]. We also demonstrate that when the underlying random walk process has steps that occur with finite speed, the solution to the fractional diffusion equation is actually a truncated Lévy distribution. Taking account of the diffusing front in the dynamics of a Lévy process was first suggested by Shlesinger, West and Klafter [21] in the context of turbulent diffusion.

2 Brief Review of Long-time Memory

The theory of the influence of long-time memory on stochastic phenomena was most recently developed in the physical sciences in the study of anomalous diffusion, where the mean square value of a random variable does not increase linearly in time. The model for this phenomenon has relied heavily on the workhorse of statistical physics, the random walk model. In continuous form we can write a simple random walk as the stochastic differential equation

$$\frac{d}{dt}X(t) = \xi(t), \tag{1}$$

where $\xi(t)$ is a stochastic quantity. The interpretation of (1) requires a specification of the statistics of the random function $\xi(t)$ along with its correlational properties. Minimally we require the correlation of the random fluctuations at two points in time, which we choose to be the stationary quantity

$$\Phi_\xi(t - t') = \frac{\langle \xi(t)\,\xi(t')\rangle}{\langle \xi^2\rangle} \tag{2}$$

for which the equation of evolution for the probability density takes on two distinct forms. If $P(x,t)\,dx$ is the probability that $X(t)$ has a value in the interval $(x, x+dx)$ at time t, its evolution is determined by [7]

$$\frac{\partial}{\partial t}P(x,t) = \langle \xi^2\rangle \int_0^t \Phi_\xi(t-t')\,dt' \frac{\partial^2}{\partial x^2}P(x,t) \tag{3}$$

when $\xi(t)$ is a Gaussian random function and by [7]

$$\frac{\partial}{\partial t}P(x,t) = \langle \xi^2\rangle \int_0^t \Phi_\xi(t-t')\,dt' \frac{\partial^2}{\partial x^2}P(x,t') \tag{4}$$

when $\xi(t)$ is a dichotomous random process. Note that (4) is a convolution in time whereas (3) is not.

If we assume that the asymptotic form of the correlation function (2) has the form

$$\lim_{t\to\infty} \Phi_\xi(t) \propto \pm\frac{1}{t^\beta} \tag{5}$$

with the power-law index in the interval

$$0 < \beta < 2, \tag{6}$$

then from (3) we obtain the second moment of the random walk variable

$$\langle X^2(t) \rangle \propto t^{2H} \tag{7}$$

where the Hurst exponent, H, is related to the index in the correlation function

$$H = 1 - \beta/2. \tag{8}$$

It is well known that (7) depicts anomalous diffusion when $H \neq 1/2$. When $1/2 < H \leq 1$ the random walk is persistent, meaning that a walker prefers to continue in the direction of motion rather than changing direction, and therefore the diffusion is faster than normal. When $0 \leq H < 1/2$ the random walk is anti-persistent, meaning that a walker prefers to change direction with each step rather than to continue in the same direction, and therefore the diffusion is slower than normal. Finally normal diffusion occurs when $H = 1/2$ and is usually referred to as Brownian motion. When $0 \leq \beta \leq 1$ the positive sign from the correlation function (5) must be used in the equation of evolution (3) to keep the second moment of the random walk variable positive definite. When $1 < \beta \leq 2$ the negative sign from the correlation function (5) must be used in (3) for the same reason.

The above model, using the equation of evolution for the probability density (3), is fully equivalent to the fractional Brownian motion (*fBm)* of Mandelbrot. However, there are several problems with this interpretation of the equation. In the subdiffusive case, $0 \leq H < 1/2$, the correct limiting behavior is obtained using the negative tail of the correlation function (5), as we said. On the other hand, the correlation function has the constraint

$$\Phi_\xi(0) = 1, \tag{9}$$

which implies that the correlation function crosses the line $\Phi_\xi(0) = 0$ at least once before achieving its asymptotic value of zero. It is not yet clear when the asymptotic value of the correlation function loses its memory of the initial condition and crosses the abscissa. Note that the early, positive part of the time evolution of $\Phi_\xi(t)$, would, by itself, generate normal diffusion, that is, at early times, $\langle X^2(t) \rangle \propto t$. This behavior would be enough to have ordinary diffusion overcome a sub-diffusional process.

In the case of superdiffusion., $0 \leq \beta < 1$, the above problem with the correlation function does not exist. However, it would be somewhat surprising that a "microscopic" variable such as ξ would be Gaussian. It seems more physical that the microscopic variable would be strongly non-Gaussian leaving the Central Limit Theorem to produce "macroscopic" Gaussianity. For example if ξ is dichotomous, rather than (3) we obtain (4) for the evolution of the probability density and consequently obtain a strong deviation from the Gaussian diffusion process predicted by the theory of *fBm* when ξ has long-time correlations.

Thus, we have a problem with specifying the dynamics underlying *fBm* using stochastic differential equations. This problem was resolved in the context of DNA sequences through the use of tertiary structures, that is, by taking into account the influence of folding on the statistics of random walkers, see Allegrini *et al.* [22]. Herein we adopt a fractional-difference approach to a self-contained derivation of *fBm* which avoids the above problems.

Let us now consider the change in the statistics that arises between the first and the second exact diffusion equation, (3) and (4), respectively. We know that under the assumption that the random walkers are constrained to traverse a step with finite speed, say unity, that the two positions x and x' are related through the two times t and t' by

$$P(x, t-t') = \frac{1}{2}\int_{-\infty}^{\infty} \delta\left(|x-x'| - t'\right) P(x', t)\, dx' \qquad (10)$$

where the delta function constrains the step size due to the finite speed of a step. The relation (10) leads from the exact equation (4) to the approximate equation [7]

$$\frac{\partial}{\partial t} P(x,t) = const. \int_{-\infty}^{\infty} \frac{dx' P(x', t)}{\left(\Delta + |x-x'|\right)^{\beta+2}} \qquad (11)$$

where $\Phi_\xi(t) \propto (\Delta + t)^{-\beta}$, which by further neglecting the constant Δ yields the equation of evolution for an α-stable Lévy process [16].

It is well established that the Lévy distribution, and the corresponding rescaling of the random walk process as well, accurately fit the results of numerical calculations. However, the resulting process when Δ is not neglected is not quite equivalent to a Lévy process, but rather to a truncated Lévy process. A truncated Lévy process does not possess any probability in the tail region beyond that which could travel ballistically in the time t. Instead the probability from the tail region is contained in symmetrically placed peaks that travel ballistically in opposite directions. A similar solution is obtained to the diffusion equation, (3), when the correlation function is exponential. The latter equation is equivalent to the telegrapher's equation [8], the solution to which is a truncated Gaussian with symmetric ballistic peaks [23].

3 Fractional Differences and Derivatives

Let us now turn our attention to the formal derivation of fractional derivatives using the continuum limit of finite difference equations. The traditional definition of the derivative of a continuous function is expressed in terms of a limiting procedure. Consider the Taylor expansion of the continuous function $X(t)$ given by

$$X(t+\tau) = X(t) + \tau DX(t) + \frac{\tau^2}{2!} D^2 X(t) + \cdots + \frac{\tau^n}{n!} D^n X(t) + \cdots \tag{12}$$

where D is the derivative operator and writing the upshift operator, E_τ, as

$$E_\tau X(t) = X(t+\tau) \tag{13}$$

we obtain by comparing the two equations for $X(t+\tau)$

$$E_\tau X(t) = e^{\tau D} X(t). \tag{14}$$

We therefore have the symbolic equivalence between the upshift and derivative operators

$$E_\tau = e^{\tau D} \tag{15}$$

and we can write

$$E_\tau \equiv 1 - \Delta_- = e^{\tau D} \tag{16}$$

so that the right-difference operator becomes

$$\Delta_- = 1 - E_\tau = 1 - e^{\tau D}. \tag{17}$$

In this way we have the definition of the derivative operator

$$\lim_{\tau \to 0} \frac{\Delta_-}{\tau} = -D \tag{18}$$

where we have expanded the right hand side of (17) in powers of τ. It is clear from this discussion that if t is the time, then D is the time derivative operator and when applied to a continuous function of time yields

$$\frac{d}{dt} X(t) = - \lim_{\tau \to 0} \frac{\Delta_- X(t)}{\tau}. \tag{19}$$

We can do a similar construction for the downshift operator

$$E_\tau^{-1} X(t) = X(t-\tau) \tag{20}$$

so that from the corresponding Taylor expansion we obtain the symbolic representation

$$E_\tau^{-1} \equiv 1 - \Delta_+ = e^{-\tau D} \tag{21}$$

so that the left-difference operator becomes

$$\Delta_+ = 1 - E_\tau^{-1} = 1 - e^{-\tau D}$$

and the limit of vanishing τ again yields the derivative operator

$$\lim_{\tau \to 0} \frac{\Delta_+}{\tau} = D. \tag{22}$$

So (18) and (22) are equivalent mathematical expressions of the derivative operator. To be on the safe side a physicist often writes the derivative operator in the symmetric form

$$\lim_{\tau \to 0} \frac{\Delta_+ - \Delta_-}{2\tau} = D \tag{23}$$

a perspective which will be of value to us later.

3.1 Differences and Derivatives of a Fractional Kind

We now extend the definition of integer-order derivatives to that of fractional order by generalizing the definition of the limit of finite differences. Consider (17) raised to the αth power

$$\Delta_+^\alpha = \left(1 - e^{-\tau \frac{d}{dt}}\right)^\alpha \tag{24}$$

so that we can write for the left-side fractional time derivative

$$\lim_{\tau \to 0} \frac{\Delta_+^\alpha}{\tau^\alpha} X(t) = \lim_{\tau \to 0} \frac{\left(1 - e^{-\tau \frac{d}{dt}}\right)^\alpha}{\tau^\alpha} X(t) = \frac{d^\alpha}{dt^\alpha} X(t) \tag{25}$$

where we have replaced D with the time-derivative operator. In a similar way we can write

$$\lim_{\tau \to 0} \frac{\Delta_-^\alpha}{\tau^\alpha} X(t) = \lim_{\tau \to 0} \frac{\left(1 - e^{\tau \frac{d}{dt}}\right)^\alpha}{\tau^\alpha} X(t) = \left(-\frac{d}{dt}\right)^\alpha X(t) \tag{26}$$

for the right-side fractional derivative. We can, using (25) and (26), write the more general formal expression for the left-side and right-side fractional derivative operators as

$$\mathcal{D}_{\pm}^{\alpha} X(t) = \lim_{\tau \to 0} \frac{\Delta_{\pm}^{\alpha} X(t)}{\tau^{\alpha}} \tag{27}$$

which is mentioned by Miller and Ross [24] and is discussed at length by Samko, *et al.* [25]. It shall require some effort to put the fractional derivatives defined by (27) into the more traditional form of an integration over a kernel dependent on the index α. Thus, we need to establish the correspondence between the fractional derivative defined by (27) and one or more fractional derivatives usually expressed in terms of integral equations. Let us examine some properties of fractional differences and what they become in the continuum limit.

Consider the right-side fractional difference between $X(t)$ and $X(t-\tau)$ for an arbitrary time interval, τ,

$$\Delta_{-}^{\alpha} X(t) = \left(1 - E_{\tau}^{-1}\right)^{\alpha} X(t). \tag{28}$$

Note that the following discussion can be repeated with minor modifications in terms of the left-side fractional difference $\Delta_{+}^{\alpha} X(t)$. As it stands, (28) is just a formal definition without content, or at least without operational content. To bring this equation alive we determine how to represent the right-side fractional difference operator in terms of more familiar functions. This operator can be defined in a natural way for integer α using the binomial theorem,

$$\left(1 - E_{\tau}^{-1}\right)^{\alpha} = \sum_{k=0}^{\alpha} \binom{\alpha}{k} (-1)^{k} E_{\tau}^{-k} \tag{29}$$

where, of course, this procedure is only defined when operating on an appropriate function. The binomial coefficients for a positive integer α are given by

$$\binom{\alpha}{k} = \frac{\Gamma(\alpha+1)}{\Gamma(k+1)\,\Gamma(\alpha+1-k)} = \frac{\alpha!}{k!\,(\alpha-k)!} \tag{30}$$

where $\Gamma(\cdot)$ denotes the gamma function. Beran [13] notes that for negative integers the gamma function has simple poles so that the binomial coefficient is zero if $k > \alpha$ in (30) and α is an integer.. Therefore the binomial expansion in (29) may be extended for any real number to an infinite series since the divergent values of the gamma functions cut-off the series for $k > \alpha$. For non-integer values of α the binomial expansion is actually an infinite series, which is to say that the binomial coefficients do not cut the series off after some finite value. Therefore, the right-side fractional difference (28) is defined by

$$\Delta_-^\alpha X(t) = \sum_{k=0}^{\infty} \binom{\alpha}{k} (-1)^k X(t - k\tau) \quad , \quad \alpha > 0 \tag{31}$$

which rather than depending only on the value of the function in the vicinity of t, depends on the entire history of the function.

One of the properties of the fractional difference operator easily obtained is that of the semi-group. We multiply two operators like (31) together to obtain

$$\begin{aligned} \Delta_-^\alpha \Delta_-^\beta X(t) &= \sum_{k=0}^{\infty} \binom{\beta}{k} (-1)^k \Delta_-^\alpha X(t - k\tau) \\ &= \sum_{k=0}^{\infty} \binom{\beta}{k} \sum_{k'=0}^{\infty} \binom{\alpha}{k'} (-1)^{k+k'} X(t - [k + k']\tau) \end{aligned}$$

where the operators are applied sequentially. Using the properties of binomial coefficients one of the series may be summed to obtain

$$\Delta_-^\alpha \Delta_-^\beta X(t) = \sum_{j=0}^{\infty} \binom{\alpha + \beta}{j} (-1)^j X(t - j\tau) . \tag{32}$$

The right hand side of (32) is the series representation of the double indexed difference so that we have

$$\Delta_-^\alpha \Delta_-^\beta X(t) = \Delta_-^{\alpha+\beta} X(t) \tag{33}$$

thereby establishing the semi-group property for this definition of a fractional difference.

Thus, we can show that the right-side fractional derivative also satisfies the semi-group property, since dividing (33) by $\tau^{\alpha+\beta}$ and taking the limit $\tau \to 0$ yields

$$\lim_{\tau \to 0} \frac{\Delta_-^{\alpha+\beta}}{\tau^{\alpha+\beta}} X(t) = \lim_{\tau \to 0} \frac{\Delta_-^\alpha}{\tau^\alpha} \lim_{\tau \to 0} \frac{\Delta_-^\beta}{\tau^\beta} X(t)$$

and assuming the same arguments with - replaced with + we have

$$\mathcal{D}_\pm^{\alpha+\beta} X(t) = \mathcal{D}_\pm^\alpha \mathcal{D}_\pm^\beta X(t) . \tag{34}$$

Equation (34) demonstrates that the definition of the right-side and left-side fractional derivatives (27) involving the continuous limit of the right-side and left-side fractional difference in time satisfies the semi-group property.

3.2 Fourier Representation

We restrict the following discussion to functions that are periodic over an interval 2π, however, the argument can be generalized to non-periodic functions [25]. Let $f(t)$ be a 2π-periodic function in $\mathcal{R}^1$ such that the Fourier series for the function is

$$f(t) = \sum_{\omega=-\infty}^{\infty} \hat{f}_{\omega} e^{i\omega t} \tag{35}$$

with the Fourier coefficient denoted by a caret over the function. The values of the Fourier coefficients are given by

$$\hat{f}_{\omega} = \frac{1}{2\pi} \int_0^{2\pi} f(t) e^{-i\omega t} dt \tag{36}$$

and we assume that the constant term at $\omega = 0$ vanishes. Now let us consider the Fourier transform, denoted by the operator $\mathcal{F}$, of the right-side fractional derivative (27)

$$\begin{aligned} \mathcal{F}\left\{\mathcal{D}_{-}^{\alpha} X(t)\right\} &= \frac{1}{2\pi} \int_0^{2\pi} e^{-i\omega t} dt \lim_{\tau \to 0} \frac{\Delta_{-}^{\alpha} X(t)}{\tau^{\alpha}} \\ &= \lim_{\tau \to 0} \frac{1}{2\pi \tau^{\alpha}} \int_0^{2\pi} e^{-i\omega t} \Delta_{-}^{\alpha} X(t) \, dt. \end{aligned} \tag{37}$$

Using the series representation of the fractional difference (31) in (37) and substituting the Fourier series for the variable we obtain

$$\mathcal{F}\left\{\mathcal{D}_{-}^{\alpha} X(t)\right\} = \lim_{\tau \to 0} \frac{1}{\tau^{\alpha}} \sum_{k=0}^{\infty} \binom{\alpha}{k} \frac{(-1)^k}{2\pi} \int_0^{2\pi} dt e^{-i\omega t} \sum_{\omega'=-\infty}^{\infty} e^{i\omega'(t-k\tau)} \hat{X}_{\omega'}$$

where the integral over time, t, yields a Kronecker delta function so that the integral is zero if $\omega \neq \omega'$ and

$$\mathcal{F}\left\{\mathcal{D}_{-}^{\alpha} X(t)\right\} = \lim_{\tau \to 0} \frac{1}{\tau^{\alpha}} \sum_{k=0}^{\infty} \binom{\alpha}{k} (-1)^k e^{-i\omega k\tau} \hat{X}_{\omega} \tag{38}$$

when $\omega = \omega'$. The summation in (38) can now be expressed as a binomial expansion and therefore we obtain

$$\mathcal{F}\left\{\mathcal{D}_{\mp}^{\alpha} X(t)\right\} = \lim_{\tau \to 0} \frac{1}{\tau^{\alpha}} \left(1 - e^{\mp i\omega\tau}\right)^{\alpha} \hat{X}_{\omega}$$

where we use (24) for Δ_{+}^{α} in the definition of $\mathcal{D}_{+}^{\alpha}$. In the $\tau \to 0$ limit we expand the exponential and the linear term cancels the divisor so that we obtain the exact results

$$\mathcal{F}\left\{\mathcal{D}_{\mp}^{\alpha} X(t)\right\} = (\pm i\omega)^{\alpha} \hat{X}_{\omega} . \tag{39}$$

Thus, we see that the effect of the fractional derivative operating on an appropriately defined function is to multiply the Fourier amplitude of that function by either $(i\omega)^{\alpha}$ or its complex conjugate. In this way the Fourier representation of the right-side and left-side fractional derivatives are

$$\mathcal{D}_{\mp}^{\alpha} X(t) = \sum_{\omega=-\infty}^{\infty} (\pm i\omega)^{\alpha} \hat{X}_{\omega} e^{-i\omega t}, \tag{40}$$

see, for example, ref.[[25]], where it is clear that for non-integer α the fractional derivative operators are non-local in time. Exactly how non-local the fractional derivatives are remains to be seen.

4 Stochastic Fractional-Difference Equations

Random time series are traditionally modeled in the physical sciences using simple random walks [8]. If ξ_j is a random variable intended to represent a step taken at the discrete time j then the random walk variable that denotes the total distance traveled in a time $t = N\tau$, or after N such steps, is

$$X(t) = \sum_{j=1}^{N} \xi_j . \tag{41}$$

An alternative way to express (41) is in terms of the right-difference equation

$$X(t) - X(t - \tau) = \xi_N$$

or if we simplify notation by setting $\tau = 1$ and define the down-shift operator for the unit time interval without the subscript, we can rewrite the right-side difference using the discrete indices for arbitrary step j,

$$\left(1 - E^{-1}\right) X_j = \xi_j. \tag{42}$$

We point out that (42) is the discrete analog of Brownian motion and is the physical scientist's guide into the world of stochastic processes [8,10]. We now take the plunge and following Hosking [11] define the right-side fractional difference process

$$\left(1 - E^{-1}\right)^{\alpha} X_j = \xi_j \tag{43}$$

where again α is not an integer. We shall show that the process defined by (43) is the discrete time analogue of *fBm* in the same sense that (42) is the discrete time analog of ordinary Brownian motion. Let us see how this analogy works.

We treat (43) as a dynamical equation that has a solution, in the same sense that an ordinary finite difference equation has a solution. We solve equation (43) by inverting the right-side fractional-difference operator to obtain

$$X_j = \left(1 - E^{-1}\right)^{-\alpha} \xi_j. \tag{44}$$

Now using the expression for the inverse of the right-fractional-difference operator developed in the preceding section in (44) yields

$$X_j = \sum_{k=0}^{\infty} \frac{(k+\alpha-1)!}{k!\,(\alpha-1)!} E^{-k} \xi_j = \sum_{k=0}^{\infty} \frac{(k+\alpha-1)!}{k!\,(\alpha-1)!} \xi_{j-k}. \tag{45}$$

We wish to determine the asymptotic form of the coefficients in (45) using Stirling's approximation of a factorial, so that the ratio of the asymptotic expansions of two gamma functions is

$$\frac{\Gamma(z+\alpha)}{\Gamma(z+\beta)} \approx z^{\alpha-\beta} \ , \quad |\arg(z+\alpha)| < \pi, z \to \infty \ ,$$

so we can rewrite (45) as

$$X_j \approx \sum_{k=0}^{\infty} \frac{k^{\alpha-1}}{(\alpha-1)!} \xi_{j-k} \tag{46}$$

for $k \to \infty$, since $k \gg \alpha$. Thus, the strength of the contributions to (46) decrease asymptotically with increasing time lag, k, as an inverse power law as long as $\alpha < 1$. We can, in fact, show that α always lies in the interval $-1/2 \leq \alpha \leq 1/2$ since any power of $\left(1 - E^{-1}\right)^q$ in the original right-side fractional-difference equation (43) can be written as $q = \alpha + n$ where n is an integer and α is in the desired range. Therefore, we can always transform the variable of interest from Y_j to $X_j = \left(1 - E^{-1}\right)^n Y_j$ and the discussion proceeds in terms of the new variable X_j with the appropriate right-side fractional difference.

4.1 The Spectrum

The spectrum of the time series (45) is obtained by taking its discrete Fourier transform and averaging the square of the modulus of the Fourier amplitude over an ensemble of realizations of the random fluctuations driving the difference equation. Recalling that the discrete convolution of two functions is the product of their Fourier coefficients, we can write the Fourier amplitudes from (45) as

$$\hat{X}_\omega = \hat{\theta}_\omega \hat{\xi}_\omega \tag{47}$$

where we have denoted the coefficient in the series (45) as θ_k and its corresponding Fourier coefficient as $\hat{\theta}_\omega$. We define the spectrum for the process by

$$S(\omega) = \left\langle \left| \hat{X}_\omega \right|^2 \right\rangle \tag{48}$$

where the brackets denote an average over an ensemble of realizations of the ξ-fluctuations driving the system. Substituting (47) into (48) and assuming that the random fluctuations have a white noise spectrum of unit strength and are delta correlated in time we obtain

$$S(\omega) = \left| \hat{\theta}_\omega \right|^2 . \tag{49}$$

The Fourier coefficient $\hat{\theta}_\omega$ is calculated by taking the discrete Fourier transform of the coefficients θ_k, so that

$$\hat{\theta}_\omega = \sum_{k=0}^{\infty} \frac{(k+\alpha-1)!}{k!\,(\alpha-1)!} e^{-ik\omega} = \left(1 - e^{-i\omega}\right)^{-\alpha} . \tag{50}$$

Rearranging terms in (50) and substituting that expression into (49) we obtain

$$S(\omega) = \frac{1}{\left[2\sin(\omega/2)\right]^{2\alpha}} \tag{51}$$

for the spectrum of the fractional-difference process driven by white noise. Therefore, we obtain the inverse power-law spectrum

$$S(\omega) \approx \frac{1}{\omega^{2\alpha}} \ , as \quad \omega \to 0 \tag{52}$$

for the right-side fractional-difference process driven by white noise as first obtained by Hosking [11].

We see from the infinite series representation of the right-side fractional-difference process that the statistics of X_j are the same as those of ξ_j, since (44) is a linear equation relating the two. Thus, since the statistics of ξ_j are Gaussian, so too are the statistics of the observed process, X_j. However, whereas the ξ-spectrum is flat, characteristic of white noise, the X-spectrum is an inverse power law (52), characteristic of fractal stochastic processes. From these analytical results we conclude that the process defined by the right-side fractional-difference stochastic equation is analogous to *fBm*. The analogy is complete if we set $\alpha = H - 1/2$ so that the spectrum (52) reads

$$S(\omega) \approx \frac{1}{\omega^{2H-1}} \ , as \quad \omega \to 0 \ . \tag{53}$$

In the language of random walks the inverse power law in (53) for $1 \geq H \geq 1/2$ or equivalently in (52) for $0 < \alpha \leq 1/2$, implies that the random walker has a tendency to continue in the direction she is going. This means there is a **persistence** to the process; given a step in a particular direction that step is remembered and the likelihood of the next step being in the same direction is greater than that of reversing directions. Analogously, for $1/2 \geq H > 0$ or equivalently $0 > \alpha \geq -1/2$, the spectrum increases as a power law in frequency. This increase implies that the random walker prefers to change her mind with each step. There is **anti-persistence** in the process; given a step in a particular direction that step is remembered and the likelihood of the next step being in the same direction is less than that of reversing directions. Hosking [11] recognized that fractional-difference processes exhibit long-term persistence and anti-persistence. This behavior has been observed in a number of physiological processes [3,4,5].

This type of time series was observed by Peng *et al.* [26] in their study of the successive increments of the cardiac beat-to-beat intervals of humans. They found that the time series for healthy individuals was anti-persistent, that being, the spectrum was an inverse power law with $H \approx 0$, giving rise to non-trivial long-time correlations in the interbeat interval increments. Their's was the first explicit identification of long-time anticorrelations in a fundamental biological variable. The inverse power-law spectrum for healthy individuals was compared with a group having a certain heart disease (dilated cardiomyopathy), whose interbeat increment spectrum was found to be flat, that is, $H \approx 1/2$, in the low-frequency region of the spectrum. The long-time correlations, so evident in the data for healthy individuals, is lost in the case of this disease, where there is no correlations in the interbeat intervals for the diseased individuals. The use of the spectral slope of physiological phenomena having fractal properties, as an indicator for disease, was first hypothesized by West and Goldberger [27,28].

4.2 The Correlation Function

The kind of memory manifest by the inverse power-law spectrum is quite different from the slower decay of exponential memory in autoregressive processes. To understand this difference in the memory let us examine the behavior of the autocorrelation coefficient directly. To evaluate the autocorrelation coefficient we first need to calculate the covariance coefficient of X_j

$$Cov_k = \langle X_j X_{j+k} \rangle \tag{54}$$

where we have used the fact that the mean is zero. Now comparing (54) with the expression for the spectrum (51) we can write the covariance coefficient as

$$Cov_k = \frac{1}{2\pi} \int_0^{2\pi} d\omega S(\omega) \cos[k\omega] \tag{55}$$

which is just the inverse Fourier transform of the spectrum. Substituting the expression for the spectrum (51) into the integral for the covariance (55) and using $\phi = \omega/2$ yields

$$Cov_k = \frac{1}{2^{\alpha-1}\pi} \int_0^{\pi/2} d\phi \sin^{-2\alpha}\phi \cos[2k\phi]. \tag{56}$$

From Gradshteyn and Ryzhik [29] [pg. 372] we obtain the integral

$$\begin{aligned} \int_0^{\pi/2} d\phi \sin^{-2\alpha}\phi \cos[2k\phi] &= \frac{2^\alpha \pi \cos[k\pi]}{(1-\alpha) B(1+k-\alpha, 1-k-\alpha)} \\ &= \frac{2^\alpha \pi (-1)^k \Gamma(1-\alpha)}{\Gamma(1+k-\alpha)\Gamma(1-k-\alpha)} \end{aligned}$$

which when substituted into (56) yields

$$Cov_k = \frac{(-1)^k \Gamma(1-\alpha)}{\Gamma(1+k-\alpha)\Gamma(1-k-\alpha)} \tag{57}$$

so that the correlation coefficient becomes

$$r_k = \frac{Cov_k}{VarX} = \frac{\Gamma(1-\alpha)\Gamma(k+\alpha)}{\Gamma(1+k-\alpha)\Gamma(\alpha)}. \tag{58}$$

Here again we use Stirling's approximation for the gamma functions to write the asymptotic relation for the correlation coefficient

$$r_k \approx \frac{\Gamma(1-\alpha)}{\Gamma(\alpha)}|k|^{2\alpha-1} = \frac{\Gamma(1-\alpha)}{\Gamma(\alpha)}|k|^{2H-2} \quad as \quad |k| \to \infty \tag{59}$$

in terms of the Hurst index H. The hyperbolic, inverse power-law form of the correlation coefficient (59) implies that the fractional-difference process is self-similar as defined by Mandelbrot [15].

Here we have been concerned with self-similar processes in which the second moment is finite. This can be seen from (59), where we have the asymptotic result $r_k = 0$ in the limit $k \to \infty$ for $0 < H \leq 1$. A more direct method of determining if the second moment of the process is finite is by direct calculation. The second moment of the time series can be expressed in terms of the integral over the spectrum

$$\langle X_j^2 \rangle = \frac{1}{2\pi}\int_0^{2\pi} d\omega S(\omega)$$

or alternatively using the covariance coefficient

$$\langle X_j^2 \rangle = Cov_{k=0} = \frac{\Gamma(1-2\alpha)}{\Gamma(1-\alpha)^2} \tag{60}$$

independently of the step number j. Thus, the right-side fractional-difference process driven by white noise is stationary, and a similar argument can be constructed for the left-side fractional-difference process driven by white noise.

5 Fractional Diffusion Equations

In this section we establish the connection between the Riemann-Liouville (RL) fractional derivative of the probability density in space with the fractional diffusion equation whose solution is a Lévy stable process. To do this we sketch the relationship between the RL fractional derivative and the limit of the fractional-difference operators. These arguments are distilled from a number of rigorous mathematical discussions in ref.[[25]]. We define the function $f(x)$ as the right-side RL fractional integral

$$f(x) =_{-\infty} \mathcal{I}_{-}^{\alpha} \varphi(x) = \frac{1}{\Gamma(\alpha)} \int_{x}^{\infty} \frac{\varphi(\zeta)\, d\zeta}{(\zeta - x)^{1-\alpha}} \tag{61}$$

and the inverse operation

$$\varphi(x) =_{-\infty} \mathcal{D}_{-}^{\alpha} f(x) = -\frac{1}{\Gamma(1-\alpha)} \frac{d}{dx} \int_{x}^{\infty} \frac{\varphi(\zeta)\, d\zeta}{(\zeta - x)^{\alpha}} \tag{62}$$

for the right-side RL fractional derivative, see Samko *et al.* [25] pages 33-35..

The right-side fractional difference of a suitably behaved function $f(x)$ can now be constructed using the definitions (31) and (61) as follows

$$\Delta_{-}^{\alpha} f(x) = \sum_{j=0}^{\infty} \left(\begin{array}{c} \alpha \\ j \end{array} \right) (-1)^{j} \int_{-\infty}^{\infty} K_{\alpha}(x - \zeta - j\delta)\, \varphi(\zeta)\, d\zeta \tag{63}$$

where δ is the spatial step size and the kernel is given by

$$K_{\alpha}(x) = \left\{ \begin{array}{ll} \frac{x^{\alpha-1}}{\Gamma(\alpha)} & for \quad x \geq 0 \\ 0 & for \quad x < 0 \end{array} \right\} \tag{64}$$

and the lower limit in (63) is accounted for in (64).Here again it is convenient to use the Fourier series, but this time in space rather than time, to obtain

$$\mathcal{F}\left\{\Delta_{-}^{\alpha} f(x)\right\} = \sum_{j=0}^{\infty} \left(\begin{array}{c} \alpha \\ j \end{array} \right) (-1)^{j} \hat{K}_{\alpha}^{(j)}(k)\, \hat{\varphi}(k). \tag{65}$$

The Fourier coefficient of the kernel, $\hat{K}_{\alpha}^{(j)}$, is dependent on the step index j, and we have again used the fact that the convolution of two functions is the product of their Fourier coefficients. Let us write the Fourier coefficient of the kernel explicitly

$$\hat{K}_{\alpha}^{(j)} = \int_{-\infty}^{\infty} K_{\alpha}(x - \zeta - j\delta)\, e^{-ik(x-\zeta)} d\zeta \tag{66}$$

so that introducing (64) and the scaled variable $y = -ik(x - \zeta - j\delta)$ into (66) and integrating, we obtain

$$\hat{K}_{\alpha}^{(j)} = \frac{e^{-ikj\delta}}{(ik)^{\alpha}}. \tag{67}$$

Substituting (67) into (65) and summing the series gives us

$$\mathcal{F}\left\{\Delta_{-}^{\alpha} f(x)\right\} = \left(\frac{1-e^{-ik\delta}}{ik}\right)^{\alpha} \hat{\varphi}(k) . \tag{68}$$

Now define the spatial right-side fractional derivative in terms of the limit as the spatial increment, δ, vanishes so that

$$\mathcal{D}_{-}^{\alpha} f(x) = \lim_{\delta \to 0} \frac{\Delta_{-}^{\alpha}}{\delta^{\alpha}} f(x) . \tag{69}$$

Thus, inserting (69) into (68) and interchanging the limit operation with taking the Fourier transform we can write

$$\mathcal{F}\left\{\lim_{\delta \to 0} \frac{\Delta_{-}^{\alpha}}{\delta^{\alpha}} f(x)\right\} = \lim_{\delta \to 0} \left(\frac{1-e^{-ik\delta}}{ik\delta}\right)^{\alpha} \hat{\varphi}(k) = \hat{\varphi}(k) . \tag{70}$$

The inverse Fourier transform of (70) yields the relation between the right-side RL fractional derivative and the limit of the right-side fractional-difference

$$\lim_{\delta \to 0} \frac{\Delta_{-}^{\alpha}}{\delta^{\alpha}} f(x) = \varphi(x) = {}_{-\infty}\mathcal{D}_{-}^{\alpha} f(x) \tag{71}$$

or equivalently from the definition of the function $\varphi(x)$ in (62)

$$\mathcal{D}_{-}^{\alpha} f(x) \equiv {}_{-\infty}\mathcal{D}_{-}^{\alpha} f(x) . \tag{72}$$

Thus, the right-side RL fractional derivative is equivalent to the limit of the right-side fractional-difference divided by the appropriate fractional power of the vanishing spatial increment and we have for the right-side RL fractional derivative

$$\mathcal{D}_{-}^{\alpha} f(x) = -\frac{1}{\Gamma(1-\alpha)} \frac{d}{dx} \int_{x}^{\infty} \frac{\varphi(\zeta)\, d\zeta}{(\zeta - x)^{\alpha}} . \tag{73}$$

Following a similar argument for the left-side RL fractional derivative we obtain

$$\mathcal{D}_{+}^{\alpha} f(x) = \frac{1}{\Gamma(1-\alpha)} \frac{d}{dx} \int_{-\infty}^{x} \frac{\varphi(\zeta)\, d\zeta}{(x - \zeta)^{\alpha}} . \tag{74}$$

So now we have independent definitions of the right-side and left-side RL fractional derivatives as the continuous limit of right-side and left-side fractional-differences, respectively, and the integral equations.

5.1 Fractional Fokker-Planck Equations

Let us now consider the evolution of a probability density in a one-dimensional space, where $P(x,t \mid x_0)\,dx$ is the probability of making a transition from the initial position x_0 to the interval $(x, x+dx)$ at time t. What we want is the evolution equation for the probability density when the process is fractal. Let us introduce the function $R(x)$ and generalize the result given by Wang and Uhlenbeck [30] using the expression involving the continuum limit of the time derivative

$$\begin{aligned} I &= \int dx R(x) \frac{\partial}{\partial t} P(x,t \mid x_0) \\ &= \lim_{\Delta t \to 0} \int dx \frac{R(x)}{\Delta t} \{P(x,t+\Delta t \mid x_0) - P(x,t \mid x_0)\}. \end{aligned} \tag{75}$$

The probability density is assumed to satisfy the discrete form of the BCSK chain condition, see Montroll and West [8] for a complete discussion, so that we can rewrite (75) as

$$I = \lim_{\Delta t \to 0} \int dx \frac{R(x)}{\Delta t} \left\{ \int dy P(x,t+\Delta t \mid y,t)\, P(y,t \mid x_0) - P(x,t \mid x_0) \right\}. \tag{76}$$

In the standard procedure one would Taylor expand the function $R(x)$ about the point $x = y$, however, now we use the generalized Taylor expansion of Osler [31] in terms of fractional derivatives. In this way we have

$$R(x) = R(y) + (x-y)^{\alpha} \left[\mathfrak{D}^{\alpha} R(x)\right]_{x=y} + \frac{(x-y)^{2\alpha}}{2!} \left[\mathfrak{D}^{2\alpha} R(x)\right]_{x=y} + \cdots \tag{77}$$

where we must define the fractional derivatives used in (77). Here we define discrete-difference operators to be

$$\Delta^{\alpha} \equiv \frac{\left(\Delta_{+}^{\alpha} - \Delta_{-}^{\alpha}\right)}{2\sin\left[\alpha\pi/2\right]} \tag{78}$$

in analogy with the symmetric Grünwald-Letnikov fractional derivatives [32,33], so that the α-derivative operators in the expansion (77) are

$$\mathfrak{D}^{\alpha} = \lim_{\delta \to 0} \frac{\Delta^{\alpha}}{\delta^{\alpha}} = \lim_{\delta \to 0} \frac{\left(\Delta_{+}^{\alpha} - \Delta_{-}^{\alpha}\right)}{2\delta^{\alpha}\sin\left[\alpha\pi/2\right]} = \mathfrak{D}_{+}^{\alpha} + \mathfrak{D}_{-}^{\alpha} \tag{79}$$

and the 2α-derivative operators in the expansion are

$$\mathfrak{D}^{2\alpha} = \lim_{\delta\to 0} \frac{\Delta^{2\alpha}}{\delta^{2\alpha}} = \lim_{\delta\to 0} \frac{\left(\Delta_+^\alpha - \Delta_-^\alpha\right)^2}{4\delta^{2\alpha}\sin^2\left[\alpha\pi/2\right]} = \left(\mathfrak{D}_+^\alpha + \mathfrak{D}_-^\alpha\right)^2. \tag{80}$$

Note that (78) has a form analogous to (23) for the symmetric time-derivative operator.

It is convenient to define the drift

$$A_\alpha(y) = \lim_{\Delta t\to 0} \int dx \frac{(x-y)^\alpha}{\Delta t} P(x-y, \Delta t) \tag{81}$$

and diffusion function

$$B_{2\alpha}(y) = \lim_{\Delta t\to 0} \int dx \frac{(x-y)^{2\alpha}}{2\Delta t} P(x-y, \Delta t), \tag{82}$$

where we have assumed that the probability is translationally invariant, that is, $P(x, t_1 \mid y, t_2) = P(x-y, t_1 - t_2)$. Substituting the generalized Taylor expansion (77) into (76) and using the averages defined by (81) and (82) we obtain

$$\int dx \left\{ \left[A_\alpha(x)\,\mathfrak{D}^\alpha R(x) + B_{2\alpha}(x)\,\mathfrak{D}^{2\alpha} R(x)\right] - R(x)\frac{\partial}{\partial t} \right\} P(x, t \mid x_0) = 0. \tag{83}$$

It is straightforward to show that the RL fractional derivatives are adjoints of one another in the sense

$$\int dx g(x)\,\mathfrak{D}^\alpha f(x) = \int dx f(x)\,\mathfrak{D}^\alpha g(x)$$

so that we have that the fractional derivatives in (83) are self-adjoint. Therefore integrating (83) by parts we obtain

$$\int dx R(x) \left\{ \left[\mathfrak{D}^\alpha A_\alpha(x) + \mathfrak{D}^{2\alpha} B_{2\alpha}(x) - \frac{\partial}{\partial t}\right] P(x, t \mid x_0) \right\} = 0 \tag{84}$$

and since the function $R(x)$ is essentially arbitrary the only way for equation (84) to be satisfied is for the terms in the curly brackets to vanish. Thus, we obtain one version of the fractional Fokker-Planck equation

$$\frac{\partial}{\partial t} P(x, t \mid x_0) = \mathfrak{D}^\alpha\left[A_\alpha(x) P(x, t \mid x_0)\right] + \mathfrak{D}^{2\alpha}\left[B_{2\alpha}(x) P(x, t \mid x_0)\right]. \tag{85}$$

We can generalize the above argument by introducing the fractional time derivative of the probability density

$$\frac{\partial^\gamma}{\partial t^\gamma} P(x,t \mid x_0) = \lim_{\Delta t \to 0} \frac{1}{\Delta t^\gamma} \{P(x,t+\Delta t \mid x_0) - P(x,t \mid x_0)\}$$

into (76). In this way the drift coefficient generalizes to

$$A_{\alpha,\gamma}(y) = \lim_{\Delta t \to 0} \int dx \frac{(x-y)^\alpha}{\Delta t^\gamma} P(x-y, \Delta t) \tag{86}$$

and the diffusion coefficient is replaced with

$$B_{2\alpha,\gamma}(y) = \lim_{\Delta t \to 0} \int dx \frac{(x-y)^{2\alpha}}{2\Delta t^\gamma} P(x-y, \Delta t) \tag{87}$$

from which (85) can be rewritten as

$$\frac{\partial^\gamma}{\partial t^\gamma} P(x,t \mid x_0) = \mathfrak{D}^\alpha [A_\alpha(x) P(x,t \mid x_0)] + \mathfrak{D}^{2\alpha} [B_{2\alpha}(x) P(x,t \mid x_0)] \tag{88}$$

which in form resembles the equation derived by Zaslavsky *et al.* [18]. However, the fractional spatial derivatives in (88) are not the same as those obtained in this earlier work, in fact these authors use only the right-side RL fractional derivatives in their equation of evolution for the probability density. It seems to the present authors that the symmetric derivatives defined by (79) and (80) are more physically reasonable.

5.2 Fractional Diffusion and Lévy Statistics

To determine the statistical properties of the solutions to the fractional diffusion equation (88) let us construct an explicit form for the fractional derivatives (79) and (80). Again using the properties of Fourier series

$$\mathcal{F}\{\mathfrak{D}^\alpha f(x)\} = \frac{(ik)^\alpha - (-ik)^\alpha}{2\sin[\alpha\pi/2]} \hat{f}(k) = i|k|^\alpha sgn(k) \hat{f}(k) \tag{89}$$

where $(ik)^\alpha = |k|^\alpha \exp[i\pi\alpha sgn(k)/2]$ and similarly

$$\mathcal{F}\{\mathfrak{D}^{2\alpha} f(x)\} = \frac{[(ik)^\alpha - (-ik)^\alpha]^2}{4\sin^2[\alpha\pi/2]} \hat{f}(k) = -|k|^{2\alpha} \hat{f}(k). \tag{90}$$

Thus, defining $f = A_{\alpha,\gamma}P$ for the α-derivative and $f = B_{2\alpha,\gamma}P$ for the 2α-derivative, we can, noting the inverse Fourier transforms

$$\mathcal{F}^{-1}\left\{i\left|k\right|^{\alpha} sgn\left(k\right)\right\} = -\cos\left[\alpha\pi/2\right] sgn\left(x\right)\frac{\Gamma\left(1+\alpha\right)}{\left|x\right|^{1+\alpha}} \tag{91}$$

and

$$\mathcal{F}^{-1}\left\{\left|k\right|^{2\alpha}\right\} = -\sin\left[\alpha\pi\right]\frac{\Gamma\left(1+2\alpha\right)}{\left|x\right|^{1+2\alpha}}, \tag{92}$$

suppressing the initial conditions, rewrite the fractional diffusion equation (88) as

$$\begin{aligned}\frac{\partial^{\gamma}}{\partial t^{\gamma}}P\left(x,t\right) &= \cos\left[\alpha\pi/2\right]\Gamma\left(1+\alpha\right)\int_{-\infty}^{\infty}\frac{sgn\left(x-x'\right)}{\left|x-x'\right|^{1+\alpha}}A_{\alpha,\gamma}\left(x'\right)P(x',t)dx' \\ &\quad +2\sin\left[\alpha\pi\right]\Gamma\left(1+2\alpha\right)\int_{-\infty}^{\infty}\frac{B_{2\alpha,\gamma}\left(x'\right)}{\left|x-x'\right|^{1+2\alpha}}P\left(x',t\right)dx' \end{aligned} \tag{93}$$

which to our knowledge is a fractional diffusion equation that has not before appeared in the literature.

Consider a $\gamma = 1$ process without drift, with constant fluctuation strength

$$A_{\alpha,1}\left(x'\right) = 0 \quad \text{and} \quad B_{2\alpha,1}\left(x'\right) = b \tag{94}$$

and with the initial condition $x_0 = 0$ so that the fractional diffusion equation (93) reduces to

$$\frac{\partial}{\partial t}P\left(x,t\right) = 2b\sin\left[\mu\pi/2\right]\Gamma\left(1+\mu\right)\int_{-\infty}^{\infty}\frac{P\left(x',t\right)}{\left|x-x'\right|^{1+\mu}}dx' \tag{95}$$

where $\mu = 2\alpha$. Equation (95) is the integro-differential equation obtained by West and Seshadri [16] for the evolution of Lévy stable processes. Note further that this form of the equation of evolution for the probability density could not be obtained from the fractional diffusion equation constructed by Zaslavsky *et al.* [18] because of the lack of symmetry.

We denote the characteristic function for the above probability density by the Fourier transform

$$\hat{\phi}\left(k,t\right) = \mathcal{F}\left\{P\left(x,t\right)\right\}. \tag{96}$$

Using (96) in the fractional diffusion equation (95) we can write the equation of evolution for the characteristic function

$$\frac{\partial}{\partial t}\overset{\wedge}{\phi}(k,t) = -b\,|k|^{\mu}\,\overset{\wedge}{\phi}(k,t)\,. \tag{97}$$

The solution to (97) with the initial condition $\overset{\wedge}{\phi}(k,t=0)=1$ is

$$\overset{\wedge}{\phi}(k,t) = \exp\left[-bt\,|k|^{\mu}\right] \tag{98}$$

whose Fourier transform is the centro-symmetric α-stable Lévy distribution for $0<\mu\leq 2$ and $b>0$,

$$P(x,t) = \frac{1}{2\pi}\int_{-\infty}^{\infty} dk\exp\left[ikx - bt\,|k|^{\mu}\right]. \tag{99}$$

The Lévy probability density does not have a simple analytic form except for a few selected values of the Lévy index μ, among these are the Gaussian for $\mu=2$ and the Cauchy for $\mu=1$, see [8,34,35] for more discussion. Another important feature of the Lévy stable distribution is that, except for $\mu=2$, it does not possess finite moments at all orders. This can be seen from the fact that, for $t>0$, the distribution has an inverse power-law asymptotic expansion [8],

$$P(x,t) = \frac{\mu}{\pi}\sum_{l=0}^{\infty}\frac{(-1)^{l}}{l!}\frac{\Gamma\left([l+1]\,\mu\right)\sin\left[(l+1)\,\mu\pi/2\right]}{|x|^{(l+1)\mu+1}}\,. \tag{100}$$

Thus, all θ-moments defined by

$$\left\langle |x|^{\theta};t\right\rangle = \int_{-\infty}^{\infty} dx'\,|x'|^{\theta}\,P(x',t) \tag{101}$$

are finite for $\theta<\mu$ and infinite for $\theta\geq\mu$. In particular the variance is infinite for $\mu<2$. Thus, a Lévy process describes anomalous diffusive phenomena in the sense that Fick's law is violated for such processes. Using the scaling properties of the Lévy distribution we can express (101) as

$$\begin{aligned}\left\langle |x|^{\theta};t\right\rangle &= t^{\theta/\mu}\int_{-\infty}^{\infty}\frac{dx'}{t^{1/\mu}}\left|\frac{x'}{t^{1/\mu}}\right|^{\theta} P\left(x'/t^{1/\mu}\right) \\ &= t^{\theta/\mu}\left\langle |x|^{\theta};1\right\rangle.\end{aligned} \tag{102}$$

Thus, we see that self-similarity is present for those moments that are finite. In the Gaussian case, where $\mu=2$, the second moment of the distribution increases linearly in time as it should for a normal diffusion process.

Of course, like in all diffusion equations based on simple random walk models we have implicitly assumed that the random walkers move with infinite speed. This is manifest in the present case through the infinite central moments of the Lévy processes. Shlesinger, West and Klafter [21] took the influence of finite velocities into account by shifting from a random walk to a random flight perspective, that is, by including the finite time it takes to execute a step in the process. In this way the central moments of a Lévy process are finite for finite times, because the Lévy process is truncated. This truncation was also achieved in another way, if we again choose (94) for the values of the parameters, but this time with $1 \leq \mu \leq 2$ in (93), we have the fractional diffusion equation derived by Allegrini *et al.* [7] for a dichotomous random walk process with fluctuations that have an inverse power-law correlation function, that is, the correlation function for the velocity of the steps for long times becomes $\Phi(t) \sim t^{1-\mu}$. The general solution to this equation consists of a central region that is a truncated Lévy distribution and two oppositely propagating fronts whose amplitudes are given by the velocity correlation function, that is, for unit velocity of each step of the dichotomous process the Green's function for this process is

$$G(x,t) = \left(\frac{1}{2\pi}\int_{-\infty}^{\infty} dk e^{ikx} e^{-bt|k|^{\mu}}\right) \Theta(t - |x|) + \frac{1}{2}\Phi(t)\left[\delta(t-x) + \delta(t+x)\right]. \tag{103}$$

This solution is similar in form to that obtained for the case of normal diffusion where the correlation function is an exponential rather than an inverse power law. In the case of the exponential correlation function the diffusion equation can be written as the telegrapher's equation and the integral in (103) is expressed in terms of complex Bessel functions that reduce to a Gaussian at long times. Note that the form of the propagating fronts in the solution to the telegrapher's equation remains the same as (103) with the coefficient of the delta functions being given by an exponential rather than the inverse power law, since that is the correlation function in the former case [23].

6 Some Applications of These Ideas

One application of the fractional difference approach is to faithfully model both the long and short time properties of a time series. The fractional-difference random walk process, like *fBm*, has an asymptotic power-law correlation coefficient. However, this process can be generalized to include appropriate short-time behavior by considering the superposition of two processes

$$Z_k = \sum_{j=0}^{k} (X_j + \xi_j) \tag{104}$$

where X_j is the solution to a fractional-difference random walk and ξ_j is a discrete white noise process. Assuming these two processes are statistically independent and zero centered, the second moment of the combined process can be written

$$\langle Z_k^2 \rangle = \left\langle \left(\sum_{j=0}^{k} X_j \right)^2 \right\rangle + \left\langle \left(\sum_{j=0}^{k} \xi_j \right)^2 \right\rangle \tag{105}$$

and the brackets denote an average over an ensemble of realizations of both types of fluctuations. The first average on the right hand side of (105) is evaluated using the correlation coefficient (59) to obtain

$$\left\langle \left(\sum_{j=0}^{k} X_j \right)^2 \right\rangle = \sum_{j=0}^{k} \sum_{j'=0}^{k} r_{j-j'} \approx A k^{2H} \tag{106}$$

where the terms in the series are dominated by the longest time, k, and A is a known constant. The second term is diffusive and of known strength so we can write the asymptotic expression

$$\langle Z_k^2 \rangle \approx A k^{2H} + Bk, \tag{107}$$

where here B is also a known constant.

The superposition process in (107) can be used to explain the statistics and correlation properties of DNA sequences using newly developed model based on random walk ideas. The DNA walk [36,37] yields the result that protein-coding DNA sequences (cDNA, exons) are remarkably like Brownian motion, having Gaussian statistics and zero correlation length. On the other hand, non-coding sequences (introns) have long-range correlations. The variance of a DNA sequence of length k is given in earlier models by k^{2H} where $H \approx 0.5$ for exons and $H > 0.5$ for introns. Voss [38], using a spectral theory finds $H > 0.5$ for both types of DNA sequences, suggesting that all DNA sequences are described by anomalous diffusive processes. Buldyrev *et al.* [39] argue that *fBm* can account for the second moment properties of DNA sequences. Thus, a joint process of the form (104) with second moment (107) can be explained in terms of the fractional difference equations presented herein.

A problem still remains, however, and that is the proper description of the statistics of DNA sequences and not just their second moments. Allegrini *et al.* [40] proposed a model based on the superposition of a dynamical (intermittent chaotic) process and a stochastic (random walk) process that leads to a form of the variance shown in (107) and resolves the controversy. They refer to their model as the copying mistake map (CMM) model. To mimic the statistical properties of cDNA the parameter values fit to the data in the CMM model requires that $B >> A$ in (107), explaining why the detection of the long-range correlation is so difficult for these kinds of sequences, and why the short-time dynamics is essentially dominated by the properties of standard diffusion. In the CMM model the diffusion stemming from the deterministic map without copying mistakes is non-Gaussian, it is a truncated Lévy process, that is, a Lévy process with propagating fronts such as those introduced in ref.[[20]], and which we described using the fractional diffusion equation of the last section. Thus, the fractional diffusion equations provide a description of DNA sequences analogous to the CMM model developed previously.

Araujio *et al.* [41] demonstrated that a generalized Lévy-walk gives the proper statistical description of DNA sequences. Allegrini *et al.* [22] proved that the generalized Lévy-walk of Araujio *et al.* and the CMM model of Allegrini *et al.* [40] are equivalent.

References

1. H.G. Stanley, Editor, ***Statphys 16***, North-Holland, Amsterdam (1986); H.G. Stanley and N. Ostrowsky, Editors, ***Random Fluctuations and Pattern Growth: Experiments and Models***, Kluwer Academic Publishers, Dordrecht, NATO Scientific Affairs Division (1988)
2. N.M. Novak and T.G. Dewey, Editors, ***Fractal Frontiers***, World Scientific, Singapore (1997); S. Harefall and M.E. Lee, Editors, ***Chaos, Complexity and Sociology***, SAGE, Thousand Oaks, California (1997); P.M. Iannaccone and M. Khokha, Editors, ***Fractal Geometry in Biological Systems***, CRC Press, Boca Raton (1996); T.F. Nonnenmacher, G.A. Losa, E.R. Weibel, Editors, ***Fractals in Biology and Medicine***, Birkhauser Verlag, Basel (1994).
3. B.J. West and W. Deering, "Fractal Physiology for Physicists: Lévy Statistics", Phys. Rept. **246** (1&2), 1-100 (1994); B.J. West and W. Deering, *The Lure of Modern Science: Fractal Thinking, Studies in Nonlinear Phenomena in Life Science Vol.* **3**, World Scientific, River Edge, New Jersey (1995); B.J. West, *Fractal Physiology and Chaos in Medicine, Studies in Nonlinear Phenomena in Life Science Vol.***1**, World Scientific,

River Edge, New Jersey (1990).
4. S.B. Lowen and M. C. Teich, "Estimation and Simulation of Fractal Stochastic Point Processes", Fractals **3**, 183-210 (1995); M.O. Vlad, B. Schonfirch and M.C. Mackey, "Self-similar potentials in random media, fractal evolutionary landscapes and Kmura's neutral theory of molecular evolution", Physica **229** A, 343-64 (1996).
5. J.B. Bassingthwaighte, L.S. Liebovitch, and B.J. West., *Fractal Physiology*, Oxford University Press, Oxford (1994).
6. E. Ott, *Chaos in Dynamical Systems*, Cambridge University Press, Cambridge (1993); N.B. Abraham, A.M. Albano, A. Passamante, P.E. Rapp and R. Gilmore, *Complexity and Chaos*, World Scientific, Singapore (1993).
7. P. Allegrini, P. Grigolini and B.J. West, "Dynamical approach to Lévy processes", Phys. Rev. E **54**, 4760-67 (1996); B.J. West, P. Grigolini, R. Metzler and T.F. Nonnenmacher, "Fractional diffusion and Lévy stable processes", Phys. Rev. E **55**, 99-106 (1997).
8. E.M. Montroll and B.J. West, "An Enriched Collection of Stochastic Processes", in *Fluctuation Phenomena*, Eds. E.W. Montroll and J. Lebowitz, North-Holland (1979); 2nd Edition, North- Holland Personal Library (1987).
9. B.B. Mandelbrot, *Fractals, Form and Chance*, W.H. Freeman, San Francisco (1977).
10. E.W. Montroll and M.F. Shlesinger, "On the wonderful world of random walks", in *Nonequilibrium Phenomena II: From Stochastics to Hydrodynamics*, edited by E.W. Montroll and J.L. Lebowitz, pp 1-121, North Holland, Amsterdam (1984).
11. J.T.M. Hosking, "Fractional Differencing", Biometrika **68**, 165-176 (1981).
12. C.W.J. Granger, "Long memory relationships and the aggregation of dynamic models", J. Econometrics **14**, 227 (1980).
13. J. Beran, *Statistics of Long-Memory Processes, Monographs on Statistics and Applied Probability* **61**, Chapman & Hall, New York (1994).
14. R.T. Baillie, "Long memory precesses and fractional integration in econometrics", J. Econometrics **73**, 5-59 (1996).
15. B.B. Mandelbrodt, "Une classe de processus stochastiques homothetiques a soi; application a la loi climatotogique de H.E. Hurst.", Comptes Rendus Acad. Sci. Paris **260**, 3274-77 (1965);B.B. Mandelbrodt and J.W. van Ness, "Fractional Brownian motions, fractional noises and applications.", SIAM Rev. **10**, 422-37 (1968).
16. B.J. West and V. Seshadri, "Linear systems with Lévy fluctuations",

Physica **113**A, 203-216 (1982).
17. A. Compte, "Stochastic foundations of fractional dynamics", Phys. Rev. E **53**, 4191-93 (1996).
18. G.M. Zaslavsky, "Anomalous transport and fractional kinetics", H.K. Moffatt et al., Editors, *Topological Aspects of Dynamics of Fluids and Plasmas*, ppsl 4581-91, Kluwer Academic Publishers, Netherlands (1992).
19. H.C. Fogedby, "Aspects of Lévy flights in a quenched random force field", in *Lévy Flights and Related Topics in Physics*, editors M.F. Shlesinger, G.M. Zaslavsky and U. Frisch, Springer, New York (1995).
20. J. Klafter, G. Zumofen, and M.F. Shlesinger, "Lévy description of anomalous diffusion in dynamical systems", in *Lévy Flights and Related Topics in Physics*, editors M.F. Shlesinger, G.M. Zaslavsky and U. Frisch, Springer, New York (1995).
21. M.F. Shlesinger, B.J. West and J. Klafter, "Lévy dynamics for enhanced diffusion: an application to turbulence", Phys. Rev. Lett. **58**, 1100-03 (1987).
22. P. Allegrini, M. Buacci, P. Grigolini and B.J. West, "Fractional Brownian motion as a nonstationary process: An alternative paradigm for DNA sequences", Phys. Rev. E **57**, 1-10 (1998).
23. P.M. Morse and H. Feshbach, *Methods of Theoretical Physics*, McGraw-Hill, New York (1953).
24. KS. Miller and B. Ross, *An Introduction to the Fractional Calculus and Fractional Differential Equations*, John Wiley and Sons, New York (1993).
25. S.G. Samko, A.A. Kilbas and O.I. Marichev, *Fractional Integrals and Derivatives*, Gordon and Breach Publishers, USA (1993).
26. C.-K. Peng, J. Mietus, J.M. Hausdorff, S. Havlin, H.E, Stanley and A.L. Goldberger, "Long-range anticorrelations and Non-Gaussian Behavior of the Heartbeat", Phys. Rev. Lett. **70**, 1343-46 (1993).
27. B.J. West, V. Bhargava and A.L. Goldberger, "Beyond the Principle of Similitude: Renormalization in the Bronchial Tree", J. Appl. Physiol. **60**, 1089-98, (1986).
28. A.L. Goldberger and B.J. West, "Fractals a Contemporary Mathematical Concepts with Applications to Physiology and Medicine", Yale J. Biol. and Med. **60**, 104-18 (1987).
29. I.S. Gradshteyn and I.M. Ryzhik, *Table of Integral, Series, and Products*, Academic Press, San Diego (1992).
30. M.C. Wang and G.E. Uhlenbeck, "On the theory of Brownian motion II", Rev. Modern Phys. **17**, 323-342 (1945).
31. T.J. Osler, "An intergral analogue of Taylor's series and its use in com-

puting Fourier Transforms", Math. Comp. **26**, 449-460 (1972).
32. A.K. Grunwald, "Uber "begrenzte" Derivationen und deren Anwendung", Z. angew. Math. und Phys. **12**, 441-80 (1867).
33. A.V. Letnikov, "Theory of differentiation with an arbitrary index", (Russian) Math. Sb. **3**, 1-66 (1868).
34. V.M. Zoloterev, *One-dimensional Stable Distributions, Translations of Mathematical Monographs* Volume **65**, American Mathematical Society, Providence, R.I. (1986).
35. G. Samorodnitsky and M.S. Taqqu, *Stable Non-Gaussian Random Processes*, Chapman & Hall, New York (1994).
36. C.K. Peng, S. Buldyrev, A.L. Goldberg, S. Havlin,.F. Sciortino, M. Simons and H. E. Stanley, "Long-range correlations in nucleotide sequences", Nature **356**, 168 (1992).
37. H.E. Stanley, S. Buldyrev, A.L. Goldberg, Z.D. Goldberger, S. Havlin, R.N. Mantegna, S.H. Ossadnik, C.K. Peng and M. Simons, "Statistical mechanics in biology: how ubiquitous are long-range correlations", Physica A **205**, 214 (1994).
38. R. Voss, "Evolution of long-range fractal correlations and 1/f-noise in DNA base sequences", Phys. Rev. Lett. **68**, 3805 (1992).
39. S.V. Buldyrev, A.L. Goldberger, S. Havlin, C.-K. Peng, M. Simons and H.E. Stanley, "Generalized Lévy-walk model for DNA necleotide sequences", Phys. Rev. E **47**, 4514 (1993).
40. P. Allegrini, M. Barbi, P. Grigolini and B.J. West, "Dynamical model for DNA sequences", Phys. Rev. E **52**, 5281-96 (1995).
41. M. Araujio, S. Havlin, G.H. Weiss and H.E. Stanley, "Diffusion of walkers with persistent velocities", Phys. Rev. **43**, 5240 (1991).
42. G.M. Zaslavsky, M. Edelman and B.A. Niyazov, "Self-similarity, renormalization, and phase space nonuniformity of Hamiltonian chaotic dynamics", Chaos **7**, 159 (1997).

CHAPTER V

FRACTIONAL KINETICS OF HAMILTONIAN CHAOTIC SYSTEMS

G.M. ZASLAVSKY
Courant Institute of Mathematical Sciences, New York University, 251 Mercer St., New York, NY 10012
and Department of Physics, New York University, 2-4 Washington Place, New York, NY 10003

Contents

1 Introduction

It was found fairly recently that numerous physical processes do not satisfy the principle of the short-time memory randomness and that long power-like tails in distributions of different time-scales can frequently occur in physical systems. The corresponding kinetic description for such cases involves new notions, such as Lévy processes, Lévy random walks, fractal Brownian motion, etc. [1]. Numerous applications of the new concept in kinetics were discussed in [2] where a notion of "strange kinetics" has been used. It seems that the strange kinetics can be applied to the wide class of phenomena related to chaotic dynamics. The corresponding "fractional" kinetic equation (FFPK) was derived in works [3,4,5] and applied to the Hamiltonian dynamical systems with few degrees of freedom and intrinsic chaotic dynamics [6]. For the case when the fractional kinetics can be applied, there is a fractal support in the phase space-time description of the system. The origin of the fractal support comes from a strong nonuniformity of the phase space where special zone(s) of particle trapping exist. These zones are called "traps" or "quasi-traps". The latter notion is related to the Hamiltonian systems where the absolute traps do not exist. The (quasi) traps can have fractal structure, and so can the set of time intervals that the trajectory spends in the traps [6,7,8]. This is how we lose the universality of the kinetic description of a system with chaotic dynamics, because it depends now on the local space-time properties of the support or, more generally speaking, on the phase space topology. The topology of dynamical systems does not persist when a critical parameter exhibits a finite change [6,9,10] and only a special condition of the new approximation can bring us again to the universal kinetics. A brief review of the problem of fractional kinetics is the subject of this chapter.

2 Mapping the Dynamics

Dynamical systems can be described by the properties of their trajectories in the phase space. In this article we will deal with Hamiltonian systems although some of the statements can have more general applications. Despite an arbitrary smoothness of the trajectories, they display different kinds of fractal features which are crucial for the large scale asymptotic properties of the system. In this section, we show how the fractal characteristics arise from smooth dynamics.

Typically, a trajectory in the phase space is $(\mathbf{r}(t), \mathbf{p}(t))$ with position $\mathbf{r}$ and momentum $\mathbf{p}$ at the time instant t. These functions can not be presented in an explicit way for general situations because the dynamics are chaotic and the trajectories carry properties of randomness. An alternative way to describe the dynamics of a system is to map a trajectory according to some rules and to reduce desirable information about the system behavior. Here are some examples of such mappings.

a. Poincaré Map. This is the most typical map due to its convenience and effectiveness. The Poincaré map can be defined as a set of points in phase space $\{p_j, q_j\}$ obtained from trajectory intersections with a hypersurface in the phase space at time instants $\{t_j\}$. For 1 1/2 or 2 degrees of freedom and finite motion, the closure of the set $\{p_j, q_j\}$ is an invariant curve if the motion is quasi-periodic, and is randomly distributed points (stochastic sea) if the motion is chaotic. For a generic situation, the Poincaré map forms a stochastic sea with implanted islands of invariant KAM curves. Sometimes, the stochastic sea can be reduced to very thin channels of chaotic dynamics, called stochastic layers and webs.

b. Poincaré Recurrences (or simply recurrences). Consider a domain A and a point $(p_0, q_0) \in A$. A trajectory with initial coordinates (p_0, q_0) will definitely escape from A after a finite time and then come back again in a finite time. Such escapes and returns will repeat an infinite number of times. Let $\{t_j^-\}$ be a set of the escape from A instants and $\{t_j^+\}$ be a set of the entrance to A instants. The set

$$\{\tau_j^{(\mathbf{rec})}\}_A = \{t_j^- - t_{j-1}^-\}_A \tag{2.1}$$

is the set of recurrences. It can be fractal or multifractal [7,8,11,12]. The notion of fractal time was discussed in [1,13]. Numerous properties of systems can be characterized by the probability distribution function $P_{\mathbf{rec}}(\tau; A)$.

c. Exit Times. Similarly to (1.1), one can introduce a set of exit (escape) times $\{\tau_j^{(\mathbf{esc})}\}$:

$$\{\tau_j^{(\mathbf{esc})}\}_A = \{t_j^- - t_{j-1}^+\}_A \tag{2.2}$$

and the corresponding probability distribution function $P_{\mathbf{esc}}(\tau; A)$. Properties of distributions of recurrences and escapes are strongly correlated [14,15].

d. Zeno Map. The Zeno map is a set of positions $\{x_j\}$ and corresponding instants attributed to the Zeno paradox about Achilles who can never overtake a turtle. The example demonstrates that not every map generates sufficient information to solve a problem.

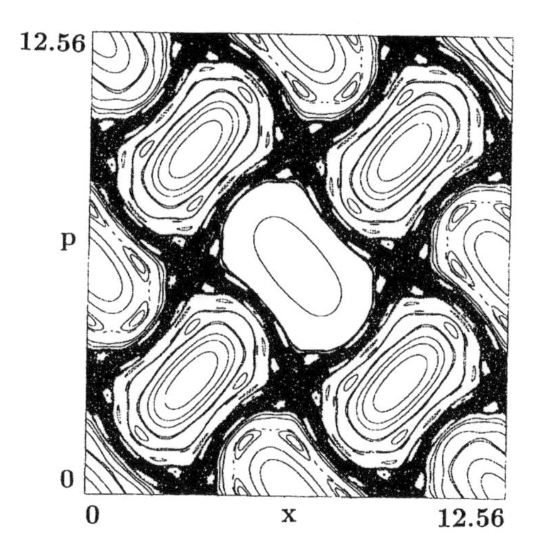

Figure 1: Example of the phase space for the web map

One should consider different ways of mapping dynamical systems while keeping in mind the final goal, i.e. obtaining necessary probabilistic properties about the system's evolution. In this article we focus on kinetics and transport, i.e. time evolution of moments. We can expect that, after choosing a specified mapping method, a fractal or multifractal support arises for the system's trajectories in the phase space. Important questions are: How intrinsic are fractal properties for the selected way of mapping? Why do we need this complication and is it unavoidable? In the next section we demonstrate that natural chaotic trajectories of real dynamical systems possess a space-time (multi)fractal structure for a typical situation. Here "real" means typical physical systems rather than specially prepared models; "space" means the phase space; and "fractal" situation means an actually multifractal one with a very narrow spectral function of dimensions.

3 Topological Zoo (Singular Zones)

In this section we want to provide a motivation for introducing a fractal support in space-time for chaotic trajectories. As it was mentioned above, a typical structure of the phase space of a Hamiltonian system with finite chaotic motion consists of the so-called stochastic sea and islands imbedded into the sea. In

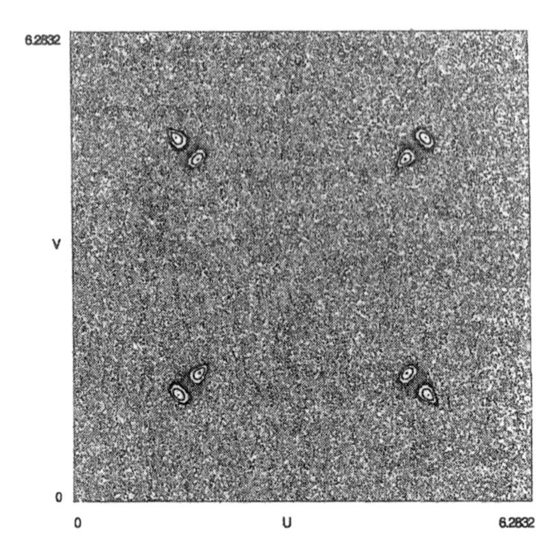

Figure 2: Hierarchy of islands for the web map: a) general phase portrait

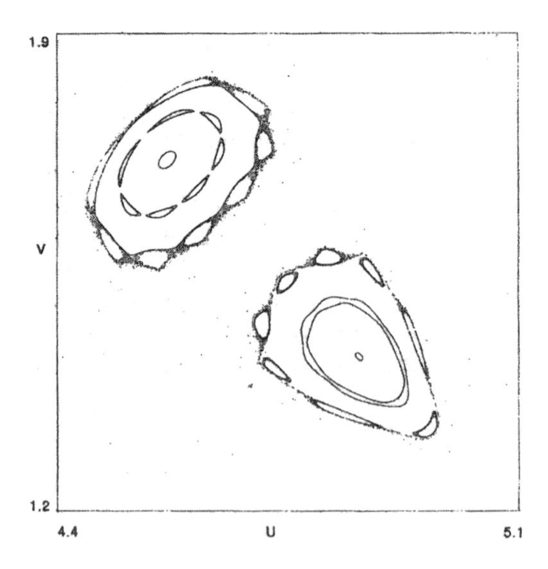

Figure 3: Hierarchy of islands for the web map: b) magnification of islands

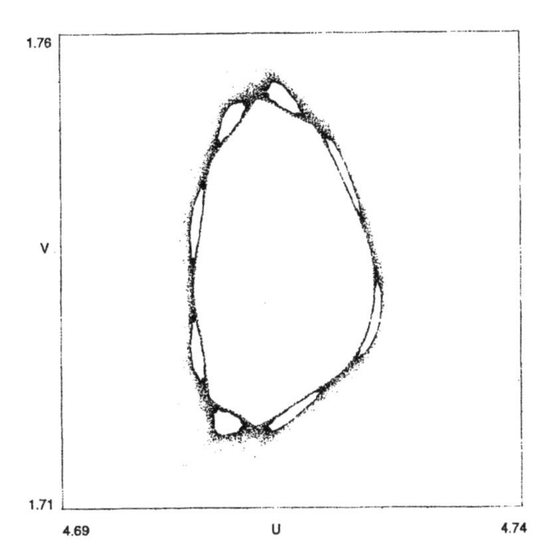

Figure 4: Hierarchy of islands for the web map: c) next magnification of an island from 2.b)

fact, there are also other structures, such as cantori or stable/unstable manifolds. As a result, we see what is presented in Figs. 1,2 [6] where we plot the mapping of trajectories for the web-map

$$u_{n+1} = v_n, \quad v_{n+1} = -u_n - K \sin v_n, \quad (u_n, v_n = \text{mod } 2\pi) \tag{3.1}$$

for two different values of the parameter K. Figure 1 simply demonstrates how complicated the phase portrait of the system can be. In contrast to Fig. 1, Fig. 2 shows up an elusive simplification. The dark area near the islands in Fig. 2 corresponds to the stickiness [16], the theory of which does not yet exist. This zone can be considered a singular one, in which a particle is trapped for a long time. The "trap" has a fine structure (Fig. 3) which displays a new set of islands with again sticky boundaries. This hierarchy of islands continues to infinity, creating a fractal or multifractal structure of the trap (see Fig. 4). The remaining part of the stochastic sea outside the islands appears uniform, although it is only a problem of resolution to see more structures.

The second example in Fig. 5–7 is taken for the standard map

$$p_{n+1} = p_n - K \sin x_n, \quad x_{n+1} = x_n + p_{n+1}, \quad (x, p, \text{ mod } 2\pi) \tag{3.2}$$

where the trapping zone looks as a set of strips (Fig. 5,6). The strips disappear if we change the value of K so that it becomes close to the bifurcation point

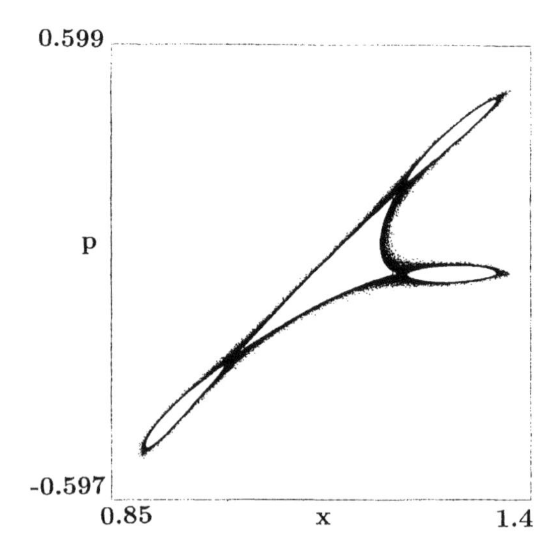

Figure 5: Stickiness near a bifurcation

(Fig. 7). It is a fairly small change of the parameter K which makes such a big difference in the topology.

The third example in Fig. 8 is taken for the perturbed pendulum model

$$\ddot{x} + \sin x = -\epsilon \sin(x - \nu t) \tag{3.3}$$

where one can see the stochastic layer with a strip-shaped trapping zone along the layer. The strip, in its turn, is composed of multiple strips. There is a fast propagation of particles along the strips, which we call flights.

Despite the given three examples corresponding to three different models, all of them can occur in any model depending on parameter values. Dark domains in Figs. 1–8 correspond to a nonuniformity of the phase space and, consequently, to a nonuniformity of the level of randomness of different parts of trajectories. This situation leads to the absence of uniformity of the kinetics for different time scales and to different strategies in the statistical description of chaotic dynamics. We expect a loss of universality of kinetics for chaotic dynamical systems, in general, and a "restricted universality" for systems within the same topological class of dynamics.

For the conclusion of this section we discuss, in a qualitative way, how one can distinguish different topological classes of dynamics. Depending on the time of observation of the system (say 10^{11}), we can neglect some sin-

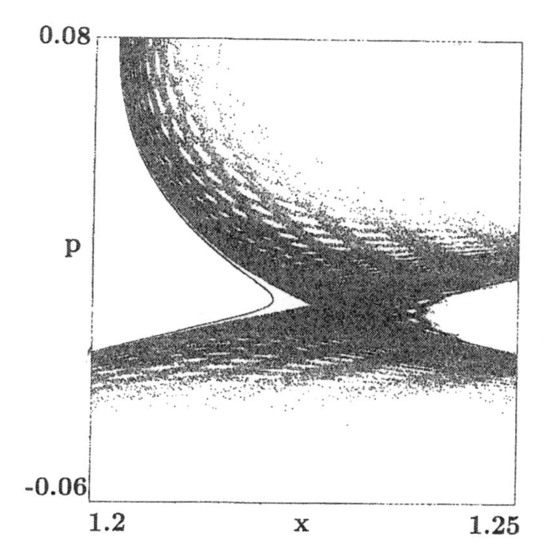

Figure 6: Magnification of a part from Fig. 3a

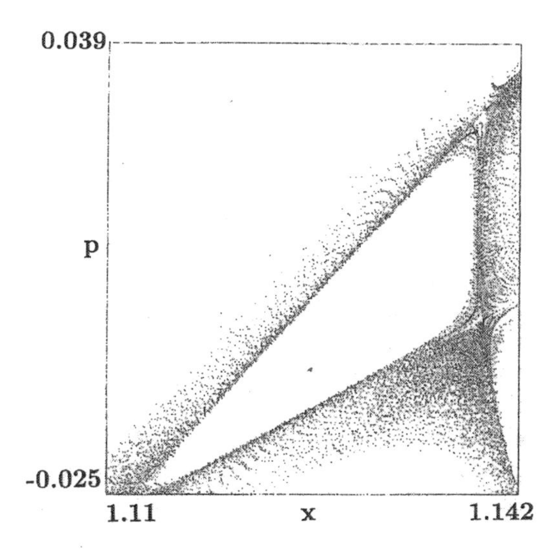

Figure 7: Boundary layer near the very vicinity of the bifurcation

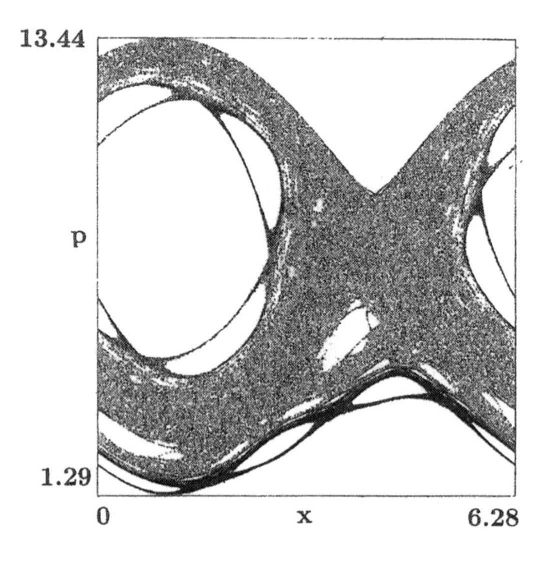

Figure 8: Sticky strips in the stochastic layer

gular zones, i.e. the zones of stickiness that have too small probability to capture a trajectory. Only a few singular zones are responsible for the observable asymptotics of the trajectories and their characteristic time windows can overlap. Assuming, for example, that the dynamics in each singular zone can be characterized by a kind of self-similarity, one can consider a general situation of motion as multi-fractal in space and time. To avoid such a high level of complexity of motion, let us introduce parameter windows in such a way that for a given parameter window there are well separated time scales of the particle residence in singular zones, i.e. characteristic times of trapping into different domains of the phase space strongly differ. This gives an opportunity to select a parameter window and a corresponding time window for which an intermediate asymptotics can be considered, and the fractional kinetic equation can be introduced. Such a case is more similar to fractal than to multi-fractal dynamics. Some of the windows are described in [6,8,9,10]. In the following two sections we describe interesting topological cases related to the stickiness.

4 Self-Similar Hierarchy of Islands

In Figs. 1–4 of Section 3 we showed an example of the hierarchical structure of islands for the web-map (3.1). A corresponding value of K, found in [17], is

$$K^* = 6.349972\ldots \tag{4.1}$$

the value (4.1) provides an infinite set of islands-around-islands with the proliferation number $q = 8$. Let the sequence of the islands be written in the form of a "word" $\{m_j\} = \{m_1, m_2, \ldots\}$ where m_j is the number of islands of the j-th generation around an island of the $(j-1)$-th generation. Typically, the sequence $\{m_j\}$ depends on the value of K and all possibilities can occur in $\{m_j\} = \{m_1, m_2, \ldots\}$ up to some minor restrictions. For the case (4.1) all m_j are equal $m^* = 8$.

Let us introduce an area ΔS_j of an j-th generation island and a period T_j of the last invariant curve for the same island. It was shown in [17] that there exists a self-similarity

$$\begin{aligned} T_{k+1} &= \lambda_T T_k \ , \qquad \lambda_T > 1 \\ \delta S_{k+1} &= \lambda_S \delta S_k \ , \qquad \lambda_S < 1 \end{aligned} \tag{4.2}$$

where

$$\delta S_k = q \Delta S_k \tag{4.3}$$

and λ_T, λ_S are some scaling parameters. The relation between λ_T, λ_S can be established for a given dynamical system.

The situation of the existence of equations (4.2), which appear for a special choice of the parameter $K = K^*$, is nontrivial. Nevertheless, one can consider different regular sequences $\{m_j\}$ for different values of K and determine a spectral function of scalings $F(\lambda)$ in analogy to the multifractal spectral function [18]. Then, the sequence of periods $\{T_j\}$ will have a much more complicated form than (4.2). At the same time, sequence $\{T_j\}$ defines how long a trajectory spends in the vicinity of the boundary of an island when it is being trapped in the singular zone. This brief comment explains the origin of (multi)fractal dynamics in the set of boundary layers of the sequence of islands $\{m_j\}$. To simplify the situation, it is convenient to select a "window" of K, which corresponds to the case of constant values of λ_T, λ_S as in (4.1),(4.2). Then we have to introduce a kinetic equation for the specific dynamics when trajectories randomly switch from islands of one hierarchy to islands of another. This indicates the fractal support of dynamics in space and time simultaneously. In Fig. 9, we show a structure of the islands hierarchy described above. It

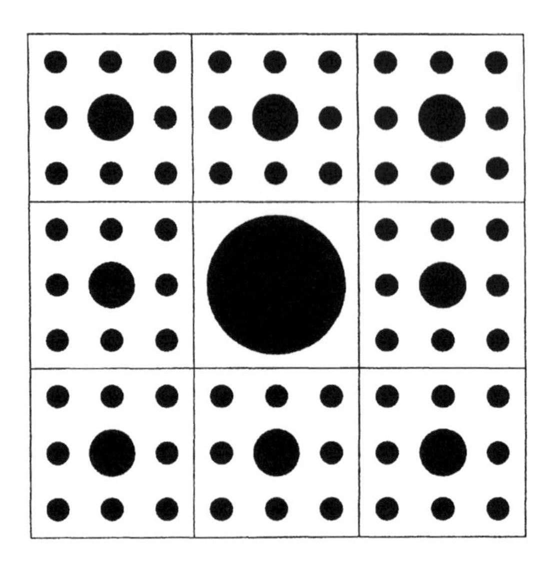

Figure 9: Scheme of the island hierarchy

somewhat resembles the Sierpinsky carpet. The crucial difference between this structure and the Sierpinsky carpet is that each circle (island) in Fig. 9 carries an additional parameter which is the time span that trajectories spend inside the circle.

5 Quasi-Traps

In this section we formalize some statements of the previous one using the notions of fractal time and quasi-traps (since traps do not exist in Hamiltonian dynamics).

The concept of fractal time is based on the fractal properties of a time set $\{t_j\}$ where the instants t_j are related to a characteristic set of events, such as the length of steps in a random walk [1,13]. The application of the fractal time concept to the Hamiltonian chaotic dynamics needs to be modified [6,7,8,11,12].

Let the finite phase space Γ_0 (square in Fig. 10) be filled by a chaotic trajectory almost uniformly, except for a relatively small domain $\Gamma_1 \subset \Gamma_0$ (a smaller square in Fig. 10) where the trajectory is almost uniform except of a

Figure 10: Sketch of the quasi-trap

smaller domain $\Gamma_2 \subset \Gamma_1$ and so on. We assume the existence of a set

$$\Gamma_0 \supset \Gamma_1 \supset \dots \tag{5.1}$$

with a space-scaling constant λ_Γ:

$$\lambda_\Gamma = \Gamma_{n+1}/\Gamma_n < 1, \qquad (\forall n). \tag{5.2}$$

It follows from (5.2) that

$$\Gamma_n = \lambda_\Gamma^n \Gamma_0. \tag{5.3}$$

Now introduce a time T_0 that the trajectory spends in $\Gamma_0 - \Gamma_1$, a time T_1 that the trajectory spends in $\Gamma_1 - \Gamma_2$, and so on. Assume the existence of a time-scaling constant λ_T:

$$\lambda_T = T_{n+1}/T_n > 1, \qquad (\forall n) \tag{5.4}$$

which means that the smaller the domain Γ_j, the larger the residence time of a particle in the domain. This is a way to construct a quasi-trap, which corresponds to the possibility to approach a singular zone (or a set in the phase space) where the Lyapunov exponent approaches zero. Particularly, the sequence (1.2) for the islands hierarchy of the web-map with parameter (4.1) can be considered as an example to apply the described construction.

Let us make some simple transformations. It follows from (5.4) that

$$T_n = \lambda_T^n T_0 \tag{5.5}$$

or

$$n = \ln(T_n/T_0)/\ln \lambda_T . \tag{5.6}$$

The substitution of (5.6) into (5.3) gives

$$\Gamma_n/\Gamma_0 = (T_0/T_n)^{\mu_\Gamma} \tag{5.7}$$

with

$$\mu_\Gamma = |\ln \lambda_\Gamma|/\ln \lambda_T . \tag{5.8}$$

Consider an asymptotic formula for the escape probability density

$$\psi(t; \Delta\Gamma) \sim P_{\text{esc}}(t) \sim 1/t^\gamma \tag{5.9}$$

where we introduce an exponent γ and the location of the domain $\Delta\Gamma$ will be specified below. The phase volume G_n of particles that pass the domain Γ_n during the escape time $t > T_n$ is of the order [19]:

$$G_n \sim \int_{T_n}^{\infty} t\psi(t; \Gamma_n)dt \sim 1/T_n^{\gamma-2} \tag{5.10}$$

in accordance with (5.9). These particles guarantee the recurrence after T_n. The expression (5.10) should be of the same order Γ_n up to a small correction of order $\Gamma_{n+1} \ll \Gamma_n$. Comparing (5.10) and (5.17) with $G_n \sim \Gamma_n$ we get in the limit $n \to \infty$

$$\gamma = 2 + \mu_\Gamma = 2 + |\ln \lambda_\Gamma|/\ln \lambda_T . \tag{5.11}$$

This result coincides with the result in [6] obtained by the Renormalization Group approach. A qualitative way of obtaining the expression (5.12) shows how the quasi-trap can work in Hamiltonian systems. The described construction of the quasi-trap corresponds to the situation considered in [6] when islands-around-islands make a fractal type hierarchy.

6 Fokker-Planck-Kolmogorov Equation (FPK)

Before considering the kinetic equation with a fractal support, we would like to remind the reader about the phenomenological derivation of the FPK equation. Some modernization of the original Kolmogorov scheme [20] will be used.

Let $P(x,t)$ be the probability density to have a particle coordinate x at an instant t. The Markov process equation can be written in the form

$$W(x_3,t_3|x_1,t_1) = \int dx_2 W(x_3,t_3|x_2,t_2)W(x_2,t_2|x_1,t_1) \tag{6.1}$$

where $W(x,t|x',t')$ is the probability density of having a particle at the position x at the time t if at the time $t' \leq t$ the particle was at the point x'. The function W depends on the initial condition, while P does not. A typical assumption for W

$$W(x,t|x',t') = W(x,x';t-t') \tag{6.2}$$

corresponds to the regular kinetic equation derivation scheme. For an infinitesimal time interval $\Delta t = t - t'$ the typical expansion is

$$W(x,x_0;t+\Delta t) = W(x,x_0;t) + \Delta t \frac{\partial W(x,x_0;t)}{\partial t} \tag{6.3}$$

where we omit all terms of higher order in Δt. Equation (6.3) is written, providing the existence of the limit

$$\lim_{\Delta t \to 0} \frac{1}{\Delta t}\{W(x,x_0;t+\Delta t) - W(x,x_0;t)\} = \frac{\partial W(x,x_0;t)}{\partial t}. \tag{6.4}$$

From (6.3) and (6.1),(6.2) we can derive the equation

$$\frac{\partial P(x,t)}{\partial t} = \lim_{\Delta t \to 0} \frac{1}{\Delta t} \left\{ \int_{-\infty}^{\infty} dy\; W(x,y;\Delta t)P(y,t) - P(x,t) \right\} \tag{6.5}$$

where new notation has been introduced

$$P(x,t) \equiv W(x,x_0;t) \tag{6.6}$$

to connect probabilities P and W. It stresses a typical assumption that only large time asymptotics

$$t \gg t_c \tag{6.7}$$

will be considered which means that large difference $|x - x_0|$ asymptotics is expected. The value t_c in (6.7) is a characteristic time of "interaction", or "phase mixing", or something else, which defines a short-scale time of the process. It is the existence of t_c that makes a difference between $W(x,x_0;t)$ in (6.6) which is responsible for the long-time behavior of the probability distribution, and between $W(x,y;\Delta t)$ in (6.5) which is responsible for a short-time behavior of

W due to the limit $\Delta t \to 0$. This separation of time scales is crucial to derive the kinetic equation given below.

The initial condition for $W(x, y; \Delta t)$ can be taken in the form

$$\lim_{\Delta t \to 0} W(x, y; \Delta t) = \delta(x - y) \tag{6.8}$$

which reflects no transition for zero interval. For a finite small Δt the condition (6.8) yields the expansion

$$W(x, y; \Delta t) = \delta(x - y) + A(y; \Delta t)\delta'(x - y) + \frac{1}{2}B(y; \Delta t)\delta''(x - y) \tag{6.9}$$

where $A(y; \Delta t), B(y; \Delta t)$ are some functions. The prime denotes derivative with respect to the argument, and we restrict ourselves by the second order terms only. These functions can be expressed as moments of the transfer probability W:

$$A(y; \Delta t) = \int dx(y - x)W(x, y; \Delta t) \equiv \langle\langle \Delta y \rangle\rangle$$

$$B(y; \Delta t) = \int dx(y - x)^2 W(x, y; \Delta t) \equiv \langle\langle (\Delta y)^2 \rangle\rangle. \tag{6.10}$$

Function W trivially satisfies the normalization condition

$$\int dx W(x, y; \Delta t) = 1 \ . \tag{6.11}$$

Simultaneously with (6.11), function W should satisfy another normalization condition

$$\int dy W(x, y; \Delta t) = 1 \ . \tag{6.12}$$

which is nontrivial and which gives after substitution of (6.9) into (6.12)

$$A(y; \Delta t) = -\frac{1}{2}\frac{\partial B(y; \Delta t)}{\partial y}. \tag{6.13}$$

Equation (6.13) can be rewritten as

$$\langle\langle \Delta y \rangle\rangle = \frac{1}{2}\frac{\partial}{\partial y}\langle\langle (\Delta y)^2 \rangle\rangle \tag{6.14}$$

which is known as Landau relation [21] obtained from the detailed balance principle. It is interesting to emphasize that we can get the important connection (6.14) directly (and simply) using the expansion (6.9).

The next step is very formal. Let us assume that there exist the limits

$$\lim_{\Delta t \to 0} \frac{1}{\Delta t} \langle\langle \Delta x \rangle\rangle \equiv \mathcal{A}(x)$$

$$\lim_{\Delta t \to 0} \frac{1}{\Delta t} \langle\langle (\Delta x)^2 \rangle\rangle \equiv \mathcal{B}(x). \tag{6.15}$$

These are very important conditions which will not work for the fractal case. Then substitution of (6.8)-(6.10) in (6.5) gives

$$\frac{\partial P}{\partial t} = -\frac{\partial}{\partial x}(\mathcal{A}P) + \frac{1}{2}\frac{\partial^2}{\partial x^2}(\mathcal{B}P) \tag{6.16}$$

which is the FPK-equation and x replaces y. The special relation (6.14) transforms the FPK-equation (6.16) into a form of the conservation law for the flux of probability

$$\frac{\partial P}{\partial t} = \frac{1}{2}\frac{\partial}{\partial x}\frac{\langle\langle (\Delta x)^2 \rangle\rangle}{\Delta t}\frac{\partial P}{\partial x}\,. \tag{6.17}$$

The expansion (6.9) in δ-functions permits one to get fast results and to make generalizations for the fractal situation.

7 Fractional Generalization of the Fokker-Planck-Kolmogorov Equation (FFPK)

In this section we derive the FFPK equation in a formal way. The relation of this equation to the wandering process in fractal space-time will be discussed in the next section. A phenomenological derivation of the FFPK equation was proposed in [3] and then specified in relation to some physical problems in [4,5]. Here we follow, mainly, these results. A special comment should be made to the FFPK derivation.

The conventional way to describe kinetics consists in defining a set of events or states $\{\xi\}$ and a simplified scheme of transitions between different states. This scheme determines which transitions are available (or have a high probability) and which ones can be neglected because of their small probability. A "Markov tree" construction is just another name for the transitions scheme. A set of islands in the phase space of Hamiltonian systems is so complicated that it allows a variety of different Markov trees. Some of them were described in [16,22]. Let us introduce transitional paths which involve particle jumps between islands of different hierarchical order. This scheme has its specific self-similar structure which was described in Section 4. The construction of the corresponding Markov tree results from the dynamics. The dynamics dictates which kind of self-similarity should be applied to the process under consideration.

To introduce a process of wandering along a fractal space-time set $\{t; x\}$, assume

$$\frac{\partial^\beta W(x, x_0; t)}{\partial t^\beta} = \frac{\partial^\beta P(x,t)}{\partial t^\beta} \tag{7.1}$$

instead of (6.4), where β is some constant

$$0 < \beta \leq 1 \tag{7.2}$$

and $\partial^\beta / \partial t^\beta$ is the fractional derivative of order β (see auxiliary material on fractional derivatives in the Appendix).

In analogy to expansion (6.9) for an infinitesimal $|x - y|$ its fractional generalization can be considered

$$W(x, y; \Delta t) = \delta(x-y) + A(y; \Delta t)\delta^{(\alpha)}(x-y) + B_1(y; \Delta t)\delta^{(\alpha_1)}(x-y) + \ldots \tag{7.3}$$

with appropriate fractal space dimension characteristics

$$0 < \alpha \leq 2\,, \qquad \alpha < \alpha_1 \leq 2 \tag{7.4}$$

(definition of fractional derivatives of δ-function see in the Appendix). Restrictions (7.2),(7.4) represent the fact that fractal set $\{t, x\}$ is imbedded into regular time t and space x and will be discussed later.

In analogy to (3.10) and (3.15) new functions $\mathcal{A}, \mathcal{B}_1$ can be introduced

$$\mathcal{A}(x) \equiv \lim_{\Delta t \to 0} \frac{A(x; \Delta t)}{(\Delta t)^\beta}$$

$$\mathcal{B}_1(x) \equiv \lim_{\Delta t \to 0} \frac{B_1(x; \Delta t)}{(\Delta t)^\beta} \tag{7.5}$$

They characterize new local properties of the wandering process in the limit $\Delta t \to 0$. The limit (7.5) appears as a result of fractal dimensionality of the set $\{t, x\}$ along which particle wandering is performed.

To complete derivation of the FFPK equation let us use the formula (see the Appendix)

$$(g^{(\alpha)}(x) \cdot f(x)) \equiv \int dx\; g^{(\alpha)}(x) f(x) = \left(g(x) \cdot \frac{\partial^\alpha}{\partial(-x)^\alpha} f(x) \right) \tag{7.6}$$

where $g(x), f(x)$ are arbitrary functions to which fractional calculus operations can be applied. Starting from Eq. (7.7) and replacing $W(x, y; t + \Delta t)$ by (6.1), we have

$$\frac{\partial^\beta P(x,t)}{\partial t^\beta} = \lim_{\Delta t \to 0} \frac{1}{(\Delta t)^\beta} \left\{ \int dy [W(x, y; t + \Delta t) - \delta(x - y)] P(y, t) \right\} \tag{7.7}$$

where the notation (6.6) has been used. After substituting (7.1) and (7.3) into (7.7) we obtain

$$\frac{\partial^\beta P}{\partial t^\beta} = \frac{\partial^\alpha}{\partial(-x)^\alpha}(\mathcal{A}P) + \frac{\partial^{\alpha_1}}{\partial(-x)^{\alpha_1}}(\mathcal{B}_1 P) \tag{7.8}$$

where Eq. (7.6) has been used. Equation (7.8) is the fractional generalization of the Fokker-Planck-Kolmogorov (FFPK) equation. For $\alpha = 1$, $\alpha_1 = 2$, Eq. (7.8) reduces to the regular FPK equation. Different intermediate asymptotics for the FFPK equation can be obtained for large t contrary to the FPK equation for which the detailed balance principle works [21].

In fact, different comments should be attached to the derivation of (7.8). By definition, the fractional derivative is one-directed (see the Appendix). It is well-defined for the time derivative $\partial^\alpha/\partial x^\alpha$ since we cannot simply replace $x \to -x$. That is why the following more general operator should be considered:

$$L_x^{(\alpha)} = \mathcal{A}^+ \partial^\alpha/\partial x^\alpha + \mathcal{A}^- \partial^\alpha/\partial(-x)^\alpha \ . \tag{7.9}$$

Constants $\mathcal{A}^\pm$ can be considered as functions of the corresponding "phases". The correct choice of $\mathcal{A}^\pm$ depends on the physical situation. For the case of symmetric wandering, a symmetrized derivative was introduced in [23]:

$$\frac{\partial^\alpha}{\partial|x|^\alpha} = -\frac{1}{2\cos(\pi\alpha/2)}\left[\frac{\partial^\alpha}{\partial x^\alpha} + \frac{\partial^\alpha}{\partial(-x)^\alpha}\right] , \qquad (\alpha \neq 1). \tag{7.10}$$

Then the equation (7.8) can be replaced by the equation

$$\frac{\partial^\beta P}{\partial t^\beta} = \frac{\partial^\alpha}{\partial|x|^\alpha}(\mathcal{A}P) + \frac{\partial^{\alpha_1}}{\partial|x|^{\alpha_1}}(\mathcal{B}_1 P) , \qquad 1 < \alpha < \alpha_1 \leq 2 \tag{7.11}$$

or, if we neglect the second term with $\mathcal{B}_1$, we get simply

$$\frac{\partial^\beta P}{\partial t^\beta} = \frac{\partial^\alpha}{\partial|x|^\alpha}(\mathcal{A}P). \tag{7.12}$$

The physical meaning of $\mathcal{A}$ and $\mathcal{B}_1$ can be obtained similarly to (6.10). Let $\mathcal{B}_1 \equiv 0$ in the expansion (7.3). Then multiplying (7.3) by $|x-y|^\alpha$ and integrating with respect to x, we get

$$\langle\langle|\Delta x|^\alpha\rangle\rangle \equiv \int dx|x-y|^\alpha W_A(x,y;\Delta t) = \Gamma(1+\alpha)\mathcal{A}(y;\Delta t) \tag{7.13}$$

where W_A is the function defined in (7.3) and taken with $B_1 \equiv 0$. In the same way, one can define

$$\langle\langle|\Delta x|^{\alpha_1}\rangle\rangle \equiv \int dx|x-y|^{\alpha_1} W(x,y;\Delta t) = \Gamma(1+\alpha_1)\mathcal{B}_1(y;\Delta t) \tag{7.14}$$

where W is defined by (7.3) and the condition $\alpha_1 > \alpha$ is used.

Integrating (7.1) with respect to y and applying a normalization condition

$$\int dy\ W(x,y;t) = 1 \tag{7.15}$$

we obtain a relation

$$\frac{\partial^{\alpha}}{\partial(-x)^{\alpha}}\mathcal{A}(x;\Delta t) + \frac{\partial^{\alpha_1}}{\partial(-x)^{\alpha_1}}\mathcal{B}_1(x;\Delta t) = 0 \tag{7.16}$$

which is a fractional analog of (6.13).

In this section, we assumed that the wandering is performed in the fractal space-time and we used corresponding fractal supports in the expansions (7.1) and (7.3). A reason for this is that the orders (α, β) of fractional derivatives follow from a specific type of chaotic dynamics in which the condition of existence of the limit (6.15) for normal kinetics should be replaced by the new one

$$\lim_{\Delta t \to 0} \frac{\langle\langle|\Delta x|^{\alpha}\rangle\rangle}{(\Delta t)^{\beta}} = \mathcal{A} \tag{7.17}$$

with nontrivial α and β.

We need to make an important comment about the limit $\Delta t \to 0$: we can fix the value Δt_{min} and consider a situation when all typical values of Δt satisfy the condition

$$\Delta t_{\text{min}}/\Delta t_{\text{max}} \to 0\,. \tag{7.18}$$

The case (7.18) is the most reasonable one for different physical situations including maps that generate chaotic dynamics.

8 Renormalization Group for Kinetics (RGK)

In Section 4, we described a self-similar space-time structure of a singular zone with a hierarchy of islands for the web-map. The self-similarity is introduced by the equations (4.2),(4.3). We can also add an equation for the Lyapunov exponents σ_k near the boundary of the k-th generation island

$$\sigma_{k+1} = \lambda_{\sigma}\sigma_k\ , \qquad \lambda_{\sigma} < 1 \tag{8.1}$$

with an appropriate value of λ_{σ}. Typically, one can expect

$$\lambda_{\sigma} = 1/\lambda_T \tag{8.2}$$

up to a small correction.

Let us define a "flight" as a portion of a trajectory that corresponds to almost regular dynamics, i.e. to the dynamics with a very small Lyapunov exponent. One can say that a flight is also a sticky part of a trajectory. We consider a flight in a generalized way, i.e. the flight can be a propagating part of the trajectory or a rotating-around-island(s) part of the same trajectory. Denote by ℓ the length of a flight. Then the existence of the self-similarity means that

$$\ell_{k+1} = \lambda_\ell \ell_k \ , \qquad \lambda_\ell > 1 \tag{8.3}$$

where ℓ_k is a flight length during the time interval $\sim T_k$ when the trajectory sticks to the boundary of a k-th generation island. It was shown in [5] that

$$\lambda_S = \lambda_\ell^{-2} \ . \tag{8.4}$$

From (8.1)-(8.4) and (4.2),(4.3), we define a number of scaling constants $\lambda_T, \lambda_S, \lambda_\ell, \lambda_\sigma$ which are not independent, and only two of them, say λ_ℓ, λ_T, are enough to introduce a Renormalization Group transform $\hat{R}_K$ for kinetics (RGK).

We start from a more general formulation of the renormalization group transform $\hat{R}_K$ which specifies two relevant variables: action s and time t

$$\hat{R}_K : s_{n+1} = \lambda_s s_n \ , \qquad t_{n+1} = \lambda_t t_n \tag{8.5}$$

where λ_s, λ_t are scaling constants that will be discussed later. More flexibility in the definition of (8.5) is necessary if dynamical systems are considered. Specifically we have in mind that

$$\lambda_s \le 1 \ , \qquad \lambda_t \ge 1 \tag{8.6}$$

and there exist cut-off values

$$s_{\max} = s_0 \ , \qquad t_{\min} = T_0 \ . \tag{8.7}$$

Inequalities (8.6) with definitions (8.7) indicate the direction of the $\hat{R}_K$-transformation process which proceeds to arbitrarily small actions s_n with arbitrarily small area S_n of the corresponding island and to arbitrarily big value $t_n = T_n$ of the last invariant curve period of the same island.

In (8.5) λ_s is a scaling parameter for the action s and in the simple situation the action coincides (or is proportional) to the island area S. This and (8.4) give

$$\lambda_s = \lambda_S = \lambda_\ell^{-2} \, . \tag{8.8}$$

The Time scaling parameter λ_t is simply

$$\lambda_t = \lambda_T \, . \tag{8.9}$$

The relations (8.8),(8.9) complete the definitions (8.5) of the RGK.

It is important to note that time and action variables are transformed simultaneously in (8.5). The flights can be roughly considered as almost periodically modulated regular motion inside the singular zone. The modulation of the n-th order is due to the rotations around the n-th order island. There is a slow drift of the flight towards the trap for rotations around the $(n \pm 1)$-th order islands, which generates $(n \pm 1)$-th order modulation. When this happens, there is a jump in the value of s from s_n to $s_{n\pm 1}$ and from T_n to T_{n+1}

$$s_{n\pm1} = \lambda_s^{\pm 1} s_n \ , \qquad T_{n\pm1} = \lambda_T^{\pm 1} T_n \ . \tag{8.10}$$

Equation (8.10) defines a fractal Markov tree where the fractional kinetics is performed, and Equation (8.5) defines the RGK which acts along with processes (8.10).

Let us consider an infinitesimal time-derivation of the distribution function

$$\delta_t P(\ell, t) \equiv W(\ell, \ell_0, t + \Delta t) - W(\ell, \ell_0; t) \tag{8.11}$$

where notation (6.6) has been used and the coordinate $|x|$ is replaced by the flight length ℓ. This deviation can be expressed through the all available paths with their probabilities, in accordance with the basic equation (6.1) or its simplified version (6.5). Thus

$$\delta_t P(\ell, t) = \sum_{\Delta\ell} \{\overline{P(\ell + \Delta\ell, t) - P(\ell, t)}\} \equiv \delta_\ell P(\ell, t) \tag{8.12}$$

where the sum over $\Delta\ell$ means summation over all paths within an infinitesimal interval $\Delta\ell$ that gives the infinitesimal evolutional change $\delta_t P$, and the overline means averaging over the paths. As an example one can recall the conventional averaging over phases. Equation (8.12) can be rewritten in the form

$$(\Delta t)^\beta \frac{\partial^\beta P}{\partial t^\beta} = \sum_{\Delta\ell} \frac{\partial^\alpha}{\partial \ell^\alpha} (\overline{(\Delta\ell)^\alpha \mathcal{A}'} \ P) \tag{8.13}$$

where $\mathcal{A}'$ is some expansion constant and the choice of a term with power α depends on the first nonvanishing value after averaging over phases.

Now apply the RGK transform (8.5) n times to the equation (8.12)

$$\hat{R}_K^n(\delta_t P) = \hat{R}_K^n(\delta_\ell P) \ . \tag{8.14}$$

Using (8.13) we get

$$[\hat{R}_K^n(\Delta t)^\beta] \frac{\partial^\beta P}{\partial t^\beta} = \sum_{\Delta\ell} \frac{\partial^\alpha}{\partial \ell^\alpha} (\overline{[\hat{R}_K^n(\Delta\ell)^\alpha] \mathcal{A}' P}) \ . \tag{8.15}$$

From Eqs. (8.5),(8.8),(8.9) we can obtain explicit expressions:

$$\hat{R}_K^n(\Delta t)^\beta = \lambda_T^{\beta n}(\Delta t)^\beta\,, \qquad \hat{R}_K^n(\Delta \ell)^\alpha = \lambda_\ell^{\alpha n}(\Delta \ell)^\alpha\,. \tag{8.16}$$

Substitution of (8.16) into (8.15) gives

$$(\Delta t)^\beta \frac{\partial^\beta P}{\partial t^\beta} = \left(\frac{\lambda_\ell^\alpha}{\lambda_t^\beta}\right)^n = \sum_{\Delta\ell} \frac{\partial^\alpha}{\partial \ell^\alpha}(\overline{(\Delta\ell)^\alpha \mathcal{A}'}\; P) \tag{8.17}$$

or in the more convenient form

$$\frac{\partial^\beta P}{\partial t^\beta} = \left(\frac{\lambda_\ell^\alpha}{\lambda_t^\beta}\right)^n \sum_{\Delta\ell} \frac{\partial^\alpha}{\partial \ell^\alpha} \overline{\frac{(\Delta\ell)^\alpha}{(\Delta t)^\beta}\mathcal{A}'}\; P\,. \tag{8.18}$$

Formula (8.18) is the renormalization group equation for the fractional kinetics.

Starting from $\ell_{\max} = \ell_0$ and $t_{\min} = T_0$ (see (8.7) and (8.6)) we have $(\lambda_\ell^\alpha/\lambda_t^\beta)^n \to 0$ or $\to \infty$, unless

$$\lambda_\ell^\alpha = \lambda_t^\beta\,. \tag{8.19}$$

Equation (8.19) defines a nontrivial "fixed point" for (8.18). It gives

$$\mu_0 \equiv \beta/\alpha = \ln\lambda_\ell/\ln\lambda_t = |\ln\lambda_S|/2\ln\lambda_T \tag{8.20}$$

and the "fixed point kinetic equation"

$$\frac{\partial^\beta P}{\partial t^\beta} = \frac{\partial^\alpha}{\partial \ell^\alpha}\sum_{\Delta\ell} \overline{\frac{(\Delta\ell)^\alpha}{(\Delta t)^\beta}\mathcal{A}'}\; P \tag{8.21}$$

which coincides with (3.25) if one restricts oneself by only the first term in the right hand side of (3.25), and puts

$$\mathcal{A} = \sum_{\Delta\ell} \overline{\frac{(\Delta\ell)^\alpha}{(\Delta t)^\beta}\mathcal{A}'}. \tag{8.22}$$

The kinetic equation obtains the form

$$\frac{\partial^\beta P}{\partial t^\beta} = \frac{\partial^\alpha}{\partial \ell^\alpha}(\mathcal{A}P) \tag{8.23}$$

that coincides with (7.12).

Solution (8.20) corresponds to a specific window of parameters and time, where corresponding values α, β are applicable. For different windows, one can expect different pairs (α, β). Each of α and β is proportional to the generalized fractal dimension of the corresponding support. The fix-point condition (8.19) defines only ratio (8.20). Nevertheless, this information is sufficient to get a transport exponent. Multiplying (8.23) by ℓ^α and integrating it, we obtain the moment equation

$$\langle \ell^\alpha \rangle \sim \mathcal{A} t^\beta \tag{8.24}$$

or, approximately,

$$\langle \ell^2 \rangle \sim \langle |x|^2 \rangle \sim t^{2\beta/\alpha} = t^{2\mu_0} \,. \tag{8.25}$$

This is an asymptotic result that is valid for

$$t^\beta / \ell^\alpha \ll 1 \,. \tag{8.26}$$

It will be discussed more in the next section.

9 Solutions for the FFPK

The formal solution for (8.23) can be written as a series expansion [23]. For this goal rewrite the Eq. (8.23) with a source

$$\frac{\partial^\beta P(x,t)}{\partial t^\beta} = \frac{\partial^\alpha P(x,t)}{\partial |x|^\alpha} + \frac{t^{-\beta}}{\Gamma(1-\beta)} \delta(x) \,, \qquad (t \to 0) \tag{9.1}$$

where the expression for the source is taken to satisfy the normalization condition

$$\int_{-\infty}^{\infty} P(x,t) dx = 1 \,. \tag{9.2}$$

Let us introduce the Laplace-Fourier transform

$$P(q,u) = \int_0^\infty dt \int_{-\infty}^\infty dx \; e^{-ut+iqx} P(x,t) \tag{9.3}$$

which reduces (9.1) to a form

$$u^\beta P(q,u) + |q|^\alpha P(q,u) = u^{\beta-1} \tag{9.4}$$

equivalent to (9.1). The form (9.4) is convenient for different approximations and comparisons.

If we apply only Fourier transformation with respect to x

$$P(q,t) = \int_{-\infty}^{\infty} dx\, e^{iqx} P(x,t) \ , \tag{9.5}$$

then from (9.1) follows

$$\frac{\partial^\beta}{\partial t^\beta} P(q,t) + |q|^\alpha P(q,t) = \frac{t^{-\beta}}{\Gamma(1-\beta)} \ , \qquad (t>0) \ . \tag{9.6}$$

The solution of (9.6) can be written in a series form [24]

$$P(q,t) = \sum_{m=0}^{\infty} \frac{(-1)^m}{\Gamma(m\beta+1)} (|q|^\alpha t^\beta)^m \ . \tag{9.6}$$

It is easy to prove that (9.6) satisfies the equation (5.2) by applying the derivative $\partial^\beta/\partial t^\beta$ to (5.3) and using the formula (see the Appendix)

$$\frac{\partial^\beta}{\partial t^\beta} t^\gamma = \frac{\Gamma(\gamma+1)}{\Gamma(\gamma+1-\beta)} t^{\gamma-\beta} \ . \tag{9.7}$$

Result (9.6) can be written in the form

$$P(q,t) = E_\beta(-|q|^\alpha t^\beta) \tag{9.8}$$

where the Mittag-Leffler function of order β has been introduced

$$E_\beta(z) = \sum_{m=0}^{\infty} \frac{z^m}{\Gamma(m\beta+1)} \ . \tag{9.9}$$

This function was also used with a similar goal in [25,26]. Recall that for $\beta = 1$

$$E_1(z) = e^z \tag{9.10}$$

and (9.8) is reduced to the characteristic function of a Lévy process [27].

Let us apply inverse Fourier transformation in q to (9.6). It gives the solution in a series form

$$P(x,t) = \frac{1}{\pi |y|} \frac{1}{t^{\mu/2}} \sum_{m=1}^{\infty} \frac{(-1)^m}{|y|^{m\alpha}} \frac{\Gamma(m\alpha+1)}{\Gamma(m\beta+1)} \cos\left[\frac{\pi}{2}(m\alpha+1)\right] \tag{9.11}$$

where

$$y = x/t^{\mu/2} \tag{9.12}$$

and

$$\mu = 2\mu_0 = 2\beta/\alpha \ . \tag{9.13}$$

It follows the asymptotic solution from (9.11)

$$P(x,t) \sim \frac{1}{\pi} \frac{t^\beta}{|x|^{\alpha+1}} \frac{\Gamma(1+\alpha)}{\Gamma(1+\beta)} \sin\frac{\pi\alpha}{2} \tag{9.14}$$

where

$$|x| \gg t^{\mu/2} = t^{\beta/\alpha} \ . \tag{9.15}$$

In particular, from (9.14) we conclude that the moments $\langle |x|^\delta \rangle$ with $\delta > 0$ are bounded if

$$0 < \delta < \alpha < 2 \ . \tag{9.16}$$

This restriction evidently disappears for $\alpha = 2$ when all positive moments are finite.

10 Generalization and Physical Utilization

Fractional analysis is a convenient tool to apply to chaotic dynamics. Nevertheless, in the form described above, we use a too formal approach and it is worthwhile to have more discussions on the physical reasons of applying fractional techniques. The main comment starts from the idea of a convenient mapping of the trajectories with a chaotic and nonuniform behavior in the phase space, as it was described in Section 2. After a mapping, or coarse-graining, or renormalization we can get fairly "wild" distributions with fractal or multifractal support. This is a typical situation in statistical physics. In dynamical systems analysis we have to add a new, temporal, variable to the phase space.

It is conventional to write down a corresponding distribution function $P(x,t)$ in the form

$$P(x,t) = P_n(x,t) + P_s(x,t) \tag{10.1}$$

where n corresponds to the normal part and s to the singular one. The same splitting can be performed in the q-space

$$P(q,t) = \int e^{iqx} P(x,t) dx = P_n(q,t) + P_s(q,t) \ . \tag{10.2}$$

Considering the asymptotics

$$|x| \to \infty , \qquad t \to \infty , \tag{10.3}$$

i.e. $q \to 0$ in (10.2), we can assume that

$$P(q,t) = \sum_{m=0}^{n} \frac{1}{m!} q^m \left. \frac{\partial^m P(q,t)}{\partial q^m} \right|_{q=0} + A_m |q|^\alpha + \text{remainder} \tag{10.4}$$

where A_m is a constant and

$$n < \alpha < n+1 \tag{10.5}$$

with the definition

$$P_s(q,t) = A_m |q|^\alpha . \tag{10.6}$$

The normal part of (10.4) can be rewritten using definition (10.2) as

$$P_n(q,t) = \sum_{m=0}^{n} \frac{(iq)^m}{m!} \langle x^m \rangle . \tag{10.7}$$

This expression, together with (10.6) and (10.5) simply states that $P(x,t)$ possesses finite first n moments and infinite $(n+1)$-th moment because of

$$\left. \frac{\partial^{n+1} P(q,t)}{\partial q^{n+1}} \right|_{q=0} = i^{n+1} \langle x^{n+1} \rangle$$

$$= A_m \alpha(\alpha - 1) \dots (\alpha - n) |q|^{\alpha - 1} = \infty , \qquad q = 0 . \tag{10.8}$$

As a result of this simple introduction we can consider the distribution function with an arbitrary power-law tail as a function with corresponding infinite moments, and use for such a function an expansion of the (10.4)-type.

It is clear now that the fractional kinetic equation should be written for the singular part only, i.e. equation (8.23), for example, should be considered as

$$\frac{\partial^\beta P_s}{\partial t^\beta} = \frac{\partial^\alpha}{\partial \ell^\alpha} (A P_s) \tag{10.9}$$

with replacement of x in (10.1) by ℓ, and

$$P_s = P - P_n \tag{10.10}$$

i.e. the fractional derivatives in (10.9) are applied to the renormalized distribution (10.10) rather than to the full distribution $P(x,t)$. A corresponding generalization can be done with respect to the time-variable. In fact, one can consider different expansions in (q,u) instead of (10.8) with a singular term of the type

$$P_s(q,u) = A_1|q|^{\alpha_1}u^{\beta_1} + A_2|q|^{\alpha_2}u^{\beta_2} + \dots \tag{10.11}$$

with parameters α_j, β_j depending on the physical situation. In Sections 7.8 it was, for example,

$$\alpha_1 = \alpha\ , \quad \beta_1 = 0\ , \quad \alpha_2 = 0\ , \quad \beta_2 = \beta \tag{10.12}$$

but one can imagine more sophisticated cases with coupled $|q|$ and u. It is important that we lift the restrictions $\alpha, \beta < 1$ and permit an arbitrary integer part for α, β. Corresponding integers n_α, n_β should be adjusted to the expansion of $P(q,u)$ in the power series on q and u.

Similar generalizations should be performed for the infinitesimal differences (8.11) and (8.12).

An actual intervention of physics comes with definitions of the infinitesimal variations $\delta_\ell P, \delta_t P$ in (8.11),(8.12) and the following assumption in (8.13) on the existence of limits

$$\lim_{\Delta t \to 0} \frac{\delta_t P}{(\Delta t)^\beta} = \frac{\partial^\beta P}{\partial t^\beta}$$

$$\lim_{\Delta \ell \to 0} \frac{\delta_\ell P}{(\Delta \ell)^\alpha} = \frac{\partial^\alpha (AP)}{\partial \ell^\alpha}\ . \tag{10.13}$$

Using the expansion (10.1) we can rewrite the limits (10.13) for only the singular part P_s which gives

$$\lim_{\Delta t \to 0} \frac{\delta_t P_s}{(\Delta t)^\beta} = \frac{\partial^\beta P_s}{\partial t^\beta}$$

$$\lim_{\Delta \ell \to 0} \frac{\delta_\ell P_s}{(\Delta \ell)^\alpha} = \frac{\partial^\alpha (AP_s)}{\partial \ell^\alpha} \tag{10.14}$$

assumed to exist. It becomes evident that β and α are Hölder exponents because of the property

$$\delta_t P_s = |P_s(t+\Delta t, \ell) - P_{n_\beta}(t,\ell)| = O(\Delta t^\beta) \tag{10.15}$$

and similar expression for $\delta_\ell P_s$, where n_β is the integer part of β. Indeed, consider the expansion

$$P(\ell, t+\Delta t) = \sum_{m=0}^{n_\beta} \frac{(\Delta t)^m}{m!} P^{(m)}(\ell, t)$$

$$+C_\beta(\Delta t)^\beta = P_n(\ell,t;\Delta t) + \delta P_s(\ell,\Delta t) \tag{10.16}$$

where C_β is some coefficient. From (10.16)

$$|\delta_t P_s| = |P(\ell,t+\Delta t) - P_n(\ell,t;\Delta t)| = O((\Delta t)^\beta)\,. \tag{10.17}$$

Equation (10.17) coincides with the Hölder condition because P_n is a polynomial of Δt of order n_β. Therefore β is the Hölder exponent with respect to time. The same can be shown for α with respect to the flights length.

The described meaning of the variations $\delta_t P, \delta_\ell P$ was introduced at the beginning of the derivation of the FFPK [3,4]. More rigorous considerations of the finite differences and series can be found in [31,32].

11 Log-periodic Oscillations

The FFPK equation defines the transport properties of the considered system depending on initial conditions and some other possible restrictions since the equation is valid for specified time and parameter windows. It was shown in Sections 8 and 9 that, although the moments of the solution possess the form (8.24), it can be a special connection between exponents α,β if the renormalization group $\hat{R}_K$ is valid for the considered situation. Here we would like to show that the renormalization transform (8.5) leads to wider possibilities than (8.24) if we consider solutions of the fix-point equation (see (8.18),(8.19))

$$\lim_{n\to\infty}(\lambda_\ell^\alpha/\lambda_T^\beta)^n = 1\ . \tag{11.1}$$

when the renormalization transform

$$R_K:\ \ell_{n+1}\to\lambda_\ell\ell_n;\qquad t_{n+1}=\lambda_T t_n;\qquad (\lambda_\ell,\lambda_T>1) \tag{11.2}$$

takes place, i.e. kinetics and transport are invariant under the transform (11.2).

Let us consider, as an example, solutions of (11.1) in the complex plane of β. It gives

$$\beta_j = \frac{1}{2}\alpha\mu + 2\pi i j/\ln\lambda_T\ ,\qquad (j=0,\pm1,\dots) \tag{11.3}$$

where μ is the same as (9.13), or, using (8.20),

$$\mu = 2\beta/\alpha = 2\ln\lambda_\ell/\ln\lambda_T = |\ln\lambda_S|/\ln\lambda_T\,. \tag{11.4}$$

The expression (11.3) defines complex exponents β_j with

$$\mathrm{Re}\ \beta_j = \frac{1}{2}\alpha\mu \tag{11.5}$$

i.e. the same real part of β for all j. This permits to write down the transport equation in the form

$$\langle \ell^\alpha \rangle \sim \sum_j \mathcal{A}_j t^{\beta_j} = \mathcal{A}_0 t^{\alpha\mu/2} \left\{ 1 + 2 \sum_{j=1}^{\infty} (\mathcal{A}_j/\mathcal{A}_0) \cos\left(\frac{2\pi j}{\ln \lambda_T} \ln t \right) \right\} \tag{11.6}$$

where $\mathcal{A}_j$ are some amplitudes and we use the condition $\mathcal{A}_j = \mathcal{A}^*_{-j}$ to make the expression $\langle \ell^\alpha \rangle$ real. The expression in the braces is periodic in $\ln t$ with a period

$$T_{\ln} = \ln \lambda_T \, . \tag{11.7}$$

The obtained result is a consequence of the existence of the discrete renormalization group (11.2) and it reflects a general property of critical phenomena (see more in the review article [33]). We can also write an analog to (8.25) which has the form

$$\langle \ell^2 \rangle \sim \mathcal{D}_0 t^\mu \left\{ 1 + 2 \sum_{j=1}^{\infty} (\mathcal{D}_j/\mathcal{D}_0) \cos\left(\frac{2\pi j}{\ln \lambda_T} \ln t \right) \right\} \tag{11.8}$$

where

$$\mathcal{D}_0 = \mathcal{A}_0^{2/\alpha}$$

and $\mathcal{D}_j$ are coefficients of the Fourier expansion which should be obtained from (11.6).

Results of this section were obtained in [34]. Log-periodic oscillations for the anomalous transport were observed in [6,9,17]. The log-periodicity with the period (11.7) was confirmed in simulations [6,9]. Figure 11 taken from [17] displays the property of log-periodicity for the web-map. Some other applications of the log-periodicity were known before: for the phase transition [35], Weierstrass random walk [36], geometric structures [37], branching processes [38], dynamical systems [39,40,41], geophysics [33], et al.

Acknowledgements

This work was supported by the U.S. Department of Navy, Grant N00014-96-1-0055 and by the U.S. Department of Energy, Grant DE-FG02-92ER54184.

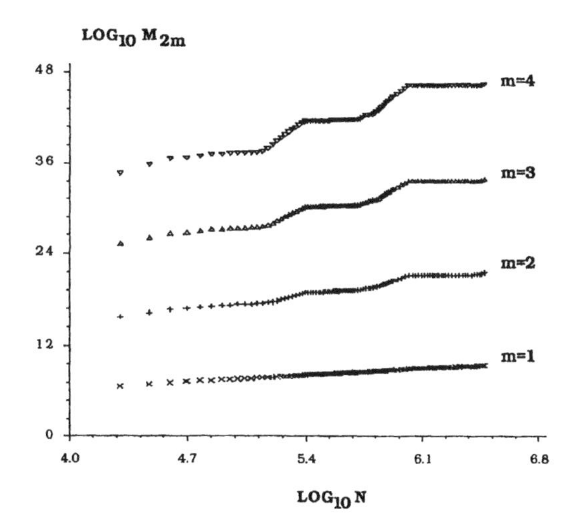

Figure 11: Log-periodicity for the web-map

Appendix: Some Basic Formulas of the Fractional Calculus

Here we describe some basic formulas and the relations that are used in the text. The following material is taken from [23]. More information on these and related subjects can be found in [28,29,24]. Particularly, different applications and properties of the Mittag-Leffler function are in [30,31].

Fractional integration of order β is defined by the operator I_β

$$I_\beta f(t) = \frac{1}{\Gamma(\beta)} \int_{-\infty}^{t} f(\tau)(t-\tau)^{\beta-1} d\tau \,, \qquad (\beta > 0) \tag{A.1}$$

or in a compact form as a convolution

$$I_\beta f(t) = f(t) \star \frac{t_+^{\beta-1}}{\Gamma(\beta)} = f(t) \star K_\beta^+(t) \tag{A.2}$$

where

$$K_\beta^+(t) = \frac{t_+^{\beta-1}}{\Gamma(\beta)} = \frac{1}{\Gamma(\beta)} \begin{cases} t^{\beta-1}, & t > 0 \\ 0, & t < 0 \end{cases} \tag{A.3}$$

is the kernel of the operator of fractional integration.

Fractional differentiation of order β can be defined in a similar way

$$\frac{d^\beta}{dt^\beta} f(t) = f(t) \star r_\beta^+(t) \;, \qquad (\beta > 0) \tag{A.4}$$

where $r_\beta^+(t)$ is the kernel of the operator of fractional derivation

$$r_\beta^+(t) = \frac{d^n}{dt^n} K_{n-\beta}^+(t) \tag{A.5}$$

with $n = [\beta]$ as the least integer larger than β. The derivative of order n in (A.5) is considered in a generalized sense and $r_\beta^+(t)$ is a singular distribution.

For $t \neq 0$, when derivatives in (A.5) can be considered in a classical sense, there exists a standard pointwise correspondence

$$r_\beta^+(t) = K_{-\beta}^+(t) \;, \qquad (t \neq 0) \tag{A.6}$$

and therefore

$$\frac{d^\beta}{dt^\beta} = I_{-\beta} \tag{A.7}$$

with

$$I_\beta = \frac{d^{-\beta}}{dt^{-\beta}} \;. \tag{A.8}$$

Using (A.7), (A.8), the fractional derivative can be written in the same form as (A.2)

$$\frac{d^\beta}{dt^\beta} f(t) = f(t) \star \frac{t_+^{-\beta-1}}{\Gamma(-\beta)} \tag{A.9}$$

or in the integral form

$$\frac{d^\beta}{dt^\beta} f(t) = \frac{1}{\Gamma(-\beta)} \int_{-\infty}^{t} f(\tau) \frac{d\tau}{(t-\tau)^{\beta+1}} \;. \tag{A.10}$$

The expression (A.10) has a symbolic character since the singularity is not integrable in the vicinity of the upper limit for $\beta \geq 0$.

To avoid the difficulty, the integral should be regularized. In particular for

$$0 < \beta < 1 \tag{A.11}$$

a regularized expression for the fractional derivative has the form

$$\frac{d^\beta}{dt^\beta} f(t) = \frac{1}{\Gamma(1-\beta)} \int_{-\infty}^{t} \frac{df(\tau)}{d\tau} \frac{d\tau}{(t-\tau)^\beta} \;. \tag{A.12}$$

Another convenient form of the regularized definition of a fractional derivative is

$$\frac{d^\beta}{dt^\beta} f(t) = \frac{1}{\Gamma(-\beta)} \int\limits_{-\infty}^{t} [f(\tau) - f(t)] \frac{d\tau}{(t-\tau)^{\beta+1}} , \qquad (0 < \beta < 1) . \qquad (A.13)$$

Let, for example,

$$f(t) = 1_+(t)\phi(t) \qquad (A.14)$$

where $\phi(t)$ is an everywhere continuously differentiable function and

$$1_+(t) = \begin{cases} 1, & t > 0, \\ 0, & t < 0. \end{cases} \qquad (A.15)$$

Then

$$\frac{df(t)}{dt} = \phi(0)\delta(t) + 1_+(t)\frac{d\phi(t)}{dt} \qquad (A.16)$$

and (A.12) takes the form

$$\frac{d^\beta}{dt^\beta}(1_+(t)\phi(t)) = \phi(0)\frac{t^{-\beta}}{\Gamma(1-\beta)} + \frac{1}{\Gamma(1-\beta)} \int\limits_0^t \frac{\phi'(\tau)}{(t-\tau)^\beta} d\tau , \qquad (t > 0) . \quad (A.17)$$

In particular, for $\phi(\tau) \equiv 1$

$$\frac{d^\beta}{dt^\beta} 1_+(t) = \frac{t^{-\beta}}{\Gamma(1-\beta)} , \qquad (t > 0) . \qquad (A.18)$$

For $\beta \to 1$ the right-hand side of (A.18) weakly converges to δ-function and we get the well-known result

$$\lim_{\beta \to 1} \frac{d^\beta}{dt^\beta} 1_+(t) = \delta(t) . \qquad (A.19)$$

Let us put in (A.17) $\phi(t) = t^\delta$ $(t > 0)$. Then

$$\frac{d^\beta}{dt^\beta}(1_+(t)t^\delta) = \frac{d^\beta}{dt^\beta} t_+^\delta = \frac{\delta}{\Gamma(1-\beta)} \int\limits_0^t \tau^{\delta-1}(1-\tau)^{\gamma-1} d\tau . \qquad (A.20)$$

The last integral is equal to

$$\int\limits_0^t \tau^{\beta-1}(t-\tau)^{-\beta} d\tau = \tau^{\beta-\delta} \frac{\Gamma(\delta)\Gamma(\gamma)}{\Gamma(\delta+\gamma)} . \qquad (A.21)$$

Thus from (A.20), (A.21) we have

$$\frac{d^\beta}{dt^\beta} t^\delta = \frac{\Gamma(\delta+1)}{\Gamma(1+\delta-\beta)} t^{\delta-\beta} \; . \tag{A.22}$$

It follows from (A.22) and (A.3) that

$$\frac{d^\beta}{dt^\beta} \frac{d^{-\beta}}{dt^{-\beta}} = \text{identity} \; . \tag{A.23}$$

One can prove by a similar way the universal relation [28]

$$\frac{d^\beta}{dt^\beta} \frac{d^\gamma}{dt^\gamma} = \frac{d^\gamma}{dt^\gamma} \frac{d^\beta}{dt^\beta} = \frac{d^{\beta+\gamma}}{dt^{\beta+\gamma}} \tag{A.24}$$

for arbitrary β, γ.

Consider now three kinds of the fractional derivatives in coordinate space. The first one is similar to (A.9)

$$\frac{d^\alpha}{dx^\alpha} g(x) = g(x) \star \frac{x_+^{-\alpha-1}}{\Gamma(-\alpha)} = g(x) \star r_\alpha^+(x) \tag{A.25}$$

where

$$r_\alpha^+(x) = \frac{x^{-\alpha-1}}{\Gamma(-\alpha)} \; , \qquad (x > 0) \tag{A.26}$$

and $r_\alpha^+(x)$ is defined in the same way as in (A.5) up to the replacement of t by x and β by α, i.e. $r_\alpha^+(x) = 0$ for $x < 0$. Similarly we can introduce

$$\frac{d^\alpha}{d(-x)^\alpha} g(x) = g(x) \star \frac{x_-^{-\alpha-1}}{\Gamma(-\alpha)} = g(x) \star r_\alpha^-(x) \tag{A.27}$$

with

$$r_\alpha^-(x) = r_\alpha^+(-x) = \frac{x^{-\alpha-1}}{\Gamma(-\alpha)} \; , \qquad (x < 0) \tag{A.28}$$

which is zero for $x > 0$.

Let

$$[g(x) \cdot f(x)] = \int_{-\infty}^{\infty} g(x) f(x) dx \tag{A.29}$$

denote the scalar product of $g(x)$ and $f(x)$. Then there exists a formula

$$\left[g(x) \cdot \frac{d^\alpha}{dx^\alpha} f(x)\right] = \left[f(x) \cdot \frac{d^\alpha}{d(-x)^\alpha} g(x)\right] \tag{A.30}$$

which is equivalent to integration by parts.

The Fourier transform

$$g(q) = \int_{-\infty}^{\infty} g(x)\, e^{iqx}\, dx \tag{A.31}$$

will be denoted as

$$g(x) \xrightarrow{F} g(q) . \tag{A.32}$$

The next formulae are valid

$$\frac{d^\alpha}{dx^\alpha} g(x) \xrightarrow{F} (-iq)^\alpha g(q) , \tag{A.33}$$

$$\frac{d^\alpha}{d(-x)^\alpha} g(x) \xrightarrow{F} (iq)^\alpha g(q) . \tag{A.34}$$

Now introduce the operator of the symmetrized fractional derivative using the Fourier transforms (A.33), (A.34)

$$\frac{d^\alpha}{d|x|^\alpha} g(x) \xrightarrow{F} -|q|^\alpha g(q) . \tag{A.35}$$

Then, combining (A.35) and (A.29)-(A.31), obtain

$$\frac{d^\alpha}{d|x|^\alpha} = -\frac{1}{2\cos(\pi\alpha/2)} \left[\frac{d^\alpha}{dx^\alpha} + \frac{d^\alpha}{d(-x)^\alpha} \right] , \qquad (\alpha \neq 1) . \tag{A.36}$$

Finally, we will show that for any $\alpha > 0$ and $f(x)$ as a probability density function, the following identities are valid

$$\int_{-\infty}^{\infty} dx \frac{d^\alpha}{dx^\alpha} f(x) = \int_{-\infty}^{\infty} dx \frac{d^\alpha}{d(-x)^\alpha} f(x) = \int_{-\infty}^{\infty} dx \frac{d^\alpha}{d|x|^\alpha} f(x) \equiv 0 . \tag{A.37}$$

Indeed, in correspondence to (A.35) we have

$$\int_{-\infty}^{\infty} dx \frac{d^\alpha}{d|x|^\alpha} f(x) = -|q|^\alpha f(q) \Big|_{q=0} , \qquad (\alpha > 0) . \tag{A.38}$$

If $f(x)$ is a probability density, $f(q)$ is its characteristic function and $f(q = 0) = 1$. Then the right-hand side of (A.38) is zero.

References

1. E.W. Montroll and M.F. Shlesinger, in *Studies in Statistical Mechanics*, edited by J. Lebowitz and E.W. Montroll (North-Holland, Amsterdam, 1984), Vol. 11, p. 1.
2. M.F. Shlesinger, G.M. Zaslavsky, and J. Klafter. *Nature* **363**, 31 (1993).
3. G.M. Zaslavsky, in *Topological Aspects of The Dynamics of Fluids and Plasmas*, edited by H.K. Moffatt, G.M. Zaslavsky, P. Compte, and M. Tabor (Kluwer, Dordrecht, 1992), p. 481.
4. G.M. Zaslavsky, *Chaos* **4**, 25 (1994).
5. G.M. Zaslavsky, *Physica* D **76**, 110 (1994).
6. G.M. Zaslavsky, M. Edelman, and B.A. Niyazov, *Chaos* **7**, 159 (1997).
7. G.M. Zaslavsky, *Chaos* **5**, 653 (1995).
8. G.M. Zaslavsky and M. Edelman, *Phys. Rev.* E **56**, 5310 (1997).
9. S. Benkadda, S. Kassibrakis, R.B. White, and G.M. Zaslavsky, *Phys. Rev.* E **55**, 4909 (1997).
10. R.B. White, S. Benkadda, S. Kassibrakis, and G.M. Zaslavsky, *Chaos* **8**, No. 4 (1998).
11. V. Afraimovich, *Chaos* **7**, 12 (1997).
12. V. Afraimovich and G.M. Zaslavsky, *Phys. Rev.* E **55**, 5418 (1997).
13. M. Shlesinger, *Ann. Rev. Phys. Chem.* **39**, 269 (1988).
14. M. Kac, *Probability and Related Topics in Physical Sciences, Proc. of Summer Seminar, Boulder, Colorado, 1957* (Interscience, NY, 1958).
15. J.D. Meiss, *Phys. Rev.* A **34**, 2375 (1986); *Rev. Mod. Phys.* **64**, 795 (1992).
16. J.D. Meiss and E. Ott, *Phys. Rev. Lett.* **55**, 2741 (1985).
17. G.M. Zaslavsky and B.A. Niyazov, *Physics Reports* **283**, 73 (1997).
18. H.G.E. Hentschel and I. Procaccia, *Physica* D **8**, 435 (1983); P. Grassberger and I. Procaccia, *ibid.* **13**, 34 (1984).
19. J. Meiss, *Chaos* **7**, 39 (1997).
20. A.N. Kolmogorov, *Uspekhi Mat. Nauk* **5**, 5 (1938).
21. L.D. Landau, *Zhurnal Eksp. i Teor. Fiz* **7**, 203 (1937).
22. J.D. Hanson, J.R. Cary, and J.D. Meiss, *J. Stat. Phys.* **39**, 327 (1985).
23. A. Saichev and G.M. Zaslavsky, *Chaos* **7**, 753 (1997).
24. A.I. Saichev and W.A. Woyczńński, *Distributions in the Physical and Engineering Sciences,* (Birkhäuser, Boston, 1997), Vol. 1.
25. R. Hilfer, *Fractals* **3**, 549 (1995).
26. R. Hilfer and L. Anton, *Phys. Rev.* E **51**, R848 (1995).
27. P. Lévy, *Theorie de l'Addition des Variables Aleatoires* (Gauthier-Villiers, Paris, 1937).

28. I.M. Gelfand and G.E. Shilov, *Generalized Functions* (Academic, New York, 1964), Vol. 1.
29. K.S. Miller and B. Ross, *An Introduction to the Fractional Differential Equations* (John Wiley & Sons, NY, 1993).
30. F. Mainardi, *Chaos, Solitons, and Fractals* **7**, 17 (1996).
31. F. Mainardi, in *Fractals and Fractional Calculus in Continuum Mechanics* (Springer, New York 1997), p. 291, Eds. A. Carpinteri and F. Mainardi.
32. S.G. Samko, A.A. Kilbas, and O.I. Marichev, *Fractional Integrals and Derivatives and Their Applications* (Nauka i Tekhnika, Minsk, 1987). Translation by Harwood Academic Publishers).
33. D. Sornette, *Phys. Reports* **297**, 239 (1998).
34. S. Benkadda, S. Kassibrakis, R. White, and G.M. Zaslavsky, *Phys. Rev. E* (1998) (to be published).
35. T. Niemeijer and J. van Leeuwen, *Phase Transitions and Critical Phenomena*, Vol. 6. Eds. C. Domb and M. Green (Academic Press, London, 1976), p. 425.
36. B.D. Hughes, M.F. Shlesinger, and E.W. Montroll, *Proc. Natl. Acad. Sci. USA* **78**, 3287 (1981).
37. B.B. Mandelbrot, *The Fractal Geometry of Nature*, San Francisco, 1982.
38. M.F. Shlesinger and B. West, *Phys. Rev. E* **67**, 2106 (1991).
39. R. MacKay, *Renormalisation in Area-Preserving Maps*, Vol. 6 of *Advanced Series in Nonlinear Dynamics*, World Scientific, Singapore, 1993.
40. R. MacKay, J. Meiss, and I. Percival, *Physica D* **13**, 55 (1984).
41. J. Hanson, J. Cary, and J. Meiss, *J. Stat. Phys.* **39**, 327 (1985).

CHAPTER VI

POLYMER SCIENCE APPLICATIONS OF PATH-INTEGRATION, INTEGRAL EQUATIONS, AND FRACTIONAL CALCULUS

JACK F. DOUGLAS
National Institute of Standards and Technology, Polymers Division
Gaithersburg, Maryland 20899

Contents

1 Introduction

There are numerous applications in materials science and condensed matter physics involving complicated shaped boundaries and boundary data. Often these complex-shaped boundaries can be idealized as "fractal" rather than as smooth boundaries[1],[2], leading to challenging mathematical problems associated with the solution of partial differential equations or other functional equations involving such irregular boundaries. Formal treatment of transport properties involving rough boundaries is naturally made using path-integration methods since the formulation of the problem, at least, is straightforward. The treatment of this kind of problem requires the introduction of new analytical methods or numerical implementation of path averages by Monte-Carlo sampling. In particular, the analytic treatment of the configurational and transport properties of polymer chains frequently involve fractional order differential operators of various kinds. Both initial and boundary value fractional differential equations arise from the path-integral formulation of this type of problem, providing insight into the physical origin and geometrical interpretation of non-integer order differential operators under more general circumstances.

This broad class of problems is first illustrated by some polymer science applications where the configurational shapes of polymers are idealized as Brownian paths such that the chain contour position replaces the time coordinate of the Brownian paths. The path-integral descriptions of the interacting polymer models considered in this chapter (polymers interacting with surfaces of varying dimension and the friction coefficient of polymer chains) are converted into equivalent integral equation representations which are expressible in terms of fractional differential equations. These equations are then solved formally using the semi-group properties of the fractional order operators and eigenfunction expansion methods. Notably, the fractional order operators are not introduced arbitrarily, as in recent modeling of transport and relaxation in disordered media [3]−[7], but are obtained directly from the path-integral formulation. The interpretation of the fractional order exponent powers in terms of fractal geometry is also considered. Applications involving the configurational paths of polymers are followed by the modeling of relaxation in condensed materials. The relaxation modeling is similar to the interacting polymer problems where the "polymers" correspond to the phase space evolution of the dynamical system describing the condensed material.

2 Path-Integration Model of Flexible Polymers

The path-integral formulation of polymer science problems is well established in the physics literature, but this kind of modeling is probably less familiar in the mathematical community. In this section the physical motivation for a path-integral description of flexible polymer chains and the relation between the parameters in this model and the more familiar Brownian motion model is briefly reviewed.

The synthesis of commercially important polymers often involves the end-to-end addition of small groups of atoms to form long chain molecules containing many thousands of these "monomers". As a leading approximation, the polymer configuration may be identified as a sequence of independent random variables describing the position vector of each successive monomer unit with respect to the previous one. The length of each monomer unit is taken to be identical, but the orientation of each monomer depends on the orientational angle of rotation about each bond and the angle describing the relative monomer orientation in the plane containing successive bond position vectors. These orientational correlations generally depend on temperature through potentials governing these bond orientations,which complicates matters, but these interactions are usually short-ranged along the chain so that the polymer adopts a "random coil" conformation whose large scale dimensions can be reasonably described by a Gaussian distribution. The polymer chain is then appropriately modeled as a random walk in which the average length of each segment (the "Kuhn length" ℓ) is defined to absorb molecular details reflecting the short-ranged correlations within the chain [8]-[13]. An image of a representation "random coil" chain is shown in Fig. 1

The Brownian motion of a particle in solution can similarly be described by a succession of independent steps taken with random orientations and the properties of these particles involve an averaging over paths which can be conveniently framed in terms of Wiener path-integration. It is then possible to represent the configurations of flexible polymer chains in the same notation where time t is replaced by chain contour length n and the diffusion coefficient D of the Brownian particle is replaced by $\ell/2d$ where d is the spatial dimension. Based on this correspondence between static polymer configurations and the dynamic trajectories of Brownian motion, we then write the mean-square end-to-end distance between the ends of the flexible polymer chain $\langle|\mathbf{R}(N) - \mathbf{R}(0)|^2\rangle_o \equiv \langle|\mathbf{R}|^2\rangle_o$ as a Wiener path-integral (See Refs. [10]-[12] for further details) where the 'o' subscript denotes an "ideal" chain without interaction. The translational invariance of the end-to-end distance distribution function $G_o(|\mathbf{R}(N) - \mathbf{R}(0)|)$ allows us to take one end of the chain to be the

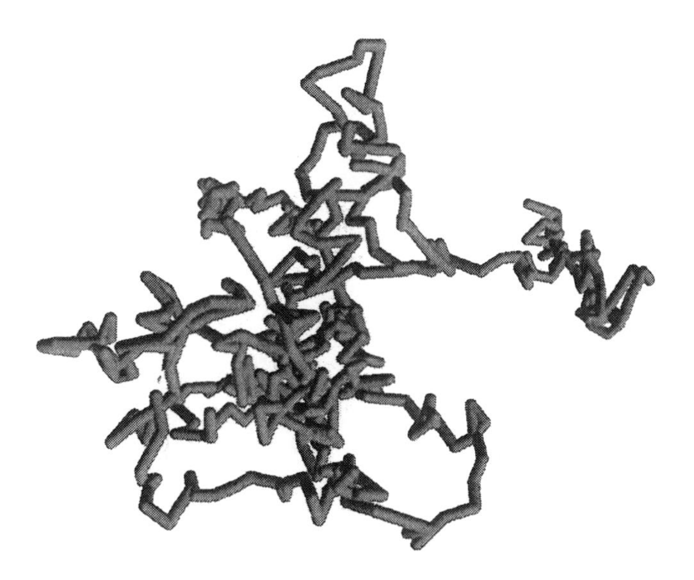

Figure 1: "Random coil" polymer chain. Each cylindrical section ("statistical segment length" ℓ) of the polymer representation connects the endpoints of a random walk having 50 elementary walk steps.

origin $\mathbf{R}(0) = \mathbf{0}$ without loss of generality so that $\langle|\mathbf{R}|^2\rangle_o$ equals,

$$\langle|\mathbf{R}|^2\rangle_o = \int d\mathbf{R}\ G_o(\mathbf{R}, N)|\mathbf{R}|^2/Q_o \tag{1}$$

$$Q_o \equiv \int d\mathbf{R}\ G_o(\mathbf{R}, N), \tag{2}$$

$$G_o(\mathbf{R}, N) = \int_{\mathbf{R}(0)=0}^{\mathbf{R}(N)=\mathbf{R}} D[\mathbf{R}(\tau)] \exp(-H_o/k_BT) \tag{3}$$

$$H_o = (d/2)\int_0^N d\tau\ |d\mathbf{R}(\tau)/d\tau|^2 \tag{4}$$

where k_BT denotes the thermal energy units. The evaluation of the path-integral Eq.(2) leads to the Gaussian distribution function [10]-[16],

$$G_o(\mathbf{R}, N) = (d/2\pi N\ell)^{d/2} \exp(-d|\mathbf{R}|^2/2N\ell) \tag{5}$$

which upon insertion into Eq.(1) gives,

$$\langle |\mathbf{R}|^2 \rangle_o = n\ell^2 = N\ell \tag{6}$$

It is experimentally difficult to measure $\langle |\mathbf{R}|^2 \rangle_o$ of polymers directly, but the closely related radius of gyration R_g can be determined by scattering techniques [10]-[16]. The radius of gyration of a Gaussian chain (second moment of the scattering structure factor) equals,

$$R_{g,o}^2 = n\ell/6 = \langle |\mathbf{R}|^2 \rangle_o/6 \tag{7}$$

In the discussion below, we define the Kuhn length as the unit of length $\ell = 1$ so that $\langle |\mathbf{R}|^2 \rangle_o = n$ and $N = n$. The relation Eq.(7) indicates that the average "size" of a flexible polymer chain should be proportional to the square root of its molecular weight M since $n \propto M$. For polymers dissolved in a solution of small molecules this situation is well realized at a special temperature ("theta-temperature") and the Gaussian chain model also provides a model for chains within a "melt" of other polymer chains [8]-[16]. More generally, the interaction of the polymer chain with boundaries and with itself (since no two monomers of the polymer chain can occupy the same physical space [8]-[16] - "the excluded volume problem") leads to generalizations of the Gaussian chain model (See below).

Polymer chains can be constructed to have different topologies for specific applications and the Gaussian chain model plays an important practical role in characterizing these polymer chain architectures [10]-[14],[15],[17]. For example, the ends of several linear polymer chains can be combined at a common junction to form a "star" polymer, their ends joined to form a "ring" polymer ("Brownian bridge") or polymer chains can be strung along another chain to create a polymer "comb". Various measures of polymer chain size can be calculated for these topologically restricted Gaussian chains and dimensionless ratios of the average chain size between the topologically constrained and unconstrained Wiener paths are invariants which are routinely employed to verify the form of newly synthesized molecules [17].

3 Surface-Interacting Polymers

The presence of a boundary near a polymer chain gives rise to correlations in the paths arising from the boundary constraint. The interaction may be attractive, in which case the chain configurations tend to lie closer to the surface or repulsive so that the paths tend to avoid the surface [18]-[22]. A surface interaction is incorporated into the Wiener path-integral model by introducing a weight on the paths which directly expresses the boundary constraint[18]-[22]. Specifically, the partition function Q_s of a Gaussian polymer terminally attached to a surface of dimension $d_{\|}$ ($d_{\|}=0$,point; $d_{\|}=1$,line; $d_{\|}=2$,plane, ...) can be written as the path-integral [18]-[23],

$$\begin{aligned} Q_s &= \int d\mathbf{R}\ G_s(\mathbf{R}, N), \\ G_s &= \int_{\mathbf{R}(0)=\mathbf{0}}^{\mathbf{R}(N)=\mathbf{R}} D[\mathbf{R}(\tau)] \exp[-(H_o + H_s)/k_B T] \end{aligned} \tag{8}$$

where H_o is defined as before by Eq.(4) and the surface interaction equals,

$$H_s/k_B T = \beta_S \int_0^n d\tau\ \delta[\mathbf{R}_\perp(\tau)] \tag{9}$$

The positon vector normal to the boundary $\mathbf{R}_\perp(\tau)$ corresponds to the components of $\mathbf{R}(\tau) = [\mathbf{R}_\perp(\tau),\ \mathbf{R}_{\|}(\tau)]$ orthogonal to the surface and $\delta(\cdot)$ is a delta-function which models the short-ranged polymer-surface interaction [18]-[23]. For simplicity in the present discussion one endpoint of the polymer chain is attached to the surface so that one chain end $\mathbf{R}(0)$ is positioned on the boundary at a point defined to be the origin. In this "penetrable" (e.g.,liquid-liquid interface) surface model the boundary constraint vanishes when the surface interaction vanishes, while in the "impenetrable" (solid interface) surface model the chain is constrained to lie on one side of the surface regardless of the polymer-surface interaction [20]. There is no difference between these models in the limit of a highly repulsive polymer-surface interaction [20]. The impenetrability constraint can be enforced by an additional step-function potential at the interface in $\mathbf{R}_\perp$ where the "impenetrability" coupling parameter is large [21].

The path-integral Eq.(8) can be evaluated using a variety of methods. For example, it is possible to expand $e^{-H_s/k_B T}$ formally as a Taylor series[23],

$$\exp(-H_s/k_B T) = 1 - H_s/k_B T + (H_s/k_B T)^2/2! + \cdots \tag{10}$$

Taking the average of this random variable with respect to unrestricted Wiener path-integration,

$$Q_s/Q_o \equiv \langle e^{-H_s/k_B T} \rangle \tag{11}$$

gives rise to the usual Feynman perturbation expansion[23],

$$Q_s/Q_o = 1 - z_s \int_0^1 dx\ x^{-d_\perp/2} + z_s^2 \int_0^1 dx' \int_0^{x'} dx\ x^{-d_\perp/2}(x'-x)^{-d_\perp/2} + O(z_s^3) \tag{12}$$

where Q_o is the partition function of the chain without the polymer-surface interaction and z_s is the dimensionless interaction parameter,

$$\begin{aligned} z_s &= (d/2\pi l^2)^{d_\perp/2}\beta_s\, n^{\phi_s}, \\ \phi_s &= (2+d_\| - d)/2, \quad 0 < \phi_s < 1 \end{aligned} \tag{13}$$

The diagramatic expansion associated with Eq.(12) and the evaluation of these integrals is discussed by Douglas et al. [23] and others [24].

The problem of a surface-interacting polymer provides one of the few examples where the Feynman expansion can be computed exactly to all orders[23],

$$Q_s/Q_o = \sum_{k=0}^{\infty}[-\Delta_s]^k/\Gamma(1+k\phi_s) \tag{14}$$

$$\begin{aligned} \Delta_s &= z_s\,\Gamma(1+\phi_s)/u_s^*, \\ u_s^* &= (2+d_\| - d)/2 \end{aligned} \tag{15}$$

$\Gamma(\cdot)$ denotes the gamma function and this expansion is recognized as the Mittag-Leffler function [23],[28]-[33] $E_\phi(-\Delta_s)$ of index ϕ_s. Eq.(14) is plotted for a range of ϕ_s values in Fig. 2.

Path-integrals having the form Eq.(8) can also be expressed in terms of equivalent differential or integral equations. Douglas [34] has shown that the partition function Q_s for a surface interacting polymer is also described by the Volterra-type integral equation,

$$\psi(x) = 1 - \Delta_s \int_0^x dt\ [(x-t)^{\phi_s - 1}/\Gamma(\phi_s)]\psi(t), \quad Q_s/Q_o \equiv \psi(x=1) \tag{16}$$

An iterative solution [34] of Eq.(16) leads to a Neumann expansion which recovers [23] the same series obtained from the path-integral, Eq.(11). Finally, we note that Q_s/Q_o can be derived starting from a diffusion equation and mixed boundary conditions involving the β_s parameter [18]-[22] so that the surface interacting polymer can be formulated from a variety of equivalent ways.

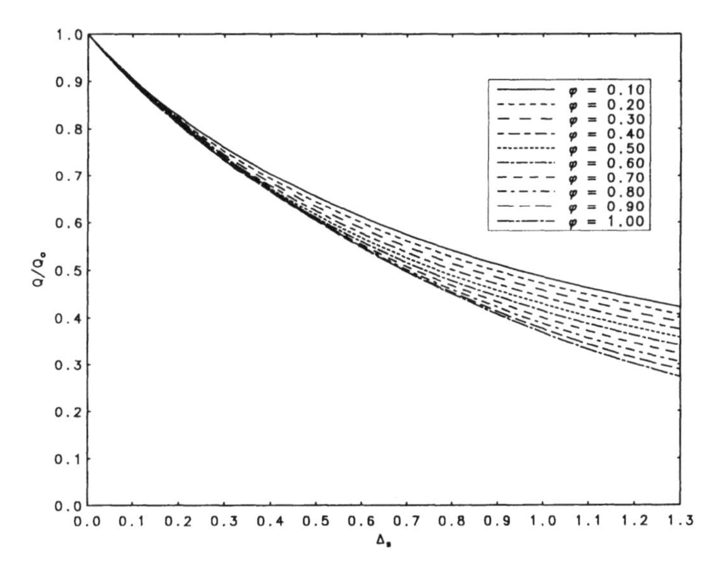

Figure 2: Partition function ratio Q_s/Q_o for surface interacting polymer. Curves are the Mittag-Leffler function $E_\phi(-\Delta_s)$ for ϕ values in the range [0.1,1].

Insight into the surface interacting polymer problem and more complicated interacting polymer problems can be obtained by noting that the integral operator in Eq.(16) is a Riemann-Liouville (RL) fractional differential operator [39]-[43] defined by,

$$I_x^\alpha f(x) = \int_0^x dt \;\; [(x-t)^{\alpha-1}/\Gamma(\alpha)] f(t) \tag{17}$$

where $f(x)$ of Eq.(17) is multiplied by a Heaviside step-function $\theta(x)$ and is chosen from a class of functions such that the operator is defined to exist (See Hille [44]-[46]).

The Riemann-Liouville operator I_x^α defines a continuous semi-group which forms the basis of an operational calculus [44]-[46]. The main properties of the fractional differential operator I_x^α are prescribed by the relations [34],[44]-[46],

$$I_x^\alpha I_x^\beta f(x) = I_x^{\alpha+\beta} f(x) \tag{18}$$

$$I_x^\alpha I_x^\beta f(x) = I_x^\beta I_x^\alpha f(x) \tag{19}$$

$$\begin{aligned} \lim_{\alpha\to 0+} I_x^\alpha f(x) &= I_x^{-\alpha} I_x^\alpha f(x) \\ &= I^o f(x) \\ &= f(x) \end{aligned} \tag{20}$$

$$x^{-\alpha} I_x^\alpha x^m = [\Gamma(m+1)/\Gamma(m+1+\alpha)]x^m \tag{21}$$

where $\alpha, \beta > 0$. The first identity defines exponent additivity, the second commutativity,
the third defines the identity operator and the left inverse operation, while the fourth defines a useful eigenvalue relation. Note that the kernel $(x-x')^{-(1-\alpha)}/\Gamma(\alpha)$ becomes a delta-function $\delta(x-x')$ in the limit $\alpha \to 0^+$, leading to the identity operator in Eq.(21). Fig. 3 shows the kernel for a range of α values and the approach to the delta-function limit for $\alpha \to 0^+$. It is pointed out that the introduction of a step-function factor $\theta(x)$ of $f(x)$ or the kernel $x^{\alpha-1}/\Gamma(\alpha)$ allows for the operators to be interpreted in terms of generalized functions for negative integer α values (i.e., extension of I_x^α to full group properties) under certain circumstances. Formal manipulations for negative α must be made with care, of course, and the results often have meaning only in a distributional sense.

The integral equation Eq.(16) can be expressed simply in terms of RL operator I_x^α as [34],

$$(1 + \Delta_s I_x^{\phi_s})\, \psi(x) = 1 \tag{22}$$

which can be formally rearranged and expanded for small Δ_s to obtain the Neumann expansion [34],

$$\psi(x) = 1/(1 + \Delta_s I_x^{\phi s}) = \sum_{k=0}^{\infty} (-\Delta_s)^k I_x^{k\phi_s} \tag{23}$$

Since, $I_x^{k\phi_s} = x^{k\phi_s}/\Gamma(1+k\phi_s)$, we trivially recover Eq.(14). The "strong coupling" expansion of Q_s/Q_o is determined by expanding Eq.(22) around the

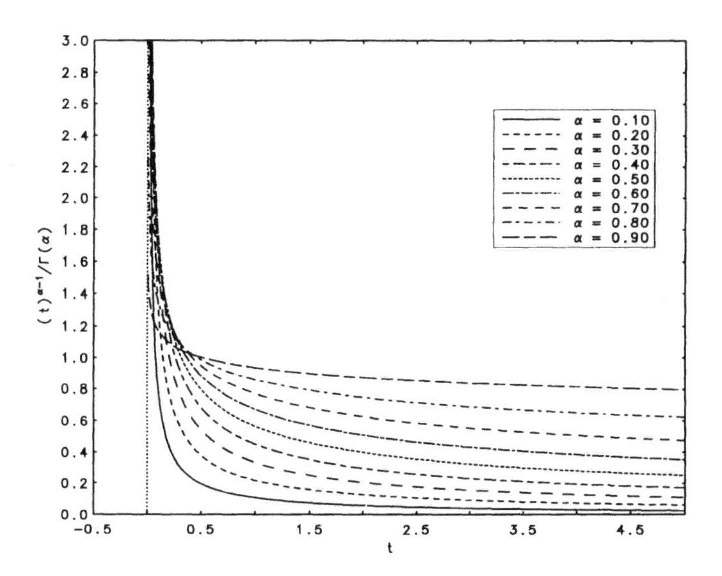

Figure 3: Kernel function $(x)_+^{\alpha-1}/\Gamma(\alpha)$ of Riemann-Liouville operator. Note that kernel approaches a delta function for $\alpha \to 0^+$ and the + subscript indicates that the function equals zero for $x < 0$.The kernel approaches a Heaviside stepfunction $\Theta(x)$ for $\alpha \to 1$.

large Δ_s limit [34]. Thus, we have formally

$$\begin{aligned}\psi(x) &= (\Delta_s I_x^{\phi_s})^{-1}/[1+(\Delta_s I_x^{\phi_s})^{-1}] \\ &= (\Delta_s I_x^{\phi_s})^{-1}\sum_{k=0}^{\infty}(-\Delta_s)^k (I_x^{\phi_s})^{-k}\end{aligned} \tag{24}$$

which gives rise to the asymptotic expansion,

$$Q_s/Q_o \sim (-1)\sum_{k=1}^{\infty}(-\Delta_s)^{-k}/\Gamma(1-k\phi_s), \quad \Delta_s \to \infty \tag{25}$$

or in leading order,

$$Q_s/Q_o \sim [\Delta_s\Gamma(1-\phi_s)]^{-1}[1+O(\Delta_s^{-1})], \quad \Delta_s \to \infty \tag{26}$$

This asymptotic expansion Eq.(26) defines a non-trivial "susceptibility" critical exponent γ_s describing the variation of Q_s/Q_o with chain length (See definition of Δ_s in Eq.(15)[34]),

$$Q_s/Q_o \sim n^{\gamma_s - 1}, \quad \gamma_s = 1-\phi_s, \; n \to \infty \tag{27}$$

A power law scaling of polymer properties with chain length in the limit of long chains is typical in polymer science and related critical phenomena problems, although it is usually difficult to calculate the critical exponents such as γ_s exactly [10],[13]. Eq.(26) implies that the fraction of all continuous (Wiener) paths is diminished by a factor $n^{-\phi_s}$ for paths initiating from a repulsive ($\Delta_s \to \infty$) surface, relative to the number of paths without the surface constraint (See Fig. 2).

The exponent ϕ_s, the "crossover exponent" in the polymer science or critical phenomena literature, has a geometric interpretation as the Hausdorff ("fractal") dimension of the contour points τ_i at which the polymer (Wiener) paths encounter the surface [47]-[52]. This geometrical interpretation of ϕ_s can be appreciated by considering the surface interaction Hamiltonian, $H_s/k_BT = \int_0^n d\tau\, \delta[\mathbf{R}_\perp(\tau)]$ which defines a counting function for the "local time" spent by the path at the boundary. In Fig. 4 we show some realizations of the random variable H_s/k_BT for random walks in $d = 2$ intersecting a line ($d_\| = 1$) from which the paths initiate, $\mathbf{R}_\perp(0) = \mathbf{0}$. An average over an ensemble of random walk trajectories (128 paths), leads to the solid line in Fig. 4, which is consistent with the expected exponent, $\phi_s(d = 2, d_\| = 1) = 1/2$. Similar geometrical interpretations of "critical exponents" such as γ_S and ϕ_S in the surface interacting polymer problem arise in many other types of interacting polymer problems.

The path-integral model of surface-interacting polymers in Eq.(8) is also interesting because it exhibits a phase transition as β_s changes sign ($n \to \infty$ limit). At this point the polymer-surface interaction changes from being predominantly repulsive to attractive and the polymer becomes confined to the surface. An exact description of this transition can be obtained from the properties of Mittag-Leffler function [28]-[34],[44] describing Q_s/Q_o in Eq.(14).

A consideration of the temperature dependence of the free-energy ΔF of the adsorbed polymer chain,

$$-\Delta F/k_BT = \ln(Q_s/Q_o) \tag{28}$$

near the transition point ($\beta_s \to 0^+$) reveals [23][34] the strength or degree of the transition (generalized "order" of transition in Ehrenfest sense [36][37][38]). The surface interaction coupling constant β_s is well approximated near the transition temperature T_c by a linear relation [20],[22],

$$\beta_s \propto (T - T_c)/T_c \tag{29}$$

and the asymptotic properties of the Mittag-Leffler function for $\beta_s < 0$ and $n \to \infty$ imply [23] [See Eq.(14)],

$$\Delta F/k_BT \sim |z_s|^{1/\phi_s}, \quad z_s \ll 0 \tag{30}$$

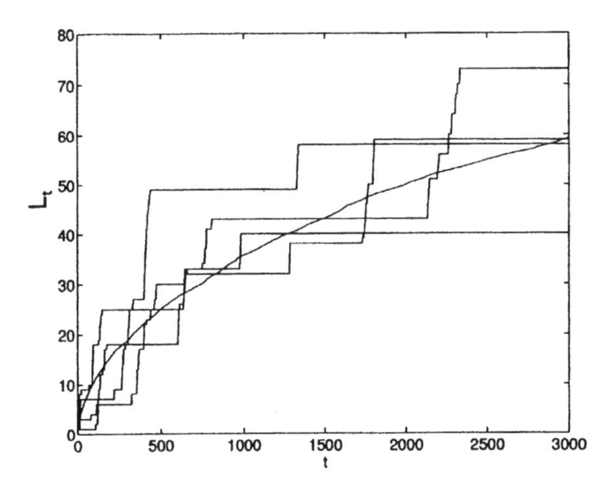

Figure 4: The local time L_t of a random walk in a plane at a designated line. Four representative random walk trajectories are considered where the initial step is positioned at the origin and successive returns to an arbitrarily designated line (containing the origin) are counted as the "local time", $L_t = \int_0^t d\tau\ \delta[X(\tau)]$. The smooth curve represents an average over 128 path realizations and the average $\langle t \rangle$ scales as $t^{1/2}$.

Equivalently, the free-energy change ΔF per unit chain length n of an adsorbed chain then scales as,

$$\Delta F/nk_BT \sim |\beta_s|^{1/\phi_s} \sim |T - T_c|^{1/\phi_s}, T < T_c \tag{31}$$

(Constant prefactors are neglected in Eqs.(29) and (30), but these constants are prescribed in Ref. [23]). The "degree" of the transition (generalized to fractional order through fractional order differential operators [36],[37]) then equals $1/\phi_s$, which indicates a direct relation between the phase transition order and the geometry [47]-[52] of path intersections through ϕ_s.

The result Eq.(14) for $Q_s/Q_o = \langle e^{-H_I/k_BT} \rangle$ is representative of a large class of interacting polymer problems in the limit of long chains. Darling and Kac [53] have considered a general class of interaction Hamiltonians ("additive functionals of Brownian motion"),

$$H_I/k_BT = \beta_I \int_0^n d\tau\ V[\mathbf{R}(\tau)] \tag{32}$$

where the positive (one-body) potential $V[\mathbf{R}(\tau)]$ is constrained by the requirement that the expansion Eq.(10) exists (See Darling and Kac [53] for a discussion of the technical restrictions on this potential). The asymptotic evaluation of the path-integral $\langle e^{-H_I/k_BT}\rangle$ leads to a remarkable "universal" functional limit theorem for long chains [34],[53]-[59],

$$Q(\beta_I)/Q_o \sim \sum_{k=0}^{\infty}(-\Delta_I)^k/\Gamma(1+k\delta), \;\; 0<\delta<1, \; n\to\infty \tag{33}$$

$$\Delta_I = \beta_I \; n^\delta g_s(n) \tag{34}$$

where $g_s(n)$ is a "slowly varying function" (i.e., $\lim_{n\to\infty} g_s(kn)/g_s(n) = 1$ for every positive k). The exponent δ arises from putting growth constraints on the allowable $V(\mathbf{R})$. For a homogeneous potential $V_{Hom}(\mathbf{R})$ we have the defining scaling property [34],

$$V_{Hom}(\lambda\mathbf{R}) = \lambda^h V_{Hom}(\mathbf{R}) \tag{35}$$

and dimensional analysis implies,

$$\delta = (2+h)/2 \tag{36}$$

The delta-function $\delta[\mathbf{R}_\perp(\tau)]$ potential in Eq.(9) has a homogeneity index $h = -d_\perp$ which leads to $\delta = \phi_s$ and thus Eq.(9) is a special case of Darling and Kac's general limit theorem.

Eq.(33) can also be deduced from Feller's "fluctuation theory of recurrent events" [60]. The contacting of the surface by the chain is a "recurrent event" and we define N_n as a random variable which counts these independent random events with "time" n (or contour length). Feller showed that the generating function $\langle e^{-N_n}\rangle$ for N_n has the general asymptotic behavior,

$$\langle e^{-N_n}\rangle \sim \sum_{k=0}^{\infty}(-\Delta_F)^k/\Gamma(1+k\delta), \;\; 0<\delta<1, \; n\to\infty \tag{37}$$

$$\Delta_F = n^\delta g_s(n) \tag{38}$$

in the case of a random walk interacting with a surface where the probability distribution for the length of time between renewals (i.e.,"loop length") has a long tail so that the average "renewal time" (first moment of renewal time distribution) is infinite. The polymer Hamiltonian H_s/k_BT in Eq.(9) is an example of such a renewal process so that Eq.(14) follows directly from Eqs.(11) and (37). The investigation of Darling and Kac was motivated by Feller's earlier derivation of Eq.(37)(See Ref.[61] for a discussion of renewal theory and random walks). The Feller and Darling and Kac theoretical frameworks provide a natural starting point for calculating the partition function of polymers interacting with fractal surfaces where the limit theorems Eqs.(33) and (37) should apply and where δ is determined by geometrical considerations [47],[50].

The limit theorems Eqs.(33) and (37) provide insight into the mathematical mechanism behind the observation of "universality" in polymer science and in a larger critical phenomena context. Many physical systems exhibit asymptotic scaling for long chains in polymeric systems or near a critical point in phase transitions (e.g., phase separation of fluid mixtures) where there is insensitivity to the form of the interaction potential. Functionals of fluctuation processes occuring in such systems are described by universal limit theorems [See Eq.(33)] having a similar mathematical origin as the "central limit theorem" governing the asymptotic Gaussian distribution for the chain end-points of long polymer chains [See Eq.(5)].

The problem of a polymer interacting with a surface also provides a novel viewpoint on the renormalization group(RG) and ϵ-expansion methods that are formally utilized in many critical phenomena and polymer science applications. As mentioned above, this is one of the few examples where a non-trivial Feynman expansion can be calculated through all orders. The series coefficients in Eq.(12) exhibit the usual poles in ϵ_s ($\phi_s = 2 - d_\perp/2 = \epsilon_s/2$) found in field theoretic calculations and the perturbative RG method can be applied to resum these expansions. Importantly, the error of the ϵ-expansion can be exactly determined in this example [23].

A discussion of the RG treatment of surface-interacting polymers [23] would involve a long technical digression which is avoided here, but contact with the results of these calculations can be easily made through the RL fractional order I_x^α operator, discussed above. The perturbative RG ϵ-expansion method is based upon a formal expansion about the "critical dimension", corresponding to the vanishing of the "crossover exponent" ϕ_s in Eq.(13). Thus, we consider the partition function ratio Q_s/Q_o of a surface-interacting chain relative to an unattached "free" chain in solution,

$$Q_s/Q_o = 1/(1 + \Delta_s I_x^{\phi_s})|_{x=1} \tag{39}$$

and take the "critical dimension" limit $\phi_s \to 0^+$ formally where we have the approximations,

$$\Delta_s \approx z_s/u_s^*, \ I_x^{\phi_s} \approx 1, \ \phi_s \to 0^+ \tag{40}$$

so that Eq.(39) reduces to the simple expression,

$$Q_s/Q_o \approx 1/(1 + z_s/u_s^*) \tag{41}$$

This is exactly the scaling function deduced from the RG analysis in Refs. [23],[34]. The origin of the pole structure in the perturbation expansion such as Eq.(14) and the asymptotic scaling of $Q_s(\Delta_s)$ for Δ_s large are readily understood as consequences of the presence of these fractional order differential operators [34].

Fractional differential equations involving Riemann-Liouville operators arise in many other polymer science applications. The treatment of many chains end-grafted onto a surface leads to integral equations similar to those arising in the "tautochrone" problem of classical mechanics (See e.g. Ref.[40]). In this polymer "brush" model the layer density profile is calculated for chains that are constrained to arrive at the surface after equal "times" (contour length) [26]. The helix-coil transition [25] of m polymer strands is also closely related to the surface-interacting polymer problem and mutually interacting "directed" polymers[24]. Modeling [27] of the diffusion-controlled growth of copolymer layers at the interface of a blend of phase separated polymers involves non-linear fractional differential equations which generalize Eq.(22).

Because of the ubiquity of random walk processes in physics, there are numerous non-polymeric problems which reduce mathematically or which are closely related to the polymer adsorption problem–the quantum mechanical density matrix for a particle interacting through a delta-function potential [62]-[66], magnetic systems with interacting defect planes [67], heat and mass transfer by diffusion from boundaries [68]-[69], electrochemical response of rough electrodes [70], rates of diffusion-limited reactions [71] and the hydrodynamics of particles moving through viscous liquids [72]-[73]. This model also arises in the description of relaxation in condensed materials and applications to this important area of polymer science are discussed in Sect.7. Eq.(14) continues to be rediscovered in other contexts. The present discussion of polymer science applications can then be seen to have a much broader physical applicability.

Generalizations of the Hamiltonian H_s/k_BT in Eq.(9) often lead to intractable mathematical problems, despite the apparent simplicity of the statement of these problems in terms of path-integration. A classical example is the "excluded volume interaction", H_{EV}[10]-[13],[16],[74],[75],

$$H_{EV}/k_BT = (\beta_2/2!)\int_0^n d\tau \int_0^n d\tau' \delta[\mathbf{R}(\tau) - \mathbf{R}(\tau')] \tag{42}$$

which is a counting function for the number of Brownian path self-interactions. A constraint ("cut-off") must be included on H_{EV} in Eq.(42) to avoid having segments unphysically interacting with themselves so that we must take $|\tau -$

$\tau'| \geq a$ where a is a positive constant on the order, [10]-[13] $a \sim O(\ell)$. The "renormalization" of the random variable H_{EV} to account for this effect has been discussed rigorously [76] and recent probablistic studies have focused on the average of H_{EV} for Wiener paths and more general random walk path processes [77]-[82].

Substantial efforts have been made developing expansions of the form Eq.(12) which are widely compared to experimental data for the swelling of polymers in small molecule solvents [10]-[13],[16],[83]. These expansions, whose evaluation is facilitated by the introduction of fractional differential operators [34], are referred to as the "two-parameter theory" or the "Edwards' model" in the polymer science literature [10]-[16]. Success has been obtained in calculating polymer properties based on perturbative calculations which start from this model in conjunction with perturbative RG methods to guide the resummation of these series expansions in term of the dimensionless excluded volume ("Fixman") parameter z_2,

$$\begin{aligned} z_2 &= (d/2\pi l^2)^{d/2}\, n^{\phi_2}, \\ \phi_2 &= (4-d)/2 \equiv \epsilon/2 \end{aligned} \tag{43}$$

The interaction parameter z_2 can be deduced simply by introducing dimensionless variables [10]-[13] in Eq.(42) and is evidently a counterpart of z_s in Eq.(13). Refs. [10]-[13],[16] review the progress in this important class of polymer science applications.

Part of the difficulty in evaluating the path-integral $\langle e^{-H_{EV}/k_B T}\rangle$ of a self-interacting polymer is associated with expressing this type of functional in terms of integral or differential equations. Attempts at obtaining such a description lead to infinite "hierarchies" of equations whose treatment requires the introduction of closure approximations of the kind well known in liquid state theory [10]-[13]. Perturbative expansion as in Eq.(10) is possible, but such expansions normally have a vanishing radius of convergence so that these expansions must be interpreted as "asymptotic". This property complicates the interpretation of these rather involved calculations. The RG theory and Borel resummation methods allow for the transformation of these perturbation expansions to obtain asymptotic approximants applicable to a wider range of the z_2 interaction, but the calculations remain formal and approximate. This difficulty in evaluating many-body functionals is common in physics and indeed the polymer excluded volume problem corresponds to the $m \to 0^+$ limit [84]-[87] of the general $O(m)$ model of phase transitions ($m = 1$, Ising model; $m = 2$, XY model; $m = 3$, Heisenberg model; $m \to \infty$, spherical model). Many phase transition models can be equivalently formulated in terms of self-interacting polymers and thus these models are similarly intractable as the Hamiltonian in Eq.(45). We briefly note a fractional calculus based method for estimating the "correlation length" exponent ν governing the size ($\langle|\mathbf{R}|^2\rangle \propto n^{2\nu}$) of swollen polymer chains to further explore the possibilities of fractional calculus methods in a more complex situation.

The perturbative expansion describing the chain swelling due to excluded volume is developed in a similar manner to the surface interacting polymer. In leading order, the distribution function governing the displacement of the chain end $\mathbf{R}(N)$ from its initial position $\mathbf{R}(0) = \mathbf{0}$ equals [10]-[13],[23],

$$\begin{aligned} G_{EV}(\mathbf{R}; z_2) &= G_o(\mathbf{R}, 1) \\ &\quad - z_2 \int_0^1 dx'(1-x')(2\pi)^{d/2} G_o(\mathbf{R}, x') G_o(\mathbf{R}, 1-x') + O(z_2^2) \end{aligned} \tag{44}$$

where the contour variable τ has been normalized by the chain length, $x' = \tau/N$ and the cut-off a in Eq.(42) has been neglected [10]-[13] ("dimensional regularization"). The leading order perturbation expansion [23] for the distribution function $G_s(\mathbf{R}_\perp, N)$ of the free-end of an end-tehered chain has the same form as Eq.(45), except that z_2 is replaced by z_s and the extra factor $(1 - x')$ in Eq.(45) which comes from folding the double integral in Eq.(42). Further details on this type of expansion are desribed in Refs. [10]-[13] and [23].

We obtain an estimate of the exponent ν for repulsively selfinteracting chains by replacing the "bare" or "unperturbed" distribution function $G_o(\mathbf{R}, N)$ [See Eq.(5)] in the integral describing the excluded volume interaction in Eq.(45) by the perturbed distribution function $G_{EV}(\mathbf{R}, z_2)$. This replacement is consistent with perturbation theory to first order in z_2. Next, the problem is simplified to an integral equation in the chain coordinate by considering the limit $|\mathbf{R}| \to 0^+$ in Eq.(45). Formal, but non-rigorous, arguments [15] indicates the exact scaling relation for self-avoiding chains is given by,

$$G_{EV}(|\mathbf{R}| \to 0^+, z_2 \to \infty) \equiv G_{EV}^*(n) \sim n^{-d\nu} \sim \langle|\mathbf{R}|^2\rangle^{-d/2} \tag{45}$$

with prefactors unspecified and unknown [Compare with the exact stable random walk result in Eq.(132)]. Inserting Eq.(45) in Eq.(45) leads to the fractional differential equation,

$$G_{EV}^*(n) = (d/2\pi n)^{-d/2} - \hat{\beta}_2 \int_0^n d\tau (n-\tau) G_{EV}^*(\tau)(n-\tau)^{-d\nu} \tag{46}$$

where the excluded volume coupling constant $\hat{\beta}_2$ absorbs unspecified constants. The operational solution of Eq.(46) is then obtained similarly to Eq.(22) for surface-interacting polymers as,

$$G^*_{EV}(n) = [1 + \hat{\beta}_2 \Gamma(\phi^*_2) I_n^{\phi^*_2}]^{-1} (d/2\pi n)^{-d/2}, \quad \phi^*_2 \equiv 2 - d\nu \tag{47}$$

The large n scaling of Eq.(47) implies $G^*_{EV}(n) \sim n^{-\phi^*_2 - d/2} \sim n^{-d\nu}$ which by scaling consistency yields,

$$\nu = (4+d)/4d, \quad 2 \leq d < 4 \tag{48}$$

$$\nu = 1/2, \quad d > 4 \tag{49}$$

The result Eq.(49) follows from the observation that $z_2 \to 0^+$ for long chains ($n \to \infty$), provided $d > 4$ [See Eq.(45)]. Eq.(48) is consistent with the leading order RG ϵ-expansion [10]-[13],[16], $2\nu = 1 + (\epsilon/8)^2 + O(\epsilon^2)$, and gives a value believed to be exact [10]-[13],[16] $\nu = 3/4$ in $d = 2$. The estimate for $d = 3$, $\nu = 7/12 \approx 0.583$, accords well with numerical estimates which are typically close to the value [10]-[13],[16], $\nu(d = 3) \approx 0.588$. Integral equations very similar to Eq.(47) arise in the theory of non-linear radiative heat conduction [69].

The exponent ν can alternatively be estimated by averaging the Hamitonian Eq.(42) in a self-consistent fashion similar to the quantum mechanical self-consistent field calculations [10]-[13],[74],[75]. This averaging procedure involves [43] the kernel of the Riesz operator, discussed in the next section, and leads to the ν estimate,

$$\nu = (3+d)/2, \quad 2 \leq d < 4 \tag{50}$$

$$\nu = 1/2, \quad d > 4 \tag{51}$$

which is originally due [8]-[13] to Flory for $d = 3$. Recent numerical lattice model estimates of ν summarized in Ref.[61] accord rather well with the estimates in Eqs.(48) and (50).

We next turn to another class of polymer applications involving the transport properties of polymer solutions. These problems can likewise be expressed in terms of path-integration and solved by fractional calculus methods. New types of fractional differential operators arise in these hydrodynamic applications, however.

4 Translational Friction of Polymer Chains

The analytic solutions of exterior boundary value problems involving complex shaped particles, such as polymer chains, is generally difficult. Progress can be obtained by recasting these problems formally in terms of path-integration [88] and obtaining approximate numerical solutions by evaluating the path-integrals describing the boundary value problem using Monte Carlo sampling of random walk paths [89]-[91],[93]. Numerical results provide an important reference point for comparison in developing analytic theories of boundary value problems involving complex surfaces.

The calculation of the friction coefficient f of a randomly coiled polymer chain undergoing Brownian motion in solution is a challenging example of this class of problems. Kirkwood and Riseman (KR)[94] established the general mathematical framework for calculating f based on the assumption that the chain is rigid and very slender so that a polymer chain is modeled as an array of hydrodynamic sources ("beads") strung along the random walk polymer chain. We start with the same physical model of a polymer chain, but proceed to calculate f starting from a formulation for objects having general shape that can be cast in terms of path-integration.

Hubbard and Douglas [97] recently derived an expression for the translational friction coefficient f of rigid Brownian particles. Stokes law for arbitrarily-shaped particles is accurately approximated by [91]-[99],

$$f = 6\pi\eta_s C_\Omega \tag{52}$$

where η_s is the solvent viscosity and C_Ω is the electrostatic capacity [89],[90], [100],[101] of the particle Ω in units such that a sphere of radius R has the capacity $C_\Omega = R$. Eq.(52) is exact for ellipsoidal particles and this approximation holds to about 1% in other cases where exact calculation of f are known[97]. The friction coefficient is experimentally related to the diffusion coefficient D of a Brownian particle of general shape, $D = k_B T/f$, which is routinely measured by dynamic light scattering techniques [102]. Within the "angular averaging" approximation used in obtaining Eq.(52), the problem of calculating f for a polymer chain reduces to the solution of Laplace's equation [97],

$$\nabla^2 \Phi(\mathbf{R}) = 0 \tag{53}$$

on the exterior of the polymer chain where $\Phi(\mathbf{R}) = 1$ for $\mathbf{R}$ on the polymer boundary polymer and $\Phi(|\mathbf{R}| \to \infty) \to 0$. The capacity C_Ω determines the decay rate of $\Phi(R)$ at a large distance $R = |\mathbf{R}|$ from the polymer,

$$C_\Omega \equiv \lim_{R\to\infty} [R\Phi(R)] \tag{54}$$

Numerical path-integration (See below for theoretical background.) calculations of C_Ω for "random coil" polymer chains are given by Douglas et al. [89],[90] and these results in conjunction with Eq.(52) are consistent with recent calculations of f based on a numerical treatment of the KR model without preaveraging [89],[90]. In this chapter, the calculation of f is pursued through analytic path-integration and fractional calculus methods. The numerical path-integration calculations have the advantage that arbitrary complexity of monomer structure (variations in monomer structure in random copolymers and in monomer shape) can be included without a substantial increase in computational effort.

The capacity of an object having general shape can be calculated by the methods of probablistic potential theory. Following Kac [103]-[104] in a different context, we consider the functional integral [103]-[105],

$$\Psi(\mathbf{R}) = \langle \exp(-\int_0^\infty d\tau \;\; V[\mathbf{R}(\tau) + \mathbf{R}]) \rangle \tag{55}$$

where we take $\mathbf{R}(0) = \mathbf{0}$ as in the previous section, and the potential $V(\mathbf{R})$ is taken to be positive and integrable,

$$V(\mathbf{R}) \geq 0, \quad \int d\mathbf{R} \;\; V(\mathbf{R}) \leq \infty \tag{56}$$

The integral of $V(\mathbf{R})$ is taken over all d-dimensional space where $d > 2$. As usual, the functional integral can be converted into an equivalent integral equation [103]-[105],

$$\Psi(\mathbf{R}) \;\; = \;\; 1 - \int d\mathbf{R}\, K(\mathbf{R},\mathbf{R}')V(\mathbf{R}')\Psi(\mathbf{R}'), \quad d > 2 \tag{57}$$

$$\begin{aligned} K(\mathbf{R},\mathbf{R}') &= [2/H_d]/|\mathbf{R}-\mathbf{R}'|^{d-2}, \\ H_d &= 4\pi^{d/2}/\Gamma(d/2-1) \end{aligned} \tag{58}$$

where $K(\mathbf{R},\mathbf{R}')$ is the Green's function of the free-space Laplacian (∇^2) operator. Treatment of the $d = 2$ case requires some modifications in this argument [104] which are important for defining capacity in $d = 2$.

The capacity C_Ω of a set Ω naturally arises in connection with the calculation of low-energy scattering lengths for certain classes of potentials in quantum theory. The "scattering length" Γ associated with $V(\mathbf{R})$ is obtained as an integral [103]-[107] of $\Psi(\mathbf{R})$,

$$\Gamma = (2/H_d) \int d\mathbf{R} \ V(\mathbf{R})\Psi(\mathbf{R}) \tag{59}$$

The relation between Γ and C_Ω requires a further specification of the potential function, $V(\mathbf{R})$. Suppose there is a bounded set Ω whose capacity we wish to know (e.g.,a "random coil" polymer) and we define $V(\mathbf{R})$ to be constant (taking the value $\beta > 0$) on this region and zero otherwise. In other words, we take $V(\mathbf{R})$ as proportional to the characteristic function $\chi(\mathbf{R})$ of the set and define,

$$\Psi(\beta, \mathbf{R}) = \langle \exp(-\beta \int_0^\infty d\tau \ \chi[\mathbf{R} + \mathbf{R}(\tau)]) \rangle \tag{60}$$

$$\begin{aligned} \chi(\mathbf{R}) &= 1 \quad , \quad \mathbf{R} \in \Omega \\ \chi(\mathbf{R}) &= 0 \quad , \quad \mathbf{R} \notin \Omega \end{aligned} \tag{61}$$

so that Eq.(57) simplifies to the form ($d = 3$),

$$\Psi(\beta, \mathbf{R}) = 1 - \hat{\beta} \int_\Omega d\mathbf{R} \ \Psi(\beta, \mathbf{R})/|\mathbf{R} - \mathbf{R}'| \tag{62}$$

where the kernel norming constant H_d is adsorbed into a redefinition of the coupling constant β,

$$\hat{\beta} \equiv \beta \ (2/H_d) \tag{63}$$

The characteristic function χ now restricts the integral in Eq.(62) to the region Ω. The scattering length Γ from Eq.(59), corresponding to the potential defined in Eq.(61), then equals,

$$\Gamma(\beta) = \hat{\beta} \int_\Omega d\mathbf{R} \ \Psi(\beta, \mathbf{R}) \tag{64}$$

The capacity C_Ω of Ω is defined through the limit [103]-[107],

$$\Gamma(\beta \to \infty) = C_\Omega \tag{65}$$

For impenetrable objects ($\beta \to \infty$) the scattering length and capacity are equivalent [103]-[107].(Subtleties regarding paths which "touch" but do not penetrate the region Ω and variations in the capacity definition are discussed by Strook [126]). Note that the function $\Psi(\beta \to \infty, \mathbf{R})$ corresponds to the probability that a path starting from $\mathbf{R}$, exterior to Ω, does not hit Ω [103]-[113]. The "capacitory potential" describing the probability of a Wiener path to hit Ω then equals,

$$\lim_{\beta \to \infty} [1 - \Psi(\beta, \mathbf{R})] = P(\mathbf{R}), \quad \mathbf{R} \notin \Omega \tag{66}$$

The limiting function $P(\mathbf{R})$ is a harmonic function on the exterior of Ω and satisfies the Dirichlet boundary condition in Eq.(53). Interestingly, the equipotential surfaces represent surfaces for which the probability for random walks initiating from these surfaces to hit Ω is constant. A basic property of C_Ω is that it vanishes for any finite collection of points or smooth (positive integer order differentiable) curves in $d = 3$. The average capacity of Wiener paths is finite in $d = 3$ because of the non-differentiability of a typical Brownian path.

The function $\Psi(\beta, \mathbf{R})$ for $\mathbf{R} \in \Omega$ also has a probabilistic interpretation in terms of the "equilibrium charge density" $\mu(\mathbf{R})$,

$$\lim_{\beta \to \infty} [\beta \Psi(\beta, \mathbf{R})] = \mu(\mathbf{R}), \quad \mathbf{R} \in \Omega \tag{67}$$

$\mu(\mathbf{R})$, when normalized, is the density of first hitting points [114]-[117] of Ω for Brownian paths initiating a large distance from Ω. (This probability is obviously conditioned on those Brownian paths which actually reach Ω). The integral of $\mu(\mathbf{R})$ over Ω equals the capacity [103]-[105],

$$C_\Omega = \int_\Omega d\mathbf{R} \;\; \mu(\mathbf{R}) \tag{68}$$

As mentioned before, the capacity C_Ω characterizes the asymptotic decay of the random walk hitting probability $P(\mathbf{R})$ of random walks launched a large distance away from Ω,

$$P(\mathbf{R}) \sim C_\Omega / |\mathbf{R}|^{d-2}, \quad |\mathbf{R}| \to \infty \tag{69}$$

C_Ω is also minimizes the functional of Ω, *(Gauss' Principle)*

$$1/C_\Omega = \frac{\int_\Omega \int_\Omega [\frac{\mu(\mathbf{R}')\mu(\mathbf{R})}{|\mathbf{R}-\mathbf{R}'|^{d-2}} d\mathbf{R} d\mathbf{R}']}{[\int_\Omega \mu(\mathbf{R}') d\mathbf{R}']^2} \tag{70}$$

This variational principle corresponds to minimizing the energy $E = 1/2C_\Omega$ of a normalized charge distribution $\mu(\mathbf{R})$ on the conductor Ω. Dirichlet's integral defines a complementary variational priciple [88],[100],[101] for C_Ω in which the integral of the square gradient of $\Phi(\mathbf{R}) = P(\mathbf{R})$ is minimized (*Dirichlet's Principle*) on the exterior of Ω so that the field energy on the exterior of Ω is minimized. Eq.(53) is the Euler equation associated with the variation of this functional acting on the exterior of Ω.

The operator in Eq.(57) is a special case of the Riesz [118] fractional order operator I_d^α,

$$I_d^\alpha f(\mathbf{R}) = \int_\Omega d\mathbf{R}' \; f(\mathbf{R}')K_\alpha(\mathbf{R},\mathbf{R}') \tag{71}$$

$$\begin{aligned} H_d(\alpha) &= \pi^{d/2}2^\alpha\Gamma(\alpha/2)\Gamma[\frac{(d-\alpha)}{2}], \\ K_\alpha(\mathbf{R},\mathbf{R}') &= |\mathbf{R}-\mathbf{R}'|^{-(d-\alpha)}/H_d(\alpha) \end{aligned} \tag{72}$$

which has the semi-group properties [44]-[46],[118], $(\alpha+\beta < d; \alpha,\beta > 0)$,

$$I_d^\alpha I_d^\beta f(\mathbf{R}) = I_d^{\alpha+\beta} f(\mathbf{R}) \tag{73}$$

$$I_d^\alpha I_d^\beta f(\mathbf{R}) = I_d^\beta I_d^\alpha f(\mathbf{R}) \tag{74}$$

$$\nabla^2 I_d^{\alpha+2} f(\mathbf{R}) = -I_d^\alpha f(\mathbf{R}), \;\; \nabla^2 = \sum_{i=1}^{d}(\partial^2/\partial x_i^2) \tag{75}$$

$$\lim_{\alpha\to 0+} I_d^\alpha f(\mathbf{R}) = f(\mathbf{R}) \tag{76}$$

As for the RL operator the limit $\alpha \to 0^+$ implies the distributional relation [119]-[124],

$$\lim_{\alpha\to 0+} |\mathbf{R}-\mathbf{R}'|^{-(d-\alpha)}/H_d(\alpha) \to \delta(\mathbf{R}-\mathbf{R}') \tag{77}$$

It is reiterated that the formal manipulation of fractional differential operators having negative α values requires caution and the functions $f(\mathbf{R})$ are restricted to a class for which the integrals exist (See Hille [44]-[46]). The value of fractional calculus to a physicist, however, often amounts to ignoring these restrictions in exploratory calculations of physical significance. Rigor can come later.

A solution of Eq.(57) can be expresed formally in terms of the I_d^α operator,

$$\Psi(\mathbf{R}) = 1 - \lambda_\Omega I_d^2 \Psi(\mathbf{R}), \;\; d > 2 \tag{78}$$

where λ_Ω equals β_Ω times $H_d(\alpha)$ and some other constants associated with the definition of the Wiener path diffusion coefficient in the polymer model [See Eq.(1)]. Eq.(78) has the formal Neumann series solution,

$$\begin{aligned} \Psi(\mathbf{R}) &= 1/[1+\lambda_\Omega I_d^2] \\ &= \sum_{k=0}^{\infty}(-\lambda_\Omega)^k [I_d^2]^k, \quad \lambda_\Omega \to 0^+ \end{aligned} \tag{79}$$

where $(I_d^2)^k(1)$ defines the kth moment of the occupation time [125],[126] of a Brownian path in the region Ω. Explicit analytic calculation of $\Psi(\mathbf{R})$ and the capacity for complicated shaped particles based on Eqs.(78) and (79) is possible in only a limited number of cases.

Some simplification occurs for polymers and other slender bodies which can be modeled as line-like paths. The integral equations defining the chain capacity then reduces to a one-dimensional integral equation defined along the chain path. To achieve this reduction, consider a string of "beads" of radius a on a polymer (Wiener path) chain where ℓ is the unit of contour distance separating the beads. Let the number of chain units be large so that the potential of the chain can be considered to be continuously distributed along its contour,

$$V_\zeta[\mathbf{R}(\tau)] = \zeta_c \int_0^n d\tau \;\; \delta[\mathbf{R} - \mathbf{R}(\tau)] \tag{80}$$

where ζ_c is the capacity of a single spherical bead $\zeta_c = a^{d-2}$ and $\delta(\cdot)$ is a delta-function concentrated along the chain contour. Inserting Eq.(80) into Eqs.(57) - (59) and adsorbing the H_d factor into the coupling parameter ζ_c as in Eq.(60) gives,

$$C_\zeta = \int d\mathbf{R} \left\langle (\zeta_c \int_0^n d\tau \;\; \delta[\mathbf{R} - \mathbf{R}(\tau)]\Psi(\mathbf{R}) \right\rangle \to \zeta_c \int_0^n d\tau \;\; \Psi[\mathbf{R}(\tau)] \tag{81}$$

$$\Psi[\mathbf{R}(\tau)] = 1 - \zeta_c \int_0^n d\tau[\Psi[\mathbf{R}(\tau)]/|\mathbf{R}(\tau) - \mathbf{R}(\tau')|^{d-2}] \tag{82}$$

We then need to configurationally average $\langle\cdots\rangle$ these equations over the Wiener path (flexible polymer) configurations,

$$\langle C_\zeta \rangle = \zeta_c \int_0^n d\tau \ \langle \Psi[\mathbf{R}(\tau)] \rangle \tag{83}$$

$$\langle \Psi[\mathbf{R}(\tau)] \rangle = 1 - \zeta_c \int_0^n \langle d\tau [\Psi[\mathbf{R}(\tau)] / |\mathbf{R}(\tau) - \mathbf{R}(\tau')|^{d-2}] \rangle \tag{84}$$

where the order of integration over chain length n is formally interchanged with the configurational average. Following Kirkwood and Riseman's [94],[127] slender body treatment of polymer friction f we introduce a "configurational preaveraging approximation",

$$\langle \Psi[\mathbf{R}(\tau)] / |\mathbf{R}(\tau) - \mathbf{R}(\tau')|^{d-2} \rangle \approx \langle \Psi[\mathbf{R}(\tau)] \rangle / \langle |\mathbf{R}(\tau) - \mathbf{R}(\tau')|^{d-2} \rangle \tag{85}$$

where the average of a product is formally factored into a product of the averages (a mean-field superposition approximation). Introduction of Eq.(85) into Eq.(84) leads to an integral equation for the average capacity $\langle C_\zeta \rangle$ which is mathematically equivalent to the Kirkwood-Riseman integral equation for a "random coil" polymer chain [94],[127] [This observation by the author provided the original impetus for the general friction-capacity relation Eq.(52)]. It is possible to calculate corrections to the approximation Eq.(85) perturbatively using the RG method [127], but the approximation Eq.(83) is sufficient for the present discussion.

The configurationally preaveraged KR integral equation for the friction coefficient f_{KR} of a random coil polymer equals [34],[94]-[96],[127],

$$f_{KR} = n\zeta \int_0^n dx \ \Psi(x), \tag{86}$$

$$\Psi(x) = 1 - (2^{\phi_H} h) \int_0^n dx \ \Psi(x) |x - x'|^{\phi_H - 1},$$

$$\phi_H = (4 - d)/2,$$

$$h = (d/2\pi)^{(d-2)/2} \beta_H n^{\phi_H}, \tag{87}$$

$$\beta_H = (\zeta / 2\pi d \eta_s)(d-1) 2^{-\phi_H} / (d-2),$$

where ζ is the Stokes friction coefficient of a spherical bead of radius a in

d-dimensions [97],[128],

$$\zeta = [d/(d-1)]H_d a^{d-2}\eta_s, \quad d > 2 \tag{88}$$

The integral equation Eq.(86) follows from Eqs.(83) to (85) by noting that $\langle|\mathbf{R}(\tau) - \mathbf{R}(\tau')|^{-(d-2)}\rangle \propto |x - x'|^{-(d-2)/2}$ and transforming the integral to the unit interval.

Despite its apparent simplicity, this equation has defied previous attempts at exact solution. Kirkwood and Riseman [94],[127] obtained an incorrect solution[127] based on a method suggested by Kac (See Ref. [94]). This "approximate" solution is often employed as being "sufficient" for practical purposes, however. An exact solution is known in the strong hydrodynamic interaction limit, $h \to \infty$ (See below).

The difficulty of solving Eq.(86), either analytically or numerically, is associated with the singular nature of the kernel $|x - x'|^{-(d-2)/2}$ which is not square-integrable for $0 \leq \phi_H \leq 1/2$ (The physically interesting case of $d = 3$ corresponds to $\phi_H = 1/2$). Unique solutions do not necessarily exist for Fredholm equations having such singular kernels or an infinity of solutions can exist[129]. Another disturbing situation which can arise with this type of kernel is that different methods of solution can lead to different, apparently unrelated, solutions [130]. We also observe that the classical Fredholm determinant method can not be applied for such singular kernels. Hilbert [131] introduced a technique("the method of pulling the poison tooth") for extending the Fredholm theory to singular kernels which are at least square-integrable, but this early "regularization" technique does not help us solve Eq.(86) in the physically interesting range $0 \leq \phi_H \leq 1/2$. The mathematical purist might also be disturbed by the RG ϵ-expansion approach which involves a formal expansion about the $\phi_H = 0$ limit where the kernel changes from being "weakly" to "strongly" singular. Despite these difficulties we note that the kernel $|x - x'|^{\phi_H - 1}$ is "compact" for $0 < \phi_H < 1$, positive, and "symmetric" [135]-[137] and, moreover, Eq.(86) is a variety of fractional differential equation (see below). These important properties allow for its formal solution by eigenvalue expansion and fractional calculus methods.

The solution of Eq.(86) is facilitated by expressing it in terms of a class of fractional differential operators introduced by Feller [138]-[145], which are a specialization of a more general class of operators introduced earlier by

Riesz[118]. The "Riesz-Feller" operators are defined by the integral relation,

$$\begin{aligned} A_x^\alpha f(x) &= \int_{-1}^{+1} dt\ [|x-t|^{\alpha-1}/H(\alpha)]f(t), \\ H(\alpha) &= 2\Gamma(\alpha)\cos(\alpha\pi/2) \end{aligned} \tag{89}$$

where $f(x)$ is defined to vanish outside the interval (-1,1) and is restricted to a class of functions for which Eq.(90)is well defined [44]-[46]. The operator A_x^α has the semi-group properties [138]-[145]:

$$A_x^\alpha A_x^\beta f(x) = A_x^{\alpha+\beta} f(x), \quad (\alpha, \beta > 0, \alpha+\beta < 1) \tag{90}$$

$$A_x^\alpha A_x^\beta f(x) = A_x^\beta A_x^\alpha f(x) \tag{91}$$

$$(D_x)^2 A_x^{\alpha+2} f(x) = -A_x^\alpha f(x), \quad D_x \equiv d/dx \tag{92}$$

$$\lim_{\alpha\to 0+} A_x^\alpha f(x) = f(x) \tag{93}$$

The operator $A_x^{-\alpha}$ should be thought of as "symbolically equivalent" to the differential operator [138]-[152], $(-D_x^2)^{\alpha/2}$. These operators have a special significance for random walk theory which explains their occurrence in Eq.(86). $A_x^{-\alpha}$ is the "generator" of symmetric stable processes of index α and the kernel of Eq.(86) is the potential function associated with these processes (See Sect. 5). Feller [138] also considers more general fractional differential operators which are the generators of "one-sided" stable processes. These generalized operators also reduce to the RL operators discussed in the previous section.

The KR integral equation Eq.(86) takes a simple form in terms of the operator A_x^α,

$$f_{KR} = n\zeta \int_{-1}^{+1} dx\ [\Psi(x)/2], \tag{94}$$

$$\begin{aligned} (1+\lambda A_x^{\phi_H})\Psi(x) &= 1 \\ \lambda &= hH(\phi_H), \\ H(\phi_H) &= 2\Gamma(\phi_H)\cos(\phi_H\pi/2) \end{aligned} \tag{95}$$

Various methods of formal solution of Eq.(94) are now possible. First, we may exploit the compactness of kernel of the operator A_x^ϕ for $0 < \phi_H < 1$ which allows for a generalization of the classical Hilbert-Schmidt [153]-[157] method of expanding the solution of Eq.(94) in terms of the eigenfunctions of the associated homogeneous eigenvalue equation,

$$\Psi_k(x) = \lambda_k^* \, A_x^{\phi_H} \, \Psi_k(x) \tag{96}$$

$$\lambda_k^* = h_k^* \, H(\phi_H) \tag{97}$$

where the λ_k^* form a discrete eigenvalue spectrum (tabulated below) in which k denotes the mode index ($k = 1, 2, 3, ...$). By virtue of the compactness property of the $A_x^{\phi_H}$ operator, the eigenfunctions $\Psi_k(x)$ form a complete set in the Hilbert space L^2(-1,1) and we may expand $\Psi(x)$ and 1 in Eq.(94) in the basis functions $\Psi_k(x)$,

$$\begin{aligned} 1 &= \sum_{k=1}^{\infty} a_k \Psi_k(x), \\ \Psi(x) &= \sum_{k=1}^{\infty} b_k \Psi_k(x) \end{aligned} \tag{98}$$

where the $\Psi_k(x)$ are taken to be orthonormal,

$$\langle \Psi_k | \Psi_m \rangle \equiv \int_{-1}^{+1} dx \; \Psi_k(x) \Psi_m(x) = \delta_{k,m} \tag{99}$$

and the coefficients a_k are obtained through the inner product relation,

$$a_k \equiv \langle \Psi_k | 1 \rangle = \int_{-1}^{+1} dx \; \Psi_k(x) \tag{100}$$

The solution $\Psi(x)$ in terms of the $\Psi_k(x)$ then has the formal expansion,

$$\Psi(x) = \sum_{k=1}^{\infty} a_k \Psi_k(x) / (1 + \lambda / \lambda_k^*) \tag{101}$$

so that the KR friction coefficient f_{KR} then equals,

$$f_{KR} = (n\zeta/2) \sum_{k=1}^{\infty} a_k^2 / (1 + \lambda / \lambda_k^*) \tag{102}$$

This solution has limited value unless the eigenvalues λ_k^* and eigenfunctions $\Psi_k(x)$ can be determined, however. Exact calculations of these quantities remains a difficult problem, but approximations can be obtained which reveal properties of the exact solution. Future work should consider the convergence properties of these formal series.

First, it is useful to focus on the eigenvalues λ_k^* which extremize the generalized "Raleigh quotient",

$$1/\lambda_k^* = \langle \Psi_k | A_x^{\phi_H} | \Psi_k \rangle \tag{103}$$

The λ_k^* can be written in terms of the characteristic values h_k^* of the "hydrodynamic interaction parameter" h [See Eq.(95)],

$$1/h_k^* = \int_{-1}^{+1} dx \int_{-1}^{+1} dx' \Psi_k(x) |x - x'|^{\phi_H - 1} \Psi_k(x') \tag{104}$$

Eq.(103) is a generalization of the vibrating string problem(Sturm-Liouville theory) and this connection will become more explicit below in the process of evaluating the eigenvalues h_k^*.

Numerical calculations show that the eigenvalue of smallest magnitude h_1^* varies differently with dimension (or equivalently ϕ_H) than the remaining eigenvalues. The estimation of h_1^* is facilitated by the observation that for $\phi_H = 1$ the solution of Eq.(95) becomes,

$$\Psi_1 = 2^{-1/2}, \; h_1^* = 1/2, \; \phi_H = 1. \tag{105}$$

This is the only eigenvalue and eigenfunction for this value of ϕ_H so that Eq.(94) has the simple solution,

$$\Psi(x) = 1/[1 + h/h_1^*] \tag{106}$$

Above we restricted ϕ_H to the open interval (0,1) where there are a countable infinity of eigenfunctions since the eigenvalue structure changes radically either as ϕ_H approaches 0 or 1. It is also important to realize that the hydrodynamic interaction parameter h in Eq.(95) is not well defined for $\phi_H = 1$ because of the Stokes paradox (vanishing of translational friction coefficient due to recurrence of random walks) and a treatment of this problem requires a modification of the kernel in Eq.(86). This physical complication is not relevant to the mathematical problem of calculating h_1^*.

The exact solution Eq.(104) for $\phi_H = 1$ suggests that a constant trial function should give a reasonable estimate of h_1^*. This simple approximation

Table 1: First Eigenvalue h_1^* of the KR Eigenvalue Equation

$\phi_H = (4-d)/2$	(numerical) h_1^*	analytical; Eq.(107) h_1^*
1.0	0.500	0.500
0.9	0.458	0.458
0.5	0.264	0.265
0.4	0.211	0.211
0.2	0.104	0.104

yields,

$$h_1^*(\phi_H) \approx \phi_H(1+\phi_H)/2^{1+\phi_H}, \quad 0 < \phi_H < 1 \tag{107}$$

Comparison of this analytic estimate of h_1^* with values obtained by numerical solution of Eq.(103) show that Eq.(107) is indeed accurate (See Table 1 and Fig. 5). Of course, refined estimates of $h_1^*(\phi_H)$ can be obtained from recursive techniques, such as Kellogg's method for estimating eigenvalues of Fredholm equations [153]-[157]. The expression Eq.(107) is sufficient for the present discussion of the qualitative nature of the $h_1^*(\phi_H)$ spectrum. Calculation of h_k^* for large k involves a generalization of Weyl's theorem [160]-[161] for the asymptotic eigenvalue spectrum of the Laplacian ∇^2 to fractional order operators, $(\nabla^2)^{\alpha/2}$. This connection is natural given that the Riesz operator [see Eqs.(73) - (77)] is the fractional order generalization of ∇^2 (See next subsection) and that A_x^ϕ involves a specialization of the these operators to one dimension, the chain contour coordinate.

It is well known that the asymptotic eigenvalue spectrum of the Laplace operator for bounded domains of rather general shape having a variety of boundary conditions can be calculated using Kac's "principle of not feeling the boundary"[162]-[163], which allows the use of the short time expansion of random walk processes to deduce high frequency properties of vibrating continua. It is less well known that Kac also calculated the equivalent of the asymptotic variation of h_k^* using this same kind of probabilistic path-integration approach [164]. Of course, this calculation involves the introduction of generalized random walks paths associated with the $A_x^{-\alpha}$ operators and the corresponding path-integral description. This problem is briefly discussed in the next section.

After some transcription to the notation of this chapter the generalization

Table 2: Eigenvalues h_k^* of the KR Eigenvalue Equation

Mode number $(k, \phi_H = 1/2)$	(numerical) h_k^*	Eq.109 h_k^*
3	0.767	0.771
5	1.04	1.05
7	1.26	1.26
9	1.45	1.45

of Weyl's theorem to fractional order operators $A_x^{\phi_H}$ is given by,

$$h_k^* \sim (k\pi/2)^{\phi_H}/H(\phi_H), \quad k \to \infty \tag{108}$$

Note the power of k directly reflects the order of the fractional order operator involved. The equivalent of the fundamental asymptotic result Eq.(108) has also been obtained by direct analytic methods by Widom [166]-[168] and Rosenblatt [169] starting from the integral equation Eq.(96)(See also ref.[169]-[172]). The usual vibrating string (Sturm-Liouville theory) scaling of the eigenvalue spectrum is recovered for the special case $\phi_H = 2$, corresponding to a second order differential operator. Interestingly, the connection of Eq.(96) with the Riesz-Feller operator was not appreciated in any of these previous calculations where fractional order operators are not mentioned.

Further numerical progress in an equivalent problem of estimating h_k^* was made by Ukai [174] who obtained the equivalent of corrections to the asymptotic result Eq.(108) using Hilbert transform methods,

$$h_k^* \sim (k\pi/2)^{\phi_H}[1 - (1 + \phi_H/2)/2k]^{\phi_H}/H(\phi_H) + O(k^{-d/2}), \quad k \to \infty \tag{109}$$

This expression leads to rather good estimates for h_k^* in comparison with numerical values calculated directly from Eq.(96) using a method in which the kernel is expanded in Chebyshev polynomials. Values of h_k^* for the physically interesting case of $d = 3(\phi_H = 1/2)$ are given in Table 2 and Fig. 5. Only the odd eigenvalues ($k \geq 3$) are tabulated since the a_k in Eq.(98) for even k vanish by symmetry, $\psi_{2k}(x) = -\psi_{2k}(x)$. Notably Eq.(109) gives a poor approximation for the leading eigenvalue and Eq.(107) should be used in this case. Improved h_k^* estimates for the lower modes can be obtained using recursive methods and Ukai[174], for example, finds $h_{k=3}^*(\phi_H = 1/2) = 0.768$ in the next order of recursion, which agrees well with the numerically calculated value 0.767 in Table 2. The corrections are smaller for higher k values and the values in Table 2 are unchanged in the higher order recursive calculations.

With some effort highly accurate analytic estimates of the eigenvalues h_k^* can be obtained.

The eigenvalue spectrum h_k^* shown in Fig. 5 for $1 \leq k \leq 10$ reveals an interesting variation as a function of d. As $\phi_H \to 0^+$ (near the "critical dimensionality" in critical phenomena jargon) the eigen-values become nearly degenerate as the kernel becomes "singular" (not integrable in the ordinary sense). This kernel $|x - x'|^{\phi_H - 1}/H(\phi_H)$ becomes a delta-function in the $\phi_H \to 0^+$ limit (See Fig. 3), as is required by the existence of the identity operator in the Riesz-Feller semi-group [See Eqs.(90)-93]. Indeed, every continuous function $f(x)$ defined on the interval (-1,1) is an eigenfunction of the $A_x^{\phi_H}$ operator in the $\phi_H \to 0^+$ limit, amounting to a extreme example of degeneracy. The spectrum also takes on curious features as $\phi_H \to 1^-$ where all the eigenvalues, but the lowest magnitude one, h_1^*, diverge to infinity. This phenomena is important for solving fractional differential equations of the form Eq.(86) which arise in applications because the Neumann perturbative series (or "Feynman" expansion" from the equivalent path-integral [171]-[173]) have radii of convergence which are determined by the smallest magnitude eigenvalue [153]-[157]. The radius of convergence of such series apparently becomes vanishingly small near the "critical dimension", $\phi_H \to 0^+$. The perturbative RG method (see below) exploits this singular situation.

Analytic calculations of the eigenfunctions $\psi_k(x)$ of the KR eigenvalue equation Eq.(96) is a more difficult matter than the eigenvalues h_k^*. Numerical calculation is possible, however, and representative eigenfunctions ($k = 1, 3, 5; \phi_H = 1/2$) are shown in Fig 6.

It is seen that the $\psi_k(x)$ rather resemble Jacobi polynomials and this superficial observation can be understood from a related eigenvalue equation derived by Pólya and Szegö[175],

$$\int_{-1}^{+1} dt \;\; K_\phi(t,x)\Phi_k(t) \;=\; \lambda_k^{(\phi)}\Phi_k(x), \;\; 0 < \phi < 1 \tag{110}$$

$$K_\phi(t,x) \;=\; (1-t^2)^{-\phi/4}[|x-t|^{\phi-1}/H(\phi)](1-x^2)^{-\phi/4} \tag{111}$$

$$\lambda_k^{(\phi)} \;=\; \Gamma(k+1-\phi)/\Gamma(k+1), \tag{112}$$

$$\lambda_k^{(\phi)} \;\sim\; k^{-\phi} \;\; as\; k \to \infty \tag{113}$$

where the eigenfunctions $\Phi_k(x)$ are proportional to ultraspherical polynomi-

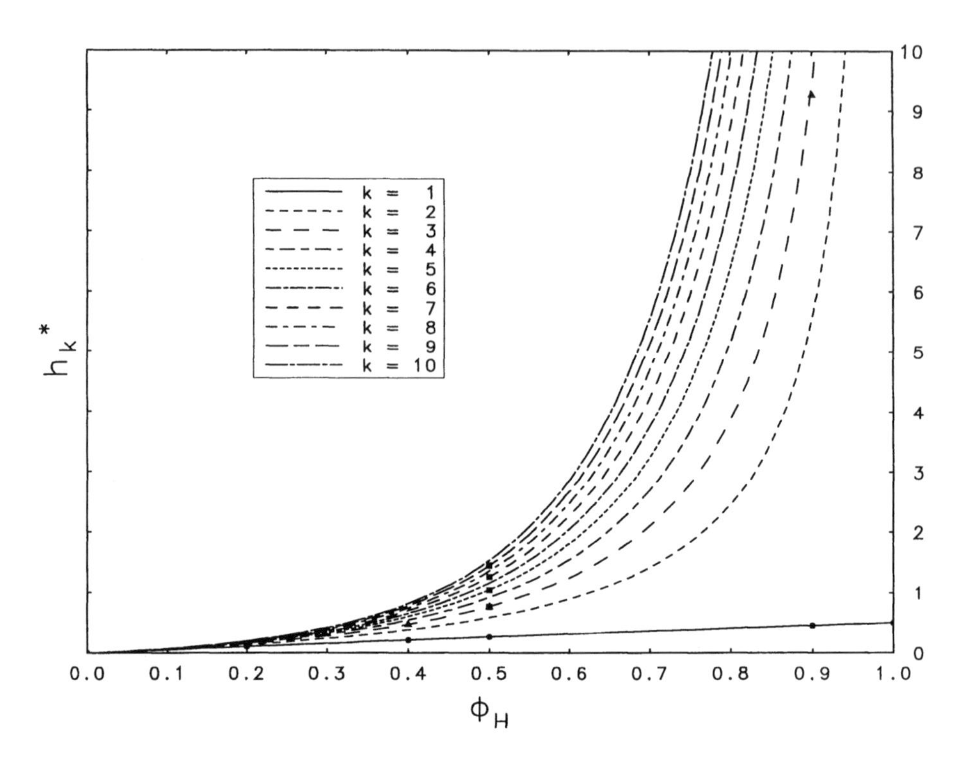

Figure 5: Leading eigenvalues h_k^* of Kirkwood-Riseman integral equation, $1 < k < 10$. h_k^* are observed to coalesce to a value of zero as the critical dimension is approached ($\phi_H \rightarrow 0^+$) while the eigenvalues rapidly spread out ($k \geq 2$) as the lower critical dimension is approached ($\phi_H \rightarrow 1^-$).The smallest magnitude eigenvalue, which is related to the fixed-point h_{RG}^* of the RG theory, exhibits a weaker dimensional variation. Solid lines exhibit analytic eigenvalue estimates from Eqs.(107) and (109) and filled symbols ($\blacksquare, \phi_H = 1/2, k = 3,5,7,9; \blacktriangle, \phi_H = 0.2, 0.4, 0.5, 0.9, k = 3$) denote eigenvalue estimates from the numerical solution of the Kirkwood-Riseman integral equation. $\bullet$ symbols show the smallest eigenvalue h_1^*.

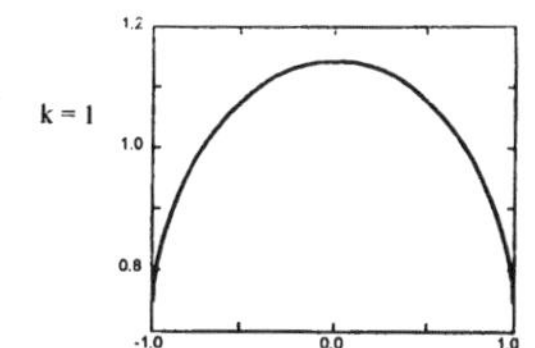

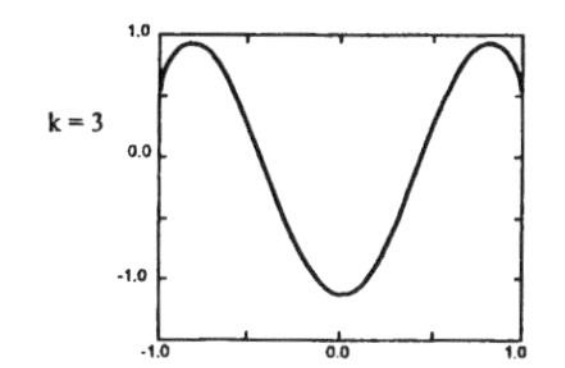

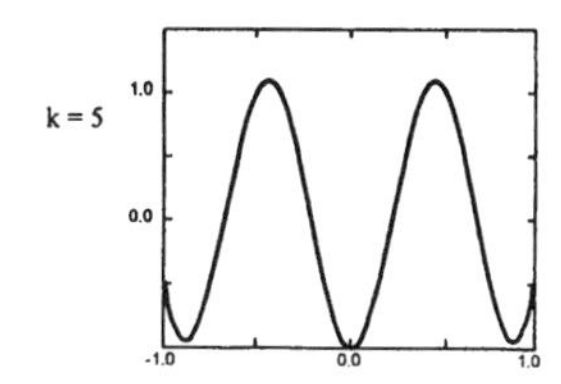

Figure 6: Representative eigenfunctions of the Riesz-Feller eigenvalue equation,Eq.(96). The even function modes ($k = 1, 3, 5; \phi_H = 1/2$) shown are important for calculating the polymer friction coefficient. Note similarity to Jacobi polynomials multiplied by their weight function.

als[175], $P_k^{(-\phi/2,-\phi/2)}(x)$,

$$\Phi_{k+1}(x) = (1 - x^2)^{-\phi/4} P_k^{(-\phi/2,-\phi/2)}(x)/N_k \tag{114}$$

which can also be expressed as Gegenbauer polynomials[175]. The constants N_k are prescribed by normalizing the $\Phi_k(x)$-functions,

$$\langle \Phi_m | \Phi_n \rangle = \delta_{m,n} \tag{115}$$

For small $\phi \approx 0$ we have the approximation, $K_\phi(t,x) \approx |x - t|^{\phi-1}/H(\phi)$, which is the kernel of the A_x^ϕ operator. The eigenfunctions $\Phi_k(x)$ are useful for constructing the operator inverse to A_x^ϕ needed in the exact solution of Eq.(94) in the limit $h \to \infty$ (see below). Good approximations to the polymer friction coefficient can be made by making a basis set transformation from the ψ_k in Eq.(96) to the Φ_k using the expansion of the kernel $|x - x'|^{\phi-1}$ in the Φ_k functions to perform this "rotation" in Hilbert space. This method, which is similar to basis set transformations commonly employed in molecular spectroscopy, will not be developed here.

As an alternative to the eigenfunction expansion method we can pursue the direct operator expansion approach as in the Riemann-Liouville semi-group calculation of the preceeding section and Ref. [34]. Formally, Eq.(94) is rewritten as,

$$\psi(x) = 1/[1 + \lambda A_x^{\phi_H}] \quad \text{or} \quad (1 + \lambda A_x^{\phi_H})\psi(x) = 1 \tag{116}$$

and $\psi(x)$ is developed as a Neumann expansion about the $\lambda \to 0^+$ limit,

$$\psi(x) = \sum_{k=0}^{\infty} (-\lambda A_x^{\phi_H})^k = 1 - \lambda A_x^{\phi_H} + O(h^2) \tag{117}$$

This expansion is analogous to Eq.(22) involving the RL operator. Directly interpreting the operator expansion yields,

$$\begin{aligned}\psi(x) &= 1 - h\int_{-1}^{+1} dt|x-t|^{\phi_H - 1} + O(h^2) \\ &= 1 - h[(1+x)^{\phi_H} + (1-x)^{\phi_H}]/\phi_H + O(h^2)\end{aligned} \tag{118}$$

Integrating $\psi(x)$ term by term in Eq.(118), using the definition of f_{KR} in Eq.(94), gives the first order perturbative expansion,

$$\begin{aligned}f_{KR} &= n\zeta[1 - C_f h + O(h^2)], \\ C_f &= 2^{\phi_H+1}/\phi_H(1+\phi_H)\end{aligned} \tag{119}$$

Notably, the leading coefficient C_f equals the analytic estimate of $1/h_1^*$ in Eq.(107). This connection is natural since h_1^* controls the radius of convergence of the series Eq.(119). Higher order perturbative calculations of f_{KR} become increasingly involved, which is the typical situation for perturbation calculations in polymer science and the theory of critical phenomena. Resummation methods are then needed to extend the perturbative expansion outside the "weak coupling" regime, $h \leq h_1^*$. Alternatively, we can develop a "strong coupling" expansion about the $h \to \infty$ limit to determine the asymptotic scaling function as a function of h. As in Eq.(25), we develop Eq.(116) in a "strong coupling" expansion,

$$\begin{aligned}\psi(x) &= (\lambda A_x^{\phi_H})^{-1}/[1 + (\lambda A_x^{\phi_H})^{-1}] \\ &= (\lambda A_x^{\phi_H})^{-1}[1 - (\lambda A_x^{\phi_H})^{-1} + ...]\end{aligned} \tag{120}$$

The first term gives the power law scaling and the residual gives the "corrections to scaling" series. If we keep just the first term in Eq.(120) we have,

$$\begin{aligned}\psi_\infty &\equiv \psi(h \to \infty) \sim (\lambda A_x^{\phi_H})^{-1} \\ &\text{or} \\ A_x^{\phi_H}\psi_\infty &\sim \lambda\end{aligned} \tag{121}$$

This weakly singular Fredholm integral equation of the first kind can be solved by inspection by noting that Eq.(110) for $k = 0$ reduces to the simple expression,

$$\int_{-1}^{+1} dx \ [|x-t|^{\phi_H - 1}/H(\phi_H)](1-x^2)^{-\phi_H/2} = \text{constant} \tag{122}$$

This directly [175],[176] leads to the exact result for the strong hydrodynamic interaction limit,

$$\psi_\infty(x) = C_\infty(\phi)(1-x^2)^{-\phi_H/2}, \quad h \to \infty \tag{123}$$

where C_∞ is a constant. This equation has a number of interpretations. For example, Eq.(103) is the equilibrium charge density [175],[177] of charges interacting by a $|x|^{\phi-1}$ potential on the interval (-1,1) in $d=1$ and it has arisen again recently in the context of the equilibrium measures of rational maps [178]. The equivalent of Eq.(123) was also found by a similar mathematical method by Auer and Gardner [179],[180] in their polymer hydrodynamics calculations in $d=3$ and in this context ψ_∞ can be interpreted as a momentum flux density [89],[90],[97]. Roughly speaking, this function describes the average distribution of frictional forces along a polymer chain due to Stokes drag and this expression implies that in $d=3$ the drag forces are stronger (on average) towards the ends of the polymer chain.

An exact expression [181] for f_{KR} in the strong hydrodynamic interaction limit ($h \to \infty$) is obtained by inserting Eq.(123) into Eq.(94),

$$f_{KR}(h \to \infty)/\eta_s = [2d/(d-1)](2\pi\langle \mathbf{R}^2\rangle/d)^{(d-2)/2} A_{KR}(\phi_H)\phi_H \tag{124}$$

$$A_{KR}(\phi_H) = [2\pi/H(\phi_H)][\Gamma(1-\phi_H/2)/\Gamma(1-\phi_H)]^2 \tag{125}$$

Notice that this result exhibits no dependence on the "bead" friction coefficient ζ and the friction coefficient of the entire polymer scales with the average chain dimensions,

$$f_{KR}(h \to \infty) \sim \langle \mathbf{R}^2\rangle^{(d-2)/2}\eta_s \tag{126}$$

in much the same fashion as the friction on a sphere [See ([91],[128])]. The prefactor in Eq.(126) is specified by Eq.(124).

The Riesz-Feller semi-group A_x^ϕ also provides a perspective on previous RG calculations for the friction coefficient and into the original calculations of Kirkwood and Riseman [34],[94],[127]. As discussed in the previous section [See Eq.(39)], the perturbative RG method involves a formal expansion about the "critical dimension" ($\phi_H = 0$) and thus we may apply this limit ($\phi_H \to 0^+$)

formally to the exact ψ operational expression for $\psi(x)$ from Eq.(116),

$$\psi(\phi_H \to 0^+) = \lim_{\phi_H \to 0+} 1/(1 + \lambda A_x^{\phi_H}) \tag{127}$$

$$\begin{aligned} \lim_{\phi_H \to 0+} A_x^{\phi_H} &\approx 1, \\ \lim_{\phi_H \to 0+} [hH(\phi_H)] &\approx h/h^*, \\ h^* &\approx \phi_H/2. \end{aligned} \tag{128}$$

This approximation implies,

$$\psi(x) \approx 1/[1 + h/h^*], \; f \approx n\zeta/[1 + h/h^*] \tag{129}$$

A second order in ϕ_H RG analysis [127] of Eq.(86) and the KR estimate [94] of f both agree with Eq.(127), except for a very slight modification of h^* to the value,

$$\begin{aligned} h^*_{RG} &= 2^{-(1+\phi_H)}\phi_H(1 + \phi_H) \\ &\approx \phi_H/2, \quad 0 < \phi_H < 1 \end{aligned} \tag{130}$$

which happens to equal the eigenvalue h_1^* estimate in Eq.(107). Kirkwood and Riseman implicitly make an assumption (less generously an error) equivalent to replacing the kernel $|x - t|^{\phi_H - 1}/H(\phi_H)$ in Eq.(86) by a delta-function, $\delta(x - t)$. Moreover, the perturbative RG method involves an expansion in the "singularity strength" ϕ_H of the kernel about the "critical dimension".

Application of the RG method to the reasonably tractable KR integral equation Eq.(86) and the integral equation for polymers interacting with a surface, Eq.(16) shows some limitations of this perturbative scheme. The Fredholm alternative and Eq.(101) implies that f_{KR} diverges as $h \to (-h_k^*)$ for each k. This important feature is not found in the corresponding perturbative RG calculations, but this deficiency leads only to a small error in the physical interesting range $h > 0$ for polymer applications. However, there are other models of interacting polymers where the equivalent of the $h < 0$ range in the present problem becomes important(e.g., collapse of a polymer due to attractive self-interaction, phase separation of high molecular weight polymer solutions) and where the perturbative RG theory is not applicable[10]-[13]. Further efforts are needed to develop resummation methods appropriate for this class of problems. The fractional calculus is very promising in connection with this problem [34] [See Eq.(3.23) of Ref.[34] where the operational approximation in Eqs.(39)and (127) of this chapter is reversed in the RG expression for $\langle|\mathbf{R}|^2\rangle$ for swollen polymer chains to obtain a description of polymer contraction using this operational analytic continuation]. The calculation of the friction coefficient of membranes modelled as multi-dimensional time generalizations of Brownian paths to describe random surfaces leads to integral equations involving Riesz operators having general α and d values. Knowledge of the eigenfunctions and eigenvalues [166]-[168], even for special cases, is very limited in this case. A generalization of Weyl's theorem to these fractional order operators in d-dimensions is given by [166],[182],[183],

$$\lambda^*_{k,\Omega} \sim [\Gamma(1+d/\alpha)/p(0)](k/V)^{\alpha/d} \tag{131}$$

where V is the volume of the region Ω and $p(0)$ is a constant defined by the $R \to 0^+$ ($R = |\mathbf{R}|$) limit of the (stable process) random walk end-to-end distribution function [166]-[167],[182],[185]

$$G(R \to 0^+, n) \sim p(0) n^{-d/\alpha} \tag{132}$$

Eq.(108) is the one dimensional analog of Eq.(132).

A previous paper [43] considered the problem of calculating the osmotic second virial coefficient between a flexible polymer and objects of general shape, such as proteins, viruses, filler particles, etc. The path-integral formulation of this problem led to integral equations involving Riesz operators. These integral equations were similar to those arising in the treatment of the friction coefficient of a polymer chain in the present section, except that non-fractal regions Ω were emphasized. It was found that the virial coefficient between a high molecular weight polymer and a smaller compact particle was proportional to the capacity of the particle and the polymer molecular weight, provided the polymer-particle interactions were repulsive. The critical binding energy of a polymer to a region (micelle, fluid droplet, membrane) depended sensitively on particle shape through the lowest magnitude eigenvalue of the Riesz operator. It was suggested that this effect has significance in biological molecular recognition phenomena since changes in the binding region shape can cause a polymer to bind or unbind to the region as required for biological function. This example points to the significance of the lower eigenvalues $\lambda^*_{k,\Omega}(k = 1, 2, 3, ...)$ of the Riesz operator in physical applications involving particle "binding" or "localization". These examples were not discussed in this chapter because of their mathematical similarity to the friction coefficient calculation.

5 Stable Processes and Subordination

The Riesz operator I_d^α and its one dimensional analog, the Riesz-Feller operators A_x^α, are fundamental in the theory of random walks. The symmetric stable processes correspond to random walks in which the transition density does not necessarily have a finite variance and these processes are conventionally specified by an index α which characterizes the smallest (not necessarily integer) moment of the transition density which is infinite. For dimensionalities $d > \alpha$ this parameter equals the Hausdorff ("fractal") dimension of the random walk paths [186]-[191]. In models of polymer chains with excluded volume interactions and for chains with some rigidity the mass scaling exponent ν is greater than the Gaussian chain value $\nu = 1/2$ [See Eq.(48)] and these exponents are often interpreted in terms of random walk paths having variable Hausdorff dimension. The Lévy flight model can be invoked as a model of polymer chain configurations which corresponds exactly to this kind of fractal ("Hausdorff") dimension variation. Of course, it remains unclear how the properties of this model quantitatively reflect those of real interacting polymers, but qualitative trends accompanying the variation of the mass scaling exponent ν are expected to be reproduced. The main purpose of this section, however, is to clarify how and why fractional order operators arise in the solution of transport problems involving fractal surfaces such as polymer chains. Some details of the relation between fractional order differential operators and general random walk processes having variable fractal dimension α are developed in this section.

The functions describing the end-to-end distribution of "stable" random walks [$\mathbf{R}(0)=\mathbf{0}$] satisfy a fractional order partial differential equation [151,152],

$$\partial G(\mathbf{R}, t)/\partial t = -(-\nabla^2)^{\alpha/2} G(\mathbf{R}, t), \quad t > 0 \tag{133}$$

which evidently reduces to the ordinary diffusion equation for $\alpha = 2$. In $d = 1$ we write Eq.(133) more explicitly in terms of the A_x^α operator [138],

$$\partial G(x, t)/\partial t = -A_x^{-\alpha} G(x, t) \tag{134}$$

where for generality we take the initial value boundary condition,

$$G(x, 0) = f(x) \tag{135}$$

Eq.(134) has the formal operational solution,

$$G(R, t) = \exp(-A_x^{-\alpha} t) f(x) \equiv T_t f(x) \tag{136}$$

where $f(x)$ is specified as a delta-function in the calculation of the Green's function associated with Eq.(134). The origin of the "generator" terminology for $A_x^{-\alpha}$ is apparent from Eq.(136).

The operator T_t, describing the propagation of stable processes defines its own semi-group [44]-[46],[138],

$$\begin{aligned} T_t T_s f(x) &= T_s T_t f(x) \\ &= T_{t+s} f(x), \quad t, s > 0 \end{aligned} \tag{137}$$

$$\lim_{t \to 0^+} T_t f(x) = f(x) \tag{138}$$

The path-integral representation of stable random paths [164,165][SeeEq.(166)] can be thought of as the "the gradual unfolding of a transition probability"[192] transformation T_t which "propagates" the path,

$$G(x, 0; t) = (T_\tau)^n \delta(x), \quad t = n\tau \tag{139}$$

where τ is the time increment (contour length increment in polymer chain context) of each step, n is the number of steps and the delta-function constraint ensures that the chain is situated at the origin 0 at $t = 0$, $G(x, 0; t = 0) = \delta(x)$. Explicitly, T_t is defined by the integral operator [44]-[46],[138],

$$T_t f(x) = \int_{-\infty}^{\infty} dy \; k_\alpha(|x - y|/t^{1/\alpha}) f(y) \tag{140}$$

where the "propagator kernel" k_α is recovered for $f(x) \equiv \delta(x)$,

$$T_t \delta(x) = k_\alpha(|x|/t^{1/\alpha}) \tag{141}$$

k_α for $\alpha = 2$ and $\alpha = 1$ correspond to Gaussian [See Eq.(5)] and Cauchy ("Lorentzian") distributions, respectively. The normalization of the density Eq.(141) gives rise to the well known time dependence for the path return probability [184]-[188],

$$k_\alpha(|x| \to 0^+, t) \sim p(0; d = 1) t^{-1/\alpha}, \quad t \to \infty \tag{142}$$

and the corresponding return probability exponent in d-dimensions equals $-d/\alpha$ [See Eq.(132)]. An explicit representation of k_α is obtained by expanding [138] Eq.(136) formally,

$$k_\alpha(|x|, t) = \sum_{k=0}^{\infty} (-t)^k A_x^{-k\alpha} \delta(x)/k! \tag{143}$$

which reduces to the explicit form,

$$
\begin{aligned}
k_\alpha(|x|,t) &= \sum_{k=0}^{\infty}(-t)^k|x|^{-k\alpha-1}/k!\; H(-k\alpha) &(144)\\
H(-k\alpha) &= 2\Gamma(-k\alpha)\cos(-k\alpha\pi/2) &(145)
\end{aligned}
$$

Using standard gamma function identities, Eq.(143) can be written in the more conventional form [138],[193]-[202],

$$
k_\alpha(|x|,t) = \frac{1}{\pi t^{1/\alpha}}\sum_{k=0}^{\infty}(-1)^{k+1}\left(\frac{t}{|x|^\alpha}\right)^{k+1/\alpha}[\Gamma(1+k\alpha)\sin(k\alpha\pi/2)/k!] \quad (146)
$$

This expression is convergent for $\alpha < 1$ and asymptotic for $1 < \alpha < 2$, but rearrangement of this series allows us to obtain a convergent expression [202] for $1 < \alpha < 2$. The Fourier transform $\mathcal{F}$ of k_α equals [182],[183],

$$
\mathcal{F}\{k_\alpha(|x|,t)\} = \exp(-t|k|^\alpha) \quad (147)
$$

These power laws in momentum or Laplace space are signatures of fractional order differential operators (See below).

Stable distributions are more generally described by a second parameter characterizing the "asymmetry" of the distribution function [138] and such distributions are associated with a generator operator more complicated than $A_x^{-\alpha}$. The distribution function from these asymmetric stable processes is defined by an operator expansion having the form Eq.(143) with the operator $A_x^{-\alpha}$ suitably modified [138]. An important special case of this general class of operators and random walks corresponds to the extreme limit of "asymmetry", the "one-sided" stable processes. In this case the random walk ($d = 1$) moves in only one time-like direction so that the probability density function is defined on the half-line $[0, \infty)$.

$A_x^{-\alpha}$ in Eq.(136) is then replaced by the RL operator $I_x^{-\alpha}$ so that k_α ("one-sided") formally equals, $x \geq 0$, $0 < \alpha < 1$,

$$
\begin{aligned}
k_\alpha(\text{"one-sided"}) &= \sum_{k=0}^{\infty}(-t)^k I_x^{-k\alpha}\delta(x)/k! \\
&= \sum_{k=0}^{\infty}(-t)^k|x|^{-k\alpha-1}/k!\;\Gamma(-k\alpha)
\end{aligned} \quad (148)
$$

or equivalently,

$$k_\alpha(\text{"one-sided"}) = \frac{1}{\pi t^{1/\alpha}} \sum_{k=0}^{\infty} (-1)^{k+1} \left(\frac{t}{|x|^\alpha}\right)^{k+1/\alpha} [\Gamma(1+k\alpha)\sin(k\alpha\pi)/k!] \tag{149}$$

The one-sided stable processes have important applications to the relaxation of polymers and other amorphous materials and this has led to accurate tabulations of k_α("one-sided")[200]-[202].

The "potential function" $V(\mathbf{R})$ is obtained from the steady-state limit of the generalized diffusion equation, Eq.(134). The classical case corresponds to Laplace's equation in d-dimensions where $V(\mathbf{R})$ equals,

$$V(\mathbf{R}) = |\mathbf{R}|^{-(d-2)}/H_d \tag{150}$$

corresponding to Wiener paths ($\alpha = 2$) and, more generally, for symmetric stable paths the solution of the fractional order Laplace equation gives the potential ("Green's function"),

$$V_\alpha(\mathbf{R}) = |\mathbf{R}|^{-(d-\alpha)}/H_d(\alpha) \tag{151}$$

$$H_d(\alpha) = \pi^{d/2} 2^\alpha \Gamma(\alpha/2)/\Gamma[(d-\alpha)/2] \tag{152}$$

The potential function is obtained as the integral of k_α,

$$V_\alpha(\mathbf{R}) = \int_0^\infty dt\ k_\alpha(|\mathbf{R}|/t^{1/\alpha}) \tag{153}$$

by the initial-value theorem of Laplace transform theory. In the case of $d = 1$ the potential function $V_\alpha(x)$ becomes,

$$V_\alpha(x) = |x|^{-(1-\alpha)}/H(\alpha) \tag{154}$$

The potential function for the "one-sided" stable processes in $d = 1$ is similarly found to equal,

$$V_\alpha(x; \text{"one-sided"}) = \Theta(x)|x|^{-(1-\alpha)}/\Gamma(\alpha) \tag{155}$$

where $\Theta(x)$ is a Heaviside step-function. The potential function is evidently the kernel of the fractional differential operator which generates the associated stable random walk process.

Bochner [151],[152] first noticed that symmetric stable processes could be constructed starting from ordinary random walks satisfying an ordinary diffusion equation in the continuum limit. A paraphrasing of his arguments into a path-integral notation gives insight into the reduction of the capacity calculation for Wiener paths and other "fractal" objects into a fractional differential equation description.

Consider the path-integral for a Gaussian chain [Eq.(1)] and replace the time variable in the path-integral by a random time variable $T(\tau)$ (specified below) and we formally have,

$$G[\mathbf{R}, T] = \int_{\mathbf{R}(0)=\mathbf{0}}^{\mathbf{R}(T)=\mathbf{R}} D[\mathbf{R}(T)] \exp[(-d/2) \int_0^T d\tau |d\mathbf{R}[T(\tau)]/d\tau|^2] \tag{156}$$

This construction implies that the original chain path $\mathbf{R}(\tau)$ is decimated so that the new random walk path $\mathbf{R}[T(\tau)]$ connects the parts of the chain path at the random time points $T(\tau)$[146]-[150],[182],[183] (See also illustrations of subordination in Ref. [1]). We specify $T(\tau)$ by considering an additional random walk process (e.g. the τ values at which an ordinary one-dimensional random walk returns to the origin.). If the second moment of the probability distribution for the time increment between these recurrent events is finite we simply obtain a coarse-grained random walk path by this procedure, but if the probability distribution for the time increments has a long tail so the second moment does not exist, then the paths are Lévy walks (described by a symmetric stable distribution function in the limit of long chains). This process of decimating random walk paths to create paths, having an altered Hausdorff ("fractal") dimension exponent α, is a type of "subordination".

An explicit construction to generate a subordinated process is to consider the "local time" variable $L(t)$ (See also Fig. 4),

$$L(t) = \int_0^t d\tau \;\; \delta[X(\tau)] \tag{157}$$

where the auxiliary random walk process $X(t)$ is taken for illustration to be an ordinary random walk on the real axis, $-\infty < X(\tau) < \infty$. At $t = 0$ we define $X(0) = 0$ and as t increases we obtain a monotone increasing random function $L(t)$ which increments by 1 at the time points τ_i where $X(t)$ passes through the origin. The time points τ_i form a "random Cantor set" of dimension $\phi = 1/2$ which specify the points of decimation of the original Wiener chain [See Eqs.(2) and (156)]. In general, the random time variable $T(t)$ in Eq.(156)

is then defined as the inverse function [55] of $L(t)$ and corresponds to a one-sided stable process ("stable subordinator") of index $\phi = 1/2$. For the $X(\tau)$ process specified we have the averages,

$$\begin{aligned} \langle L(t) \rangle &\sim t^{\phi}, \\ \langle T(t) \rangle &\sim t^{1/\phi}, \quad \phi = 1/2 \end{aligned} \tag{158}$$

and by varying the process $X(\tau)$ the exponent ϕ can be "tuned" in the interval (0,1) to obtain random walks having a range of α values. With this random time transformation the original Wiener path $\mathbf{R}(\tau)$, corresponding to $\alpha = 2$ becomes the "subordinated" path process [47],[203] $\mathbf{R}[T(\tau); \alpha]$ where $\alpha = 2\phi$. The probability distribution of the random walk end-to-end separation distribution for Wiener paths $G(\mathbf{R}/t^{1/2})$ is transformed to the form,

$$G_{\alpha}(\mathbf{R}/\langle T(t) \rangle) = G_{\alpha}(\mathbf{R}/t^{1/\alpha}), \quad \alpha = 2\phi \tag{159}$$

which describes the end-to-end separation of symmetrical stable (Lévy walk) paths. $G_{\alpha}(\mathbf{R}/t^{1/\alpha})$ can be explicitly calculated by averaging the Gaussian distribution function $G_{\alpha=2}(\mathbf{R}/t^{1/\alpha}) \equiv G_o(\mathbf{R}/t^{1/2})$ with respect to the "one-sided" stable distribution function [145],[150],

$$G_{\alpha}(\mathbf{R}/t^{1/\alpha}) = \int_0^{\infty} d\tau \ k_{\phi}(\text{"one-sided"}; x = \tau, t) G_o(\mathbf{R}/\tau^{1/2}), \quad 0 < \phi < 1 \tag{160}$$

where the distribution function $k_{\phi}(x,t)$ of the stable subordinator $T(t)$ is defined in Eq.(148). ϕ is taken to be variable in Eq.(160), corresponding to more general local time processes. Feller [150] discusses Eq.(160) in the special case of $\phi = 1/2$ where $k_{\phi}(x,t)$ reduces to an inverse Gaussian distribution and $G_{\alpha}(\mathbf{R}/t^{1/\alpha})$ becomes a Cauchy distribution.

More generally, we could start with an original stable process described by $G_{\beta}(\mathbf{R}/t^{1/\beta})$ where $0 < \beta < 2$ and subordinate this process as above to obtain the new path process and distribution G_{α} where $\alpha = \beta\phi$ so that repeated random time decimation of the Wiener path decreases [182],[183] the "dimension exponent" α. This process is evidently a kind of position-space renormalization group transformation [149] .

A similar decimation process occurs for Wiener paths having repulsive excluded volume interactions inhibiting path self-intersection [See Eq.(42)]. We can view such self-avoiding walks (SAWs) to be much like ordinary random walks moving in a fractal obstruction environment, where the path's previous and future contour history are the source of obstruction. This effect is evidently stronger in lower dimensions and it is well known that SAWs can be described as simple random walks for $d > 4$ with a renormalized step length [10]-[13] ℓ (See Sect. 3). Numerical evidence and RG calculations [10]-[13],[16] indicate that the fractal (mass scaling)dimension $(1/\nu)$ of self-avoiding paths decreases in $d = 3$ (the chains swell) and that a universal distribution function $G(\mathbf{R}/t^{\nu})$ governs the displacement of the chain ends in the long chain limit $(t \to \infty)$. $G(\mathbf{R}/t^{\nu})$ preserves the fundamental scaling property of "stable" random walks in Eq.(141) with $\alpha \leftrightarrow 1/\nu$, but the history dependence of the path process implies that the semi-group property Eq.(137) no longer holds. Such non-Markovian processes are classified as "semi-stable"[204],[205].

Since the random time decimation of the Wiener path process leads to fractional order operators in the spatial variable describing the path displacement process, it is natural to expect that the decimation of paths due to spatial constraints should lead complimentary to fractional order operators involving fractional time derivatives. Indeed, exact solution of the fractional order diffusion equation [4]-[6] with a fractional time derivative yields a change of ν and a $G(\mathbf{R}/t^{\nu})$ having a "stretched Gaussian" form,

$$G(\mathbf{R}/t^{\nu}) \sim \exp[-(const.)(\mathbf{R}/t^{\nu})^{1/(1-\nu)}], \quad \mathbf{R}/t^{\nu} \to \infty \tag{161}$$

as indicated also by formal RG calculations for SAWs [10]-[13],[16] and heuristic calculations for diffusion on fractals[6]. Interestingly, the exponent ν decreases $(\nu < 1/2)$ and the random walk motion "slows" in the case of a fixed fractal space, while ν increases $(\nu > 1/2)$ for the self-interacting polymer where the interaction occurs both forward and backward in time (the chain contour coordinate).

The geometrical path decimation origin of fractional order (RL) differential operators in time is particularly evident in the surface-interacting chain model of Sect. 2 where the boundary geometry directly determines the order of the differential operator (See Sect. 3). In this model the operator order ϕ_s is the fractal dimension of the "local time" spent spent by the path at the boundary (See Fig. 3),

$$\phi_s = (2 + d_f - d)/2, \quad 0 < \phi_s < 1 \tag{162}$$

where d_f is the boundary fractal dimension [47]-[52] (Notably ν is not changed by these exterior surface boundary constraints in this model.). The value of ϕ_s can also be made to vary by constraining the random walk to lie in a wedge [47]-[52] and ϕ_s also depends on the dimension of the random walk path [See

Eq.(170)]. Thus, we see a general pattern responsible for the generation of fractional order operators in space and time: decimation at random time points which preserve the random walk nature of the path leads to fractional order differential operators in the space variable, while decimation of the random path in space, which preserve the fractal character of the path process, lead to fractional order operators in the time coordinate (or the chain contour coordinate in a polymer context). This geometrical phenomenon, responsible for the occurrence of fractional order operators in many applications, deserves further investigation.

Subordination in a physical context is illustrated through the probabilistic calculation of the charge density of a conducting disc of unit radius. The charge density can be calculated by enclosing a two-dimensional disc in a large three dimensional sphere and launching random walks from the surface of the enclosing sphere [89],[90]. Some of the random walks hit the disc and the density of these first hits on the surface of the disc, when launching occurs from a great distance, determines the equilibrium charge density $\mu(x)$ [See Eq.(67)], $d = 3$,

$$\mu(x; \text{disc}) = N_\mu (1 - x^2)^{-1/2} \tag{163}$$

where x is the distance from the center of a unit disc and N_μ is a normalization constant. More generally for a d_s-dimensional unit disc in d-dimensions the resulting charge density equals,

$$\begin{aligned} \mu(x; \text{disc}) &= N_\phi (1 - x^2)^{-\phi_s}, \quad 0 < \phi_s < 1, \\ \phi_s &= (2 + d_s - d)/2 \end{aligned} \tag{164}$$

where ϕ_s is the dimension of the Wiener path local time at the d_s-dimensional disc boundary and N_ϕ is a constant. For $\phi_s < 0$ the capacity of the disc vanishes since the walkers miss the disc with unit probability. The random hitting process of a random walk hitting a surface ($d_s = 2, d = 3$) can also be viewed from the plane in which the disc is embedded so that it appears the random walk in the plane takes quite erratic steps between points at which the full three-dimensional walk contacts the surface [146]-[150]. Indeed, the "local time" dimension in this case equals $\phi_s = 1/2$ [See Eqs.(13) and Eqs.(159)], reflecting the large fluctuation in times between random walk surface contacts. The resulting walk in the plane corresponds exactly to the subordinated Wiener path example discussed above. The Hausdorff dimension α of these "flatland" walks equals $\phi_s = 1/2$ so that the walks in the plane are Cauchy walks [146]-[150], $\alpha = 1$. This result can be checked by calculating the charge density of the disc by launching Cauchy walks from a large concentric ring in the plane of the disc and constrained to lie within this plane. The charge density of first random walk hit points by the Cauchy paths is again given by Eq.(164) where $\alpha = 1$ replaces 2 in the definition of ϕ_s and $d = 2$. The general result for the charge density of a unit hypersphere for α-stable (Lévy walks) equals [206]-[210],

$$\mu(x; \text{hypersphere}) = N_\alpha (1 - x^2)^{-\alpha/2} \tag{165}$$

A disc viewed from a "flatland"($d = 2$) world is a "sphere" having the embedding space dimension. For a Cauchy process ($\alpha = 1$) Eq.(165) reduces to Eq.(163).

The erratic random walk process (Lévy walk) in the plane of the disc arising from the decimation of the three dimensional walk is described by a fractional differential equation whose order is dictated by the local time dimension of the Wiener path time points of intersection. The emergence of fractional order differential operators through the process of subordination is also at the heart of the calculation of the capacity (friction) of Wiener paths, but this example involves a less trivial calculation. The charge density of a Brownian chain can alternatively be calculated (preaveraging approximation) in the one-dimensional contour coordinate of the chain (the analog of the disc surface in the example above [211]) or in the full three dimensional space. Some details of this calculation of the polymer chain charge density are given utilizing path-integration and fractional calculus methods.

In the one-dimensional coordinate x of the chain the polymer is taken to occupy the interval $(-n/2, n/2)$ so that $x = 0$ corresponds to the center of the chain. The α-capacity C_α of this interval is calculated by path-integral methods starting from the functional [See Eq.(55)],

$$\psi(x) = \left\langle \left(\exp(-\beta \int_0^\infty d\tau \chi[x + x(\tau)]) \right) \right\rangle \tag{166}$$

where χ is the indicator function for the interval $(-n/2, n/2)$ and $x(t)$ is a symmetric stable process of index α in $d = 1$. Expanding this functional formally in the coupling constant β shows that it is equivalent to the integral equation,

$$\psi(x) = 1 - \beta \int_{-n/2}^{n/2} dx' \; [|x' - t|^{\alpha-1}/H(\phi)]\psi(x') \tag{167}$$

(The transformation from Eq.(166) to (167) is very similar to the treatment of Ciesielski and Taylor [125] of the corresponding path functional for $\alpha = 2$ and d-dimensions, $d > 2$.) Transforming the integral equation Eq.(167) to the interval (-1,1) and taking the index of the stable process equal to $\alpha = \phi_H$ recovers the Kirkwood-Riseman integral equation for the friction coefficient [see Eq.(94)]. The subordination idea comes into this construction by noting that ϕ_H is the Hausdorff dimension of the random walk "local time" for a process occuring in d-dimensions. Viewed from the chain coordinate the intersections appear as a random walk on this line with jumps occuring over a large range of scales (a symmetric Lévy walk on the line). This construction is entirely analogous to the disc problem discussed above.

The charge density $\mu(x)$ of the interval (-1,1) can be obtained from the solution of the fractional differential equation,

$$1 = A_x^\alpha \ \mu(x;\alpha) \tag{168}$$

where the normalization constant of $\mu(x;\alpha)$ is unspecified in this equation. As discussed in Sect. 4 the solution of Eq.(168)is obtained formally as,

$$\mu(x;\alpha) = A_x^{-\alpha} 1 \ = N_\alpha (1 - x^2)^{-\alpha/2} \tag{169}$$

where N_α is a constant and α lies in the range (0,1). This expression describes the density with which the polymer chain is hit by the random walk process. This density is insensitive to whether the hitting process is viewed from within the chain coordinates or a diffusion process in d-dimensional space.

Riesz [118] has extended this type of charge density calculation to α-potentials, corresponding to stable processes hitting a d-dimensional hypersphere. This calculation leads to the form Eq.(169) where x is the distance from the center of the unit hypersphere.

In connection with this result and the results above for hitting discs and random paths with random walks, Hawkes [47] has considered the general problem of calculating the intersection set dimensions of stable paths with domains of specified fractal dimension. In $d = 1$ he proves that the local time dimension ϕ_s of the intersection of the walk path with a domain having dimension d_f equals,

$$\phi_s = (\alpha + d_f - d)/\alpha, \quad 0 < \phi_s < 1, \tag{170}$$

Fragmentary exact results by Hawkes suggest that Eq.(170) is also true in higher dimensions [49], although in some instances it might be necessary to identify d_f as the Minkowski dimension rather than the Hausdorff "fractal" dimension [47]. For a Wiener path "surface" [16],[204],[205] we have $d_f = 2$

and for the calculation of Newtonian capacity the dimension of the probing trajectories equals 2 so that ϕ_s (Wiener path) becomes,

$$\phi_s = (4-d)/2, \quad 0 < \phi_s < 1, \tag{171}$$

Since the dimension of the hitting points vanishes for $d \geq 4$ we conclude that Wiener paths have zero capacity for $d \geq 4$ (See below). This accords with the exact results of Dvoretsky et al.[212]-[214] The dimension of the surface hitting points gives a direct indication of the subordinated random walk process occurring on Ω.

Apart from providing a physical interpretation of the origin of fractional differential equations in modeling boundary value problems having rough boundaries and a geometrical interpretation of the order of these operators, the subordination concept also affords a direct means for the numerical calculation of the Hausdorff dimension of fractal sets. The capacity can be calculated by simply hitting the fractal set with α-stable walks launched as long distance away and the resulting charge density [114]-[117] and capacity [215],[216] C_α can be determined numerically as a function of α. If we define a dimensionless capacity $\psi_\alpha^* \equiv C_\alpha / R_p^{d-\alpha}$ where R_p is a characteristic dimension of the probed set Ω (eg. radius of gyration), then C_α should decrease in a monotone fashion with a decrease of α until it vanishes [214], reflecting the increasingly penetrating nature of the probing trajectories and the associated diminished screening of these trajectories. It is rigorously known (Frostman's theorem [119]-[124],[214]) that there is a critical value of the probing trajectory dimension α for which C_α first vanishes and which determines the Hausdorff dimension of any bounded region Ω. The critical α value equals the dimension of space from which the walks are launched minus the Hausdorff dimension d_f of the bounded set Ω. C_α vanishes for $\phi_s < 0$ and RG calculations of C_α in the case of a Wiener paths ("polymers") indicate that ψ_α^* vanishes as,

$$\psi_\alpha^* \sim \phi_s + O(\phi_s^2), \quad \phi_s \to 0^+ \tag{172}$$

with an unspecified prefactor on the order of unity. Since it is probably difficult to directly determine the exact value of α where C_α vanishes, an extrapolation determination of $\alpha^*(C_\alpha = 0)$ based on Eq.(172) should be made. Apparently, this method has never been implemented numerically, despite the fact that many of rigorous results about the fractal geometry of random walks and random surfaces are based on this idea [186]-[191],[214]. Fig. 7 shows a model aggregate of spheres hit by incoming diffusive trajectories (small spheres whose lightness increase with hitting frequency) launched from a large enclosing sphere. Fig.8 illustrates the hitting of a compact object with sharp corners.

Figure 7: Random walk hit density of a model branched polymer. The lighter the shading of the small spheres the greater the density of hits (i.e.,these points are "hotter"). The larger spheres denote the polymer "monomer" units. In the limit of a large launching sphere the hit density approaches the charge density of the branched polymer in the limit of infinite random walk launches. The integral of the charge density gives the capacity.

This problem of the capacity of a cube is one of the great unsolved problems of electrostatics.

6 Interior-type Boundary Value Problems

The calculation of the charge density and capacity of α-stable processes is a natural extension of the solution of Laplace's equation with Dirichlet boundary conditions on the exterior of a region Ω. There are also many interior-type boundary value problems defined on "fractal" objects such as polymer chains where fractional order equations arise. The normal mode vibrations of the polymer chain in solution provides a good example of this type of problem

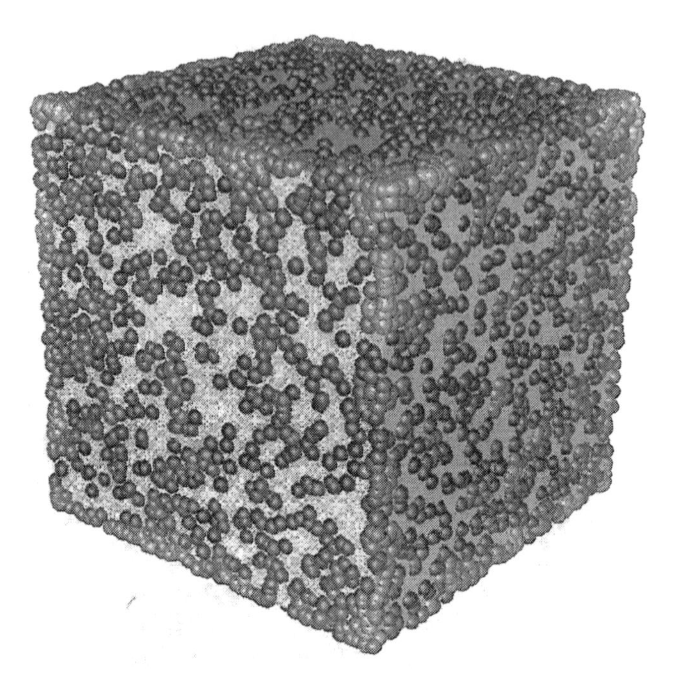

Figure 8: Random walk hit density for a unit cube. The capacity of a cube is one of the great unsolved problems of classical electrostatics. The dark patches indicate the hitting of the cube by (2000) random walks launched from an enclosing sphere. Note the increased hit density (charge density) near the sharp cube corners [See Eq.(163)-(164).]. Ref. [98] utilizes a probabilistic method to calculate $C(unit\ cube) = 0.660675 \pm 0.00001$ where 4.7 billion trajectories were sampled in the Monte Carlo calculation. This example illustrates the accuracy which can be achieved by probabilistic methods.

[94]-[96]. In the absence of hydrodynamic interactions between the polymer chain segments, the normal modes of a polymer chain are much like those for a vibrating string so that the calculation of the characteristic values which determine the time constants of these normal mode motions reduces to classical Sturm-Liouville theory [217]. This theory is attributed to Rouse in the polymer science literature [218]. The inclusion of hydrodynamic interactions (Rouse-Zimm Theory [94]-[96]), which is physically necessary for real polymers in solution, leads to fractional differential equations [219] involving the operator A_x^α, as in the calculation of the friction above within the Kirkwood-Riseman rigid chain model. Moreover, perturbative RG calculations indicate that the order of the operator involved is influenced by chain swelling which plays a

similar role to varying the spatial dimension (ϕ_H). It should be possible to expand the solution of these equations describing polymer dynamics in terms of the eigenvalues of A_x^ϕ, as in the calculation of the friction coefficient. Exact calculation of autocorrelation functions associated with the relaxation of the chain normal modes based on fractional calculus methods would be very valuable for understanding the normal mode motions of fractal objects in a fluid or elastic medium [220]. There are many open problems in this area demanding new mathematics involving fractional calculus.

Another important class of interior fractional differential boundary value problems arise from generalizations of the diffusion and Poisson equations to the interior of complex shaped regions. Kac, for example, considered the problem of α-stable walks in $d = 1$ where the paths are restricted to the interior of a finite interval [164],[165]. Paths which leave the interval are "killed" as in the solution of Laplace's equation with Dirichlet boundary conditions on the interior of a region in $d = 2$ or $d = 3$. Solution of this problem leads to an expansion in the eigenvalues and eigenfunctions of a fractional differential equation analogous to the wave equation and several authors investigated the generalization of Weyl's theorem [See Eq.(131)] describing the asymptotics of the eigenvalue spectra arising in this type of interior-type fractional differential boundary value problem. The potential function associated with solving α-stable diffusion processes on a finite interval with "killing" boundary conditions has also been calculated in $d = 1$ and Trigt describes the generalization to d-dimensions for a spherical boundary [137]. These exact solutions are consistent with Kac's conjecture for the general form of the kernel $k(x, y)$ for the α-stable generalization of the Laplace equation with an "absorbing" boundary defined on the interior of a finite interval [164],[165],[206]-[210],

$$k(x, y) = |x - y|^{\alpha - 1}/H(\alpha) + d(x, y) \tag{173}$$

where $d(x, y)$ is the "non-singular" or "regular" part of the kernel, corresponding to the fractional order Laplacian. The generalization of Kac's conjecture to d-dimensions is evident,

$$K_\Omega(\mathbf{R}, \mathbf{R}') = |\mathbf{R} - \mathbf{R}'|^{-(d-\alpha)}/H_d(\alpha) + D(\mathbf{R}, \mathbf{R}') \tag{174}$$

where Ω is the region on which $K_\Omega(\mathbf{R}, \mathbf{R}')$ is defined and $D(\mathbf{R}, \mathbf{R}')$ is a non-singular function. The asymptotic eigenvalue spectrum in leading order should be dominated by the singular part of the kernel so that $k(x, y)$ has the same leading order eigenvalues variation with mode number k as the Riesz-Feller kernel [164],[165]. The same should be true for the Riesz kernel in d-dimensions so that results obtained for the Riesz kernel [166] should also apply to the asymptotic eigenvalue spectrum of the fractional differential analog of the vibrating membrane with fixed boundary. The Riesz operator spectrum should have importance in describing the dynamics of "membranes" and "sponges" embedded in fluid or elastic media [219]-[221].

Finally, we mention a fractional differential analog of the solution of the interior Poisson equation with Dirichlet boundary condition which has polymer and hydrodynamic applications. In the case of Wiener paths we consider the classical Poisson equation,

$$\nabla^2 \hat{u} = -1 \tag{175}$$

in the interior of a region Ω where $\hat{u}$ is taken to vanish on the boundary (Dirichlet boundary condition). The solution of Eq.(175) is formally,

$$\hat{u} = I_\Omega^2 1, \; I_\Omega^2 \equiv (-\nabla^2)^{-1} \tag{176}$$

where I_Ω^2 is the integral operator associated with the K_Ω kernel in Eq.(62). The generalization of Eq.(175) to variable α is termed a "fractional Poisson equation",

$$(-\nabla^2)^{\alpha/2} \hat{u}_\alpha = 1 \tag{177}$$

with a "killing" type boundary condition or equivalently $\hat{u}_\alpha = I_\Omega^{-\alpha} 1$. Explicit calculation of $\hat{u}$ in Eq.(175) and the more general case of $\hat{u}_\alpha$ is limited to special shaped boundaries. Kac [164],[165] and Elliott [207] obtained $\hat{u}_\alpha(x)$ in $d = 1$ for the interval (-1,1),

$$\hat{u}_\alpha(x) = M_\alpha (1 - x^2)^{\alpha/2} \tag{178}$$

where M_α is a constant. Eq.(178) arises in the calculation of the mutual capacity between coaxially positioned charged conducting circular plates in $d = 3$ where [208] $\alpha = 1$. Getoor [206] has also shown that Eq.(178) describes $\hat{u}_\alpha(R)$ for a d-dimensional unit sphere where the constant M_α then depends on d and x is the distance from the center of a hyper-sphere.

Notice the curious relation between the solution of the fractional Poisson equation $\hat{u}_\alpha(x)$ in Eq.(178) (an interior boundary value problem) and the charge density of a stable process in Eq.(165) for a sphere (an exterior boundary value problem). These solutions, apart from a constant of proportionality, are simply related by a change of exponent sign. This relation reflects a more general correspondence between the Riesz kernel $K_\alpha(\mathbf{R}, \mathbf{R}')$ and the "interior kernel" $K_\Omega(\mathbf{R}, \mathbf{R}')$ in Eq.(174) for the hyper-sphere, associated with the fractional order generalization of the Laplace equation with absorbing boundary conditions. In $d = 1$ this connection is made explicit through an expansion of these kernels in ultraspherical polynomials, $0 < \phi < 1$,

$$|x-t|^{\phi-1}/H(\phi) = \sum_{k=1}^{\infty} \Phi_k^{(\phi)}(x)\Phi_k^{(\phi)}(t)/\lambda_k^{(\phi)}, \tag{179}$$

$$\lambda_k^{(\phi)} = \Gamma(k-\phi)/\Gamma(k)$$

$$k_\phi(x,t) = \sum_{k=1}^{\infty} \Phi_k^{(-\phi)}(x)\Phi_k^{(-\phi)}(t)/\lambda_k^{(-\phi)} \tag{180}$$

where the $\Phi_k^{(\phi)}$ are defined in Eq.(110). Notably, these kernels are *identical* apart from the sign of ϕ. The solution $\hat{u}_\alpha$ of the fractional Poisson equation is just the first iterate [164]-[168] of the $K_\Omega(\mathbf{R}, \mathbf{R}')$ kernel in Eq.(174).

The analytic solution of $\hat{u}(\mathbf{R})$ in Eq.(175) for complex-shaped boundaries is generally difficult. Fortunately, $\hat{u}(\mathbf{R})$ can be interpreted probabilistically in terms of the mean first-passage time $\langle T(\mathbf{R})\rangle$ of a random walk to encounter the boundary [91]. $\langle T(\mathbf{R})\rangle$ is a function of the point R on the interior of Ω where the walks initiates and $\hat{u}(\mathbf{R})$ is proportional [91] to $\langle T(\mathbf{R})\rangle$. This probabilistic interpretation of $\hat{u}(\mathbf{R})$ is also true for stable random walks [91],[206]-[210]. Recently, Monte Carlo sampling of random walk paths ($d = 2$) has been employed to calculate $\hat{u}(\mathbf{R})$ for complex boundaries (including "fractal" boundaries), showing the numerical feasibility of the probabilistic method. In Fig. 9 we show the mean first-passage time of random walkers ($\alpha = 2$) to hit an exterior boundary starting within a fractal region [88],[91].

The Poisson equation Eq.(175) has many applications[91], but the problem of Poiseuille flow through a capillary of arbitrary shape is particularly important. It is well known that the velocity field $v(x)$ in the cross-section of a long pipe in which fluid (perhaps polymeric) is flowing laminarly (low Reynolds number) satisfies the Poisson equation within the pipe cross-section so that Eq.(166) reduces the Poiseuille flow velocity field for a circular duct. The corresponding problem of slow flow of fluid through an orifice in a plane screen can be interpreted in terms of subordination ideas. In a fashion parallel to the case of a conducting disc we should expect the velocity field in the plane of the orifice to be described by subordinated random walks in the plane, a Cauchy process [146]-[150]. This intuitive view implies that the velocity field for slow flow through a circular orifice should be proportional to $v(x) \approx (1-x^2)^{1/2}$, which is directly analogous with the charge density calculation for a circular disc where $\mu(x) \approx (1-x^2)^{-1/2}$. This argument accords with the exact calculation of the velocity field of slow flow through a circular orifice [222],[223]

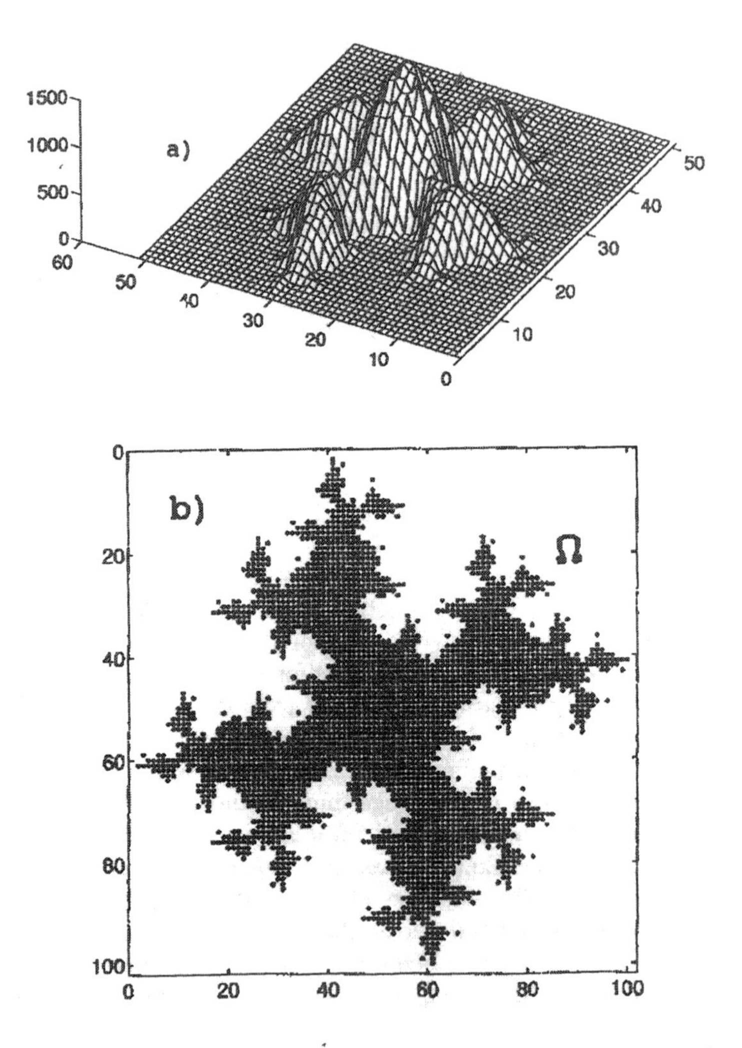

Figure 9: Poiseuille velocity field for a model duct having a fractal cross-section. a) denotes velocity field in profile and b) indicates the duct cross-section Ω. Calculations are performed using Monte Carlo path-integration [91]. The local velocity field has the probabilistic interpretation in terms of the average time $\langle T(\mathbf{R})\rangle$ taken for a random walk to hit the boundary. This time is short if the walk starts near the boundary and is maximal the farthest the way it starts from the nearest boundary point. Note the sluggish nature of the flow near the diffuse boundary. This flattening should become more accentuated for smaller α, corresponding to modeling the average local velocity field of turbulent duct flow (See text).

and probably holds for screens having complex shaped hole configurations. Elsewhere, it is suggested [91] that Eq.(169) should also describe average flow velocity of pipe flows at high Reynolds number, Re. This model seems to agree [91] well with experiments for a fully developed turbulent [$Re \sim O(10^6)$] where α was fitted [91] to the value $\alpha(\text{expt.}) \approx 0.28$. Analytic calculations by Barenblatt and Chorin [92] for the average velocity field $\langle v \rangle$ near the wall of a tube require that $\alpha = 3/ln(Re)$ for consistency so this exponent should be weakly dependent on Re. Eq.(169), with α fixed by the Barenblatt-Chorin relation, approaches a logarithmic velocity field in the $Re \to \infty$ limit. This limiting velocity field ("law of the wall") is commonly invoked as a description of highly turbulent flows [92]. Notably, Eq.(169) is not restricted to the near wall region and can be generalized to ducts of arbitrary cross section and a wide range of Re. These examples suggest that fractional differential equations should arise pervasively in solving boundary value problems having rough boundaries and in modeling systems driven far from equilibrium where fractal structures naturally arise [197]-[199].

7 Fractional Calculus and Polymer Relaxation

Fractional calculus methods have long been invoked to model relaxation processes in polymeric materials, although the modeling has generally been rather phenomenological [224]-[233]. The success of these modeling efforts has led to interesting discussions into the nature of the transport coefficients appropriate to describe these complex materials [226]. Recently, it has become appreciated that patterns of relaxation found in polymeric materials are also observed in a much broader class of materials (glasses, colloids, etc.), and this has prompted speculations about the "universality" of disordered material relaxation [234]-[236]. In this section a general model of condensed matter relaxation is developed to address this issue [237]-[239]. We start with a formally exact integral equation determining the relaxation of a Hamiltonian dynamical system corresponding to a material system near thermal equilibrium [240]. Probabilistic reasoning, utilizing Feller's "fluctuaion theory of recurrent events" [60] specifies the functional form of the memory kernel appropriate to describing homogeneous materials where the rate process governing relaxation can be "regular" or "intermittent". The relaxation integral equation obtained involves Riemann-Liouville fractional order operators I_t^{ϕ} where the "dynamic exponent" ϕ characterizes the degree of intermittency in the relaxation process. Exact solution of the relaxation integral equation leads to the Cole-Cole function [241]-[244] which is commonly reported as an empirical description of relaxation in condensed materials. An extension of the model [237]-[239] is given to include dynamical heterogeneity arising from equilibrium molecular clustering – a feature which is expected to be ubiquitous in condensed materials, especially liquids. The size distribution of the clusters is argued to be governed by Boltzmann's law under conditions of equilibrium and this leads to a relaxation integral equation involving Erdélyi-Kober fractional differential operators. These generalized operators depend on an additional parameter β which Refs. [237]-[239] argues is related to the topology of the clusters (lines, sheets, compact clumps). The class of relaxation functions obtained from this probabilistic model of relaxation in condensed materials seems to correspond well to relaxation functions observed experimentally and the limit theorems we investigate provide a rationale for the existence of this "universality". Relaxation function classes ("strong mixing", "weak mixing", "non-ergodic") are defined in terms of the asymptotic time decay and integrability properties of these relaxation functions.

In developing our probabilistic model of condensed matter relaxation we invoke minimal assumptions in order to make the description as general as possible. First, we assume that our material can be described by equilibrium thermodynamics. The existence of equilibrium implies that correlations which arise from a small perturbation or which arise from spontaneous thermal fluctuations decay with time. Specifically, autocorrelations functions for large scale observable properties of the material describing the rate of "mixing" of the dynamical system, vanish at long times [250]-[253]. Transport properties are defined through integrals of these autocorrelation functions so the integrability of these functions is also a concern in classifying relaxation in condensed materials. The existence of "mixing" (and well-defined thermo-dynamic properties), however, does not imply the existence of ergodicity in the classical sense of term [254], and weaker forms of mixing can arise from memory effects associated with dynamic intermittency and the influence of material heterogeneity on the relaxation process [237]-[239] (See below). Finally, we invoke the condition of stationarity [250]-[255] so that results of our measurements do not depend on when they are performed. This is another basic property of a material system at equilibrium.

Formal calculations based on these widely applicable conditions lead to a general functional relation describing autocorrelation functions $\Psi(t)$ for long wavelength observable properties A of our condensed material. In particular, if we normalize $A(t)$ so that its average value is zero, then the dimensionless autocorrelation function of $A(t)$,

$$\Psi(t) \equiv \langle A(t)A(0)\rangle / \langle A^2(0)\rangle \tag{181}$$

exactly obeys the integral equation [240],[256]-[258],

$$d\Psi(t)/dt = -\int_0^t d\tau\ k(|t-\tau|)\Psi(\tau), \quad \Psi(0) = 1 \tag{182}$$

where $\Psi(t) = 0$ for $t < 0$. (Of course, the existence of a finite $\langle A^2(0)\rangle$ requires A to be a square-integrable function in its spatial coordinates.) The symmetric memory kernel $[k(t,\tau) = k(\tau,t)]$is associated with the assumption of constant energy and stationarity, leading to an invariance of Eq.(182) under time inversion $(t \to -t)$ and time translation $(t \to t + a)$. Although Eq.(182) has sufficient generality for describing condensed matter relaxation processes, the specification of the memory kernel is rather difficult from first principles [237]-[239],[256]. This situation is helped somewhat from a phenomenological standpoint by the sensitivity of $\Psi(t)$ to the form of the memory kernel $k(t)$[256].

The usual method of modeling $k(t)$ involves an approach patterned after the Langevin model of Brownian motion [240]. A representative particle in the material is subjected to fluctuating forces exerted by its molecular environment [240],[259] and the time averages of these fluctuating forces is related to $k(t)$ by a fluctuation-dissipation theorem [240]. Here we follow a different strategy, since our interest is in studying classes of relaxation functions $\Psi(t)$ appropriate to describing long wavelength condensed matter relaxation.

Feller introduced a general theory of "return to equilibrium" in dynamical processes governed by a random evolution [60]. This theory assumes that the relaxation events giving rise to the relaxation process occur as independent random variables in time, which is compatible with our "mixing" and "stationarity" assumptions for condensed matter relaxation. A heuristic physical motivation for describing the dynamical evolution of a material composed of interacting particles as a random walk ("polymer") in phase space is discussed by Uhlenbeck [260]and connections between ergodic theory and random walks are well known [261]-[264]. This random walk picture gives a vivid conceptual picture of the origin of "mixing" in many-body systems and of the occurrence of relaxation event times after random (independent) time intervals. The intensity of these fluctuations has important implications for the character of the relaxation process and general "universality classes" of relaxation emerge from this approach.

We can establish a direct connection between $\Psi(t)$ and the "renewal theory" of relaxation by integrating both sides of Eq.(181) from 0 to t (i.e., multiply by I_t^α) to obtain the "survival equation" of renewal theory,

$$\Psi(t) = 1 - \int_0^t d\tau\, \mathcal{R}(t,\tau)\Psi(\tau), \tag{183}$$

$$\mathcal{R}(t,\tau) = \mathcal{R}(|t-\tau|)$$

$$\mathcal{R}(t) = \int_0^t d\tau\, k(\tau) \tag{184}$$

Eq.(183) first arose in the context of describing the decay in the relative number of charter members of an insurance group [265]-[269], where $\mathcal{R}(t)$ describes the rate at which members drop out through death and where new members are added to keep the total number of policy holders constant. This type of equation [270] can be translated to many contexts (population dynamics, replacement of industrial equipment, etc.), but here we should think of $\Psi(t)$ as the probability that the initial state of the dynamical system property $A(0)$ persists (i.e., "survives") up to time t.

The main object of modern "renewal theory" is the probabilistic modeling of the "renewal rate" $\mathcal{R}(t)$, the kernel of Eq.(183). Taking the continuum limit of Feller's classical "fluctuation theory of recurrent events" [271]-[273] provides an integral equation for the average rate of "renewal events" governing the relaxation process,

$$\mathcal{R}(t) = p(t) + \int_0^t d\tau\;\; p(|t-\tau|)\mathcal{R}(\tau) \tag{185}$$

where $p(t)$ is the probability density describing the (first-passage) time between relaxation increment events. The solution of this equation gives the average rate $\langle\mathcal{R}(t)\rangle = \mathcal{R}(t)$ of the random process governing the large-scale relaxation process. The assumption of a mixing, stationary system restricts modeling of relaxation processes to large-scale properties of equilibrium materials. Subsystems within the material do not necessarily follow these macroscopic relaxation patterns (See below).

The occurrence of universality in $\Psi(t)$ and $\mathcal{R}(t)$ in this model of condensed matter relaxation derives from the observation that the solution $\mathcal{R}(t)$ from

Eq.(185) for large t depends only on the existence of p(t) moments,

$$\langle t^n \rangle = \int_0^\infty d\tau \;\; \tau^n p(\tau) \tag{186}$$

There are functional limit theorems governing the classes of $\mathcal{R}(t)$ which can arise from this probabilistic model. Three general cases occur: (a)$\langle t^2 \rangle, \langle t \rangle < \infty$, (b)$\langle t^2 \rangle \to \infty, \langle t \rangle < \infty$, and (c)$\langle t^2 \rangle, \langle t \rangle \to \infty$, corresponding to different degrees of intensity in the fluctuations governing the relaxation process.

(a) In the case where $\langle t^2 \rangle < \infty$, the relaxation events occur with a well-defined average "period", $\langle t \rangle < \infty$, leading to a relatively rapid mixing (i.e., $\Psi(t \to \infty) \to 0$). If we take, for example, the relaxation process events to occur as a Poisson process [$p(t) = e^{-\tau/\tau_o}/\tau_o$], then Eq.(185) implies the average rate is constant,

$$\mathcal{R}(t) = 1/\tau_o \tag{187}$$

Inserting this result into Eq.(183) implies exponential decay of the relaxation function, $\Psi(t) = e^{-t/\tau_o}$. More generally, $\mathcal{R}(t)$ for any $p(t)$ with a finite second moment has the asymptotic dependence,

$$\mathcal{R}(t) \sim 1/\tau_o + C_1/t, \quad t \to \infty \tag{188}$$

where C_1 depends on $\langle t \rangle$ and $\langle t^2 \rangle$, corresponding to a rapid approach to the constant rate in Eq.(187). Exponential decay ("strong mixing") [273] is commonly found in idealized models of condensed matter relaxation [274] and more generally in the autocorrelation function decays of mathematical models of strongly mixing dynamical systems. Gordon [276] considers the problem of translational and rotational velocity correlations in a gas which provides a good example of "strong mixing" and the probabilstically defined renewal process (molecular collision events) governing the velocity relaxation process.

(b) In the case where $\langle t^2 \rangle \to \infty$ and $\langle t \rangle < \infty$, corresponding to a more intermittent relaxation process, we still have a finite relaxation rate at long times [$\mathcal{R}(t \to \infty) \sim 1/\tau_o$], but there is a slower approach of $\mathcal{R}(t)$ to its asymptotic limit[60],

$$\mathcal{R}(t) \sim 1/\tau_o + C_2 t^{\phi - 1}, \quad 0 < \phi < 1, \quad t \to \infty \tag{189}$$

where C_2 depends on the "critical index" ϕ characterizing the strength of the fluctuations in the relaxation process.

c) In the case of strongly intermittent relaxation, where even the first moment diverges $\langle t \rangle \to \infty$, the relaxation rate approaches zero at long times [60],

$$\mathcal{R}(t) \sim C_3 t^{\phi-1}, \quad 0 < \phi < 1, \quad t \to \infty \tag{190}$$

where C_3 is a known constant [60]. Geometrically, ϕ is the Hausdorff ("fractal") dimension of the time points at which relaxation events occur. $\mathcal{R}(t)$ is the derivative of the average number of renewal events for a process like that shown in Fig.4. The "local time" variable L_t in Fig.4 is a counter for those intermittent "renewal" events having their origin in a random walk path process and for this model $\mathcal{R}(t) = d\langle L_t \rangle / dt$. The density function $p(t)$ for the first passage times between relaxation events approach (are "attracted" to) the one-sided stable densities $p_\alpha(t)$ in the long time limit such that: $\alpha = 2$ for (a), $1 < \alpha = 2 - \phi < 2$ for(b), $0 < \alpha = \phi$ for(c). The index α (or ϕ) characterizes the degree of fluctuations in the occurrence of the relaxation event process. Notably, the distribution function $p_\alpha(t)$ obeys the fractional differential equation [195],

$$t^{-1} I_t^{1-\alpha} p_\alpha(t) = [1/\Gamma(1-\alpha)] \; p_\alpha(t), \quad 0 < \alpha < 1 \tag{191}$$

which curiously has a continuous eigenvalue spectrum. The $p_\alpha(t)$ distribution functions describe random walks which move only in the positive direction where the jump length variance is not generally finite. Stable distributions such as Eq.(191)can be represented in terms of Fox functions [195] and were discussed in Sect. 5

The three limiting expressions for $\mathcal{R}(t)$ indicated above can be approximated by the expression,

$$\mathcal{R}_\phi(t) \approx \Omega_o \; |t|^{\phi-1}/\Gamma(\phi), \quad 0 < \phi \leq 1 \tag{192}$$

where Ω_o is a "coupling constant" governing the relaxation rate intensity and $\Gamma(\phi)$ is a normalizing factor. Some particular expressions for Ω_o are discussed by Douglas and Hubbard [237]-[239] for models of relaxation in polymeric systems in the $\phi \to 1$ limit where $\mathcal{R}_\phi$ reduces to a constant ($\Omega_o = 1/\tau_o$). Note that the additive constant in Eq.(177) is neglected in Eq.(192).

Insertion of Eq.(192)into Eq.(182) and utilizing the definition of the Riemann-Liouville operator I_t^ϕ yields a fractional differential equation for $\Psi(t)$,

$$\Psi(t) = 1 - \Omega_o \; I_t^\phi \Psi(t) \tag{193}$$

where ϕ is a "dynamical critical index" characterizing the degree of intermittency in the relaxation process. As before [see Eq.(22)], the exact solution of

$\Psi(t)$ equals the Mittag-Leffler function E_ϕ,

$$\begin{aligned} \Psi(t) &= E_\phi(-\Delta_\tau), \\ \Delta_\tau &= (t/\tau_o)^\phi, \\ \tau_o &= (\Omega_o)^{-1/\phi} \end{aligned} \tag{194}$$

Notably, the solution of Eq.(193) is not a "stretched exponential", as indicated in Ref. [258] and a number of subsequent works reproducing this error.

The Mittag-Leffler function is a natural generalization of the exponential relaxation function observed in "strongly mixing" dynamical systems at equilibrium. Cole [243] utilized the Mittag-Leffler relaxation function to model nerve stimulation and Cole and Cole [241] introduced this function to phenomenologically describe dielectrical relaxation in a broad range of condensed materials. This function is commonly discussed in the frequency domain through its (one-sided) Fourier-transform $\chi(\omega)$,

$$\chi(\omega) = \int_0^\infty d\tau \; e^{-i\omega\tau}[-d\Psi(\tau)/d\tau] \tag{195}$$

where $d\Psi/dt$ is proportional to the decay current and $\chi(\omega)$ is the dielectric susceptibility in a dielectric relaxation context.

The fractional operators lead to a simple algebra in a Fourier or Laplace transform representation [277]-[280]. Quite generally, we can obtain an I_t^α representation ("I-transform") of any Laplace-transformable function $f(t)$,

$$\begin{aligned} \mathcal{L}\{f(s)\} = F(s) &= \textstyle\int_0^\infty e^{-st} f(t)dt, \\ f(I_t) &= [sF(s)]_{s=1/I_t} \end{aligned} \tag{196}$$

where $s > 0$ and $f(t)$ is multiplied by $\Theta(t)$ so that $f(t < 0) = 0$. Multiplication of $f(t)$ by I^α and $t^{-\alpha}$ leads to the transform relations,

$$\begin{aligned} \mathcal{L}\{I_t^\alpha f(t)\} &= s^{-\alpha} F(s), \\ \mathcal{L}\{t^\alpha f(t)\} &= \textstyle\int_s^\infty d\tau \; [(\tau - s)^{-\alpha-1}/\Gamma(-\alpha)]F(\tau) \end{aligned} \tag{197}$$

where the integral operator in Eq.(197) is the adjoint ("Weyl") operator $K_s^{-\alpha}$ of $I_s^{-\alpha}$ [39]-[43]. We then deduce from Eqs.(193)-(197) the susceptibility function $\chi_{cc}(\omega)$,

$$\chi_{cc}(\omega) = 1/[1 + (i\omega\tau_o)^{\phi}] \tag{198}$$

where we have taken $s = i\omega$. It is usual practice to plot the real and imaginary parts of $\chi(\omega) = \chi(\omega) + i\chi"(\omega)$ [the energy "storage" and "loss" contributions to $\chi(\omega)$] against each other to obtain a "Cole-Cole plot" of relaxation data. The function $\chi_{cc}(\omega)$ gives rise to a symmetric arc plot in this representation [241]-[244]. It should be apparent from Eqs.(195) and (196) that any manipulation involving fractional powers of the Laplace variables or the Fourier transform variables(ω or k) actually involves fractional differential operators,even if this is not explicitly stated.

The Mittag-Leffler function has also been obtained in various time domain studies of relaxation in condensed materials. De Oliveira Casto [281] developed a phenomenological theory giving a Mittag-Leffler relaxation function describing the discharge of capacitors. Gross [227] investigated a phenomenological model of creep mechanical relaxation and found that this function leads to the widely found Hopkinson power law creep function approximately for many materials. From a molecular modeling perspective there are many models which lead to Mittag-Leffler type relaxation. Glarum [282] (See also [283]-[285]), for example,introduced a defect diffusion model of relaxation which led to the Mittag-Leffler function index 1/2,

$$\begin{aligned} \Psi(\text{Glarum}) &= E_{1/2}(-\Delta_{1/2}) \\ &= \exp(\Delta_{1/2}^2)\,\text{erfc}(\Delta_{1/2}),\ \Delta_{1/2}(\phi = 1/2) = \Delta_{1/2} \end{aligned} \tag{199}$$

where erfc is the complementerary error function. Yonezawa and coworkers [286]-[287] have recently performed a variety of simulation studies of relaxation in model condensed materials, which give much insight into the physical significance of the "intermittency index" ϕ. Constraining the relaxation process to occur on a fractal space or the particle motion to occur as a continuous time random walk with a long pausing time distribution leads to variations in ϕ in the resulting Mittag-Leffler (Cole-Cole) relaxation functions [286]-[287]. This finding is natural since the probability of recurrence of particle motion to a previously visited point is altered in these models, thereby changing ϕ (see Fig.4). The problem of the return probability of a random walk to the original point at $t = 0$ is exactly described by the renewal equations [61] above. In this context $\Psi(t)$ describes the probability that a random walk does not return to the origin ("survives") up to time t [See Eqs. (A.3) and (A.4) of

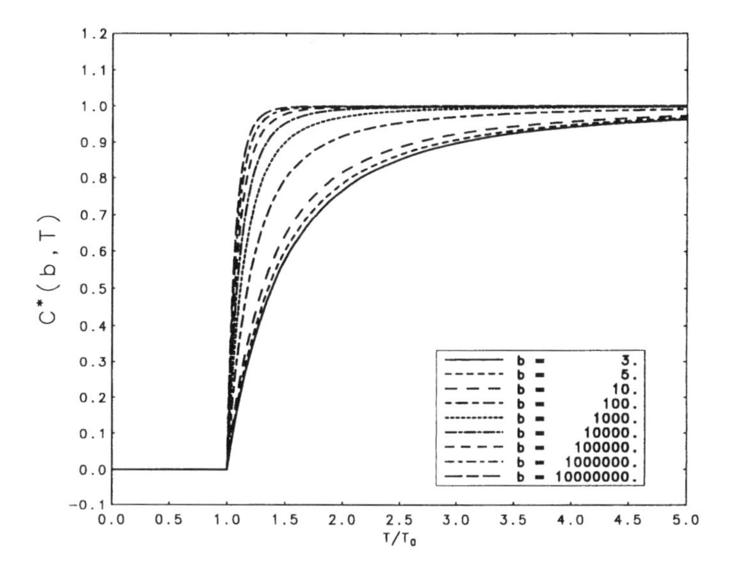

Figure 10: Capacity $C^*(T)$ of random walk on an ultrametric space with activated hopping. Note the dynamical transition at the temperature T_o in this exactly solvable model [35],[288],[289]. It is apparent that the transition becomes sharper as the branching index (coordination number of lattice tree) increases. The capacity $C^*(d)$ of a random walk on a hypercubic lattice (Fig. 11) has a similar variation to $C^*(T)$ where d corresponds to temperature in the ultrametric space diffusion model.

Ref. [288]-[289]] where the rate of return to the origin of the random walk $\mathcal{R}(t)$ obeys an integral equation of the form Eq.(183) [see Eq.(8.25b) of Ref. [61]].

A realistic model of condensed matter relaxation must also address changes in phase space structure which accompany temperature changes in condensed disordered materials. It has been suggested that diffusion in an ultrametric space should mimic this kind of complex evolution [288]-[290], and temperature can be incorporated into this model by making the hopping process thermally activated. The ultrametric space diffusion model is attractive because it allows exact calculation of the random walk recurrence properties, which are similar to the geometrical models considered by Yonezawa et al.[286]-[287]. Notably this model exhibits an "ergodic to non-ergodic transition" at a well-defined characteristic temperature where fluctuations in the random walk recurrences change their dynamic "universality class".

The temperature of this transition T_o is determined by considering an ergodic theorem [261]-[262] governing the rate of random walk exploration.

The number of distinct sites visited S_t, divided by the number of steps t taken by a random walk, defines the "capacity" C^* of the walk [261]-[264],

$$C^* = \lim_{t\to\infty} S_t/t, \quad C^* \in [0,1] \tag{200}$$

where the $*$-superscript is a reminder that the lattice quantity is under discussion. C^* is evidently a direct measure of rate at which phase space is explored. Fig. 10 shows an exact calculation of $C^*(T)$ for this ultrametric space model where temperature T is expressed in units of the transition temperature T_o at which C^* vanishes. The "dynamic transition temperature" T_o is a function of the branching index b of the ultrametric space [35],[288]-[289],

$$T_o = 1/\ln b \tag{201}$$

$C^*(T)$ is an "order parameter for the ergodicity" [35] of the random walk process. Note that the temperature variation of C^* is more rapid for higher values of the branching index b of the ultrametric space, which is reminiscent of the "fragility" concept in glass phenomenology.(The detailed calculations on which Fig.10 is based are described in Refs.[288]-[289].) Fig. 11 shows the variation [61] of C^* for simple random walks on a hypercubic lattice of dimension d where a similar transition between "transient" ($C^* > 0$) and "recurrent"($C^* = 0$) random walk motion occurs as the dimension d is lowered through the critical value $d = 2$.

It should be appreciated from this discussion that the general limit theorem leading to the $\Psi(t;\phi)$ relaxation functions occurs for a multitude of microscopic models which are "attracted" to the same "universality class" of relaxation functions. The mathematical origin of this universality is the same as in equilibrium critical phenomena, except we do not have an evident mechanism to select out particular values of the "dynamic critical exponent" ϕ. Instead, we can expect ϕ to reflect the geometry of the material's phase space [which should depend on system parameters such as temperature, pressure, etc.]. The occurrence of fractal phase space geometry (due perhaps to the "fracture" of islands of stability in sticky Cantori structures as in model dynamical systems [293]-[300]) should be sufficient to give rise to this pattern of relaxation. Probably the best we can hope to achieve in understanding this type of phenomena (i.e.,ϕ variation) is the construction of exactly solvable models where this phase space geometry change can be calculated numerically and the relaxation processes accompanying these changes monitored. In general, the parameter ϕ can be expected to be slowly varying [241]-[244] with T or constant, except in the vicinity of a phase transition.

Although many of the properties of the Mittag-Leffler function are summarized by Douglas et al. [23], there are some particular properties of this

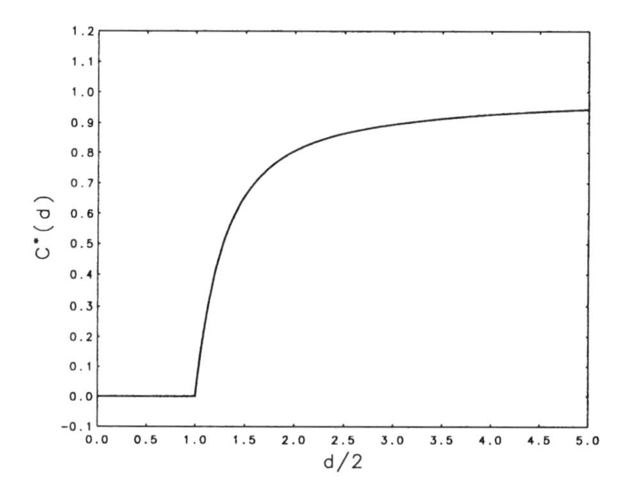

Figure 11: The capacity $C^*(d)$ of a random walk on a hypercubic lattice. The lattice capacity is a central object in the statistical mechanics of phase transitions and critical phenomena [290]. It determines the critical temperature of the spherical model, polymer adsorption, particle localization transitions, dimensions of random networks, resistivity of lattice networks and approximates many other objects of basic significance [61] (percolation thresholds, critical temperatures for phase separation, etc.). The capacity also governs the rate at which random walks explore space so this parameter is also basic to characterizing the rate of mixing in dynamical systems. $C^*(d)$ can be interpreted as the probability that a random walk escapes from its initial point to infinity and the results of this figure for a continuum Brownian chain which escapes with probabilty 0 for $d < 2$ and probabilty 1 for $d > 2$. This type of discontinuous behavoir for C^* as a function of T/T_o seems to be approached for $b \to \infty$ in Fig. 10. The capacity C_Ω dicussed in Section 4 is the continuum analog of the lattice capacity C^* and thus it can be likewise be interpreted in terms of the volume swept out by a Brownian chain per unit time.

function worth emphasizing in connection with relaxation processes in condensed materials. First, we observe that $E_\phi(-x)$ is an absolutely monotone [30]-[31] function for $0 < \phi \leq 1$,

$$d^m E_\phi(-x)/dx^m \geq 0, \quad m = 0, 1, 2, 3, ..., x \in [0, \infty) \tag{202}$$

so that these relaxation functions decay uniformly. This is a characteristic feature of large scale relaxation in condensed materials which is often taken for granted by experimentalists. We also observe that by changing temperature [and thus the magnitude of the coupling constant Ω_o in Eq.(192)] that the functional form of the relaxation function is invariant (provided ϕ does not change in this temperature interval). This "time-temperature superposition" property is a mathematical consequence of the homogeneous form of the memory kernel,

$$k(t;\phi) = d\mathcal{R}(t)/dt \sim \Omega_o \ |t|^{\phi-2}/\Gamma(\phi-1), \ \ 0<\phi\leq 1 \tag{203}$$

$$k(\lambda t;\phi) = \lambda^{\phi-2} \ k(t;\phi) \tag{204}$$

This analytic symmetry reflects the probabilistic limit theorems discussed above.

The probabilistic model of condensed matter relaxation naturally leads to a classification scheme for $\Psi(t)$ functions depending on the character of the fluctuations governing the relaxation process. The simplest case corresponds to the "strong mixing" case where $\mathcal{R}(t)$ rapidly approaches a constant value so that $\Psi(t)$ obeys the limits,

$$\begin{aligned} \lim_{t\to 0+} \Psi(t) &= 1, \\ -\lim_{t\to\infty} [\log\Psi(t)/t] &= \Omega_o \ = R(t\to\infty) \end{aligned} \tag{205}$$

The limit Eq.(205) reflects the existence of an ergodic theorem [301]-[302] governing the rate of elementary relaxation events giving rise to relaxation. Specifically, if $N(t)$ is a random variable which counts these random (and presumably independent) events, then "strong mixing" implies the ergodic theorem,

$$\mathcal{R}(t\to\infty) = \lim_{t\to\infty} \langle N(t)\rangle/t \tag{206}$$

where $\mathcal{R}(t\to\infty)$ is positive constant characterizing the rate of mixing of the dynamical system (i.e., the material).

If the relaxation process occurs more "sporadically" [301]-[302], then we have a "weak" ergodic theorem [53],[303], $0<\phi<1$,

$$\lim_{t\to\infty} \langle N(t)\rangle/t^{\phi} = \Omega_o/\Gamma(\phi) \tag{207}$$

where Ω_o is a positive constant. In this case, we have the limiting behaviour,

$$\begin{aligned} \lim_{t\to 0+} \Psi(t) &= 1, \\ -\lim_{t\to\infty} [\log\Psi(t)/t] &= 0 \end{aligned} \tag{208}$$

defining "weak mixing" relaxation functions. (The term "weak mixing" is usually used in a different technical sense in the mathematical literature). Finally, we note that if $\mathcal{R}(t)$ decays fast (e.g., exponentially fast) for large t

so that the integral of $\mathcal{R}(t)$ over $(0,\infty)$ is finite, then $\Psi(t)$ no longer decays to zero at long times [31],

$$\lim_{t\to\infty} \Psi(t) = \Psi^* = \left[1 + \int_0^\infty d\tau\, \mathcal{R}(\tau)\right]^{-1} \tag{209}$$

This "non-ergodic" limiting behavior corresponds to a material which is not in equilibrium, a commonly observed behavior in "glassy" materials. Finally, we mention the opposite extreme where $\mathcal{R}(t)$ increases with time. For example, take $\mathcal{R}(t)$ to have the homogeneous form,

$$\mathcal{R}(t) \sim t^{\phi-1}, \quad t \to \infty \tag{210}$$

where $1 < \phi \leq 2$ so the rate of the relaxation process grows with time. The relaxation function becomes oscillatory in this case, reducing to $\Psi(t) = \cos(t/\tau_o)$ in the extreme limit $\phi \to 2$. The integral [31] of $\Psi(t;\phi)$ from 0 to ∞ for any ϕ in the interval $1 < \phi \leq 2$ equals 0, meaning that the associated transport properties either vanish or diverge. This situation may well have important significance in material systems driven far out of equilibrium. Relaxation processes occuring at small scales in condensed materials (e.g., velocity autocorrelation of particle in cooled liquid [240],[256]) often exhibit a superficially similar oscillatory relaxation. Finally, it is noted that the "weak mixing" class of relaxation processes, involving fractional order operators, leads to a breakdown of ergodicity as it is classically defined [304] [assumption of $\phi = 1$ in Eq.(207)]. Variable ϕ corresponds to "quasi-mixing" [254] where ϕ quantifies the "degree of ergodicity". It is also notable that the classification scheme for relaxation functions given here exactly parallels the classification of "regenerative p-functions" in Kingman's "stochastic theory of regenerative events" [305] (a continuum limit generalization of Feller's fluctuation theory of recurrent events) where the p-functions obey limit theorems "attracting" to the Mittag-Leffler function class, as in the present development for $\Psi(t)$.

The occurrence of "weak mixing" type relaxation of the kind discussed above leads to non-trivial constitutive relations involving fractional order operators, as recognized experimentally [224]-[233]. The Mittag-Leffler function approaches a power law at long times,

$$\Psi(t;\phi) \sim (t/\tau_o)^{-\phi}, \quad t \to \infty \tag{211}$$

which can be a good approximation over appreciable time scales ("von Schweidler law" [234],[306],[307]). For example, if we introduce this approximation for the shear relaxation function into the functional relating stress $\sigma(t)$ to the strain rate $\dot{\gamma}(t)$ in the limit of linear response,then we obtain the fractional order constituitive equation relating these quantities,

$$\sigma(t) \approx s I_t^{1-\phi} \dot{\gamma} \tag{212}$$

where s is a constant. Eq.(212) reduces to Newton's constituitive relation for "liquids" and to Hooke's law for "solids" for $\phi \to 1$ and 0, respectively. Eq.(199) with $\phi \in (0,1)$ has been shown to provide a good approximation for polymer gels [308] and is useful in understanding the influence of inertial effects on the asymptotic frequency dependence of the viscosity $\eta(\omega)$ of small molecule liquids [309]. In an electrical context the counterpart to Eq.(212) relates voltage and current [6],[70].

Apart from the mathematical occurrence of fractional order operators, the "fractional" or "gel" constituitive Eq.(199) raises some interesting philosophical issues [226]. A material obeying this equation is somehow in an "intermediate state" between a liquid and solid. If we attempt to measure the viscosity of a gel obeying this equation, for example, the apparent viscosity grows indefinitely with the time [308]. On the other hand, the shear stress relaxation function seems to vanish at long times for these materials. The problem here is that the usual transport coefficients are not suitable for discussing these hybrid forms of matter. It then becomes appropriate to define new transport coefficients defined through the constant of proportionality in Eq.(59)[226],[308]. The index ϕ provides a useful and intuitive quantification of the physical state ("relative firmness") of complex materials such as foods, plastics, gels, etc. without appeal to ideal Newtonian liquids or Hookean solids [226].

A shortcoming of our probabilistic approach to modeling condensed matter relaxation is that it does not account for heterogeneities in those materials which develop transiently through interparticle interactions. It is now recognized that cooled liquids develop large scale heterogeneities which come to have a significant influence on their properties. Douglas and Hubbard [237]-[239] suggested that this clustering phenomenon, having its origin in topological interchain interactions, is responsible for the universal "entanglement" phenomena observed in high molecular weight and concentrated polymer solutions and for characteristic features of stress relaxation in glass forming liquids where the clustering arises from interparticle attractions. Douglas and Hubbard [237]-[239] extended the probabilistic model of condensed matter relaxation by making arguments for the form of this clustering process, based on the assumption of thermal equilibrium (Boltzmann's law assumed to govern the cluster size distribution). According to this cluster model the energy of the living-polymer-like transient clusters is extensive in their size where the geometry of these structures can be string, sheet, or clump-like depending on the particulars of the interparticle interactions. String-like structures were suggested to form in "fragile" glass-forming liquids, sheets in "strong" glasses, while amoeba-like clumps were suggested to occur in "entangled" polymer fluids.

The Douglas-Hubbard model indicated that the presence of this material heterogeneity governed by Boltzmann's Law should give rise to stretched exponential stress relaxation [237]-[239],

$$\Psi(t) \approx \exp[-\Omega_o \; t^{1-\beta}/(1-\beta)], \quad t \to \infty \tag{213}$$

even for systems which were locally "strongly mixing" where β depends on the spatial dimension of the excitation involved,

$$\beta(\text{linear}) = 2/3, \; \beta(\text{sheets}) = 1/2, \; \beta(\text{clumps}) = 2/5 \tag{214}$$

This corresponds to a memory kernel having the form,

$$k(t,\tau) = \Omega_o \; \tau^{-\beta}\delta(t-\tau) \tag{215}$$

where δ is a delta-function distribution.

Rigorous calculations [310] of the survival probability of a random walk in a medium with randomly placed absorbing obstructions exactly leads to "stretched exponential" decay functions of the form Eq.(200) where d is the spatial dimension rather than the cluster dimension as in the present modeling. The calculations of Donsker and Varadhan [310] clearly indicate that material disorder can lead to stretched exponential relaxation and the cluster model of Douglas and Hubbard [237]-[239] is patterned after this same idea. Importantly, the fluid inhomogeneities in the stress-relaxation model persist on timescales greater than the relaxation process, but to finite times so the fluid can remain in equilibrium.

An interesting implication of the exact Donsker-Varadhan calculations for relaxation in a disordered medium is the breaking of the time-translation symmetry of the memory kernel [see Eq.(172)]. This analytic symmetry is linked to conservation of energy in the derivation of Eq.(182). Of course, this is only an apparent symmetry breaking which derives from the disorder average [237]-[239]. There is no paradox if we consider our material to be composed of an ensemble of dynamical systems comprised of different local environments on the timescale of the relaxation process under discussion (This is nothing but the classical "distribution of relaxation times" idea.). The wide occurrence of stretched exponential relaxation in condensed materials suggests that material heterogeneity arising from long-lived particle clustering is a rather widespread phenomenon in condensed materials. Douglas and Hubbard introduce the dynamic index β as a measure of fluid "heterogeneity" where β is related to the geometry of the clusters involved. Although unstated in the original work, it was also understood that other values of β can occur for clusters having a fractal structure. As a final note, it is mentioned that string-like fluid "excitations" have been observed in molecular dynamics simulations of cooled Lenard-Jones particles which form a glass at low temperatures [311],[312].

It is natural to consider relaxation in condensed materials to involve both intermittency in the rate of the relaxation process (arising from collective particle motions in dense environments) and from material heterogeneity and these effects are modeled by the generalized relaxation kernel,

$$\mathcal{R}(t,\tau) = \Omega_o \; |t-\tau|^{\phi-1}\tau^{-\beta}/\Gamma(\phi), \quad 0 < \phi, \beta \leq 1 \tag{216}$$

where the ϕ and β "critical indices" characterize the "degree" of intermittency and heterogeneity, respectively. Inserting $\mathcal{R}(t,\tau)$ into the relaxation integral equation Eq.(209) yields,

$$\Psi(t;\phi,\beta) = 1 - \Omega_o \int_0^t d\tau \;\; [|t-\tau|^{\phi-1}\tau^{-\beta}/\Gamma(\phi)]\Psi(\tau;\phi,\beta) \tag{217}$$

This integral equation involves the replacement of the Riemann-Liouville operator of Eq.(193) by the more general Erdélyi-Kober fractional order operator [245]-[249],

$$I_t^{\delta,\phi} f(t) = t^{-(\phi+\delta)} \int_0^t d\tau \;\; [|t-\tau|^{\phi-1}\tau^{\delta}\Gamma(\phi)] f(\tau) \tag{218}$$

The Erdélyi-Kober operator is related to the RL operator I_t^{α} as,

$$I_t^{\delta,\phi} f(t) = t^{-(\phi+\delta)} I_t^{\phi}[t^{\delta} f(t)] \tag{219}$$

This representation is sometimes more convenient for algebraic manipulations involving these more complicated operators.

The $I_t^{\delta,\phi}$ operators obey the exponent relations [245]-[249], $\alpha > 0$, $\delta > -1/2$,

$$I_t^{\delta,\phi} I_t^{\delta+\phi,\beta} f(t) = I_t^{\delta,\phi+\beta} f(t) \tag{220}$$

$$I_t^{\delta,\phi} I_t^{\delta+\phi,\beta} f(t) = I_t^{\delta+\phi,\beta} I_t^{\delta,\phi} f(t) \tag{221}$$

$$I_t^{\delta+\phi,-\phi} I_t^{\delta,\phi} f(t) = I_t^{\delta,0} f(t) = f(t) \tag{222}$$

$$I_t^{0,\phi} f(t) = t^{-\phi} I_t^{\phi} f(t) \tag{223}$$

$$I_t^{\delta,\phi}[t^{\beta} f(t)] = t^{\beta} I_t^{\delta+\beta,\phi} f(t) \tag{224}$$

and for negative integer $-n$ values of ϕ the operator $I_t^{\delta,\phi}$ has the formal representation,

$$I_t^{\delta,-n} f(t) = t^{-\delta+n} d^n [t^\delta f(t)]/dt^n \tag{225}$$

We can obtain some insights into these operators by transforming to the unit inverval (0,1),

$$I_t^{\delta,\phi} f(t) = \int_0^1 ds \ [(1-s)^{\phi-1} s^\delta / \Gamma(\phi)] f(st) \tag{226}$$

The operators $I_t^{\delta,\phi}$ evidently involve averaging functions $f(t)$ with respect to the weight functions of the Jacobi polynomials. The singular "differential" operators correspond to negative integer δ values where the kernel of these operators reduces to generalized functions [313] (delta-function or delta-function derivatives). The kernel of Eq.(211) also becomes singular for large positive values of ϕ and δ which leads to many further important analytic consequences [313]. Many of the special functions of mathematical physics may be built up from $I_t^{\delta,\phi}$ operators acting on simple functions [40],[313] and much mathematical analysis simply involves exploring the group properties of these operators.

Exact solution of the integral equation Eq.(217) by the usual semi-group methods gives [?],[237]-[239],

$$\Psi(t;\phi,\beta) = \sum_{k=0}^{\infty} a_k(\phi,\beta)[z_\Omega(\phi,\beta)]^k \ , \ a_0(\phi,\beta) = 1; \tag{227}$$

$$a_k(\phi,\beta) = \prod_{m=1}^{k} \Gamma(1+m\hat{\phi}-\phi)/\Gamma(1+m\hat{\phi}), \quad k>0 \tag{228}$$

$$\hat{\phi} = \phi - \beta, \quad 0 < \hat{\phi}, \phi, \beta \leq 1) \tag{229}$$

$$z_\Omega(\phi,\beta) = \Omega_o \tau^{\hat{\phi}} = (t/\tau^*)^{\hat{\phi}}, \quad \tau^*(\phi,\beta) = \Omega_o^{-1/\hat{\phi}} \tag{230}$$

The relaxation function $\Psi(t;\phi,\beta)$ reduces to a Mittag-Leffler function for $\beta = 0$ and a stretched exponential function for $\phi = 1$, but $\Psi(t;\phi,\beta)$ becomes a hybrid form of these functions otherwise. Some representative plots of these relaxation curves are given in Figs.12-14. The function $\Psi(t;\phi,\beta)$ arises in an electrochemistry context in the description of the dropping mercury electrode [314]-[316]. In this context $\Psi(t,\phi,\beta)$ the case is known as the Koutecky equation [237]-[239],[314]-[316] in the case $\hat{\phi} = 3/14$.

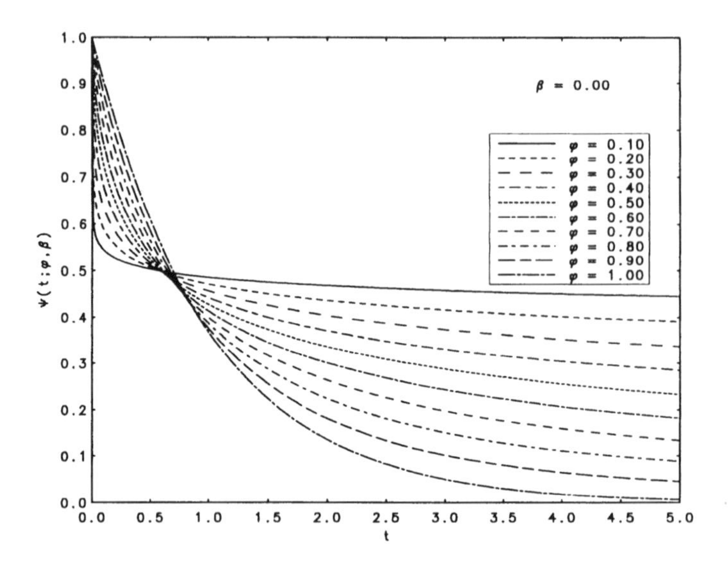

Figure 12: Relaxation function $\Psi(t;\phi,\beta)$ for a range of ϕ and β values. The indices ϕ and β model hetrogeneity in time and space, respectively (see text): $\beta = 0, \phi \in [0,1]$ The time constant τ^* in Eq.(227) is taken to equal 1.

The Mittag-Leffler function exhibits the same asymptotic dependence as $\Psi(t;\phi,\beta)$ with $\hat{\phi} = \phi$, but there are significant differences between these functional forms for $\hat{\phi} \neq \phi$ which are apparent when these functions are examined over wide time intervals. These differences are particuarly evident in the frequency domain through "Cole-Cole" plots of the real and imaginary [See Eqs.(195)-(198)] contributions of the Fourier transform [241]-[244] of $d\Psi/dt$. A decrease of the index ϕ.causes the symmetric arc of the Debye (exponential relaxation function) to be symmetrically "squashed", while an increase of β causes the Cole-Cole arc to become "skewed". The Havriliak-Negami function [317]-[319], which is widely used to correlate frequency domain relaxation data for condensed materials, depends on two parameters which have a similar effect to the variation of ϕ and β for the $\Psi(t;\Phi,\beta)$ function [318]. The Cole-Davidson class [318] of Havriliak-Negami functions resemble the commonly utilized "stretched exponential [319] class of functions which corresponds to $\Psi(t;\Phi,\beta)$ for $\Phi = 1$.

The relaxation function $\Psi(t;\phi,\beta)$ is readily shown to have the asymptotic

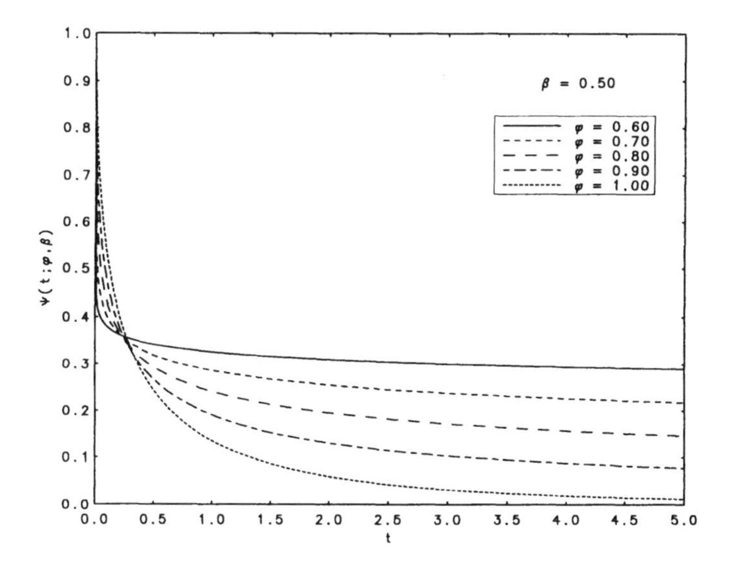

Figure 13: Relaxation function $\Psi(t;\phi,\beta)$ for a range of ϕ and β values. The indices ϕ and β model hetrogeneity in time and space, respectively (see text): $\beta = 0.50, \phi \in [0.5,1]$ The time constant τ^* in Eq.(227) is taken to equal 1.

scaling [237]-[239], $0 < \hat{\phi} < 1$,

$$\Psi(t;\phi,\beta) \sim t^{\hat{\phi}}, \quad t \ll \tau^* \tag{231}$$

$$\Psi(t;\phi,\beta) \sim t^{-\hat{\phi}}, t \gg \tau^* \tag{232}$$

where the prefactor coefficients are unspecified. Jonscher [234] has emphasized that the relaxation decay in Eq.(213) is found in a wide range of materials–dielectrics, semi-conductors, materials with covalent and ionic bonding and structures having structures ranging from crystalline to amorphous. This type of decay is also characteristic of transient photocurrent decay in amorphous materials [320], where the rescaling $\tau^*(\phi,\beta) = \Omega_o^{-1/\hat{\phi}}$ has been observed (See also Ref. [236]).

Although the limiting stretched exponential ($\phi \to 1; 0 < \beta < 1$) relaxation functions obey the conditions of "weak mixing" given in Eq.(207), the integral of $\Psi(t;\phi = 1,\beta)$ from 0 to ∞ is finite for this class of functions,

$$\int_0^\infty d\tau \;\; \Psi(\tau;\phi = 1,\beta) < \infty \tag{233}$$

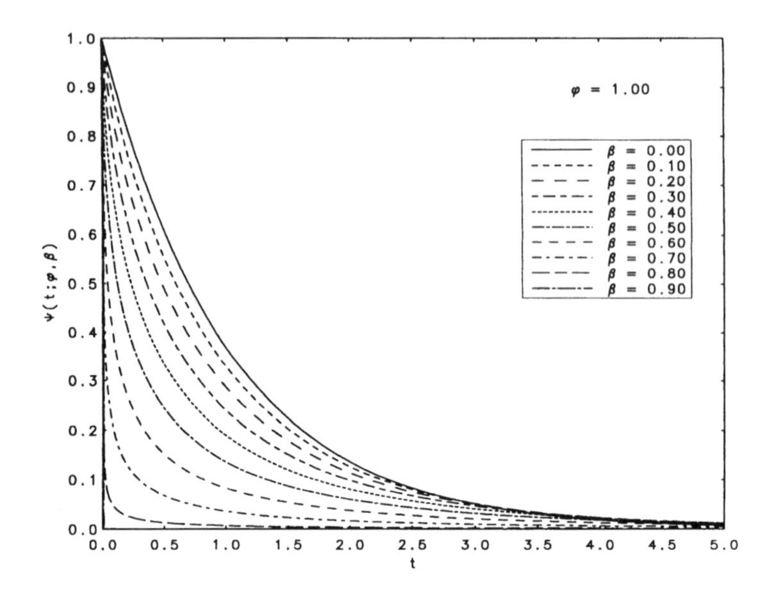

Figure 14: Relaxation function $\Psi(t;\phi,\beta)$ for a range of ϕ and β values. The indices ϕ and β model hetrogeneity in time and space, respectively (see text): $\beta \in [0,1], \phi = 1$ The time constant τ^* in Eq.(227) is taken to equal 1.

making this class of functions suitable for modeling liquid relaxation (i.e., viscosity is finite if $\Psi(t)$ is the shear relaxation function.) This property cannot be discussed for the Havriliak-Negami function [317], which is not actually derived from a theoretical model. Given these advantages it would be interesting to test this new class of relaxation functions against accurate dielectric data to determine the system dependence of ϕ and β parameters. The "stretched exponential" or "Williams-Watts" function, of corresponding to $\Psi(\Phi \to 1, \beta)$, is widely used to correlate relaxation data in complex liquids [236].

The slow decay of the $\Psi(t;\phi,\beta)$ function for $\phi \neq l$ and $\beta \neq 0$ requires a refinement of the relaxation function classification introduced above. We observe that the integral of $\Psi(t;\phi;\beta)$,

$$\int_0^\infty d\tau \ \ \Psi(\tau;\phi,\beta=0)d\tau \to \infty, \quad 0<\phi<1 \tag{234}$$

diverges while the integral of $\Psi(t, \phi = 1, \beta)$ over $[0, \infty)$ is finite,

$$0 < \int_0^\infty d\tau \; \Psi(\tau; \phi = 1, \beta) d\tau < \infty, \quad 0 < \beta < 1. \tag{235}$$

We have argued that the case of ϕ variable and $\beta = 0$ corresponds to homogeneous materials and the case of β variable and $\phi = 1$ corresponds to inhomogeneous materials. Thus, we further divide the "weak mixing" class of relaxation functions into "homogeneous weak mixing" and "inhomogeneous weak mixing" subclasses, according to whether the integral of $\Psi(t; \phi, \beta)$ over $[0, \infty)$ is finite or not. This provides a complete classification scheme for condensed matter relaxation functions.

8 Conclusion

The polymer science applications of this chapter illustrate a common problem in materials science and condensed matter physics applications. In many real materials, typified by polymeric materials, we are confronted with boundaries which can be idealized as "fractals" [1],[2]. Path-integration provides a natural language for stating boundary value problems in cases where the boundary is not differentiable in the classical (positive integer order) sense. The path-integral formulation of classical boundary value problems (e.g., Dirichlet problem for Laplace's equation, Stokes drag for a Brownian particle with stick boundary condition) with "fractal" boundaries commonly leads to fractional differential equations. These fractional differential initial and boundary value problems are obtained by converting the path-integral description to the equivalent integral equation formulation. The order of the fractional differential operators obtained reflect the geometry of boundary, the space in which the boundaries (polymers) are embedded, and the type of random walk process governing transport. It is observed that spatial inhomogeneity in the presence of the polymer tends to give rise to fractional order differential operators in the time (or contour) coordinate, while temporal hetrogeneity tends to give rise to fractional order differential operators in the space variable. The fractional calculus then arises from an analytic description of diffusion processes associated with fractal boundaries [1],[2].

The recasting of polymer science applications involving path-integration in terms of integral equations and fractional calculus also illuminates many aspects of perturbative renormalization group theory, which can be viewed as another analytic method adapted to the treatment of geometrical systems having fractal structure. For example, the pole structure of the interaction perturbation theory in the dimensionally dependant parameter Φ directly reflects the presence of the fractional order operators, which in turn have an origin in fractal geometry. The "critical dimensionality" corresponds to the point where the order of the fractional order operator vanishes and the kernel of the operator describing the interaction changes from being "weakly singular" to "singular"(not integrable in $\mathcal{L}^1$ sense). Thus, the ϵ-expansion perturbation theory is an expansion in the singularity strength of the interaction kernel associated with these operators. This point of view helps us appreciate the significance of the RG "fixed-point" which determines the critical exponents and amplitudes in the "strong coupling" regime for a large class of physical models. In the case of the friction coefficient the RG fixed-point was found to correspond to the lowest eigenvalue of a fractional differential equation [Eq.(96)] and a similar correspondence was discussed in connection with the fractional differential equations arising in surface interacting polymers. It seems likely that the type of connection is quite general [321]. Fractional order operators were long ago utilized as a method for controlling singularities in early field theoretic renormalization calculations so this perspective is not entirely new [322]-[331].

The fractional calculus is also a natural tool in characterizing relaxation in complex dynamical systems such as polymeric materials. The large number of polymer chain conformations gives rise to complexity in the configurational dynamics of these materials. This is reflected in the intermittency in the chain segment motions and in the tendency for the particles to cluster and thus move in a collective fashion. The general theory of relaxation, involving Erdélyi-Kober fractional order operators, is relevant for describing long wavelength relaxation proceses in condensed materials such as polymers. Basic properties of condensed matter relaxation, such as time-temperature superposition, monotone decay of relaxation functions, etc. become apparent from a consideration of the types of "mixing" which should occur for statistical mechanical systems at equilibrium. A general classification scheme relaxation functions is introduced as a byproduct of this investigation.

Acknowledgements

Fern Hunt, Marjorie McClain and Robert Lipman of the Information technology Laboratory at NIST are thanked for their help in making the figures. Fern Hunt is also thanked for numerous discussions intermittency in dynamical systems and its ramifications for relaxation processes.

References

1. B.B.Mandelbrot, The Fractal Geometry of Nature (Freeman,San Francisco,1982). See also Ref.[2]
2. S.J. Taylor, Math. Proc.Camb.Phil.Soc. **100**, 13 (1986).
3. R.R. Nigmatullin, Phys.Stat.Solidi B **133**, 425 (1986).
4. W. Wyss, J.Math.Phys. **27**, 2782 (1986).
5. W.R. Schneider and W. Wyss, J.Math.Phys. **30**, 134 (1989).
6. M. Giona and H.E. Roman, J.Phys.A **25**, 2093 (1992).
7. B. Ross, Math.Mag. **50**, 115 (1977).
8. P.J. Flory, Statistical Mechanics of Chain Molecules (Wiley-Interscience, New York,1969).
9. P.J. Flory, Brit.Polym.J. **8**, 1 (1976).
10. K.F. Freed, Renormalization Group Theory of Macromolecules (Wiley-Interscience,NY,1987).
11. K.F. Freed, Adv.Chem.Phys. **22**, 1 (1972).
12. K.F. Freed, Ann.Prob. **9**, 537 (1981).
13. See Ref. [74],[75].
14. H. Yamakawa, Modern Theory of Polymer Solutions(Harper and Row, NY,1971).
15. P.-G. De Gennes, Scaling Concepts in Polymer Physics(Cornell University Press,Ithaca,NY,1979).
16. J.Des Cloizeaux and G. Jannink, Polymers in Solution (Clarendon Press, Oxford,1990).
17. J.F. Douglas, J. Roovers and K.F. Freed, Macromolecules **23**, 4168 (1990).
18. K.F. Freed, J.Chem.Phys. **79**, 3121 (1983).
19. A.M. Nemirovsky and K.F.Freed, J.Chem.Phys. **83**, 4166 (1985).
20. J.F. Douglas, A.M. Nemirovsky and K.F.Freed, Macromolecules **19**, 204 (1986).
21. Z.-G.Wang, A.M. Nemirovsky and K.F.Freed, J.Chem.Phys. **85**, 3068 (1986).
22. J.F. Douglas and K.F. Freed, Macromolecules **30**, 1813 (1997).
23. J.F. Douglas, S.-Q. Wang and K.F. Freed, Macromolecules **19**, 2207 (1986); **20**, 543(1987). See also Ref. [34].
24. S.M.Bhattachargee, Physica A **186**, 183 (1992).
    This work considers the closely related problem of the second virial coefficient of directed polymers.
25. The m-strand helix-coil transition corresponds to m-directed random walks where the chain "contacts" are regions of chain ordering. The

virial coefficient of the m-strands equals the integral (I_n) of the Mittag-Leffler function where the contact index ϕ eqauals $\phi = 1 - (m-1)(d/2)$ for random coil chains($a = 2$). See also M.E. Fisher [J.Stat. Phys. **34**, 667 (1984)] for applications of of directed polymer models to describe melting.

26. S.D. Milner, Z.G. Wang, Macromolecules **22**, 489 (1989).
27. G.H. Fredrickson and S.T. Milner, Macromolecules **29**, 7386 (1996).
28. G.H. Hardy, Divergent Series (Clarendon, Oxford,1949).
29. H. Bateman, Higher Transcendental Functions, Vol.3, Chapt.18, ed., A. Erdélyi(McGraw-Hill, NY, 1955).
30. H. Pollard, Bull.Amer.Math.Soc. **54**, 1115 (1948).
31. A. Friedman, J. D'Analyse Math. **11**, 381 (1963).
32. C.G. Fry and H.K. Hughes, Duke Math.J. **9**, 791 (1942).
33. H. Matison, Duke Math.J. **4**, 9 (1938).
34. J.F. Douglas, Macromolecules **22**, 1786(1989); Macromolecules **24**, 3163 (1991).
35. J.F. Douglas, Comp.Mat.Sci. **4**, 292 (1995).
36. R. Hilfer, Phys.Rev.Lett. **68**, 190 (1992); Phys.Rev.E **48**, 2466 (1993).
37. R. Hilfer, Physica Scripta **44**, 321 (1991);
38. J.F. Nagle, Proc.Roy.Soc.Lond.337569(1974). This is probably the first mention of a fractional order phase transition, although there is no use of the fractional order operators in this work.
39. B. Ross, Fractional Calculus and Its Applications Lect. Notes Math. 457 (Springer-Verlag, NY,1975).
40. K.B. Oldham and J. Spanier, The Fractional Calculus (Academic Press, NY,1974).
41. G.G. Samko, A.A. Kilbas and O.I. Marichev, Fractional Integrals and Derivatives,Theory and Applications(Gordon and Breach Sci.Publ., Langhorne, PA,1993).
42. B. Ross, "The Development,Theory and Applications of the Gamma-Function and a Profile of Fractional Calculus", Ph.D. thesis, Mathematics, New York University, 1974
43. J.F. Douglas, Adv.Chem.Phys. **102**, 121 (1997).
44. E. Hille, Functional Analysis and Semi-Groups Amer.Math.Soc. Colloq. Publ.,Vol.31 (Amer.Math.Soc., NY, 1948).
45. E. Hille, Kungl.Fys.Sälls.Lund Förh. **21**, 130 (1951)
46. E. Hille, Ann.Math. **40**, 1 (1939);
    Trans.Amer.Math.Soc. **57**, 246 (1945).
47. J.F. Douglas, Macromolecules **22**, 3707 (1989).
48. J.Hawkes, Z. Wahr. **19**, 90 (1971).

49. R.F. Bass and D. Khoshnevisan, Prob.Theor. and Related Fields **92**, 465 (1992).
50. J.F. Douglas, Macromolecules **21**, 3515 (1988).
51. S.J. Taylor, Z. Wahr. **6**, 170 (1996).
52. J.-F. Le Gall, Prob.Theor. and Related Fields **76**, 587 (1987). This work considers the rigorous calculation of the local time of Brownian motion in a wedge and the results accord with the heuristic discussion in Ref. [47].
53. D.A. Darling and M. Kac, Trans.Amer.Math.Soc. **84**, 444 (1957). See also Ref [54]-[59]
54. K.L. Chung and M. Kac, Mem.Amer.Math.Soc. **6**, 11 (1951).
55. N.H. Bingham, Z. Wahr. **17**, 1 (1971).
56. H. Kesten, Ann.Math.Soc. **103**, 82 (1962).
57. Y. Kasahara, Jap.J.Math.Phys. **1**, 67 (1975).
58. C. Stone, Ill.J.Math. **7**, 638 (1963).
59. J. Lamperti, J.Math.Mech. **7**, 433 (1958).
60. W. Feller, Trans.Amer.Math.Soc. **67**, 98 (1949).
61. J.F. Douglas and T. Ishinabe, Phys.Rev.E **51**, 1791 (1995).
62. M.J. Goovaerts,A. Babcenco and J.T. Devreese, J.Math.Phys. **14**, 554 (1973).
63. M.J. Goovaerts and J.T. Devreese, J.Math.Phys. **13**, 1070 (1972).
64. D.Bauch, Nuovo Cimento **85**, 118 (1985).
65. H.Taitlebaum, Physica A **190**, 295 (1992).
66. S.V.Lawande and K.V. Bhagwat, Phys.Lett.A **131**, 8 (1988).
67. S.-Q. Wang and K.F. Freed, J.Math.A 19,L-637(1986).
68. J.C. Jaeger, Proc.Roy.Soc.Edin. **61**, 223 (1943).
69. W.E. Olmsted and R.A. Handelsam, SIAM J. Appl.Math. **30**, 180 (1976);
    R.A. Handelsam and W.E. Olmsted,SIAM Rev. **18**, 275 (1976).
70. R. de Levie and A. Vogt, J. Elect.Anal.Chem. **278**, 25 (1990).
71. B.P. Lee, J.Phys.A **27**, 2633 (1994).
72. J.W. Strutt("Lord Raleigh"), Phil.Mag. **21**, 697 (1911).
73. E.J. Hinch, J. Fluid Mech. **72**, 499 (1975).
74. S.F. Edwards, Proc.Phys.Soc.Lond. **85**, 1656 (1965); **88**, 265 (1966); J.Phys.A **10**, 1670 (1975).
75. M.J. Westwater, Comm.Math.Phys. **72**, 131 (1980); **79**, 53 (1981).
76. S.R.S. Varadhan, Appendix to "Euclidean Quantum Field Theory" by K. Symanzik in Local Quantum Field Theory,ed., R. Jost(Academic Press, NY, 1969).
77. J. Rosen, Comm.Math.Phys. **88**, 327 (1983);

Ann.Prob. **14**, 1245 (1986).
78. E. Dynkin, J.Funct.Anal. **62**, 3 (1985); Ann.Prob. **16**, 1 (1988).
79. J. Le Gall, Ann.Prob. **16**, 991 (1988); J.Funct.Anal. **88**, 299 (1990).
80. X.-Y. Zhou, Prob.Th.and Related Fields **91**, 375 (1992).
81. R.L. Wolpert, J.Funct.Anal. **30**, 329 (1978); **30**, 341 (1978).
82. R.J. Adler and J.S. Rosen, Ann.Prob. **21**, 1073 (1993).
83. J.F. Douglas and K. F. Freed,Macromolecules **17**, 1854(1984); **17**, 2334 (1984); **18**, 201 (1985).
84. P.G. De Gennes, Phys.Lett.A **36**, 339 (1972).
85. V.J. Emery, Phys.Rev.B **11**, 239 (1975).
86. A.L. Kholodenko and K.F. Freed, J.Chem.Phys. **80**, 900 (1983).
87. See Ref. [10].
88. J.F.Douglas and A. Friedman, "Coping With Complex Boundaries" in Mathematics in Industrial Problems, Pt. 7 (Springer-Verlag,N.Y.,1994).
89. J.F. Douglas,H.-X. Zhou and J.B. Hubbard, Phys.Rev.E **49**, 5319 (1994).
90. See Ref. [97].
91. F.Y. Hunt, J.F. Douglas and J. Bernal,J.Math.Phys. **36**, 2386 (1995).
92. G.I. Barenblatt and A.J. Chorin, Proc.Natl.Acad.Sci. **93**, 6749 (1996).
93. S.Torquato, Phys.Rev.Lett. **64**, 2644 (1990).
94. J.G. Kirkwood and J. Riseman, J.Chem.Phys. **16**, 565 (1948).
95. B.H. Zimm, J.Chem.Phys. **24**, 269 (1956).
96. B.H. Zimm, G.M. Roe and L.F. Epstein,J.Chem.Phys. **24**, 279 (1956).
97. J.B. Hubbard and J.F. Douglas,Phys.Rev.E47,R-2983(1993).
98. J.A. Given, J.B. Hubbard and J.F. Douglas, J.Chem.Phys.**106**, 3761 (1997).
99. J.F. Douglas and E.J. Garboczi, Adv.Chem.Phys. **91**, 85 (1995).
100. G. Polya and G. Szegö, Isoperimetric Inequalities in Mathematical Physics,(Princeton Univ.Press,Princeton,NJ,1951).
101. See Ref. [217] M. Schiffer, Bull.Amer.Math.Soc. **60**, 303 (1954).
102. W. Brown,ed., Dynamic Light Scattering, (Clarendon Press,Oxford, 1993).
103. M. Kac, Rocky Mount.J.Math. **4**, 511 (1974).
104. M. Kac, Amer. Math. Monthly **77**, 586 (1970).
105. J.F. Douglas, Adv.Chem.Phys. **102**, 121 (1997).
106. M. Kac and J. M. Luttinger, Ann.Inst.Four.(Grenoble) **25**, 317 (1975).
107. J.M. Luttinger, J.Stat.Phys. **15**, 215 (1976).
108. F. Spitzer, Z.Wahr. **3**, 110 (1965).
109. G. Louchard, J.Math.& Phys. **44**, 177 (1965); Duke Math.J. **33**, 13 (1966).
110. A.M. Berezhkovskii, J.Stat.Phys. **76**, 1089 (1994).

111. S.C. Port, J.Math.&Mech. **15**, 805 (1966); Ann.Prob. **20**, (1992).
112. T.Jaroszewicz and P.S. Kurzepa, Ann.Phys. **216**, 226 (1992).
113. P. Erdös and S.J. Taylor, Acta Sci.Math.(Szeged) **11**, 137 (1960); **11**, 231 (1960).
114. H.P. McKean, Jr., J.Math.(Kyoto Univ.) **4**, 617 (1965).
115. W.E. Pruitt and S.J. Taylor, Trans.Amer.Math.Soc. **146**, 299 (1969).
116. S.C. Port, Ann.Math.Stat. **39**, 365 (1968).
117. K.L. Chung, Ann.Inst.Four. **23**, 313 (1973); **25**, 131 (1975).
118. M. Riesz, Acta Math. **81**, 1 (1949); Acta Szeged (Sci. Sect.)**9**, 1 (1938-1940). See Ref. [244] for the complete calculation of the equilibrium charge density associated with charges interacting through the Riesz potential on a hypersphere.
119. O. Frostman, Medd.Lunds Mat.Sem. **3**, 1 (1935); Kungl.Fys. Lund **20**, 3 (1949).
120. L. Lithner, Arkiv Mat. **4**, 31 (1959).
121. G. Björck, Arkiv Mat. **3**, 255 (1955).
122. H. Wallin, Arkiv Mat. **4**, 527 (1961).
123. B. Fuglede, Math.Scand. **8**, 287 (1960).
124. J. Deny, Acta Math. **82**, 107 (1950).
125. Z. Ciesielski and S.J. Taylor, Trans.Amer.Math.Soc. **103**, 434 (1962).
126. D.W. Strook, J. Math.& Mech. **16**, 829 (1967); Comm.Pure Appl.Math. **20**, 775 (1967).
127. S.-Q. Wang,J.F. Douglas and K.F. Freed, J.Chem.Phys.**85**,3674 (1986); **87**, 1346 (1987).
128. H. Brenner, J.Fluid Mech. **111**, 197 (1981).
129. E. Hille and J. Tamarkin, Ann.Math. **31**, 479 (1930).
130. T.E. Hull, SIAM J.Math. **7**, 290 (1959).
131. D. Hilbert, Gött.Nach., pg.49 (1904). See also Ref. [132]-[134].
132. R.G. Newton, Scattering Theory of Waves and Particles, (Springer, NY, 1982), pg.253.
133. T. Carleman, Math.Zeits. **9**, 196 (1921); **15**, 111 (1922).
134. N. Zeilon, Arkiv Mat., Astron., Fys. **18**, 1 (1924).
135. H. Widom, Trans.Amer.Math.Soc. **98**, 430 (1961).
136. M. Tsuji, Jpn.J.Math. **23**, 1 (1954).
137. C.J. van Trigt, J.Math.Phys. **14**, 863 (1973).
138. W. Feller, Comm.Sem. Lunds Univ.Mat.Sem.Suppl. **72**, 73 (1952).
139. K. Yosida, Math.Rev. **14**, 561 (1953).
140. A.V. Balakrishnan, Trans.Amer.Math.Soc. **91**, 330 (1959); Pacific J.Math. **10**, 419 (1960).
141. J. Elliot and W. Feller, Trans.Amer.Math.Soc. **82**, 392 (1956).

142. D. Ray, Trans.Amer.Math.Soc. **82**, 452 (1956).
143. W. Feller, Ann.Math.60417(1954).
144. J. Lamperti, Stochastic Processes(Springer,NY,1977).
145. A.S. Carasso and T. Kato, Trans.Amer.Math.Soc.327867(1991).
146. E. Nelson, Trans.Amer.Math.Soc. **88**, 400 (1958).
147. D. Geman and J. Horowitz, Duke Math.J. **43**, 809 (1976).
148. R.M. Blumenthal and R.K. Getoor, J.Math.Mech. **10**, 493 (1961).
149. G. Jona-Lasinio,Nuovo Cimento **25**, 99 (1975).
150. W. Feller, An Introduction to Probability Theory and its Applications, Sec.Ed.(Wiley,NY,1971), pg.175.
151. S. Bochner, Proc.Nat.Acad.Soc. **35**, 368 (1949).
152. S. Bochner, Harmonic Analysis and the Theory of Probability (Univ.Calif.Press, Berkeley, Cal.,1955).
153. G.F. Carrier, J.Math.&Phys.(MIT) **27**, 82 (1948).
154. F. Smithies, Integral Equations (Cambridge Univ.Press, 1958).
155. C. Baker The Numerical Treatment of Integral Equations (Clarendon Press,Oxford,1977).
156. W.L. Lovitt, Linear Integral Equations(Dover,NY,1950).
157. V.Volterra, Theory of Functionals and Integral and Integro-Differential Equations(Dover,NY,1959).
158. F.G.Tricomi, Integral Equations (Dover,NY, 1985).
159. H. Weyl, Göttinger Nachrichten, p.110(1911); Math.Ann. **71**, 441 (1912); Bull.Amer.Math.Soc. **56**, 115 (1950).
160. D. Ray, Tran.Amer.Math.Soc. **77**, 299 (1954).
161. H.P. McKean and I.M. Singer, J.Diff.Geom. **1**, 43 (1967).
162. M. Kac, Amer.Math.Month. **73**, 1 (1966).
163. Z.Ciesielski, Bull.Acad.Sci.Math.(Poland) **12**, 265 (1964); **13**, 147 (1965); **14**, 435 (1966).
164. M. Kac, Mich.Math.J. **3**, 141 (1955).
165. M. Kac,"Some Remarks on Stable Processes with Independent Increments" in Probability and Statistics, ed., U. Grenander(Wiley,NY,1959), pg.130.
166. H. Widom, Trans.Amer.Math.Soc. **98**, 430 (1961); **100**, 252 (1961); **106**, 391 (1963); **109**, 278 (1963); Arch.Rat.Mech. **17**, 215 (1964).
167. H. Widom, Probability, Statistical Mechanics and Number Theory, Adv. Math. Suppl. Stud., Vol.9, ed., G. Rota(Academic Press,NY,1986), pg.63.
168. V.S. Vladimirov, Theor.Prob.Appl. **1**, 101 (1956). This work develops a Monte Carlo method for calculating the eigenvalue of the Riesz operator defined on bounded Ω in R^d.

169. M. Rosenblatt, J. Math. &Mech. **12**, 619 (1963).
170. J.B. Reade, Proc.Edin.Math.Soc. **22**, 137 (1979).
171. T.G. Ostrom, Bull.Amer.Math.Soc. **55**, 343 (1949).
172. M. Rosenblatt, Trans.Amer.Math.Soc. **71**, 120 (1951).
173. M. Kac, Proc.Second Berkeley Symp.Math.Stat.Prob.,ed.,J. Neyman (Univ.Calif.Press,Berkeley,Cal.,1951), pg.189. Refs. [171]-[173] concern the relation between path-integrals and integral equations.
174. S. Ukai, J.Math.Phys. **12**, 83 (1971).
175. G. Polya and G. Szegö, J.Reine Angew.Math. **165**, 4 (1931).
176. G. Szegö, Orthogonal Polynomials(A.M.S.Colloq.Publ.23, NY, 1959).
177. R.L. Spencer, Amer.J.Phys. **58**, 385 (1990).
178. A.O. Lopes, Math.Zeit. **202**, 261 (1989).
179. P. Auer and C. Gardner, J.Chem.Phys. **23**, 1545 (1955).
180. M. Fixman, Macromolecules **14**, 1710(1981).
181. J.F. Douglas and K. F. Freed, Macromolecules **27**, 6088(1994).
182. R.M. Blumenthal and R.K. Getoor, Trans.Amer.Math.Soc. **95**, 263 (1960).
183. R.M. Blumenthal and R.K. Getoor, Pacific J.Math. **9**, 399 (1959).
184. W.E. Pruitt and S.J. Taylor, Z.Wahr. **12**, 267 (1969).
185. R.N. Mategna and H.E. Stanley, Nature **376**, 46 (1995). This work considers evidence for a random walk return probability of a physical system(stock market prices) modeled by a stable process.
186. S.J. Taylor, J.Math.&Mech. **16**, 1229 (1967).
187. S.C. Port, Trans. Amer.Math.Soc. **313**, 805 (1989).
188. M.Kac, Probability and Related Topics in Physical Sciences (Interscience, NY,1959), Chapt.4.
189. R.M. Blumenthal and R.K. Getoor, Ill.J.Math. **6**, 308 (1962).
190. M.T. Barlow and S.J. Taylor, Proc.Lond.Math.Soc. **64**, 125 (1992).
191. R.K. Getoor, Z.Wahr. **4**, 248 (1965).
192. S. Chandrasekhar, Rev.Mod.Phys. **15**, 1 (1943),pg.32.
193. V.M. Zolotarev, One-Dimensional Stable Distributions Trans. Math. Monogr.,Vol.65 (Amer.Math.Soc., Providence, RI, 1986).
194. H. Bergstrom, Arkiv.Mat. **2**, 375 (1952).
195. W.R. Schneider, "Generalized One-Sided Stable Distributions" in Stocahstic Processes in Mathematical Physics, eds. S. Albervio, P. Blanchard, and L. Streit (Springer Verlag,NY,1985).
196. R.N.Mantegna, Phys.Rev. E **49**, 4677 (1994).
197. M.F. Shlesinger, U. Frisch and G. Zaslavsky, eds., Lévy Flights and Related Phenomena in Physics(Springer, Berlin, 1995).
198. M.F. Schlesinger, G.M. Zaslavsky and J. Klafter, Nature **363**, 31 (1993).

199. C. Tsallis, S.V.F. Levy, A.M.C. Souza and R. Maynard, Phys. Rev.Lett. **75**, 3589 (1995).
200. M. Dishon, G.H. Weiss, and J.T. Bendler, J.Res.Nat.Bur.Stds. **90**, 27 (1985).
201. M. Dishon, J.T. Bendler and G.H. Weiss, J.Res.Nat.Bur.Stds. **95**, 433 (1990).
202. V. Uchaiken and G. Gusarov, J.Math.Phys. **38**, 2453 (1997).
203. H.P. McKean, Jr.,Duke J.Math. **22**, 229 (1955).
204. J. Lamperti, Trans.Amer.Math.Soc. **104**, 62 (1962).
205. C. Stone, Ill.J.Math. **7**, 731 (1963).
206. R.K. Getoor, Trans.Amer.Math.Soc. **101**, 75 (1961).
207. J. Elliot, Ill.J.Math. **3**, 200 (1959);
See Ref. [141].
208. E. Reich, Quart.Appl.Math. **11**, 341 (1953).
209. Y. Nozaki, Jpn.J.Math. **29**, 92 (1959).
210. V. Seshdri and B.J. West, Proc.Nat.Acad. **79**, 4501 (1982).
211. R. Pemantle, Y. Peres, J.W. Shapiro, Prob.Theor. and Related Fields **106**, 379 (1996).
212. A. Dvoretsky, P. Erdös and S. Kakutani, Acta Sci.Math. B (Szeged) **12**, 75 (1950).
213. S.Kakutani, Proc.Acad.(Tokyo) **20**, 648 (1944).
214. S.J.Taylor, Proc.Camb.Phil.Soc. **51**, 265 (1955).
215. N.S. Landkof, Foundation of Modern Potential Theory (Springer-Verlag, NY,1977). There are actually very few analytic calculations of the generalized analytic capacity C_α for $d > 1$. The exact result for the hypersphere may be found in Landkof and Kruglikov, but there seem to be no further examples.
216. Y.I. Kruglikov, Math.USSR(Sbornik) **58**, 185 (1987).
217. R. Bellman, Ill.J.Math. **2**, 577 (1958). See Refs. [135]-[136] and [164]-[165].
218. P. E. Rouse, J.Chem.Phys. **21**, 1272 (1953).
219. The integral equations of the Rouse-Zimm theory (Refs. [95]-[96]) can be directly expressed in terms of the A_x^α operator and the extension to "membrane" and "sponge" polymers [221] should involve the corresponding Riesz operators acting on higher dimensional manifolds.
220. It is noted that the "spectral dimension" does not generally govern the normal-mode asymptotic eigenvalue spectrum for fractal objects. For example, hydrodynamic forces strongly modify the normal mode motions in solution.
221. J.F. Douglas, Phys.Rev.E **54**, 2677 (1996). See extensive random surface

refs. cited in this work.
222. R.P. Roscoe, Phil.Mag. **40**, 338 (1949).
223. H. Hasimoto, J.Phys.Soc.Jpn. **13**, 633 (1958).
224. A. Gemant, Physics **6**, 363 (1935);**7**, 311 (1936);
Phil.Mag. **25**, 540 (1938).
225. R.C.L. Bosworth, Nature **157**, 447 (1946).
226. G.W.S. Blair and B.C. Veinoglou, Proc.Roy.Soc.Lond.A189,69 (1947).
227. B. Gross, J.Appl.Phys. **18**, 212 (1947); **19**, 257 (1948).
228. G.L. Slonimsky, J.Polym.Sci.C **16**, 1667 (1967).
229. W. Smit and H. DeVries, Rheol.Acta **9**, 525 (1970).
230. C. Friedrich, Rheol.Acta **30**, 151 (1991); J.Non-Newt.Fluid Mech. **46**, 307 (1993).
231. R.L. Bagley and P.J. Torvik, Rheol.Acta **27**, 201 (1983).
232. R.C. Koeller, J.Appl.Mech. **51**, 299 (1984).
233. W.G. Glöckle and T.F. Nonenmacher, J.Stat.Phys. **71**, 741 (1993).
234. A.K. Jonscher, Phys.Stat.Sol. **83**, 585 (1977);
Nature **253**, 717 (1975); **267**, 673 (1977);
Phys.Thin Films **11**, 205 (1980).
235. K.L. Ngai,A.K. Jonscher and C.T. White, Nature **277**, 185 (1979).
236. K.L. Ngai, R.W. Rendell, A.K. Rajagopal and S. Teitler, Ann. N.Y.Acad.Sci. **484**, 150 (1987).
See also: G. Williams and D.C Watts, Trans.Farad.Soc. **66**, 80 (1970);
F.Kohlrausch, Pogg.Ann.Phys. **12**, 393 (1847);
Pogg.Ann.Phys.IV- **91**, 56 (1854); **91**, 79 (1854).
237. J.F. Douglas and J.B. Hubbard, Macromolecules **24**, 3163(1991).
238. See Ref. [35].
239. See Refs. [311],[312].
240. D.A. McQuarrie, Statistical Mechanics(Harper and Row, NY, 1976), pgs.572-579.
241. K.S. Cole and R.H. Cole, J.Chem.Phys. **9**, 341 (1941); **10**, 98 (1942) ;
242. H.T. Davis, The Theory of Linear Operators (Principia Press, Bloomington,Ill.,1936).
243. K.S. Cole, "Electric Conductance in Biological Systems and Electronic Excitation in Nerves, "Symp. Quant.Biol., Vol.1, 1933.
244. H. T. David, Am.J. Math. **46**, 95 (1924); **49**, 123 (1927). For an example of the application of the Cole-Cole function see: C.Duron, J.M. Wacrenier, F. Hardouin, N.H. Tinh and H. Gasparoux, J.de Phys. 44,1983). See Ref. [317].
245. A. Erdélyi, "Fractional Integral of Generalized Functions" in Ref. [39].
246. I.N. Sneddon, "The Use in Mathematical Physics of Erdélyi-Kober Op-

erators and Some of Their Generalizations" in Ref. [39].
247. H. Kober, J.Math.(Oxford Ser.) **11**, 193 (1940).
248. I.N. Sneddon and A. Erdélyi, Canad.J.Math. **14**, 685 (1962).
249. I.N. Sneddon, The Use of Integral Transforms(McGraw-Hill, NY, 1972).
250. G. Maruyama, Mem.Fac.Sci.A(Kyushu Univ.) **4**, 45 (1949).
251. J.L. Lebowitz, "Hamiltonian Flows and Rigorous Results in Non-Equilibrium Statistical Mechanics" in Statistical Mechanics : New Concepts, New Problems, New Applications, ed., S.A. Rice, K.F. Freed and J.C. Light (University of Chicago Press, Chicago, 1972), p.41.
252. M.S.Bartlett, Proc.Camb.Phil.Soc. **49**, 263 (1953); Nature **165**, 727 (1950).
253. M. Kac, Bull.Amer.Math.Soc. **53**, 1002 (1947).
254. F. Papengelou, Z.Wahr.8.259(1967).
255. J.L. Doob, Proc.Nat.Acad.Sci. **20**, 376 (1934).
256. B.J. Berne, J.P. Boon and S.A. Rice, J.Chem.Phys.45, 1086 (1966).
257. R. Zwanzig, Lect.Theor.Phys. **3**, 106 (1961).
258. K.L. Ngai, A.K. Rajagopal and S. Teitler, J.Chem.Phys.88,5086 (1988).
259. K.S. Schweitzer, J.Chem.Phys. **91**, 5822 (1989).
260. G. Uhlenbeck, "The Boltzmann Equation", in Probability and Related Topics in Physical Sciences, ed.,M. Kac(Interscience,NY, 1957), p.187.
261. F. Spitzer, Z.Wahr. **3**, 110 (1965).
262. R.K. Getoor, Z.Wahr. **4**, 248 (1965).
263. G.D. Birkhoff, Bull.Amer.Math.Soc.38 361 (1932); Amer.Math.Mon. **49**, 222 (1942).
264. G.M. Zaslavsky, Phys.Rep. **80**, 157 (1981).
265. A.J. Lotka, Ann.Math.Stat. **10**, 1 (1931). Lotka discusses the original derivation of the "renewal equation" Eq.(173) by L. Herblot.
266. F. Taylor, Rocky Mountain J.Math. **9**, 149 (1979).
267. F. Brauer, SIAM J.Math. **6**, 312 (1975).
268. F. Brauer, Adv.Math. **22**, 32 (1976).
269. F.J.S. Wang, SIAM J.Math.Anal. **9**, 529 (1978).
270. W. Feller, Ann.Math.Stat. **12**, 243 (1941).
271. W.L. Smith, J.Roy.Stat.Soc. **20**, 243 (1958); Proc.Roy.Soc.London A **232**, 6 (1955). Roy.Soc.Dein. **64**, 9 (1957); Trans.Amer.Math.Soc. **104**, 79 (1962).
272. J.L. Doob, Trans.Amer.Math.Soc. **63**, 422 (1948).
273. M. Kac and H. Kesten, Bull.Amer.Math.Soc. **64**, 283 (1958); W. Bauer and G. F. Bertsch, Phys.Rev.Lett. **65**, 2213 (1990).
274. J.C. Maxwell, Phil.Trans.Roy.Soc.London **157**, 49 (1867);
275. P. Debye, Polar Molecules(Dover Press,NY,1945).

276. R.G. Gordon, J.Chem.Phys. **44**, 228 (1966).
277. O. Heaviside, Electronic Theory, Vol.2 (London, 1922).
278. N. Wiener, Math.Ann. **95**, 557 (1926).
279. T.P.G. Liverman, Generalized Functions and Direct Operational Methods, Vol.1(Prentice Hall, Englewood Cliffs, NJ, 1964).
280. I.M.Gelfand and G.E. Shilov, Generalized Functions, Vol.1(Academic Press, New York, 1964).
281. F.M. de Oliveira Castro, Z.Phys. **114**, 116 (1939).
282. S.H. Glarum, J.Chem.Phys. **33**, 639 (1960); **33**, 1371 (1960).
283. J.L. Skinner, J.Chem.Phys. **79**, 1955 (1983).
284. P. Bordewijk, Chem. Phys.Lett. **32**, 592 (1975).
285. J.E. Shore and R.J. Zwanzig, J.Chem.Phys. **63**, 5445 (1975).
286. S. Fujawara and F. Yonezawa, Int.J.Mod.Phys.B **10**, 3561 (1996); Phys.Rev.E **51**, 2277 (1995); Phys.Rev.Lett. **74**, 4229 (1995).
287. S. Gomi and F. Yonezawa, Phys.Rev.Lett. **74**, 4125 (1995).
288. A. Blumen, J. Klafter and G. Zumofen, J.Phys.A19,L-77(1986).
289. G.H. Köhler and A. Blumen, J.Phys.A **24**, 2807 (1991). See Ref. [294].
290. A.T. Ogielski and D.L. Stein, Phys.Rev.Lett. **55**, 1634 (1985).
291. Y. Hiwatari, J. Matsui, K. Uehara, T. Muranaka, H. Miyagawa, M. Takasu and T. Odagaki, Physica A **204**, 306 (1994).
292. T. Odagaki, Phys.Rev.Lett. **75**, 3701 (1995).
293. C.F.F. Karney, Physica D **8**, 360 (1983).
294. J.D. Meiss and E. Ott, Phys.Rev.Lett. **55**, 2741 (1985).
295. K. Kaneko and T. Konishi, Phys. Rev.A **40**, 6130 (1989).
296. G.M. Zaslavsky and M.K. Tippett, Phys.Rev.Lett. **67**, 3251 (1991).
297. G.M. Zaslavsky, Chaos **4**, 25 (1994).
298. K. Lee, Phys.Rev.Lett. **60**, 1991 (1988).
299. Y. Aizawa, Prog.Theor.Phys. **81**, 249 (1989).
300. Y.Kikuchi and Y.Aizawa Prog.Theor.Phys. **84**, 563(1990).
301. X.J. Wang, Phys.Rev.A **39**, 3214 (1989); **40**, 6647 (1989).
302. P. Gaspard and X.J. Wang, Proc.Nat.Acad.Sci. **85**, 4591 (1988).
303. J. Aaronson, Isr.J.Math. **27**, 93 (1976).
304. R. Hilfer, Chaos,Solitons and Fractals **5**, 1475(1995); Physica A **221**,89 (1995); Fractals **3**, 549 (1995).
305. J.F.C. Kingman, Z.Wahr. **2**, 180 (1964).
306. J. Curie, Ann.Chim.Phys. **18**, 203 (1899).
307. E. von Schweidler, Ann.Physik **24**, 711 (1987).
308. H.H. Winter and F. Chambon, J.Rheol. **30**, 367 (1986); H.H. Winter, Progr.Coll.Polym.Sci. **75**, 104 (1987); F. Chambon, Z.S. Petrovic, W.J. McKnight and H.H. Winter, Macro-

molecules **19**, 2146(1986).
309. R. Zwanzig, Proc.Nat.Acad.Sci. **78**, 3296 (1981).
310. M.D. Donsker and S.R.S. Varadhan, Comm.Pure&Appl.Math. **28**, 1 (1975); **28**, 279 (1975); **28**, 525 (1975).
311. C. Donati, J.F. Douglas, W. Kob, S.J. Plimpton, P.H. Poole, S.C. Glotzer, Phys.Rev.Lett.(**80**,2338(1998)).
312. H. Heckmeier, M. Mix and G. Strobl,Macromolecules **30**, 4454(1997). See Ref. [237].
313. J.F. Douglas, "The Classical Orthogonal Functions and Their Weight Families", Masters Thesis in Mathematics, Virginia Commonwealth University, 1981.
314. J. Koutecky,Coll.Czech.Chem.Comm., **18**, 597 (1953).
315. H. Matsuda and Y. Aybe, Bull.Chem.Soc.Jpn. **28**, 422 (1955).
316. K.B. Oldham, Anal.Chem. **41**, 941 (1969); See Ref. [40].
317. S. Havriliak and S.J. Negami, J.Polym.Sci. **14**, 99 (1966). For an application of the H-N function to dielectric relaxation polymers: M. Yoshihara and R. N. Work, J.Chem.Phys. **72**, 5909 (1980).
318. D.W. Davidson and R.H. Cole, J.Chem.Phys. **18**, 417 (1950); **19**, 1484 (1951); D.W. Davidson, Canad.J.Chem. **39**, 571 (1961).
319. G. Williams and D.C. Watts, Trans.Farad.Soc. **66**, 80 (1970).
320. H. Scher and E. W. Montroll, Phys.Rev. B **12**, 2455 (1975).
321. A.V. Das and C.V. Coffman, J.Math.Phys. **8**, 1720 (1967).
322. N.E. Fremberg, Proc.Roy.Soc.A **188**, 18 (1946); Kungl.Fys.Salls Lund.Forh. **15**, 265 (1946) .
323. F.C. Auluck and L.S. Kothari, Proc.Camb.Phil.Soc. **47**, 436 (1951).
324. L.S. Kothari, Proc.Phys.Soc. **67**, 17 (1954).
325. L.S. Kothari, Phys.Rev. **87**, 536 (1952).
326. S.T. Ma, Phys.Rev. **71**, 878 (1947).
327. T. Gustafson, Arkiv Mat., Astrom., Fys.A **34**, 1 (1947); Nature **157**, 734 (1946).
328. S.B. Nilsson, Ark.Fys. **1**, 369 (1949).
329. E.T. Copson, Proc.Roy.Soc.Edin. **61**, 260 (1943).
330. R.L. do Amaral and E.C. Marino, J.Phys.A **25**, 5183 (1992).
331. M. Suzuki, Y. Yamazaki, G.Igarashi, Phys.Lett.A **42**, 313 (1972).

CHAPTER VII

APPLICATIONS TO PROBLEMS IN POLYMER PHYSICS AND RHEOLOGY

H. SCHIESSEL
Theoretical Polymer Physics, Freiburg University, Rheinstr. 12, 79104 Freiburg, Germany and Materials Research Laboratory, University of California, Santa Barbara, CA 93106-5130, USA

CHR. FRIEDRICH
Freiburg Materials' Research Center, Freiburg University, Stefan-Meier-Str. 21, 70104 Freiburg, Germany

A. BLUMEN
Theoretical Polymer Physics, Freiburg University, Rheinstr. 12, 79104 Freiburg, Germany

Contents

1 Introduction

Power law relaxation is of widespread occurrence in complex materials. Thus one often encounters algebraic relaxation functions

$$\Phi(t) \propto t^{-\alpha} \tag{1}$$

with $0 < \alpha < 1$. Examples are the transport of charge carriers in amorphous semiconductors [1,2], the behavior of electrical currents at rough blocking electrodes [3], the dielectric relaxation of liquids [4] and of solids [5] and the attenuation of seismic waves [6]. Especially the microscopic and macroscopic dynamical behavior of macromolecular systems (such as linear polymers or gels) is often characterized by the algebraic patterns of Eq. (1) [7].

For such systems fractional expressions come naturally into play as a result of the superposition principle. Consider an arbitrary history of the external perturbation $\Psi(t)$, and let us denote by $\Phi_s(t)$ the response of the system to steplike external perturbations $\Psi(t) = \Theta(t)$, where $\Theta(t)$ is the Heaviside step function. Then we obtain the response of the system by the causal convolution:

$$\Phi(t) = \int_{-\infty}^{t} dt' \Phi_s(t - t') \frac{d\Psi(t')}{dt'} \tag{2}$$

since the (Boltzmann) superposition integral, Eq. (2), holds for linear systems which are homogeneous in time. Specifically, let Φ_s obey:

$$\Phi_s(t) = \frac{C}{\Gamma(1-\alpha)} \left(\frac{t}{\tau}\right)^{-\alpha}, \tag{3}$$

where $\Gamma(x)$ denotes the Gamma function. Eq. (3) is chosen in such a way as to match the forthcoming definitions. Now we find the response $\Phi(t)$ to an arbitrary $\Psi(t)$ by inserting Eq. (3) into Eq. (2):

$$\Phi(t) = \frac{C\tau^{\alpha}}{\Gamma(1-\alpha)} \int_{-\infty}^{t} dt' (t - t')^{-\alpha} \frac{d\Psi(t')}{dt'}. \tag{4}$$

The right-hand-side of Eq. (4) is nothing but a fractional integral. This is most readily seen by recalling the definition of Riemann's fractional integral [8,9]:

$$_cD_t^{-\gamma} f(t) = \frac{1}{\Gamma(\gamma)} \int_{c}^{t} dt' \frac{f(t')}{(t - t')^{1-\gamma}}, \tag{5}$$

where $\gamma > 0$. Equation (5) embraces two special cases: (*i*) The case $c = 0$ corresponds to a fractional integral of Riemann-Liouville type. (*ii*) $c \to -\infty$ leads to Weyl's version of fractional calculus. Fractional differentiation of order $\gamma > 0$ is now obtained by first picking an integer n, $n > \gamma$, then performing a fractional integration of order $n - \gamma$, followed by an ordinary differentiation of order n, i.e.

$$_cD_t^\gamma f(t) = \frac{d^n}{dt^n}\left[{}_cD_t^{\gamma-n} f(t)\right]. \tag{6}$$

We work in the following within Weyl's formalism, i.e. $c \to -\infty$, and use the shorthand notation $d^\gamma/dt^\gamma \equiv {}_{-\infty}D_t^\gamma$. In Weyl's version the composition rule for differentiation and integration obeys the simple form

$$\frac{d^\alpha}{dt^\alpha}\frac{d^\beta}{dt^\beta} = \frac{d^{\alpha+\beta}}{dt^{\alpha+\beta}} \tag{7}$$

for arbitrary α and β [9].

Using Eqs. (5) to (7) we can rewrite the causal convolution, Eq. (4) as:

$$\Phi(t) = C\tau^\alpha \frac{d^{\alpha-1}}{dt^{\alpha-1}}\frac{d\Psi(t)}{dt} = C\tau^\alpha \frac{d^\alpha \Psi(t)}{dt^\alpha}. \tag{8}$$

Thus we find that here the system's response, $\Phi(t)$, follows by the fractional differentiation of the external perturbation, $\Psi(t)$.

We consider now the same system, but exchange the roles of Φ and Ψ, *i.e.*, view Ψ as being the response to a prescribed Φ. Such a reversal can be achieved experimentally for systems in which both quantities can be controlled (for instance, the stress and the strain in a rheological experiment). In order to find the corresponding fractional expression we have simply to apply the fractional operator $d^{-\alpha}/dt^{-\alpha}$ to both sides of Eq. (8), which leads to

$$\Psi(t) = C^{-1}\tau^{-\alpha}\frac{d^{-\alpha}\Phi(t)}{dt^{-\alpha}}. \tag{9}$$

By rewriting Eq. (9) as

$$\Psi(t) = C^{-1}\tau^{-\alpha}\frac{d^{-\alpha-1}}{dt^{-\alpha-1}}\frac{d\Phi}{dt} = \frac{C^{-1}\tau^{-\alpha}}{\Gamma(1+\alpha)}\int_{-\infty}^{t} dt'\,(t-t')^\alpha \frac{d\Phi(t')}{dt'} \tag{10}$$

we can infer immediately that the response Ψ_s to a steplike perturbation $\Phi(t) = \Theta(t)$ is

$$\Psi_s(t) = \frac{C^{-1}}{\Gamma(1+\alpha)}\left(\frac{t}{\tau}\right)^\alpha \tag{11}$$

2 Applications to Microscopic Models of Polymer Dynamics

Macromolecular systems show in many cases viscoelastic behavior, which combines the characteristic features of solids (elasticity) and of liquids (viscosity). Traditional rheological methods measure the materials' properties at large length scales, and we will discuss applications of fractional calculus to this class of experiments in Sec. 3. Recent optical developments allow, however, the micromanipulation of macromolecules so that one has means at hand to measure local mechanical properties. Thus Perkins et al. [10]–[12] and Wirtz [13] have dragged individual fluorescent DNAs with optical or magnetic tweezers at one end; Amblard et al. [14] have performed similar experiments with individual magnetic beads in actin networks. In this section we focus on such a class of experiments on polymer chains and networks. We prefer here not to take all experimentally relevant factors into account, such as the excluded volume, the finite extensibility, the bending rigidity, the hydrodynamic coupling (such effects were discussed, for instance, in Refs. [14]–[17]), but to focus on the rather simple Rouse model for polymer dynamics [18,19]. Thus, using the Rouse model we demonstrate how, as a result of the superposition principle (or equivalently of simple scaling relations) fractional expressions such as Eq. (9) come into play, and we determine the local dynamics of polymer chains and networks.

2.1 Pulling a Rouse Chain at One of its Ends

In the following we show how a prototype model of polymer dynamics, the Rouse model [18,19], leads rigorously to a fractional differential equation of the form of Eq. (9). We demonstrate this for an external force which acts on one end-monomer of the chain.

The polymer is modeled as a Gaussian chain consisting of N monomers, connected by harmonic springs to a linear chain. The chain's configuration is given by the set of vectors $\{\mathbf{R}_n(t)\}$, where $\mathbf{R}_n(t) = (X_n(t), Y_n(t), Z_n(t))$ is the position vector of the nth bead at time t, $n = 0, 1, ...N-1$. The potential energy $U(\{\mathbf{R}_n(t)\})$ has to account for the elastic contributions and for the influence of the external force $\mathbf{F}(t)$, which is assumed to act on the monomer $n = 0$. Thus one has

$$U(\{\mathbf{R}_n(t)\}) = \frac{K}{2}\sum_{n=1}^{N-1}[\mathbf{R}_n(t) - \mathbf{R}_{n-1}(t)]^2 - \mathbf{F}(t)\,\mathbf{R}_0(t)\,. \tag{12}$$

In Eq. (12) K is the (entropic) spring constant $K = 3T/b^2$, where T denotes the temperature in units of the Boltzmann constant k_B, and b is the mean distance between neighboring beads (in the absence of an external perturbation). The chain's dynamics is described by N coupled Langevin equations [18,19]

$$\zeta \frac{d\mathbf{R}_n(t)}{dt} = -\frac{\partial U(\{\mathbf{R}_n(t)\})}{\partial \mathbf{R}_n(t)} + \mathbf{f}_R(n,t) \tag{13}$$

and the hydrodynamic interaction between the beads is disregarded. In Eq. (13) ζ is the friction constant and $\mathbf{f}_R(n,t)$ is the random thermal-noise force which mimics the collisions of the nth bead of the chain with the solvent molecules. The thermal noise is Gaussian, white, with zero mean, so that one has:

$$\overline{f_i(n,t)} = 0, \; \overline{f_i(n,t)\, f_j(n',t')} = 2\zeta T \delta_{ij}\delta_{nn'}\delta(t-t') \,. \tag{14}$$

In Eq. (14) i and j denote the components of the force vector, i.e. $i,j = X,Y,Z$ and the overbar stands for thermal averaging, i.e. averaging over the realizations of the Langevin forces $\mathbf{f}_R(n,t)$.

Taking the variable n to be continuous (i.e. considering the chain as an elastic string) and setting $\mathbf{F}(t) = (0, F(t), 0)$, it follows from Eqs. (12) and (13) that:

$$\zeta \frac{\partial X_n(t)}{\partial t} = K \frac{\partial^2 X_n(t)}{\partial n^2} + f_X(n,t) \,, \tag{15}$$

$$\zeta \frac{\partial Z_n(t)}{\partial t} = K \frac{\partial^2 Z_n(t)}{\partial n^2} + f_Z(n,t) \tag{16}$$

and

$$\zeta \frac{\partial Y_n(t)}{\partial t} = K \frac{\partial^2 Y_n(t)}{\partial n^2} + \delta_{n0} F(t) + f_Y(n,t) \,. \tag{17}$$

At the chain's ends the Rouse boundary conditions [18,19] hold:

$$\left.\frac{\partial X_n(t)}{\partial n}\right|_{n=0,N} = \left.\frac{\partial Y_n(t)}{\partial n}\right|_{n=0,N} = \left.\frac{\partial Z_n(t)}{\partial n}\right|_{n=0,N} = 0. \tag{18}$$

Since the X- and Z-components of the $\mathbf{R}_n$ are force-independent and follow the standard Rouse behavior [18,19] we can restrict ourselves to the behavior of the Y-component, Eq. (17). This equation contains two types of forces: the thermal noise term $f_Y(n,t)$ and the external force $F(t)$ which acts on the first monomer.

The solution of Eq. (17) with the boundary conditions, Eq. (18), can be given in the form of a Fourier series [19]

$$Y_n(t) = Y(0,t) + 2\sum_{p=1}^{\infty} Y(p,t)\cos\left(\frac{p\pi n}{N}\right) \tag{19}$$

where $Y(p,t)$, $p = 0, 1, \ldots$, denote the normal coordinates:

$$Y(p,t) = \frac{1}{N}\int_0^N dn \cos\left(\frac{p\pi n}{N}\right) Y_n(t) . \tag{20}$$

In terms of the coordinates $Y(p,t)$ Eq. (17) can be rewritten as

$$\frac{\partial Y(p,t)}{\partial t} = -\frac{p^2}{\tau_R} Y(p,t) + \frac{1}{N\zeta} F(t) + \frac{1}{\zeta}\varphi_Y(p,t) . \tag{21}$$

Here τ_R denotes the Rouse time

$$\tau_R = \frac{\zeta b^2 N^2}{3\pi^2 T} \tag{22}$$

which is the longest internal relaxation time of the harmonic chain. The symbol $\varphi_Y(p,t)$ in Eq. (21) denotes the Fourier transform of the thermal noises $\varphi_Y(p,t) = N^{-1}\int_0^N dn \cos(p\pi n/N)\, f_Y(n,t)$.

We assume that the chain is at $t = 0$ in thermal equilibrium, i.e. it has a Gaussian conformation. This can be accounted for automatically by stipulating the polymer to have been subjected to the thermal forces since $t = -\infty$. Furthermore, switching on the external force at $t = 0$, i.e. having $F(t) = F_0\Theta(t)$, the normal coordinates are given by

$$\begin{aligned} Y(p,t) &= \frac{1}{\zeta}\int_{-\infty}^{t} d\tau \varphi_Y(p,\tau)\exp\left(-p^2(t-\tau)/\tau_R\right) \\ &+ \frac{F_0}{N\zeta}\int_0^t d\tau \exp\left(-p^2(t-\tau)/\tau_R\right) . \end{aligned} \tag{23}$$

From Eqs. (19) and (23) one can obtain readily the explicit time dependence of several different quantities such as the end-to-end distance, the displacement of the chain's center of mass (CM), or that of a tagged bead (cf. for instance Refs. [20,21]).

Consider first the CM's motion. The Y-component of the trajectory of the CM is given by the 0th normal coordinate, i.e. by $Y_{CM}(t) = N^{-1}\int_0^N dnY_n(t) = Y(0,t)$. Using Eq. (23) with $p = 0$ we find for the mean averaged position of the chain's CM in the Y-direction:

$$\overline{Y_{CM}(t)} = \frac{F_0}{N\zeta}t \tag{24}$$

i.e., the chain drifts with a constant velocity $F_0/N\zeta$. From Eq. (24) one can infer that the friction constant of the overall chain is $N\zeta$, which is the sum of the friction constants of the individual monomers: Due to the neglect of hydrodynamic interactions in Eq. (13), the Rouse chain is free draining.

The behavior of a tagged bead, say one of the chain's ends, is more complicated. Using Eq. (19) with $n = 0$, i.e.

$$Y_0(t) = Y(0,t) + 2\sum_{p=1}^{\infty} Y(p,t) \tag{25}$$

we obtain from Eqs. (23) and (25) for the MSD of the 0th bead

$$\overline{Y_0(t)} = \frac{F_0}{N\zeta}t + \frac{2F_0}{N\zeta}\sum_{p=1}^{\infty}\int_0^t d\tau \exp\left(-p^2\tau/\tau_R\right) . \tag{26}$$

From Eq. (26) the short-time behavior, $t \ll \tau_R$, follows by converting the sum over p into an integral; this leads to:

$$\overline{Y_0(t)} \cong \frac{2bF_0}{\sqrt{3\pi\zeta T}}t^{1/2} = \frac{2}{\sqrt{\pi}}\frac{F_0}{\sqrt{\zeta K}}t^{1/2}. \tag{27}$$

In Eq. (27) we have omitted the first term on the rhs of Eq. (26), which is of the order of $\sqrt{t/\tau_R}$ smaller than the second term. In the long time regime $t \gg \tau_R$ one finds the second term of Eq. (26) to be of the order of τ_R/t smaller than the first one. Thus one has

$$\overline{Y_0(t)} \cong \frac{F_0}{N\zeta}t, \tag{28}$$

i.e. the bead's motion mirrors the motion of the CM of the chain, Eq. (24).

Now Eq. (27) is a nice example which shows how fractional relationships come into play in polymer dynamics. Assume the chain to very long, $N \gg 1$; since the Rouse time obeys $\tau_R \propto N^2$, the range of validity of Eq. (27)

becomes very large. One may even consider the limit $N \to \infty$, where Eq. (27) becomes exact for all positive times. Assume now an arbitrary history $F(t)$ of the external force. Then, due to the linearity of the system one finds the convolution integral

$$\overline{Y_0(t)} = \frac{1}{\sqrt{\zeta K}} \frac{1}{\Gamma(3/2)} \int_{-\infty}^{t} \frac{d\tau}{(t-\tau)^{1-3/2}} \frac{dF(\tau)}{d\tau} = \frac{1}{\sqrt{\zeta K}} \frac{d^{-3/2}}{dt^{-3/2}} \left[\frac{dF(t)}{dt} \right] \tag{29}$$

which represents Weyl's integral (cf. Eq. (5) with $c \to \infty$). Using the composition rule, Eq. (7), leads finally to:

$$\overline{Y_0(t)} = \frac{1}{\sqrt{\zeta K}} \frac{d^{-1/2} F(t)}{dt^{-1/2}}. \tag{30}$$

Thus the mean position of bead $n = 0$ (which is subjected to the external force) follows from the semiintegration of the force history $F(t)$.

We close this section by showing how Eq. (27) can also be derived using scaling arguments. Consider the tagged bead $n = 0$. After switching on the force at $t = 0$, the number $g(t)$ of monomers which move together with the tagged bead increases with t. This number follows from the Rouse time associated with a subchain of g monomers, namely from $\tau_g \approx \zeta b^2 g^2 / T$. Inverting this relation leads to:

$$g(t) \approx \frac{\sqrt{T}}{\sqrt{\zeta} b} t^{1/2}. \tag{31}$$

Eq. (31) describes the short time behavior, $t \ll \tau_R$, whereas at longer times, $t \gg \tau_R$, one has

$$g(t) \cong N, \tag{32}$$

i.e. the whole chain moves collectively. The mobility of the set of g beads decreases with time as $\mu(t) \cong (\zeta g(t))^{-1}$. The average velocity v_Y in the Y-direction of the tagged monomer is given by the velocity of the set of monomers moving together with it; hence $v_Y(t) = \mu(t) F_0 = (\zeta g(t))^{-1} F_0$, where g is given by Eq. (31) for $t \ll \tau_R$ and by Eq. (32) for $t \gg \tau_R$. The average displacement of a single bead can be estimated from the average displacement of the corresponding blob of g monomers, i.e. $\overline{Y_0(t)} = v_Y(t)\, t = (\zeta g(t))^{-1} F_0 t$. For $t \ll \tau_R$ one finds from Eq. (31)

$$\overline{Y_0(t)} \approx \frac{b F_0}{\sqrt{\zeta T}} t^{1/2}, \tag{33}$$

i.e., we recover (up to numerical coefficients of order unity) Eq. (27), the result of the previous, exact calculations. For longer times the PA drifts as a whole, and we find from Eq. (32) $\overline{Y_0(t)} \approx (N\zeta)^{-1} F_0 t$, which corresponds to Eq. (28).

In Sec. 2.2 we will discuss how Eq. (30) can be generalized to an arbitrary order α of integration, with α obeying $0 < \alpha \leq 1/2$. This is achieved by pulling a monomer which belongs to a branched structure (fractal network); then α is directly related to the connectivity of the network.

2.2 Pulling One Monomer of a Fractal Network

Consider again the scaling arguments which led us to Eq. (33). The $t^{1/2}$-subdrift displacement follows from the increase in the number $g(t) \propto t^{1/2}$ of monomers which are moving collectively: this slows down the tagged bead. It is evident that pulling a monomer of a macromolecular network (which is more tight than a linear chain) involves an even slower subdrift. Examples of such networks are the *generalized Gaussian structures* (GGS), which were considered by Sommer and Blumen [22]. We investigate here some of the dynamical properties of such GGS, especially isotropic and locally homogeneous fractals, which may be seen as prototypes for membranes, gels and polymer networks.

In order to obtain $g(t)$ for the GGS it is useful to rederive Eqs. (31) and (32) in a slightly different way. Note that Rouse dynamics leads to Eq. (17) which is a one-dimensional equation of diffusion-type. As it is well-known, its Green's function is a Gaussian, whose width increases with time as $t^{1/2}$; stated differently, the number of monomers $g(t)$ which are involved in a collective motion obeys $g(t) \propto t^{1/2}$.

Consider now the Rouse dynamics of a fractal GGS. This leads to a diffusion equation on the corresponding fractal lattice. It is well-known that the number $S(t)$ of different sites visited by a random walker during time t goes as $S(t) \propto t^{d_s/2}$ for $d_s < 2$ and as $S(t) \propto t$ for $d_s \geq 2$. Here d_s denotes the spectral dimension of the network (see, for instance, Refs. [23,24]); in the case of a regular lattice d_s equals the Euclidian dimension. Now, the dynamical process has a single parameter combination with the dimension of time, namely $\zeta b^2/T$; we thus find:

$$g(t) \approx \begin{cases} \frac{T^{d_s/2}}{\zeta^{d_s/2} b^{d_s}} t^{d_s/2} & \text{for } d_s < 2 \\ \frac{T}{\zeta b^2} t & \text{for } d_s \geq 2. \end{cases} \tag{34}$$

Eq. (34) is valid as long as the network does not move as a whole; at later times $g(t) = N$ holds. The crossover time τ_G follows by setting $g(\tau_G) = N$ in

Eq. (34). For $d_s < 2$ this leads to

$$\tau_G \approx \zeta b^2 N^{2/d_s}/T. \tag{35}$$

For a linear chain, $d_s = 1$, one recovers the Rouse time, Eq. (22). We proceed now as in Sec. 2.1. The domain which moves collectively has the velocity $v_Y(t) = (\zeta g(t))^{-1} F_0$. For $d_s < 2$ this leads to the following displacement of the tagged monomer

$$\overline{Y(t)} \approx \frac{F_0 b^{d_s}}{\zeta^{1-d_s/2} T^{d_s/2}} t^{1-d_s/2}. \tag{36}$$

The displacement due to an arbitrary force $F(t)$ is thus given by

$$\overline{Y(t)} \approx \frac{b^{d_s}}{\zeta^{1-d_s/2} T^{d_s/2}} \frac{d^{d_s/2-1} F(t)}{dt^{d_s/2-1}}. \tag{37}$$

Thus we have suceeded in deriving an expression with a fractional integral of order $\alpha = 1 - d_s/2$, where $0 < \alpha \leq 1/2$.

We close this section with a short discussion of the case $d_s \geq 2$. Sommer and Blumen [22] showed that the corresponding GGS (without an external force) is then in a collapsed state. Using a Langevin-type approach Schiessel [25] showed that in this case one has a very weak time-dependence of $\overline{Y(t)}$, which results from the fact that the external force is unable to unfold the collapsed structure. In Sec. 2.4 we show how more general perturbations which act on the whole GGS are able to unfold structures whose spectral dimensions are less than 4.

2.3 Polyampholytes in External Electrical Fields

We turn now to the dynamics of polyampholytes (PAs) in external electrical fields. This allows us to obtain relations similar to Eq. (30), with orders of integration α within the range $1/2 \leq \alpha \leq 1$. PAs are heteropolymers carrying positive and negative charges along their backbone, forming quenched (random or regular) patterns. The conformational and dynamical properties of PAs show interesting features: (*i*) The interactions between the different positive and negative charges lead in many cases to the collapse of the chains into spherical globules [26,27]. (*ii*) For PAs with a sufficiently high total charge the globular state is not stable; one has then a random necklace structure with globular and stretched parts [28,29]. (*iii*) In a sufficiently strong external field a PA globule is unstable and stretches out [29]–[32].

This plethora of static (conformational) features has its counterpart in the dynamics, and the situation is very complex. Here we restrict the analysis to an important, limiting case, namely to the weak coupling limit (WCL), in which the interaction between the charges can be neglected and where the behavior of PAs in external fields can be handled with some ease [20] [21] [29]– [36]. This WCL can be realized through weakly charged PAs. As an example consider a PA with N monomers, with fN positive $(+q)$ and fN negative $(-q)$ charges, randomly distributed along the backbone of the chain; $f \leq 1/2$ denotes the fraction of positively (as well as negatively) charged monomers. In this case the interaction between the charges can be neglected if $b > N^{1/2} f l_B$, where $l_B = q^2 / (\varepsilon T)$ denotes the Bjerrum length and ε the dielectric constant [26,27,31,32]; in water at room temperature one has $l_B \approx 7$Å .

We now follow Refs. [20,21] and model, similar to Sec. 2.1, the PA as a Rouse chain. We denote the charge on the nth bead by q_n and take it to be a quenched variable, *i.e.*, the set $\{q_n\}$ stays fixed for a given PA. The potential $U(\{\mathbf{R}_n(t)\})$ in an external electrical field $\mathbf{E}$ is then given by

$$U(\{\mathbf{R}_n(t)\}) = \frac{K}{2} \sum_{n=1}^{N-1} [\mathbf{R}_n(t) - \mathbf{R}_{n-1}(t)]^2 - \mathbf{E} \sum_{n=1}^{N-1} q_n \mathbf{R}_n(t) . \tag{38}$$

In order to calculate the dynamics of the PA one can now follow Sec. 2.1. The equations of motion are given by Eq. (13) which again decouples in Cartesian coordinates. For $\mathbf{E} = (0, E, 0)$ this leads to Eqs. (15) and (16) for the X- and the Z-component, whereas the dynamics in field direction is given by

$$\zeta \frac{\partial Y_n(t)}{\partial t} = K \frac{\partial^2 Y_n(t)}{\partial n^2} + q_n E + f_Y(n,t) . \tag{39}$$

By transforming to Rouse normal coordinates (as in Sec. 2.1) one can readily evaluate several PA properties for different charge distributions $\{q_n\}$: see Ref. [21] for an explicit calculation of the position of the CM, of that of particular beads, as well as for the evaluation of the PA's end-to-end distance. We dispense here with giving the explicit derivation of the formulas and only present a result which again leads to a fractional expression.

Consider a PA with a random, uncorrelated charge distribution, *i.e.* $\langle q_n q_m \rangle = q^2 \delta_{nm}$ (the brackets denote the average with respect to different $\{q_n\}$ realisations). At short times, $t \ll \tau_R$, and neglecting the thermal contributions, the mean-squared displacement of bead $n = 0$ obeys:

$$\left\langle \left(\overline{Y_0(t)} \right)^2 \right\rangle = \frac{8bq^2E^2\left(\sqrt{2}-1\right)}{3\zeta^{3/2}\sqrt{3\pi T}} t^{3/2} \tag{40}$$

(see Refs. [20,21] for the full expression). Hence, from Eq. (40) the mean-squared displacement of the bead follows a $t^{3/2}$-subdrift behavior. On the other hand for $t \gg \tau_R$ (and again neglecting the thermal contributions) one has

$$\left\langle \left(\overline{Y_0(t)} \right)^2 \right\rangle = \frac{q^2 E^2}{\zeta^2 N} t^2. \tag{41}$$

Here the displacement of the tagged bead mirrors the CM's motion (note that the average total charge is of the order of $q\sqrt{N}$).

Eq. (40) is again an example of a power law: After switching on the field, the displacement of the zeroth bead is given by $\widehat{Y}_0(t) \approx \left(b^{1/2} qE / \zeta^{3/4} T^{1/4}\right) t^{3/4}$, as long as $t < \tau_R$; here the hat denotes the averaging procedure $\widehat{f}(t) = \left\langle \left(\overline{f(t)} \right)^2 \right\rangle^{1/2}$. For large N the Rouse-time τ_R is very large ($\tau_R \propto N^2$) and one has

$$\widehat{Y}_0(t) \approx \frac{b^{1/2} q}{\zeta^{3/4} T^{1/4}} \int_{-\infty}^{t} \frac{d\tau}{(t-\tau)^{1-7/4}} \frac{dE(\tau)}{d\tau} \approx \frac{b^{1/2} q}{\zeta^{3/4} T^{1/4}} \frac{d^{-3/4} E(t)}{dt^{-3/4}}, \tag{42}$$

i.e., for an arbitrary history of the electrical field $E(t)$ the displacement of the bead is given by a fractional integration of $E(t)$ of order $3/4$.

What leads to an exponent of $3/4$? And can the exponent also assume other values? To clarify this we apply scaling arguments (as in Sec. 2.1) to PAs with random, uncorrelated charge distributions. In the following we consider general charge patterns, namely random, long-range correlated sequences of charges [31]. For such sequences the net charge of the PA is given by $\langle Q_{tot}^2 \rangle \cong q^2 N^{2\gamma}$ with $0 \leq \gamma \leq 1$. The case $\gamma = 1/2$ corresponds to the uncorrelated distribution discussed above, whereas for $\gamma > 1/2$ the charges are positively, and for $\gamma < 1/2$ negatively correlated. The extreme cases $\gamma = 1$ and $\gamma = 0$ are realized by polyelectrolytes ($q_n \equiv q$ or $q_n \equiv -q$) and by alternating PAs ($q_n = (-1)^n q$), respectively.

Consider now a single bead. We know from Sec. 2.1 the short- and long-time behavior of the total number g of neighboring monomers which are involved in a collective motion with this tagged bead (cf. Eqs. (31) and (32)). The excess charge Q of this collectively moving set of beads grows with time as $\langle Q^2 \rangle \cong q^2 (g(t))^{2\gamma}$, whereas its mobility decreases with time as $\mu \cong (\zeta g(t))^{-1}$. The average velocity of the tagged monomer in the Y-direction, v_Y, equals the velocity of the collectively moving set around it. Hence $\widehat{v}_Y^2(g) \cong \mu^2 \langle Q^2 \rangle E^2 \cong q^2 E^2 \zeta^{-2} g^{2\gamma-2}$, where g is given by Eq. (31) for $t \ll \tau_R$ and by Eq. (32) for $t \gg \tau_R$. The average displacement $\widehat{Y}(t)$ of a single bead, follows from the average displacement of the corresponding group of g neighbors, *i.e.* $\widehat{Y}^2(t) \cong \widehat{v}_Y^2(g)\, t^2 \cong q^2 E^2 \zeta^{-2} g^{2\gamma-2} t^2$. For $t \ll \tau_R$ one finds from Eq. (31)

$$\widehat{Y}^2(t) \cong \frac{b^{2-2\gamma} q^2 E^2}{\zeta^{1+\gamma} T^{1-\gamma}} t^{1+\gamma}. \tag{43}$$

For the uncorrelated case, $\gamma = 1/2$, this reproduces (up to a numerical constant of order unity) the field induced short-time behavior of the exact calculations, *i.e.* Eq. (40). At longer times the PA drifts as a whole, and from Eq. (32) we find that $\widehat{Y}^2(t) \cong \left(q^2 E^2/\zeta^2 N^{2-2\gamma}\right) t^2$, which for $\gamma = 1/2$ reproduces Eq. (41). Note that for polyelectrolytes, $\gamma = 1$, the drift becomes independent of N.

With Eq. (43) we have found a power law of the form $\widehat{Y}(t) \propto t^{(1+\gamma)/2}$, which leads us to a generalization of the fractional relationships Eqs. (30) and (42), namely to:

$$\widehat{Y}(t) \approx \frac{b^{1-\gamma} q}{\zeta^{(1+\gamma)/2} T^{(1-\gamma)/2}} \frac{d^{-(1+\gamma)/2} E(t)}{dt^{-(1+\gamma)/2}} . \tag{44}$$

In Eq. (44) the order of integration ranges from 1/2 (alternating case) to 1 (polyelectrolyte).

2.4 Polyampholytic Networks in External Electrical Fields

Consider a GGS where each monomer carries a charge, so that $q_n = \pm q$. Assume that the charges are distributed in such a way that $\langle Q_{tot}^2 \rangle \cong q^2 N^{2\gamma}$. Note that the cases $\gamma = 1/2$ and $\gamma = 1$ can be simply realized, whereas long-ranged (anti)correlations cannot be simply induced for many non-trivial connectivities. Using Eq. (34) one finds with $\widehat{v}_Y = QE/(g\zeta) \approx qE/\left(g^{1-\gamma}\zeta\right)$ for $d_s < 2$:

$$\widehat{Y}(t) \approx \left(\zeta b^2/T\right)^{(1-\gamma)d_s/2} \frac{qE}{\zeta} t^{1-(1-\gamma)d_s/2}. \tag{45}$$

The equilibrium value of the typical size R of the PA in the external field is reached at τ_G; one has for $\gamma = 1/2$

$$R \approx \widehat{Y}(\tau_G) \approx \frac{b^2 qE}{T} N^{(4-d_s)/(2d_s)}, \tag{46}$$

a relation which was derived by Sommer and Blumen [22].

Interestingly, the range of validity of Eq. (45) extends – depending on the value γ – to larger d_s-values. Consider, for instance, the uncorrelated case $\gamma = 1/2$. Then the exponent of t, namely $1-(1-\gamma)\, d_s/2$ is positive as long as

$d_s < 4$. One can show using more refined arguments [25] that here Eqs. (45) and (46) remain indeed valid also for $2 < d_s < 4$. Note, however, that for $d_s > 2$ $g(t)$, Eq. (34), is independent of d_s, so that our simplified argument does not work here.

This example shows that the interplay between the connectivity of the network and the charge distribution is not in every case multiplicative, *i.e.* does not subordinate (cf. the similar phenomenon discussed in Ref. [37]). In order to obtain the exponents for all d_s- and γ-values one is forced to use a more detailed analysis, which is given in Ref. [25].

We close by giving the corresponding fractional relation for an arbitrary time-dependence of $E(t)$:

$$\widehat{Y}(t) \approx \left(\zeta b^2/T\right)^{(1-\gamma)d_s/2} \frac{q}{\zeta} \frac{d^{(1-\gamma)d_s/2-1} E(t)}{dt^{(1-\gamma)d_s/2-1}}, \tag{47}$$

a relation which generalizes the fractional Eqs. (37) and (44). The order of integration extends now from 0 to -1; note, however, that Eq. (47) is only valid for sufficiently small d_s-values, namely $d_s < 2/(1-\gamma)$ [25].

2.5 Rouse Dynamics of Generalized Gaussian Structures: Connection to Macroscopic Properties

Until now we have considered the local dynamical properties of Rouse chains and of GGSs; now one may raise the question in how far the local dynamics influence the macroscopic properties of, say, a solution of GGSs. We will show in this section that the microscopic dynamics also determine the macroscopic viscoelastic properties of a GGS-solution; thus fractional force-displacement relationships translate into fractional stress-strain relationships of the whole sample.

We follow here the general treatment of viscoelasticity for polymer solutions, for suspensions etc. which is given in chapter 3.7 of Ref. [19] and apply it to GGS solutions. Let us assume to have a solution of identical GGSs and denote with c the concentration of the monomers, *i.e.* c/N is the concentration of the GGSs. The problem is now to find a microscopic expression for the stress tensor $\sigma_{\alpha\beta}$, say for the component σ_{zx} (for its definition cf. Ref. [38,39]). To this end let us consider a volume V in the CGS solution (cf. Fig. 1). We divide the volume by a hypothetical plane perpendicular to the Z-axis. The component σ_{zx} of the stress tensor is now given through the following definition:

$$\sigma_{zx} = \overline{S_{zx}}/A. \tag{48}$$

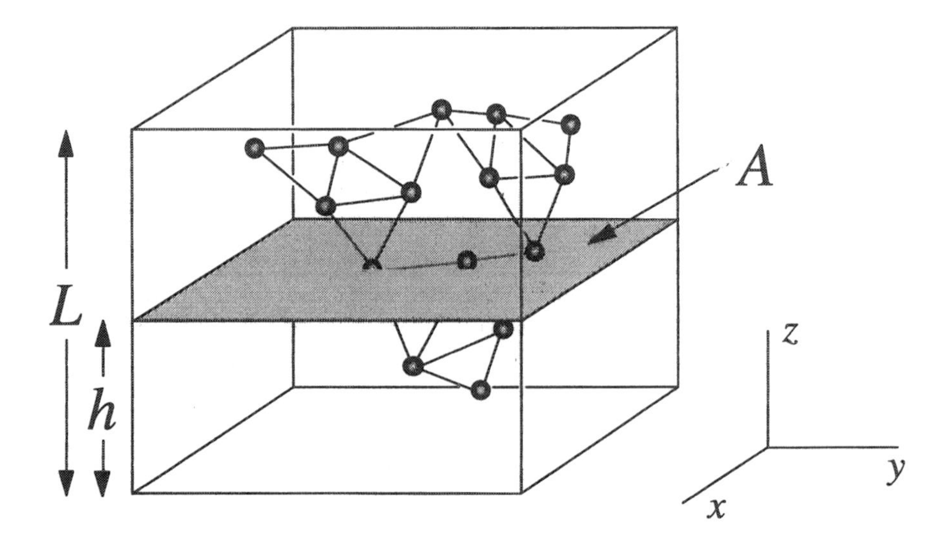

Figure 1: Microscopic derivation of the stress-tensor of a solution of GGSs (see text for details)

Here S_{zx} denotes the force which the upper part of the volume exerts on the lower part through the plane A; the dash denotes the configurational average with respect to the monomers. If we neglect the contribution of the solvent fluid to the force, S_{zx} is given by [19]

$$S_{zx}(h) = \sum_{n,m} F_{mn}^{(x)} \Theta(h - Z_m) \Theta(Z_n - h), \tag{49}$$

which depends on the height h of the dividing plane. In Eq. (49) the summations extend over all monomers in the volume V and $F_{mn}^{(x)}$ denotes the X-component of the force $\mathbf{F}_{nm}$ which monomer n exerts on monomer m. Thus σ_{zx} can be written (for short-ranged forces):

$$\sigma_{zx} = \frac{1}{AL} \int_0^L dh \overline{S_{zx}(h)} = \frac{1}{V} \sum_{n,m} \int_0^L dh \overline{F_{mn}^{(x)} \Theta(h - Z_m) \Theta(Z_n - h)}. \tag{50}$$

Eq. (50) can be transformed to the simple form [19]

$$\sigma_{zx} = -\frac{c}{N} \sum_{n=1}^{N} \overline{F_n^{(x)} Z_n} \tag{51}$$

where the summation now extends over the monomers of one arbitrary representative of the GGSs, and $\mathbf{F}_n = \sum_n \mathbf{F}_{nm}$ is the total sum of the forces acting on monomer m.

For a GGS $\mathbf{F}_n$ is the sum of all forces exerted by the monomers which are connected to the bead n via a spring (of spring constant $K = 3T/b^2$). Thus

$$F_n^{(x)} = -K \sum\nolimits_m M_{nm} X_m. \tag{52}$$

Here M_{nm} denotes the generalized Rouse matrix [40] of the GGS which can be constructed as follows. Start with all matrix elements set to zero. Then account for each bond between the monomers n and m by increasing the diagonal components M_{nn} and M_{mm} by $+1$ and M_{nm} and M_{mn} by -1. For a linear structure (*i.e.* a polymer chain) this procedure leads to a tridiagonal matrix which is the well-known Rouse matrix. Inserting Eq. (52) into (51) leads to

$$\sigma_{zx} = \frac{c}{N} \frac{3T}{b^2} \sum_{n,m} \overline{Z_n M_{nm} X_m}. \tag{53}$$

Now we switch from the $\{\mathbf{R}_n\}$ coordinates to the normal coordinates $\{\mathbf{R}(p)\}$; these two sets are related to each other through, say for the X-component, $X_n = 2\sum_p X(p)\, m_n(p)$. Here $m_n(p)$ denotes the nth component of the pth eigenvector of M, corresponding to the eigenvalue λ_p, *i.e.* $\sum_k M_{nk} m_k(p) = \lambda_p m_n(p)$. We normalize $m_n(p)$ as follows: $\sum_n m_n(p)\, m_n(q) = 2\delta_{pq}/N$. In terms of these normal coordinates Eq. (53) takes the following form:

$$\sigma_{zx} = \frac{c}{N} \frac{3T}{b^2} \sum_p 2N\lambda_p \overline{X(p)\, Z(p)} \tag{54}$$

Now let us perform a steplike deformation of the GGS solution at $t = 0$, *i.e.*, let the strain tensor be of the form $\varepsilon_{zx}(t) = \Theta(t)$ (see Ref. [39] for a definition of $\varepsilon_{\alpha\beta}$). Then for $t \geq 0$ the relaxation of the stress σ_{zx} is given by the relaxation modulus $G(t)$. Microscopically, the step deformation means that at $t = 0$ each monomer gets displaced by $Z_n(t = 0)$ in the X-direction. Solving the corresponding Langevin equation (see Ref. [25]) it follows that

$$\overline{X(p,t)\, Z(p,t)} = \frac{T}{2KN\lambda_p} \exp\left(-2K\lambda_p t/\zeta\right) \tag{55}$$

Thus we find for the relaxation modulus of the CGS solution

$$G(t) = \frac{c}{N} T \sum_p \exp\left(-2K\lambda_p t/\zeta\right) \tag{56}$$

Now the behavior for $t < \tau_G$ can be calculated by converting the sum over p into an integral; this leads to

$$G(t) \cong cT \int_0^\infty d\lambda n(\lambda) \exp(-2K\lambda t/\zeta) \cong cT (\zeta/K)^{d_s/2} t^{-d_s/2}. \tag{57}$$

Here $n(\lambda)$ denotes the spectral density of the eigenvalues of M which for $N^{-2/d_s} < \lambda < 1$ is given by $n(\lambda) \cong n^{d_s/2-1}$, cf. Refs. [23,24]. This result was also found by Cates [41].

For an arbitrary strain history $\varepsilon_{zx}(t)$ the stress $\sigma_{zx}(t)$ is given by the causal convolution

$$\sigma_{zx}(t) = \int_{-\infty}^{t} d\tau G(t-\tau) \frac{d\varepsilon_{zx}(\tau)}{d\tau}. \tag{58}$$

Using Eq. (57) we find the following fractional stress-strain relationship

$$\sigma_{zx}(t) \cong cT (\zeta/K)^{d_s/2} \frac{d^{d_s/2}\varepsilon_{zx}(t)}{dt^{d_s/2}}. \tag{59}$$

3 Applications to Rheological Constitutive Equations

In the last subsection we have shown how the (fractional) stress-strain relationship, Eq. (59) which describes the macroscopic mechanical properties of GGS solutions emerges from the microscopic dynamics of its constituents. The viewpoint of the current section is different: We attempt to find stress-strain relationships which properly *describe* the rheological properties of wide classes of materials. This phenomenological approach will lead us to rheological constitutive equations (RCE) with fractional derivatives. We note that the description of viscoelastic properties of materials by means of fractional calculus has a long history, which is discussed in detail in Ref. [42].

In the next section we describe the classical approach to viscoelasticity by exemplarily introducing the Maxwell model; furthermore we show that this model may be generalised by simply replacing the ordinary derivatives in its RCE by fractional ones. A word of caution is here necessary; arbitrarily replacing in RCE ordinary derivatives by fractional ones has its pitfalls: as a general procedure it can easily lead to physically meaningless results: It is therefore indispensable to have procedures at hand which automatically guarantee that the resulting fractional expressions are physically correct. To achieve this we start from mechanical analogues, a topic which is the subject of investigation of Sec. 3.2. Using this method large classes of physically meaningful fractional RCEs can be formulated; in Sec. 3.3 we present a list of RCEs based on fractional elements (FEs) for which all important material functions are known in analytical, closed form. Finally, we demonstrate in Sec. 3.4 the usefulness of fractional RCEs, by applying them to several viscoelastic materials, by determining the corresponding response functions and by comparing these to the experimental results.

3.1 Viscoelasticity: Classical Approach and its Fractional Generalization

The usual phenomenological viscoelastic models are based on springs and dashpots [38,39]. The springs, Fig. 2(a), obey Hooke's law

$$\sigma(t) = E\varepsilon(t) \tag{60}$$

whereas for dashpots, Fig. 2(b), Newton's law holds:

$$\sigma(t) = \eta \frac{d\varepsilon\,(t)}{dt}. \tag{61}$$

Here σ and ε are the stress and strain of the corresponding mechanical model and E and η denote the spring constant and the viscosity. Now the stress σ and the strain ε in Eqs. (60) and (61) may be interpreted as being components of the stress and strain tensors, say σ_{zx} and ε_{zx} of a macroscopic object. Then Hooke's law represents the RCE of an ideal solid, whereas Newton's law corresponds to an ideal fluid.

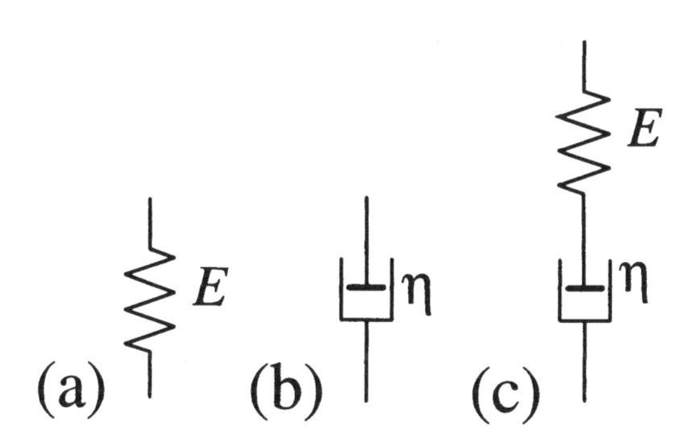

Figure 2: (a) The spring and (b) the dashpot are the structural parts of ordinary viscoelastic models. An example is (c) the Maxwell model where the spring and the dashpot are arranged in a sequential manner.

In general, real objects (such as macromolecular materials) show a behavior which combines characteristic features of solids and liquids, *i.e.* they are viscoelastic. In order to derive RCEs for viscoelastic materials one often starts from mechanical arrangements which combine a small number of springs and dashpots in suitable ways; by this method one arrives at several standard viscoelastic models. To give an example, let us combine a spring and a dashpot in series (Fig. 2(c)); this leads to the classical Maxwell model [38,39]. Here the two components obey the stress-strain relations $\sigma_1(t) = E\varepsilon_1(t)$ for the spring and $\sigma_2(t) = \eta d\varepsilon_2(t)/dt$ for the dashpot. Due to the sequential construction of the arrangement the stresses are equal for both structural parts, $\sigma_1 = \sigma_2 = \sigma$ whereas the strains add, $\varepsilon_1 + \varepsilon_2 = \varepsilon$. This leads to

$$\sigma(t) + \tau \frac{d\sigma(t)}{dt} = \tau E \frac{d\varepsilon(t)}{dt} \tag{62}$$

which is the RCE of the ordinary Maxwell model.

The *relaxation modulus* $G(t)$, *i.e.*, the stress response of the system to a shear jump $\varepsilon(t) = \Theta(t)$ follows from Eq. (62) to be exponential, namely

$$G(t) = E \exp(-t/\tau)\, \Theta(t) \,. \tag{63}$$

Thus the Maxwell model leads to an exponential stress decay; real materials, however, show in many cases a more general behavior. This is also the case for microscopic models of polymer dynamics, such as the Rouse model: In Sec. 2.5 we showed that – as a result of the linear superposition of normal modes – the stress relaxation is non-exponential, obeying an algebraic decay, namely $G(t) \propto t^{-\gamma}$ (with $0 < \gamma < 1$; cf. Eq. (57)). Such algebraic patterns occur for wide classes of polymeric materials over many decades in time. It is thus natural to ask how the classical viscoelastic models have to be extended in order to be able to reproduce such decay patterns.

A direct approach to create fractional RCE is to replace the regular derivatives of ordinary RCE by fractional derivatives ($d^{\gamma_i}/dt^{\gamma_i}$) of non-integer order ($0 < \gamma_i < 1$) [43]. In the case of the Maxwell model, Eq. (62), this procedure leads to

$$\sigma(t) + \tau^{\gamma_1} \frac{d^{\gamma_1}\sigma(t)}{dt^{\gamma_1}} = \tau^{\gamma_2} E \frac{d^{\gamma_2}\varepsilon(t)}{dt^{\gamma_2}} \tag{64}$$

with $0 < \gamma_1 < 1$ and $0 < \gamma_2 < 1$. The relaxation modulus of this generalised Maxwell model can be calculated analytically using different methods [43]–[45]. Interestingly, Eq. (64) is physically meaningful only for γ_i-values in a restricted range: An analysis of the RCE shows (see Sec. 3.3) that $G(t)$ behaves like $t^{\gamma_1-\gamma_2}$ for short times, $t \ll \tau$. Thus for $\gamma_1 > \gamma_2$ the relaxation function increases, which is in general not reasonable [43,44], and one has to require that $\gamma_1 \leq \gamma_2$.

This example shows that constructing fractional RCEs through direct replacement ($d/dt \rightarrow d^{\gamma_i}/dt^{\gamma_i}$) is unsatisfactory: Only after an in depth analysis of the solutions of such fractional RCE can one decide, *a posteriori*, whether the solutions are physically meaningful or not. In the next section we describe another method which is easy to implement and which has the big advantage that it guarantees automatically that the fractional RCEs obtained through it are physically correct.

3.2 Mechanical Analogues to Fractional Rheological Equations

Let us start by constructing a simple RCE which is physically sound. We assume a given, linear system, which displays the following response function:

$$G(t) = \frac{E}{\Gamma(1-\gamma)} \left(\frac{t}{\tau}\right)^{-\gamma} \tag{65}$$

with $0 < \gamma < 1$. We already know a physical model which leads to such a decay pattern, namely solutions made of Rouse-type GGS, cf. Eq. (57). By comparing Eq. (65) with Eqs. (3) and (8) we obtain the fractional RCE corresponding to this state:

$$\sigma(t) = E\tau^{\gamma} \frac{d^{\gamma}\varepsilon(t)}{dt^{\gamma}}. \tag{66}$$

Note that Eq. (66) is an interpolation between Hookes law, Eq. (60) ($\gamma = 0$) and Newton's law, Eq. (61) ($\gamma = 1$).

Schiessel and Blumen [46,47], Schiessel *et al* [45,48], Heymans and Bauwens [49] and Heymans [50] have demonstrated that the fractional relation, Eq. (66), can be realized physically through hierarchical arrangements of springs and dashpots, such as ladders, trees and fractal networks. In the limit of an infinite number of constitutive elements these arrangements obey Eq. (66). We sketch here two examples, namely the ladder structure introduced in Ref. [46] and the fractal arrangement discussed in Ref. [47]. In Fig. 3 we display a ladder-like structure consisting of springs (with spring constants E_0, E_1, E_2,...) along one of the struts and dashpots (with viscosities η_0, η_1, η_2,...) along the rungs of the ladder. We have shown in Ref. [46] that for equal spring constants and viscosities, *i.e.* for $E = E_0 = E_1 = ...$ and $\eta = \eta_0 = \eta_1 = ...$, this arrangement obeys Eq. (66) with $\tau = \eta/E$ and $\gamma = 1/2$. Moreover, the sequential structure of Fig. 3 also allows to attain other γ-values, by choosing E_k and η_k with k

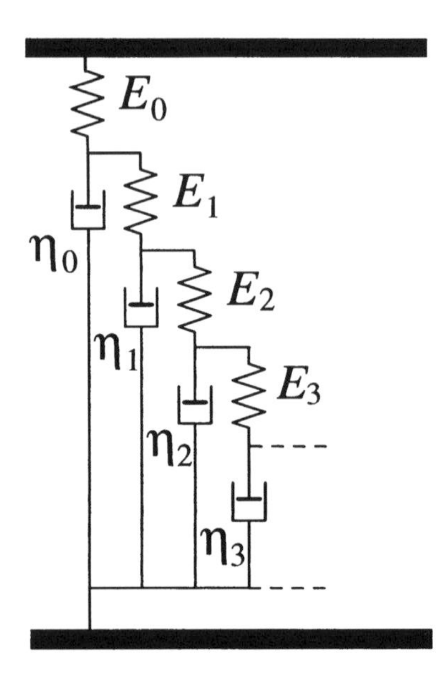

Figure 3: Spring-dashpot ladder: A sequential realisation of the fractional element.

in a suitable way. Thus it suffices to let the spring constants and viscosities obey [46]

$$E_k \propto k^{1-2\gamma} \text{ and } \eta_k \propto k^{1-2\gamma} \tag{67}$$

in order to have an arrangement which obeys Eq. (66) with $0 < \gamma < 1$.

In Ref. [51] we have shown that the linear Rouse-model and the simple ladder model (all E and η are equal, $\gamma = 1/2$) are closely related. Indeed, the similarity between Eqs. (30) and (66) is not accidental. The physical situation of the Rouse chain is, in fact, very close to the ladder arrangement in Fig. 3: In both cases one pulls at the end of an object which is connected by springs and exposed to a velocity dependent friction.

Here an interesting question arises, namely whether it is possible to construct spring-dashpot arrangements which correspond to the GGSs discussed in Sec. 2.2. Such fractals were indeed analysed by us in Ref. [47]. As an example consider the Sierpinski-like structure depicted in Fig. 4. Such a model obeys $\gamma = 1 - d_s/2$ (cf. Ref. [47]), *i.e.*, the connectivity of the network determines the order of derivation in Eq. (66).

As a next step we introduce now the concept of *fractional element* (FE); we define a FE as being a mechanical arrangement which obeys Eq. (66) (without

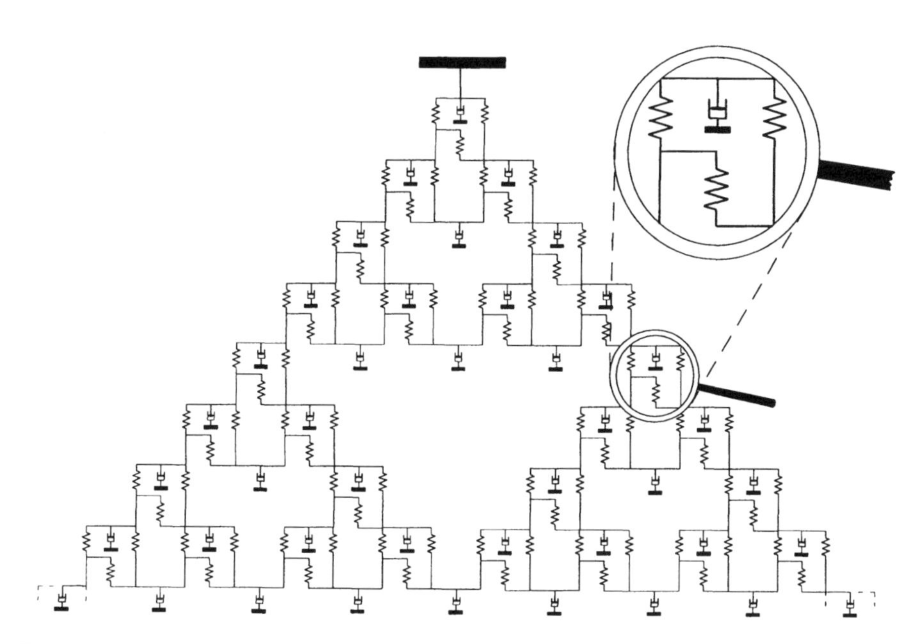

Figure 4: Sierpinski-like network of springs and dashpots: Here the connectivity of the structure controls its rheological properties.

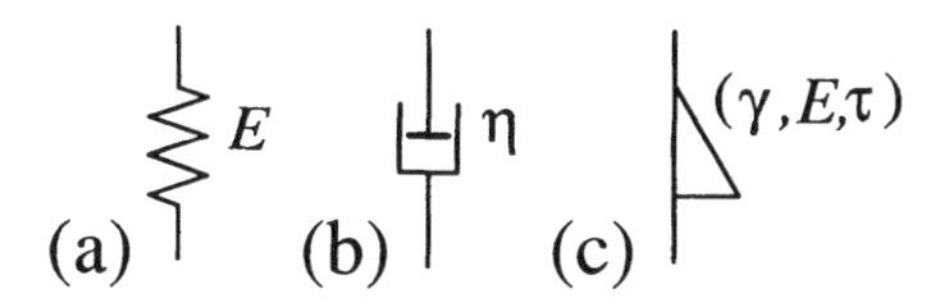

Figure 5: Single elements: (a) elastic, (b) viscous and (c) fractional element.

caring about its specific inner structure). A FE is specified by a triple (γ, E, τ) and we symbolize it by a triangle, as shown in Fig. 5(c), where also its classical counterparts, the spring and the dashpot, are depicted. In the following we will treat a FE on the same footing as a spring or a dashpot, namely as being an elementary building block of mechanical arrangements.

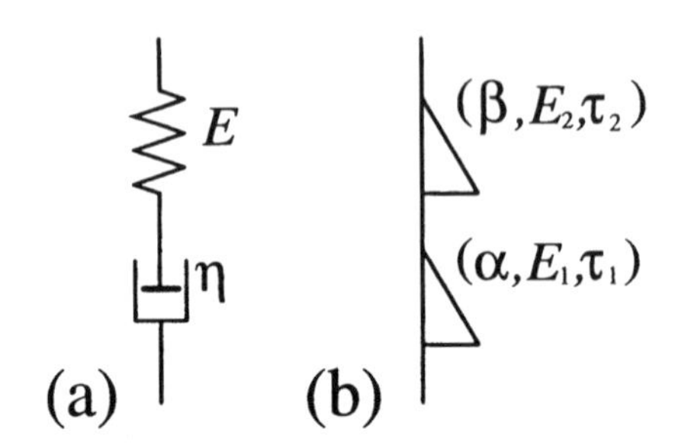

Figure 6: (a) The Maxwell model and (b) its fractional generalisation.

Now we have a method at hand to carry out a proper generalization of the classical Maxwell model, which we depict again in Fig. 6(a): We simply replace the two classical elements by FEs with the parameters (α, E_1, τ_1) and (β, E_2, τ_2), cf. Fig. 6(b). The FEs obey the stress-strain relations $\sigma_1(t) = E_1\tau_1^\alpha d^\alpha \varepsilon_1(t)/dt^\alpha$ and $\sigma_2(t) = E_2\tau_2^\beta d^\beta \varepsilon_2(t)/dt^\beta$. Due to the sequential construction of the arrangement one finds

$$\sigma(t) + \frac{E_1\tau_1^\alpha}{E_2\tau_2^\beta}\frac{d^{\alpha-\beta}\sigma(t)}{dt^{\alpha-\beta}} = E_1\tau_1^\alpha\frac{d^\alpha\varepsilon(t)}{dt^\alpha}. \tag{68}$$

Let us assume without loss of generality $\alpha > \beta$. Equation (68) can be further simplified by setting

$$\tau = \left(E_1\tau_1^\alpha/E_2\tau_2^\beta\right)^{1/(\alpha-\beta)} \text{ and } E = E_1(\tau_1/\tau)^\alpha. \tag{69}$$

This leads to

$$\sigma(t) + \tau^{\alpha-\beta}\frac{d^{\alpha-\beta}\sigma(t)}{dt^{\alpha-\beta}} = E\tau^\alpha\frac{d^\alpha\varepsilon(t)}{dt^\alpha}. \tag{70}$$

which is the RCE of the fractional Maxwell model [43,45,52]. Now compare Eq. (70) with Eq. (64): The parameter γ_1 has to be identified with $\alpha - \beta$ and γ_2 with α. Note that due to the mechanical construction, our fractional Maxwell model obeys *automatically* $\gamma_1 \leq \gamma_2$. A further advantage of our method is that the schematic representation of the model, Fig. 6(b), is much easier to grasp than the corresponding RCE, Eq. (70). This advantage increases with the number of FEs involved, as the next examples will show.

3.3 Overview over Exactly Solvable Fractional Models

Here we list FE based models for which all relevant response functions are known analytically, namely the relaxation modulus, the creep compliance, the complex modulus and the complex compliance. Now, the relaxation modulus $G(t)$ was already introduced. The *shear creep compliance* $J(t)$ is the response of the strain to the shear $\sigma(t) = \Theta(t)$. $G(t)$ and $J(t)$ are the so-called step response functions. On the other hand, the so-called harmonic response functions are defined as follows: The *complex shear modulus* $G^*(\omega)$ describes the response of the stress to a harmonic strain excitation $\varepsilon(t) = \exp(i\omega t)$, *i.e.* $\sigma(t) = G^*(\omega)\exp(i\omega t)$; from $G^*(\omega)$ the *storage* and the *loss moduli*, $G'(\omega) = \mathrm{Re}(G^*(\omega))$ and $G''(\omega) = \mathrm{Im}(G^*(\omega))$, follow. *The complex shear compliance* $J^*(\omega)$ is the response of the strain to a harmonic stress excitation $\sigma(t) = \exp(i\omega t)$; the *storage* and *loss compliances* are defined by $J'(\omega) = \mathrm{Re}(J^*(\omega))$ and $J''(\omega) = -\mathrm{Im}(J^*(\omega))$.

In the context of fractional RCEs, the dynamical response functions are extremely useful: (*i*) In many rheological experiments they are measured directly (cf., for instance, Ref. [42] and subsection 3.4) and (*ii*) for a given fractional RCE they can be easily derived. This can be inferred from the behavior of fractional derivatives under Fourier transformation. The Fourier transform

$$\mathcal{F}\{f(t)\} = \tilde{f}(\omega) = \int_{-\infty}^{\infty} dt f(t)\exp(-i\omega t) \tag{71}$$

turns the operation d^γ/dt^γ into a simple multiplication [8,9]

$$\mathcal{F}\left\{\frac{d^\gamma f(t)}{dt^\gamma}\right\} = (i\omega)^\gamma \tilde{f}(\omega). \tag{72}$$

For a given fractional RCE the complex modulus follows directly from

$$G^*(\omega) = \tilde{\sigma}(\omega)/\tilde{\varepsilon}(\omega) \tag{73}$$

by using the multiplication rule (72); the complex compliance is then given by $J^*(\omega) = 1/G^*(\omega)$.

Fractional Element

A single FE leads to Eq. (66), which is the simplest fractional RCE. The corresponding mechanical analogue is depicted in Fig. 5(c). The relaxation modulus is given by Eq. (65); the creep compliance follows in analogy to the discussion in Sec. 1, cf. Eq. (11):

$$J(t) = \frac{E^{-1}}{\Gamma(1+\gamma)} \left(\frac{t}{\tau}\right)^{\gamma}. \tag{74}$$

Fourier transforming Eq. (66) leads to $\widetilde{\sigma}(\omega) = E(i\omega\tau)^{\gamma}\,\widetilde{\varepsilon}(\omega)$, from which

$$G^*(\omega) = E(i\omega\tau)^{\gamma} \tag{75}$$

and

$$J^*(\omega) = E^{-1}(i\omega\tau)^{-\gamma} \tag{76}$$

follow.

Fractional Maxwell Model

In Sec. 3.2 we have already derived the RCE of the fractional Maxwell model, cf. its RCE, Eq. (70), and the mechanical analog, Fig. 6(b). Fourier transforming Eq. (70) and using Eqs. (72) and (73) leads to the complex modulus:

$$G^*(\omega) = \frac{E(i\omega\tau)^{\alpha}}{1+(i\omega\tau)^{\alpha-\beta}}. \tag{77}$$

The complex compliance $J^*(\omega) = 1/G^*(\omega)$ obeys

$$J^*(\omega) = E^{-1}(i\omega\tau)^{-\alpha} + E^{-1}(i\omega\tau)^{-\beta}. \tag{78}$$

Eq. (78) mirrors the fact that for serially arranged elements the compliances of the FEs, Eq. (76), simply add; similarly, the creep compliance is the sum of the compliances of the corresponding two FEs:

$$J(t) = E^{-1}\frac{(t/\tau)^{\alpha}}{\Gamma(1+\alpha)} + E^{-1}\frac{(t/\tau)^{\beta}}{\Gamma(1+\beta)} \tag{79}$$

for all $0 \le \beta \le \alpha \le 1$, a result also reported in Ref. [43] (see also the discussion in Refs. [44,45]).

On the other hand, the determination of the relaxation function $G(t)$ is a difficult task; it can be performed using a power series ansatz [43], a Laplace-Mellin [44] or a Fourier-Mellin method [45]. For the detailed calculations we refer the interested reader to the original works; here we restrict ourself to the presentation and discussion of the results. The relaxation function is:

$$G(t) = \frac{E}{\alpha - \beta} H_{12}^{11} \left[\frac{t}{\tau} \middle| \begin{matrix} \left(\frac{-\beta}{\alpha-\beta}, \frac{1}{\alpha-\beta}\right) \\ \left(\frac{-\beta}{\alpha-\beta}, \frac{1}{\alpha-\beta}\right); (0,1) \end{matrix} \right] \tag{80}$$

or, written differently

$$G(t) = E \left(\frac{t}{\tau}\right)^{-\beta} E_{\alpha-\beta,1-\beta}\left(-\left(\frac{t}{\tau}\right)^{\alpha-\beta}\right). \tag{81}$$

Equations (80) and (81) are two equivalent representations of $G(t)$: Here the $H_{pq}^{nm}(x)$ denote the Fox H-functions [53], which are defined via modified Mellin-Barnes integrals; a detailed discussion of the H-functions can be found in Ref. [45] and in chapter 8 of this book. The $E_{\kappa,\mu}(x)$ stand for the Mittag-Leffler functions [54], which are defined via power series [43,45].

The H-functions may be written in terms of alternating power series, a fact which is convenient for computations. For fractional RCEs we often encounter the form $H_{12}^{11}(x)$ (see Eq. (80) and the other models, discussed below). The power series around $x = 0$ and $x = \infty$ are given by

$$H_{12}^{11}\left[x \middle| \begin{matrix} (a, A) \\ (a, A); (0,1) \end{matrix}\right] = A^{-1} \sum_{k=0}^{\infty} \frac{(-1)^k x^{(k+a)/A}}{\Gamma\left(1 + \frac{k+a}{A}\right)} \tag{82}$$

and

$$H_{12}^{11}\left[x \middle| \begin{matrix} (a, A) \\ (a, A); (0,1) \end{matrix}\right] = A^{-1} \sum_{k=0}^{\infty} \frac{(-1)^k (1/x)^{(k+1-a)/A}}{\Gamma\left(1 - \frac{k+1-a}{A}\right)}. \tag{83}$$

A combination of Eqs. (82) and (83) allows the numerical evaluation of $H_{12}^{11}(x)$ over the complete range of x-values, $x > 0$. Furthermore, from the power series the asymptotic behavior for small and for large arguments can be immediately derived:

$$A\, H_{12}^{11}\left[x \middle| \begin{matrix} (a, A) \\ (a, A); (0,1) \end{matrix}\right] \cong \frac{1}{\Gamma(1 + a/A)} x^{a/A} \tag{84}$$

for $x \ll 1$ and

$$A\, H_{12}^{11}\left[x \middle| \begin{matrix} (a, A) \\ (a, A); (0,1) \end{matrix}\right] \cong \frac{1}{\Gamma\left(1 + \frac{a-1}{A}\right)} x^{\frac{a-1}{A}} \tag{85}$$

for $x \gg 1$.

Now, the behavior of $G(t)$, Eq. (80), for the fractional Maxwell model can be inferred readily. In the parameter range $0 \leq \beta < \alpha < 1$ $G(t)$ obeys at short times, $t \ll \tau$, cf. Eq. (84):

$$G(t) \cong \frac{E}{\Gamma(1-\beta)} \left(\frac{t}{\tau}\right)^{-\beta} \tag{86}$$

whereas at long times ($t \gg \tau$) one has asymptotically, cf. Eq. (85)

$$G(t) \cong \frac{E}{\Gamma(1-\alpha)} \left(\frac{t}{\tau}\right)^{-\alpha}. \tag{87}$$

Equation (80) also includes the special cases $\beta = 0$ and/or $\alpha = 1$; a discussion of these cases can be found in Ref. [45].

Fractional Kelvin-Voigt Model

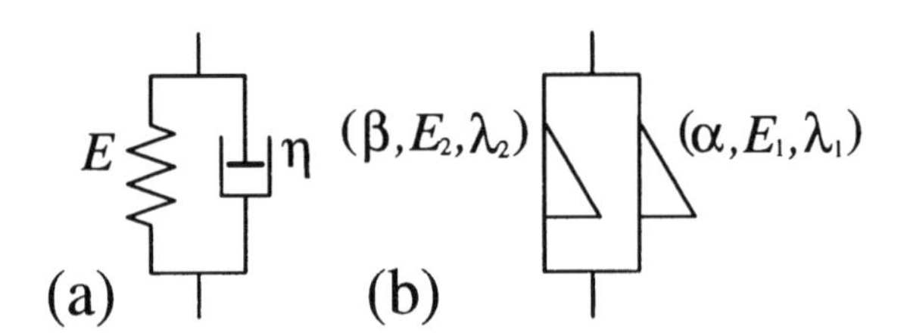

Figure 7: The (a) ordinary Kelvin-Voigt model and (b) its fractional generalisation.

By arranging the spring and the dashpot in parallel one arrives at the standard Kelvin–Voigt model, depicted in Fig. 7(a); its generalisation with two FEs is shown in Fig. 7(b). Because of the parallel construction the individual stresses add. Following the procedure used for the fractional Maxwell model we find the following RCE

$$\sigma(t) = E\tau^{\alpha} \frac{d^{\alpha}\varepsilon(t)}{dt^{\alpha}} + E\tau^{\beta} \frac{d^{\beta}\varepsilon(t)}{dt^{\beta}} \tag{88}$$

where the parameters τ and E are defined as in Eq. (69). By Fourier transforming Eq. (88) we obtain from Eqs. (72) and (73) the complex modulus

$$G^{*}(\omega) = E(i\omega\tau)^{\alpha} + E(i\omega\tau)^{\beta}, \tag{89}$$

a result that mirrors the additivity of the moduli in parallel arrangements. Similarly, one finds for the relaxation modulus

$$G(t) = E\frac{(t/\tau)^{-\alpha}}{\Gamma(1-\alpha)} + E\frac{(t/\tau)^{-\beta}}{\Gamma(1-\beta)}. \tag{90}$$

On the other hand, the parallel arrangement of the Kelvin-Voigt model causes the complex compliance

$$J^*(\omega) = E^{-1}\frac{(i\omega\tau)^{-\beta}}{1+(i\omega\tau)^{\alpha-\beta}} \tag{91}$$

to be more involved than for the Maxwell model. For the calculation of the function $J(t)$ methods similar to the determination of $G(t)$ for the fractional Maxwell model can be used. This leads to [45]

$$J(t) = \frac{E^{-1}}{\alpha-\beta}H_{12}^{11}\left[\frac{t}{\tau}\left|\begin{matrix}\left(\frac{\alpha}{\alpha-\beta},\frac{1}{\alpha-\beta}\right) \\ \left(\frac{\alpha}{\alpha-\beta},\frac{1}{\alpha-\beta}\right);(0,1)\end{matrix}\right.\right] \tag{92}$$

or, equivalently, to

$$J(t) = E^{-1}\left(\frac{t}{\tau}\right)^{\alpha} E_{\alpha-\beta,1+\alpha}\left(-\left(\frac{t}{\tau}\right)^{\alpha-\beta}\right). \tag{93}$$

From Eq. (84) we obtain for $t \ll \tau$ the short time behavior of $J(t)$:

$$J(t) \cong \frac{E^{-1}}{\Gamma(1+\alpha)}\left(\frac{t}{\tau}\right)^{\alpha} \tag{94}$$

whereas for $t \gg \tau$ we have asymptotically, cf. Eq. (85):

$$J(t) \cong \frac{E^{-1}}{\Gamma(1+\beta)}\left(\frac{t}{\tau}\right)^{\beta}. \tag{95}$$

A discussion of the special cases $\alpha = 1$ and/or $\beta = 0$ can be found in Ref. [45].

Note the symmetry of the step response functions, $G(t)$ and $J(t)$, in the fractional Maxwell model and in the fractional Kelvin–Voigt model, the two possible arrangements of two FEs. Depending on the construction (sequential for Maxwell or parallel for Kelvin–Voigt) one of these response functions is simply a sum of two algebraic terms (*i.e.* $J(t)$ in the Maxwell model, Eq. (79), and $G(t)$ in the Kelvin–Voigt model, Eq. (90)). Then the other response function is a Mittag–Leffler function (*i.e.* $G(t)$ for the Maxwell model, Eq. (81), and $J(t)$ for the Kelvin–Voigt model, Eq. (93)). Furthermore, one finds in each case a crossover between two algebraic regimes: For a sum of two algebraic terms this is obvious; in the case of the Mittag–Leffler function this follows from its expression through power series (*vide supra* and Eqs. (84) and (85)).

Fractional Zener Model

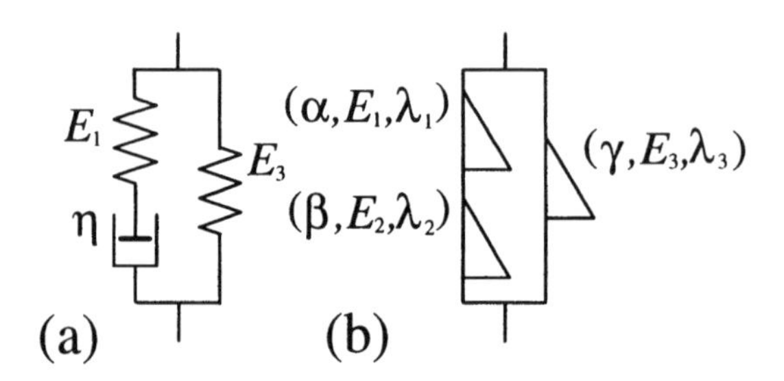

Figure 8: (a) The Zener model and (b) its fractional generalisation.

The so-called Zener or standard solid model [38,39] depicted in Fig. 8(a) involves three elements: It consists of a Maxwell model in parallel with a spring. The most general fractional version of the Zener model is displayed in Fig. 8(b) and consists of three FEs. Without loss of generality, we again require $0 \leq \beta < \alpha \leq 1$ and, of course, $0 \leq \gamma \leq 1$. For this arrangement one finds the following RCE [45]:

$$\sigma(t) + \tau^{\alpha-\beta} \frac{d^{\alpha-\beta}\sigma(t)}{dt^{\alpha-\beta}} = E\tau^{\alpha} \frac{d^{\alpha}\varepsilon(t)}{dt^{\alpha}} + E_0\tau^{\gamma} \frac{d^{\gamma}\varepsilon(t)}{dt^{\gamma}} + E_0\tau^{\gamma+\alpha-\beta} \frac{d^{\gamma+\alpha-\beta}\varepsilon(t)}{dt^{\gamma+\alpha-\beta}}. \tag{96}$$

with τ and E given by Eq. (69) and $E_0 = E_3(\tau_3/\tau)^{\gamma}$. This RCE was given by Tschoegl [39] for the special case $\beta = \gamma = 0$; it was extended to arbitrary $0 \leq \beta < \alpha$ (with $\gamma = 0$) by Friedrich and Braun [55] (cf. also the paper by Metzler et al. [56]).

Using Eqs. (72) and (73) the calculation of the complex modulus $G^*(\omega)$ is straightforward and results in:

$$G^*(\omega) = \frac{E(i\omega\tau)^{\alpha}}{1 + (i\omega\tau)^{\alpha-\beta}} + E_0(i\omega\tau)^{\gamma}. \tag{97}$$

In Eq. (97) the moduli of the Maxwell model, Eq. (77) and of the single FE, Eq. (75), add; this is the direct consequence of the parallel arrangement of Fig. 8(b).

We obtain the resulting relaxation modulus from Eq. (97) by simply adding the moduli of the Maxwell model, Eq. (80), and of the single FE, Eq. (65), *i.e.*

$$G(t) = \frac{E}{\alpha-\beta} H_{12}^{11}\left[\frac{t}{\tau}\middle| \begin{pmatrix} -\frac{\beta}{\alpha-\beta}, \frac{1}{\alpha-\beta} \\ -\frac{\beta}{\alpha-\beta}, \frac{1}{\alpha-\beta} \end{pmatrix}; (0,1) \right] + E_0 \frac{(t/\tau)^{-\gamma}}{\Gamma(1-\gamma)}. \tag{98}$$

Depending on the parameters, up to four algebraic time regimes may show up. This can be seen by comparing on the RHS of Eq. (98) the generalised Maxwell term (and its power–law behaviors, Eqs. (86) and (87)) with the second term. For instance, in the case $0 \le \gamma \le \beta < \alpha$ and for $E \gg E_0$, we find three time regimes

$$G(t) \sim \begin{cases} t^{-\beta} & \text{for} \quad t \ll \tau \\ t^{-\alpha} & \text{for} \quad \tau \ll t \ll \tau_1 \\ t^{-\gamma} & \text{for} \quad \tau_1 \ll t \end{cases} \tag{99}$$

where $\tau_1 \approx (E/E_0)^{1/(\alpha-\gamma)}\, \tau$.

The complex compliance $J^*(\omega) = 1/G^*(\omega)$, *i.e.*

$$J^*(\omega) = \frac{(i\omega\tau)^{-\alpha} + (i\omega\tau)^{-\beta}}{E + E_0(i\omega\tau)^{\gamma-\alpha} + E_0(i\omega\tau)^{\gamma-\beta}} \tag{100}$$

shows a more involved pattern. The creep compliance is known analytically only for the special cases $\gamma = \alpha$ or $\gamma = \beta$, cf. Ref. [45]; we present here the result for $\gamma = \beta$:

$$J(t) = \frac{C_1^{\alpha-2\beta}}{\alpha-\beta} \frac{E}{E_0^2} H_{12}^{11}\left[C_1 \frac{t}{\tau}\middle| \begin{pmatrix} \frac{\alpha}{\alpha-\beta}, \frac{1}{\alpha-\beta} \\ \frac{\alpha}{\alpha-\beta}, \frac{1}{\alpha-\beta} \end{pmatrix}; (0,1) \right] + \frac{(E+E_0)^{-1}}{\Gamma(1+\beta)} \left(\frac{t}{\tau}\right)^{\beta}. \tag{101}$$

Here we set $C_1 = (E_0/(E+E_0))^{1/(\alpha-\beta)}$. Equation (101) was given in Ref. [55] for the special case $\gamma = \beta = 0$. Note that Eq. (101) expresses $J(t)$ as the sum of two terms which have the same functional dependences as the compliances of the Kelvin–Voigt model, Eq. (92) and of the single FE, Eq. (74), respectively. This similarity follows from the correspondence between the fractional Zener model and the fractional Poynting-Thomson model, which we discuss below.

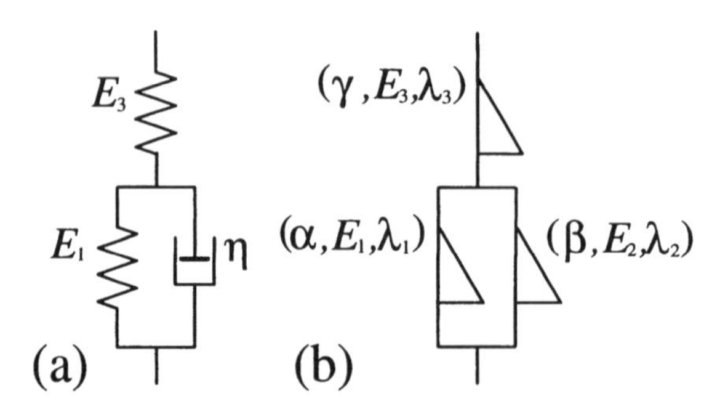

Figure 9: (a) The ordinary and (b) the fractional Poynting-Thomson model.

Fractional Poynting-Thomson Model

The Poynting–Thomson model and its generalisation based on FEs are shown in Figs. 9(a) and (b), respectively. The stress–strain relation of the fractional model obeys

$$\sigma(t) + \frac{E}{E_0}\tau^{\alpha-\gamma}\frac{d^{\alpha-\gamma}\sigma(t)}{dt^{\alpha-\gamma}} + \frac{E}{E_0}\tau^{\beta-\gamma}\frac{d^{\beta-\gamma}\sigma(t)}{dt^{\beta-\gamma}} = E\tau^{\alpha}\frac{d^{\alpha}\varepsilon(t)}{dt^{\alpha}} + E\tau^{\beta}\frac{d^{\beta}\varepsilon(t)}{dt^{\beta}} \tag{102}$$

where τ, E and E_0 are defined as for the fractional Zener model. Due to the serial arrangement of the Poynting–Thomson model we obtain immediately its creep compliance as being the sum of the compliances of its subunits, a Kelvin–Voigt element, Eq. (92) and a FE, Eq. (74):

$$J(t) = \frac{E^{-1}}{\alpha-\beta}H^{11}_{12}\left[\frac{t}{\tau}\left|\begin{matrix}\left(\frac{\alpha}{\alpha-\beta}, \frac{1}{\alpha-\beta}\right) \\ \left(\frac{\alpha}{\alpha-\beta}, \frac{1}{\alpha-\beta}\right); (0,1)\end{matrix}\right.\right] + \frac{E_0^{-1}}{\Gamma(1+\gamma)}\left(\frac{t}{\tau}\right)^{\gamma}. \tag{103}$$

We are able to calculate the relaxation modulus when we restrict ourselves to $\gamma = \alpha$ or to $\gamma = \beta$. Interestingly, the Poynting–Thomson and the Zener model lead then to the same RCEs. In order to distinguish between the material constants of the two models we introduce the superscripts P for Poynting-Thomson and Z for Zener. Thus for $\gamma = \beta$ the RCE of the Poynting–Thomson model, Eq. (102), takes the form

$$\sigma(t) + \frac{E^P\left(\tau^P\right)^{\alpha-\beta}}{E^P + E_0^P}\frac{d^{\alpha-\beta}\sigma(t)}{dt^{\alpha-\beta}} = \frac{E^P E_0^P}{E^P + E_0^P}\left[\left(\tau^P\right)^{\alpha}\frac{d^{\alpha}\varepsilon(t)}{dt^{\alpha}} + \left(\tau^P\right)^{\beta}\frac{d^{\beta}\varepsilon(t)}{dt^{\beta}}\right] \tag{104}$$

whereas the RCE of the Zener model, Eq. (96), for $\gamma = \alpha$ reads:

$$\sigma(t) + \left(\tau^Z\right)^{\alpha-\beta} \frac{d^{\alpha-\beta}\sigma(t)}{dt^{\alpha-\beta}} = \left(E^Z + E_0^Z\right)\left(\tau^Z\right)^{\alpha} \frac{d^{\alpha}\varepsilon(t)}{dt^{\alpha}} + E_0^Z\left(\tau^Z\right)^{\beta} \frac{d^{\beta}\varepsilon(t)}{dt^{\beta}}. \tag{105}$$

By comparing the corresponding terms of Eqs. (104) and (105) we find as transformation rules:

$$\tau^P = \left(\left(E^Z + E_0^Z\right)/E_0^Z\right)^{1/(\alpha-\beta)} \tau^Z, \tag{106}$$

$$E^P = \left(\left(E^Z + E_0^Z\right)/E_0^Z\right)^{(\alpha-2\beta)/(\alpha-\beta)} \left(E_0^Z\right)^2/E^Z \tag{107}$$

and

$$E_0^P = \left(\left(E^Z + E_0^Z\right)/E_0^Z\right)^{(\alpha-2\beta)/(\alpha-\beta)} E_0^Z. \tag{108}$$

These relations connect the two models and are valid for $\gamma = \beta$; for $\gamma = \alpha$ one has simply to exchange α and β in Eqs. (106) to (108), cf. Ref. [45]. Using Eq. (103) and the relations (106) to (108) one recovers Eq. (101) of the fractional Zener model.

This duality can now be invoked to establish the relaxation modulus $G(t)$ in the Poynting–Thomson model; one has only to replace the material constants in the corresponding function of the fractional Zener model, Eq. (98). For $\gamma = \beta$ this leads to [42]

$$G(t) = \frac{C_2^{\alpha-2\beta}}{\alpha-\beta}\frac{E_0^2}{E} H_{12}^{11}\left[\frac{t}{C_2\tau}\,\middle|\,\begin{matrix}\left(\frac{-\beta}{\alpha-\beta}, \frac{1}{\alpha-\beta}\right) \\ \left(\frac{-\beta}{\alpha-\beta}, \frac{1}{\alpha-\beta}\right); (0,1)\end{matrix}\right] + \frac{EE_0}{E+E_0}\frac{(t/\tau)^{-\beta}}{\Gamma(1-\beta)} \tag{109}$$

with $C_2 = \left(E/\left(E+E_0\right)\right)^{1/(\alpha-\beta)}$.

3.4 Application to Experimental Data

In Sec. 2 we demonstrated that some special interconnected structures, called GGS, lead directly to power law relaxation, or, equivalently to fractional rheological constitutive equations. The unifying characteristics of different materials displaying such behaviors is the fact that their internal structural units are interconnected and that these units move cooperatively. From this point of view, it is interesting to test the applicability of the derived models for materials whose rheological properties are dominated by their strong connectivity.

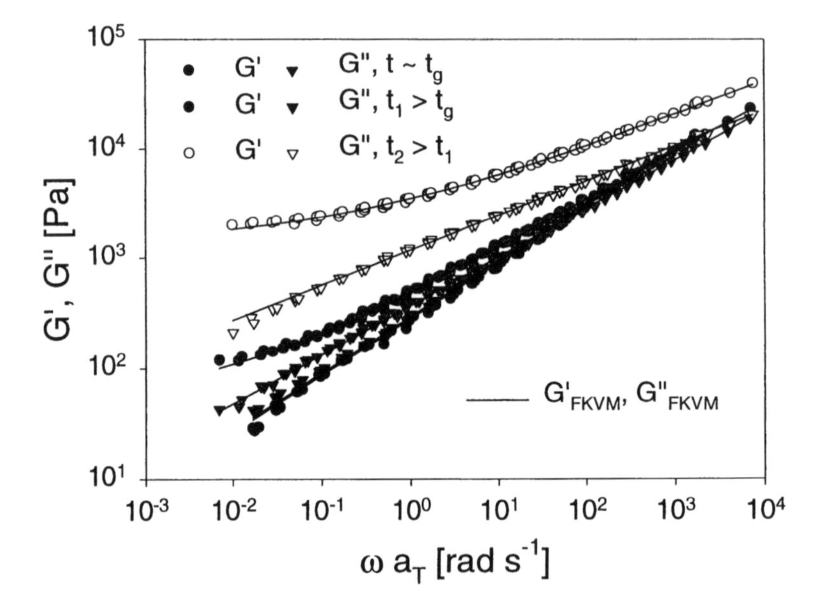

Figure 10: Storage modulus $G'(\omega a_T)$ and loss modulus $G''(\omega a_T)$ of crosslinked polydimethylsiloxane. The shade of the symbols indicates different degrees of crosslinking. The degree depends on the time of the reaction: t_g is the time where the gel point is reached; t_1 and t_2 are later times. The lines represent the fit to the data.

Table 1: Material parameters used in Fig. 10.

time	$S = E\tau^\alpha$, Pa s$^\alpha$	α, –	E_0, Pa
t_g	407	0.51	0
$t_1 > t_g$	572	0.45	52
$t_2 > t_1$	2461	0.31	1340

Gels are obtained either by polymerization of multifunctional monomers or by crosslinking of already formed strands [57]. Such gels correspond closely to GGS. We display in Fig. 10 the dynamical moduli G' and G'' as a function of the reduced frequency, ωa_T, for polydimethylsiloxane (PDMS) at different crosslinking stages [58]. Thus the situation at which the gel structure is attained for the first time is given in black symbols, whereas the situation at two later times is indicated in gray and white. As to be expected, we find at the gelation point in a double-logarithmic plot a linear dependence of both moduli over the whole experimentally accessible domain. This behavior can be fully represented by a single FE, Fig. 5(c). The corresponding expression is Eq. (66). On the other hand, the post-gel regime is characterized by the appearance of an elastic component, which leads to a plateau at low frequencies. In this domain one has to add to the FE a Hookian spring in parallel. This construction is a special case of the Kelvin–Voigt model, Eq. (88), where the order of one of the fractional derivatives is zero, say $\beta = 0$. As also shown in Fig. 10 this model reproduces the data very well; the parameters used are listed in Table 1.

A second example based on interconnected structures is a polymeric matrix (here polypropyleneoxide, PPO) in which a dibenzoylsorbitol (DBS) network is formed [59]. Such networks emerge due to the self-organizing tendency of DBS molecules in certain temperature ranges. Fig. 11 shows such a network. The PPO is the bright background and the grayish threads are fibrils of DBS. The dark points are DBS molecules which are not built in the fibrils and, therefore, form spheres. The interconnection of the fibrils to form a network is the main structural factor leading to the observed power law relaxation in the G' and G'' functions in Fig. 12. The data are best represented by a FE with parameters $\alpha = 0.038$ and $S = E\tau^\alpha = 119000$ Pa s$^\alpha$. In this frequency range the stress in the matrix is always relaxed and the observed behavior is exclusively due to the DBS network.

The third example also pertains to a branched structure [60]. Here a few branches, which are subunits of a syndiotactic polypropylene synthesised with a metallocene catalysator are randomly distributed along the backbone. From

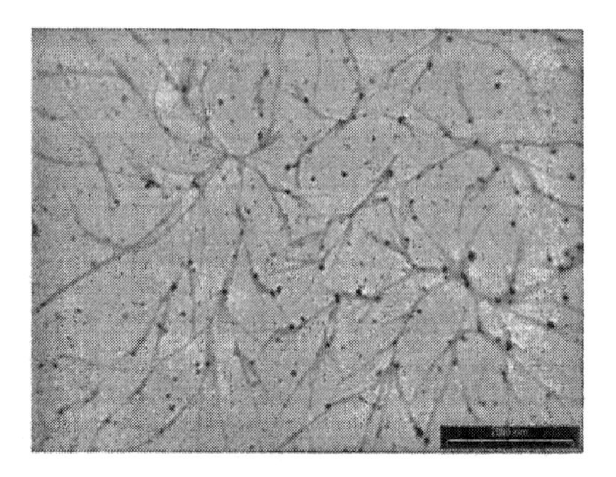

Figure 11: Transmission electron micrograph of a polypropyleneoxide (bright matrix) with dibencoyl sorbitol fibrils (darker network).

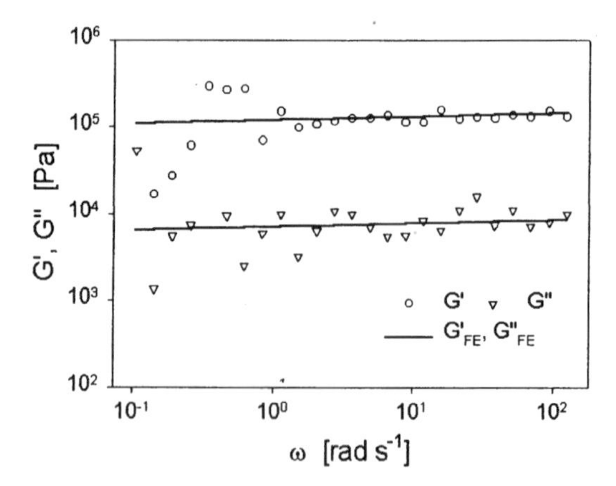

Figure 12: Storage modulus G' (ωa_T) and loss modulus G'' (ωa_T) of polypropyleneoxidcontaining 1 % of dibencoyl sorbitol at $T = 30°C$. The lines represent fits based on a fractional element.

Table 2: Material parameters used in Fig. 14–16.

polymer	τ, s	E, 10^{-5}Pa	α	β
EB64	0.062	3.47	0.69	0.10
EB80	0.048	5.68	0.56	0.06
EB88	0.057	7.24	0.52	0.05

a conformational point of view, such a polymer is classified as "long chain branched" [61]. Such materials flow and, therefore, show a terminal relaxation region, characterized by $G' = \omega^2$ and $G'' = \omega^1$. This behavior is also displayed in Fig. 13. In our case we were able to approach this terminal region experimentally by judiciously combining different techniques (here: dynamic mechanical measurements and creep measurements). On the other hand, in the plateau and in the intermediate ranges we are able to reproduce the observed relaxation by a fractional Maxwell model (FMM). The description of the whole curve by constitutive equations with fractional derivatives is also possible [42,62,63], but beyond the scope of this chapter. The results obtained through the FMM are depicted in Fig. 13 by continuous and dashed lines. Note that the fit of the model to the data is very good in the frequency range considered. For the fit we have taken $E = 4.36 \times 10^5$ Pa, $\tau = 0.242$ s, $\alpha = 0.610$ and $\beta = 0.090$.

Similar features are observed for a poly(ethen-co-1-butene) (EB) copolymer [60]. The storage and loss moduli of three samples with different amounts of ethene incorporation (64, 80 and 88 mol %) are displayed in Figs. 14–16. Only in this range of compositions such a behavior can be observed; we also attribute it to long chain branching. In all three cases the data are well described by the FMM; the corresponding material parameters are given in Table 2.

The fourth example relates to filled polymers. Here we have to use mechanical models composed of three fractional elements. The experimental results are taken from Ref. [64]. Here silicagel nonoparticles are immersed in a polystyrene (PS) matrix. These particles have an average diameter of about 200 nm and are clusters of primary structural units. The primary units are silica spheres of about 15 nm. Such fillers possess an enormous specific interfacial area, which attracts a considerable amount of matrix polymer. This polymeric material can connect the particles, and hence may create a network at quite high particle concentrations. Fig. 17 shows a s.e.m. image of such a compound with 1 vol % of filler. One can recognize the aggregates sitting in the centre of the "nests" formed by the immobilized polymer. The bright rim of the nest is formed from polymer material which disentangles during the process of frac-

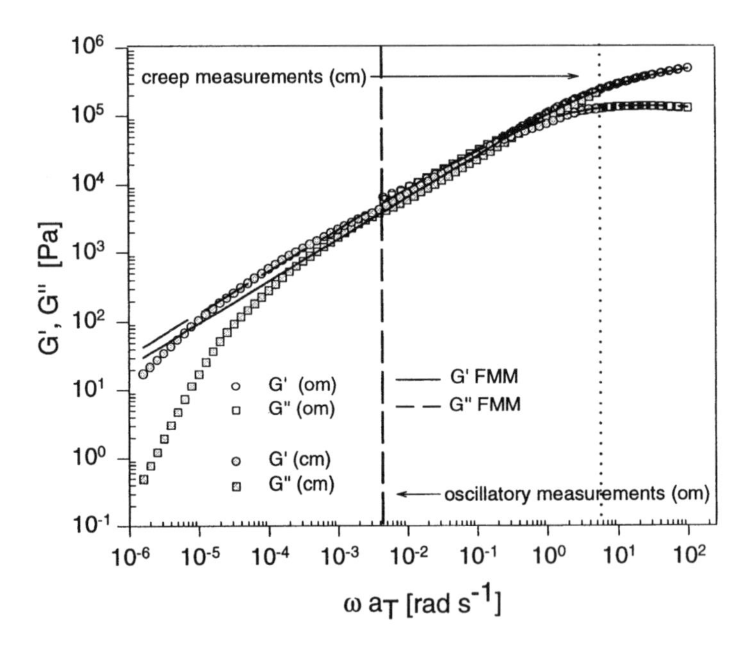

Figure 13: Storage modulus $G'(\omega a_T)$ and loss modulus $G''(\omega a_T)$ of syndiotactic polypropylene determined by oscillatory and creep measurements. The vertical lines indicate the frequency range of the corresponding technique of measurements and the reference temperature is $T = 190°C$. The lines represent fits based on the fractional Maxwell element.

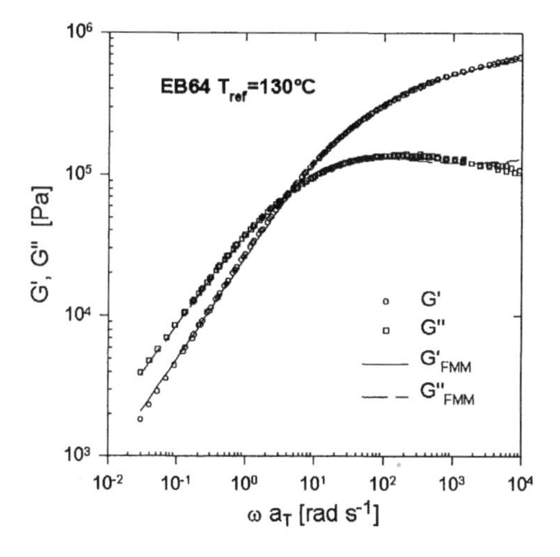

Figure 14: Storage modulus $G'(\omega a_T)$ and loss modulus $G''(\omega a_T)$ of ethene–co–1–butenecopolymer with 64 mol % ethene. The lines represent the fits based on the fractional Maxwell element.

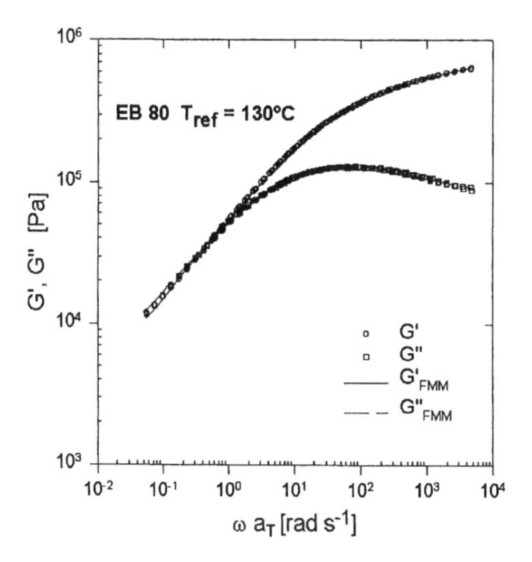

Figure 15: Same as Fig. 14, but with 80 mol % ethene.

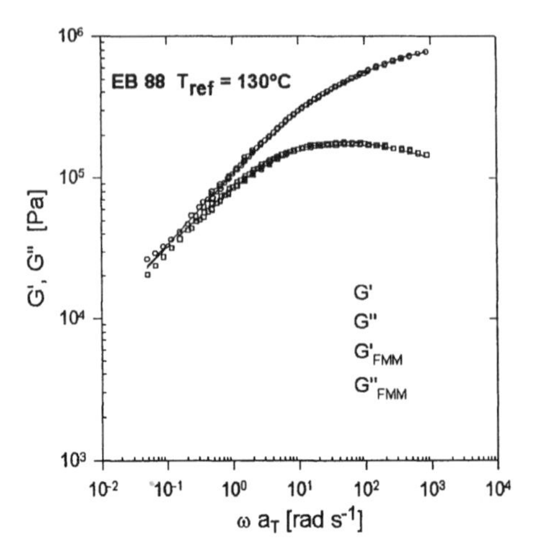

Figure 16: Same as Fig. 14 but with 88 mol % ethene.

Table 3: Material parameters used in Fig. 18.

conc., vol %	t, s	E, 10^{-5}Pa	α	β	E_0, Pa
3	0.087	1.1	0.98	0.20	20
4	0.110	1.51	0.97	0.20	44
5	0.102	1.82	0.93	0.20	532
7	0.089	3.31	0.77	0.17	4460

ture. Fig. 18 shows the storage moduli, G', of polystyrene filled with different amounts of silicagel as a function of the reduced frequency ωa_T. The material functions of these compounds follow a power law in a range intermediate between the plateau region ($\omega a_T > 10$ rad/s) and the low frequency plateau. For modelling purposes we thus have to add a Hookian element in parallel to the two FEs of the FMM. Such a model renders the data very well, as can be seen from the good agreement obtained in Fig. 18. The corresponding material parameters are given in Table 3. We hasten to note that similar results were found by Metzler et al. [56] in describing through RCEs the rheological behavior of filled polymers.

As a last example we display in Fig. 19 the creep behavior of a polypropylene exposed to a constant shear load at room temperature [65]. Under these conditions the material is in the transitional region between the entanglement

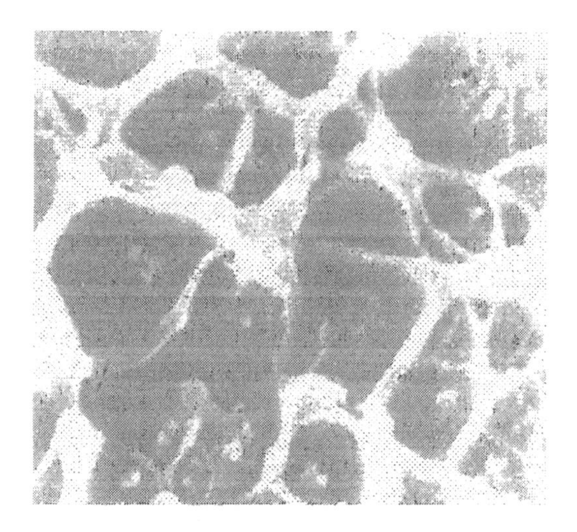

Figure 17: Scanning electron micrograph of a polystyrol filled with 1 vol% silicagel R974.The image represents an area of 1 mm^2.

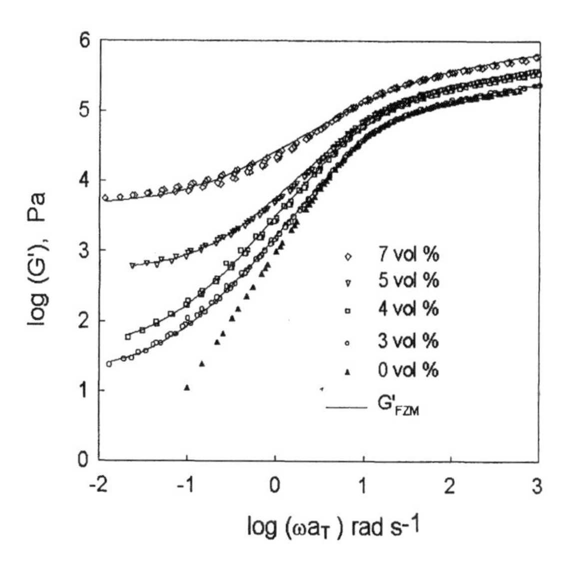

Figure 18: Storage modulus $G'(\omega a_T)$ of silicagel filled polystyrene at $T = 190°C$. The lines represent fits based on the fractional Zener model.

plateau and the glass region. Here only some segments of the polymeric backbone are involved in the relaxational processes. Now, it is well known that such relaxational processes are highly cooperative and, therefore, may lead to a power law behavior for $J(t)$. To describe the experimental results by an appropriate combination of FEs we may view the data as arising as a combination of two distinct patterns. The first pattern is an S-shaped transition from a lower plateau around $J_0 \approx 1.5 \times 10^{-10}$ Pa^{-1} to an upper plateau at $J_1 \approx 4.5 \times 10^{-10}$ Pa^{-1}, whereas the second pattern is a steep increase in the compliance for times longer than 10 s. The S-shaped transition is well represented by the fractional Zener model (FZM), Fig. 8(b). The creep behavior of the FZM is given by Eq. (101) which is valid only under restricted conditions. The presence of the second plateau demands $\gamma = 0$ and we can use Eq. (101) when we set $\beta = 0$. In this case Eq. (101) degenerates to the following expression, which was already suggested elsewhere [55]:

$$J(t) = J_1 - (J_1 - J_0)\, E_{\alpha,1}\left[-(t/\tau_1)^{\alpha}\right], \tag{110}$$

with $J_0 = E^{-1}$ and $J_1 = (E + E_0)^{-1}$. The generalized Mittag-Leffler function $E_{\alpha,1}$ can be calculated based on Padé-approximants [55]. The second pattern can be represented by a simple FE whose creep behavior is given by $J_2 t^{\delta}$. The description of the whole data set is possible by combining these elements in series. Using $J_0 = 1.50 \times 10^{-10}$ Pa^{-1}, $J_1 = 4.21 \times 10^{-10}$ Pa^{-1}, $J_2 = 0.43 \times 10^{-10}$ Pa^{-1}, $\tau = 2.9 \times 10^{-4}$ s, $\alpha = 0.282$ and $\beta = 0.351$ we attain a very good fit. This and all the other results presented here show convincingly that models containing various numbers of FEs allow to reflect quantitatively the rheological behavior of highly interconnected materials.

4 Conclusion

In this chapter we have discussed different applications of fractional calculus to polymer physics and rheology. Fractional expressions come naturally into play for systems which respond algebraically to external perturbations. We show that such situations occur on the microscopic level both for linear polymers (Rouse chains) and for networks (generalized Gaussian structures). Furthermore, we demonstrated rigorously how such behaviors translate into the macroscopic level, which is usually accessible by rheological experiments. This explains why fractional constitutive equations are extremely successful in describing the viscoelastic properties of many polymeric materials. We presented several systems (linear, interconnected and branched structures, gels and filled polymers), for which the fractional expressions lead to excellent fits for the observed relaxation behaviors.

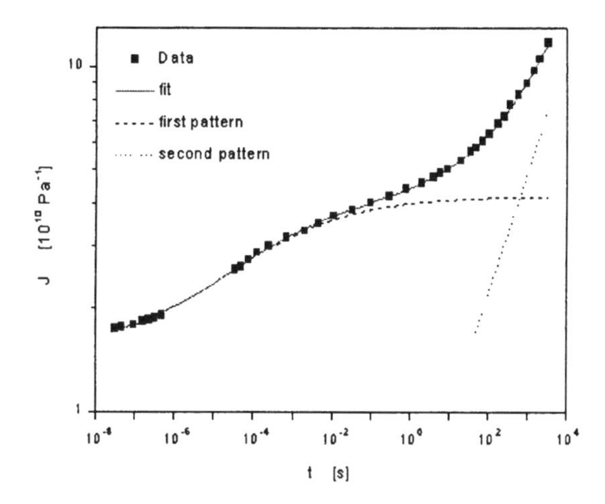

Figure 19: Creep curve of polypropylene at room temperature. The solid line represents a fit using Eq. (111). The dashed lines correspond to the fractional Zener model (left branch, first pattern) and to a simple fractional element (right branch, second pattern).

Acknowledgments

The authors thank Prof. P. Pincus for discussions. The support of the Fonds der Chemischen Industrie and of the DFG (through SFB 428, Graduiertenkolleg, and a stipend to H.S.) are thankfully acknowledged.

References

1. H. Scher and E.W. Montroll, *Phys. Rev.* B **12**, 2455 (1975).
2. H. Schnörer, H. Domes, A. Blumen, and D. Haarer, *Phil. Mag. Lett.* **58**, 101 (1988).
3. R.D. Armstrong and R.A. Burnham, *J. Electroanal. Chem.* **72**, 257 (1976).
4. K.S. Cole and R.H. Cole, *J. Chem. Phys.* **9**, 341 (1941).
5. A.K. Jonscher, *Nature* **267**, 673 (1977).
6. E. Kjartansson, *J. Geophys. Res.* **84**, 4737 (1979).
7. R. Richert and A. Blumen (eds.), *Disorder Effects on Relaxational Processes: Glasses, Polymers, Proteins* (Springer, Berlin, 1994).
8. K.B. Oldham and J. Spanier, *The Fractional Calculus* (Academic Press, New York, 1974).

9. K.S. Miller and R. Ross, *An Introduction to the Fractional Calculus and Fractional Differential Equations* (Wiley, New York, 1993).
10. T.T. Perkins, D.E. Smith, and S. Chu, *Science* **264**, 819 (1994).
11. T.T. Perkins, S.R. Quake, D.E. Smith, and S. Chu, *Science* **264**, 822 (1994).
12. T.T. Perkins, D.E. Smith, R.G. Larson, and S. Chu, *Science* **268**, 83 (1995).
13. D. Wirtz, *Phys. Rev. Lett.* **75**, 2436 (1995).
14. F. Amblard, A.C. Maggs, B. Yurke, A.N. Pergellis, and S. Leibler, *Phys. Rev. Lett.* **77**, 4470 (1996).
15. F. Brochard-Wyart, *Europhys. Lett.* **26**, 511 (1994).
16. F. Brochard-Wyart, *Europhys. Lett.* **30**, 387 (1995).
17. S. Manneville, P. Cluzel, J.-L. Viovy, D. Chatenay, and F. Charon, *Europhys. Lett.* **36**, 413 (1996).
18. P.E. Rouse, *J. Chem. Phys.* **21**, 1272 (1953).
19. M. Doi and S.F. Edwards, *The Theory of Polymer Dynamics* (Claredon Press, Oxford, 1995).
20. H. Schiessel, G. Oshanin, and A. Blumen, *J. Chem. Phys.* **103**, 5070 (1995).
21. H. Schiessel, G. Oshanin, and A. Blumen, *Macromol. Theory Simul.* **5**, 6036 (1996).
22. J.-U. Sommer and A. Blumen, *J. Phys.* A **28**, 6669 (1995).
23. D. Dhar, *J. Math. Phys.* **18**, 577 (1977); S. Alexander and R. Orbach, *J. Physique* **43**, L625 (1982); R. Hilfer and A. Blumen, *J. Phys.* A **17**, L537 (1984).
24. S. Havlin and A. Bunde in *Fractals and Disordered Systems*, eds. A. Bunde and S. Havlin (Springer, Berlin, 1991).
25. H. Schiessel, submitted to *Phys. Rev. E.*
26. P.G. Higgs and J.F. Joanny, *J. Chem. Phys.* **94**, 1543 (1991).
27. A.V. Dobrynin and M. Rubinstein, *Journal de Physique II* **5**, 677 (1995).
28. Y. Kantor and M. Kardar, *Phys. Rev.* E **51**, 1299 (1995).
29. Th. Soddemann, H. Schiessel, and A. Blumen, *Phys. Rev. E* in press.
30. H. Schiessel and A. Blumen, *J. Chem. Phys.* **105**, 4250 (1996).
31. H. Schiessel and A. Blumen, *Macromol. Theory Simul.* **6**, 103 (1997).
32. A.V. Dobrynin, M. Rubinstein, and J.-F. Joanny, *Macromolecules* **30**, 4332 (1997).
33. H. Schiessel and A. Blumen, *J. Chem. Phys.* **104**, 6036 (1996).
34. R.G. Winkler and P. Reineker, *J. Chem. Phys.* **106**, 2841 (1997).
35. D. Loomans, H. Schiessel, and A. Blumen, *J. Chem. Phys.* **107**, 2636 (1997).

36. H. Schiessel, I.M. Sokolov, and A. Blumen, *Phys. Rev. E* **56**, R2390 (1997).
37. G.H. Köhler and A. Blumen in *MATH/CHEM/COMP 1988*, ed. A. Graovac, Elsevier, Amsterdam, *Studies Phys. Theor. Chem.* **63**, 339 (1989).
38. I.M. Ward, *Mechanical Properties of Solid Polymers* (Wiley, Chichester, 1983).
39. N.W. Tschoegl, *The Phenomenological Theory of Linear Viscoelastic Behavior* (Springer, Berlin, 1989).
40. J.-U. Sommer, M. Schulz, and H. Trautenberg, *J. Chem. Phys.* **98**, 7515 (1993); J.-U. Sommer, T.A. Vilgis, and G. Heinrich, *J. Chem. Phys.* **100**, 9181 (1994).
41. M.E. Cates, *J. Physique* **46**, 1059 (1985).
42. Chr. Friedrich, H. Schiessel, and A. Blumen in *Advances in the Flow and Rheology of Non-Newtonian Fluids*, eds. D.A. Siginer, D. De Kee, and R.P. Chhabra (Elsevier, Amsterdam, in press).
43. C. Friedrich, *Rheol. Acta* **30**, 151 (1991).
44. W.G. Glöckle and T.F. Nonnenmacher, *Macromolecules* **24**, 6426 (1991).
45. H. Schiessel, R. Metzler, A. Blumen, and T.F. Nonnenmacher, *J. Phys.* A **28**, 6567 (1995).
46. H. Schiessel and A. Blumen, *J. Phys.* A **26**, 5057 (1993).
47. H. Schiessel and A. Blumen, *Macromolecules* **28**, 4013 (1995).
48. H. Schiessel, P. Alemany, and A. Blumen, *Progr. Colloid Polym. Sci.* **96**, 16 (1994).
49. N. Heymans and J.-C. Bauwens, *Rheol. Acta* **33**, 210 (1994).
50. N. Heymans, *Rheol. Acta* **35**, 508 (1996).
51. H. Schiessel and A. Blumen, *Fractals* **3**, 483 (1995).
52. T.F. Nonnenmacher in *Rheological Modeling: Thermodynamical and Statistical Approaches*, eds. J. Casas-Vazquéz and D. Jou, Springer, Berlin, *Lecture Notes in Physics* **381**, 309 (1989).
53. A.M. Mathai and R.K. Saxena, *The H-function with Applications in Statistics and Other Disciplines* (Wiley Eastern, New-Delhi, 1978).
54. A. Erdélyi (ed.), *Bateman Manuscript Project, Higher Transcendental Functions*, volume III, (McGraw-Hill, New York, 1955).
55. C. Friedrich and H. Braun, *Rheol. Acta* **31**, 309 (1992).
56. R. Metzler, W. Schick, H.-G. Kilian, and T.F. Nonnenmacher, *J. Chem. Phys.* **103**, 7180 (1995).
57. H.H. Winter and M. Mours, *Adv. Polym. Sci* **134**, 165 (1997).
58. F. Chambon and H.H. Winter, *Polym. Bull.* **13**, 499 (1985).
59. K. Fuchs, R.-D. Meier, M. Fahrländer, and Chr. Friedrich, *in prepara-*

tion.
60. A. Eckstein, J. Suhm, O. Meincke, R. Mühlhaupt, and Chr. Friedrich, *in preparation.*
61. J.D. Ferry, *Viscoelastic Properties of Polymers* (John Wiley, New York, 1980).
62. Chr. Friedrich, *Phil. Mag. Letters* **66**, 287 (1992).
63. Chr. Friedrich and H. Braun, *Colloid Polym. Sci.* **272**, 1536 (1994).
64. Chr. Friedrich and H. Dehno in: *Progress and Trends in Rheology IV* (Steinkopfverlag, Darmstadt, 1994), p. 45.
65. B.E. Read, G.D. Dean, and P.E. Tomlins, *Polymer* **29**, 2159 (1988).

CHAPTER VIII

APPLICATIONS OF FRACTIONAL CALCULUS TECHNIQUES TO PROBLEMS IN BIOPHYSICS

THEO F. NONNENMACHER, RALF METZLER
Department of Mathematical Physics, University of Ulm,
Albert-Einstein-Allee 11, 89069 Ulm/Donau, Germany

Contents

1 Introduction

Biomedical research deals with the structure and function, and all the activities of living systems such as organs, tissues or cells [67,97]. The high level of organisation necessary for live is only possible in a macroscopic (or mesoscopic) system—otherwise the order would be destroyed by microscopic fluctuations[3,4,97].

Modelling structures and functions in biomedical matter is, as elsewhere in science, to obtain simple models that capture the essential features of the structure or dynamical processes under investigation. Many observations of structures and functions of biophysical systems show macroscopic phenomena that apparently lack a characteristic length or time scale (or both). Such phenomena are usually modelled by ***allometric*** [a] laws. Power–law scaling is the most prominent example of an allometric law.

A common feature of all these objects and processes lacking a characteristic scale, which Mandelbrot called ***fractals*** [51] is, for instance, that the shape of the boundaries of biological cells is very complex (irregular) and that the dynamics going on in a protein during ion channel gating, or other dynamical processes, may also be modelled by an irregular (fractal) energy landscape[24,32,64,97]. We will discuss some examples in this review. Here, however, we already comment that, in contrast to ideal mathematical fractals like Koch curves, Cantor sets *etc*, all natural objects show scaling (fractal) properties only between limits (scaling windows) [19,49,66,67]. Over the last years, methods have been discussed in order to find the turnover points and to classify natural fractals into asymptotic [76] and bi–asymptotic [19] fractals.

This review is mainly concerned with the dynamical complexity of physical and biophysical systems. Typical examples that will be discussed here, are relaxation and diffusion processes. The standard (Debye) relaxation process, for example, is formulated via an initial value problem

$$\tau \frac{d\Phi(t)}{dt} = -\Phi(t), \quad t > 0 \quad \text{and } \Phi(0) = \Phi_0. \tag{1}$$

The solution is an exponential decay

$$\Phi(t) = \Phi_0 e^{-t/\tau} \tag{2}$$

for $t \geq 0$, and $\Phi(t) = 0$ for $t < 0$. The same initial value problem, Eq. (1) can

[a] *Greek* $\alpha\lambda\lambda o\mu\varepsilon\tau\rho o\nu$.

also be formulated by integration of (1), leading to

$$\Phi(t) - \Phi_0 = -\tau^{-1}\frac{\mathrm{d}^{-1}\Phi(t)}{\mathrm{d}t^{-1}} \equiv -\tau^{-1}\int_0^t \mathrm{d}t'\,\Phi(t'). \tag{3}$$

This integral equation incorporates the initial value Φ_0 and can be used as a starting point for a formulation of a fractional initial value problem [28,84]. In Section 3 we will come back to this point. The integral equation (3) can formally be extended into a fractional integral equation by replacing $\tau^{-1}\frac{\mathrm{d}^{-1}\Phi(t)}{\mathrm{d}t^{-1}} \to \tau^{-\beta}\frac{\mathrm{d}^{-\beta}\Phi(t)}{\mathrm{d}t^{-\beta}}$, which leads to

$$\Phi(t) - \Phi_0 = \tau^{-\beta}\,{}_0D_t^{-\beta}\Phi(t), \tag{4}$$

where we have introduced the notation $\mathrm{d}^{-\beta}/\mathrm{d}t^{-\beta} \stackrel{\mathrm{def}}{=} {}_0D_t^{-\beta}$ for the Riemann–Liouville (RL) operator (see Sec. 2). In Section 3 we will give some physical motivation and derivation of the fractional relaxation integral equation (4).

After a veritable boom of fractal geometry in mathematics and physics, it has meanwhile become a widely used characterisation method in biology and medicine in order to define and specify structural and functional parameters[67]. On the other hand, fractional calculus has always been overshadowed by this development. Despite the book by Oldham and Spanier in 1974 [69] and its intuitive relation to power–laws, the mathematical difficulties in handling fractional equations have prevented a broader use of these techniques, especially in experimental branches. There are no nice colour–plates for fractional integrals.

It is the goal of this review to present the usefulness of fractional ideas in relaxation and diffusion processes in rheology, biology, and complex systems, in general. The resulting equations exhibit a rich behaviour in dependence on only a few parameters, and they are still *linear*, *i.e.* they can be solved via integral transforms. The solution procedure reveals the wide class of Fox' H–functions as perfectly suited to solve these equations. Fox functions have been introduced, to our knowledge, as exact representations of Lévy stable distribution by Schneider [83]. Schneider and Wyss established the Fox function method to the description of anomalous diffusion [84,99]. The method was explicitly applied to relaxation problems in rheology by Glöckle and Nonnenmacher[28,29,30], and to self–similar biophysical processes [31,32]. Further applications in diffusion theory were reported by Metzler *et al* [54,56] and by Hilfer and Anton [38]. Explicit solutions of fractional wave and diffusion equations were given by Schneider and Wyss [84,99] and, more recently, by Metzler *et al* [54,56]. Many examples discussed within the context of rheology came up with Mittag–Leffler (ML) functions, representing a special subclass of Fox functions. Here, however, we demonstrate that subclasses, or irreducible classes, of Fox functions different from the ML function do represent exact analytical solutions of fractional diffusion and wave equations [18,54,56,68,84,99].

In this chapter we therefore want to present applications of fractional calculus to bio–medical processes. A large part of this chapter, however, will be devoted to the overall introduction of fractional integrals and derivatives, and their physical motivation. We will show, how they come about from easy considerations. Mathematics will be kept on a fairly modest niveau.

2 Fractional calculus

2.1 Preliminary remarks

Standard differential equations contain differentials of integer order. Fractional operators like $\mathrm{d}^\alpha/\mathrm{d}t^\alpha$, on the other hand, are still defined when α, $\mathrm{Re}\{\alpha\} \in \mathbb{R}^+$, is an arbitrary number (rational, real– or even complex–valued). Fractional calculus ideas go back to Leibniz, but it was not before the end of the last century that the first applications of the «Heaviside operational calculus» accelerated the development of the fractional calculus theory (an historical survey of this topic is given in References [58,69,79].

In general, the differentiation of a power can be obtained heuristically from the result for integer n and m:

$$\frac{\mathrm{d}^n x^m}{\mathrm{d}x^n} = \frac{m!}{(m-n)!} x^{m-n} \tag{5}$$

by introducing the gamma function replacing the factorial. Equation (5) then reads:

$$\frac{\mathrm{d}^n x^m}{\mathrm{d}x^n} = \frac{\Gamma(m+1)}{\Gamma(m-n+1)} x^{m-n}. \tag{6}$$

This allows for a generalisation to arbitrary order $\nu, \mu \in \mathbb{R}$:

$$\frac{\mathrm{d}^\nu x^\mu}{\mathrm{d}x^\nu} = \frac{\Gamma(\mu+1)}{\Gamma(\mu-\nu+1)} x^{\mu-\nu}. \tag{7}$$

This heuristic extension is consistent with the Riemann–Liouville calculus defined below. In principle, relation (5) could be used to differentiate analytic functions.

It is an intrinsic property of fractional calculus that the differentiation of a power–law (of integer, non–integer or zeroth order) results in another power–law. The connection of power–laws to real systems which will often be encountered throughout the text makes fractional equations an excellent candidate for a well–suited phenomenological description of natural, *i.e.* non–ideal systems.

2.2 Definition(s) of fractional differ–integrals

Cauchy's n–fold ($n \in \mathbb{N}$) integral

$$\begin{aligned} {}_aD_t^{-n} f(t) &= \int_a^t \mathrm{d}t_{n-1} \int_a^{t_{n-1}} \mathrm{d}t_{n-2} \cdots \int_a^{t_1} \mathrm{d}t_0 f(t_0) \\ &= \frac{1}{(n-1)!} \int_a^t \mathrm{d}t' (t-t')^{n-1} f(t') \end{aligned} \tag{8}$$

can be generalised to real order $q \in \mathbb{R}^+$ by introduction of the Gamma function:

$${}_aD_t^{-q} f(t) \stackrel{\text{def}}{=} \frac{1}{\Gamma(q)} \int_a^t \mathrm{d}t' \frac{f(t')}{(t-t')^{1-q}}. \tag{9}$$

Eq. (9) defines the fractional integral operator ${}_aD_t^{-q}$. A fractional differentiation is defined as the succession of a fractional integration and a standard differentiation ($n \in \mathbb{N}$, $n > q$) as follows:

$${}_aD_t^q f(t) \stackrel{\text{def}}{=} \frac{\mathrm{d}^n}{\mathrm{d}t^n} \left({}_aD_t^{q-n} f(t) \right). \tag{10}$$

Further on, we will make use of the *Riemann-Liouville* definition basing on the choice $a = 0$ in Eqs. (9) and (10). We will also meet *Weyl's* definition ${}_{-\infty}D_t^r$ (according to Oldham and Spanier [69]).

As an example, let us consider the Riemann–Liouville differ–integration of the simple exponential function:

$$\begin{aligned} {}_0D_t^\nu \exp t &= {}_0D_t^\nu \sum_{m=0}^\infty \frac{t^m}{\Gamma(m+1)} = \sum_{m=0}^\infty \frac{t^{m-\nu}}{\Gamma(m-\nu+1)} \\ &= \frac{t^{-\nu}}{\Gamma(1-\nu)} \, {}_1F_1\left(1, 1-\nu, t\right) \end{aligned} \tag{11}$$

which reveals the confluent hypergeometric function ${}_1F_1$. On the other hand, the Weyl differ–integration yields

$$ {}_{-\infty}D_t^{\mu} \exp t = \exp t, \tag{12} $$

for arbitrary $\nu, \mu \in \mathbb{R}$.

2.3 Power–law relations and asymptotic fractals

Often, one encounters functional relations in the frequency domain of the form $\omega^{-\alpha}F(\omega)$, for example $\omega^{-\alpha}\exp(-\omega\tau)$. Such relations, back–transformed to time space, reveal, by the generalised differentiation theorem,

$$ \omega^{-\alpha}F(\omega) \to {}_0D_t^{-\alpha}F(t) \tag{13} $$

a fractional integral. Depending on the experiment, see the discussion in Sec. 3.1, we could also find a Weyl operator in time space. Thus, fractional expressions can arise quite naturally from functional relations observed in the so–called spectral domains, e.g. in the measurement of dynamical structure factors, or in the interpretation of a time series [64] in terms of a waiting time distribution that represents a typical example of an asymptotic fractal of the form

$$ G(t) \sim (t/\tau)^{-\beta}\gamma(\beta, t/\tau) \tag{14} $$

where $\gamma(a,t)$ is the incomplete Gamma function. In Section 3.2 we will come back to this point.

2.4 Some comments on the definitions

Like the Cauchy multiple integral, the fractional integral (9) represents a convolution of a power–law with the function $f(t)$: ${}_aD_t^{-q}f(t) = t^{q-1} * f(t)/\Gamma(q)$. For $a = 0$ (Riemann–Liouville) we have a Laplace, in the case $a = -\infty$ (Weyl) it is a Fourier convolution. Therefore, Laplace or Fourier transforms will be an important tool in solving fractional equations. This is due to the convolution theorem typical for these transforms, a convolution being transformed into the product of the simple product of the convoluted functions: $f(t) * g(t) \to \tilde{f}(u)\tilde{g}(u)$. Thus it is not surprising that convolution algebra plays a central part in dealing with fractional calculus problems [33]. The convolution can be interpreted as a memory integral (see below). From this point of view, the memory kernel is the power–law t^{q-1}. A typical property of a power–law is its scale invariance. Let $y(t) = t^{\beta}$, then $y(\lambda t) = (\lambda t)^{\beta} = \lambda^{\beta}y(t)$: A scaling of the time axis results in a simple scaling of the ordinate y. The law $y(t)$ is not altered by changing scales. In the log–log plot this means a shift $\log\lambda$ on the $\log t$ axis, and a shift $\beta\log\lambda$ in the ordinate. Thus, having given only one point in the diagram, one can use this property to re–construct the whole power–law. This scale invariance, however, is the cause for the stability of systems following power–laws. As exterior disturbances are mostly of a single scale (a given wavelength for example), they cannot much affect the system.

3 Relaxation processes

Standard relaxation, often referred to as Maxwell–Debye relaxation, see Eqs. (1) leads to the exponential relaxation function (2) with a characteristic time τ. Relaxation processes showing deviations from this classical Maxwell–Debye behaviour are referred to as Non–Debye or non–exponential relaxation processes and include dielectric relaxation, stress relaxation, stress–strain relations, NMR relaxation or diffusion–controlled relaxation in liquids, liquid crystals, polymer melts and solutions, amorphous polymers, rubbers or biopolymers. The deeper physical understanding of non–exponential relaxation processes belongs to the problems in physics which have not been completely resolved, as well as other topics occurring in the description of complex materials: the glass transition, the Vogel–Tamann–Fulcher (VTF) behaviour for viscosity or the Williams–Landel–Ferry (WLF) relation for viscoelasticity or the physical aging of polymers [26,60].

Typically, Non–Debye relaxation processes are described either by a Kohlrausch–Williams–Watts (KWW) decay

$$\Phi(t) = \Phi_0 \exp\{-(t/\tau)^\alpha\} \tag{15}$$

with $0 < \alpha < 1$, or via an asymptotic power–law (Nutting law)

$$\Phi(t) = \Phi_0 \frac{1}{(1+t/\tau)^n} \tag{16}$$

with $0 < n < 1$. To our knowledge, the first formulation of a power–law relaxation can be found in 1729 when B G Buelfinger used it to describe the strain relaxation in materials like steel or stone. Later, this law became known as Bach's law of elasticity, 1888 [75].[b]

For a sufficiently large experimental window, it is, in general, possible to observe transitions from the KWW– to the Nutting–behaviour or vice versa.

[b] Concerning the KWW law several mechanisms have been shown to lead to the KWW law (15), including defect diffusion, hierarchically constrained dynamics, multipolar direct energy transfer [86], and low energy excitations with infrared divergencies [62].

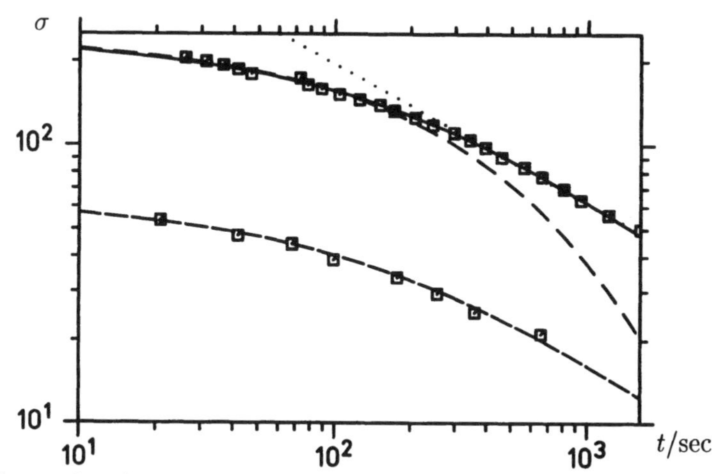

Figure 1: Stress relaxation at constant strain, for two different initial conditions: data from Ref. [85]. (- - -) stretched exponential (15), (···) Nutting law (16), and (—) Mittag–Leffler function according to Eq. (39c). The dimension of the stress σ is dyn/mm^2.

Such a situation is depicted in Figure 1, which has been presented and discussed this way for the first time in terms of Mittag–Leffler functions by Glöckle and Nonnenmacher [28,65]. This new result interpolates between both cases (15) and (16). Figure 2 shows the upper part of Figure 1 in a linear plot. Again, the interpolative property of the Mittag–Leffler function becomes clear. The Mittag–Leffler function plays a dominant rôle in fractional relaxation problems. Its appearance in rheology has also been observed by Friedrich [25], Caputo and Mainardi [13,50], and Glöckle and Nonnenmacher [28,65].

Non–exponential relaxation implies memory, *i.e.* the underlying fundamental relaxation processes are of Non–Markovian nature [47]. In Subsec. 3.1, it will be shown that a natural way to incorporate such memory effects is fractional calculus. Via the involved convolution integral in time, the present state is being influenced by all the states the system has been running through at the times $t' = 0 \ldots t$. The power–law kernel defining the fractional expression represents a particularly long memory.

3.1 Memory integrals

Glöckle and Nonnenmacher [30] showed that, within the linear response régime, the Zwanzig formalism involving the projection operator technique, leads to

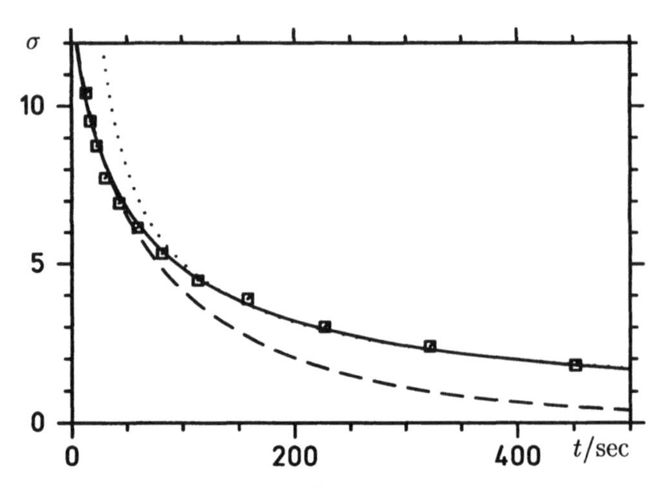

Figure 2: Upper part of Figure 1 in a linear scale.

an integral equation of the form

$$\frac{\mathrm{d}\Phi(t)}{\mathrm{d}t} = -\int_0^t \mathrm{d}\tau\, K(t-\tau)\Phi(\tau) \tag{17}$$

where the kernel $K(t)$ can be integrated in phase space. For a non–trivial K, equation (17) is called a *memory integral*, *i.e.* all instances from $\tau = 0$ to $\tau = t$ contribute to the situation at t. Several cases can be distinguished [30].

(i) The memory breaks down, *i.e.* the Markovian case is recovered, for

$$K(t) = K_0\delta(t). \tag{18a}$$

In this case, the memory equation (17) reduces to Eq. (1) and the exponential relaxation function

$$\Phi(t) = \Phi_0 \exp\{-K_0 t\} \tag{18b}$$

is found.

(ii) The opposite case with a constant memory,

$$K(t) = K_0, \tag{19a}$$

leads to an oscillating solution:

$$\Phi(t) = \Phi_0 \cos\left(\sqrt{K_0} t\right). \tag{19b}$$

(iii) For a slowly varying kernel which, for small times, behaves like

$$K(t) \sim t^{\gamma}, \tag{20a}$$

one can show [62] that the stretched exponential or KWW function

$$\Phi(t) = \Phi_0 \exp\left\{-K_0 t^{\gamma+2}\right\} \tag{20b}$$

results.

(iv) Finally, supposing

$$K(t) = K_0 t^{q-2}, \ (0 < q \leq 2) \tag{21a}$$

for $t > 0$, the fractional relaxation equation

$$\frac{\mathrm{d}\Phi(t)}{\mathrm{d}t} = -\tau^{-q} \, {}_0D_t^{1-q}\Phi(t), \ \ \tau^{-q} = K_0\Gamma(q-1) \tag{21b}$$

results. Applying ${}_0D_t^{-1}$ on both sides, leads to the fractional integral equation (4). Further application of ${}_0D_t^{q}$ brings us to the fractional differential equation

$${}_0D_t^{q}\Phi(t) - \frac{\Phi_0 t^{-q}}{\Gamma(1-q)} = -\tau^{-q}\Phi(t). \tag{22}$$

Thus we have *derived* the fractional integral and differential equations on the basis of Zwanzig's projections operator formalism for a power–law memory kernel. The solution of the fractional relaxation equation subjected to the initial condition $\Phi(0) = \Phi_0$ is given in Eq. (33*b*). In Fig. 3, this solution is graphed for various values of q.

Another physical basis to come up with a memory integral, is to start from Boltzmann's superposition principle which formally incorporates memory via the (phenomenological) causal convolution [94,96]:

$$\sigma(t) = \int_{-\infty}^{t} \mathrm{d}t' \, G(t-t') \frac{\mathrm{d}\epsilon(t')}{\mathrm{d}t'}. \tag{23}$$

The Boltzmann superposition integral, equation (23), holds for linear systems which are homogeneous in time; in it $\sigma(t)$ denotes the stress, $\epsilon(t)$ the strain and $G(t)$ the relaxation modulus, i.e. the response of the stress to a shear jump. Consider now a system whose stress decays after a shear jump in an algebraic manner, similar to equation (16). Its stress relaxation modulus obeys then:

$$G(t) = \frac{E}{\Gamma(1-\beta)} \left(\frac{t}{\tau}\right)^{-\beta} \tag{24}$$

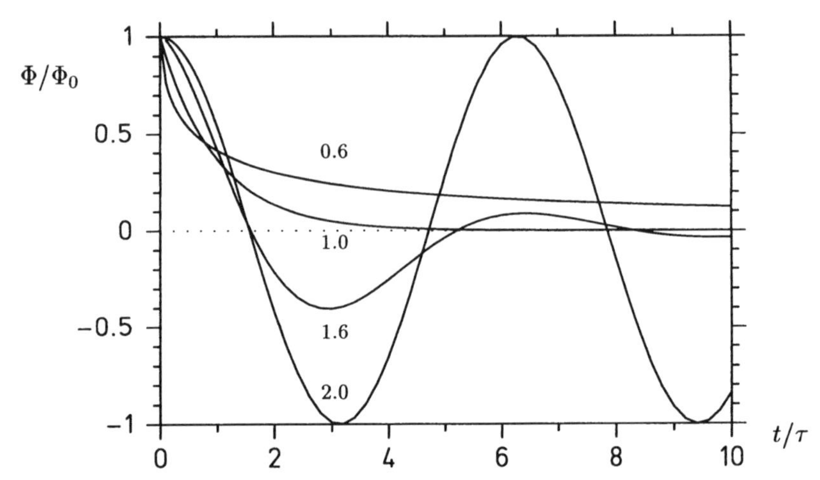

Figure 3: Fractional differentiation between zero and two: Solution 33*b* of the fractional relaxation equation 4 for various values of q in the range $0 < q \leq 2$.

where E and τ are constants and $\Gamma(x)$ is the complete Gamma function. For convenience in (24) the pre–factor is chosen in a way which matches our forthcoming definitions. Combining equations (23) and (24) one arrives at

$$\sigma(t) = \frac{E\tau^{\beta}}{\Gamma(1-\beta)} \int\limits_{-\infty}^{t} \mathrm{d}t' \, (t-t')^{-\beta} \frac{\mathrm{d}\epsilon(t')}{\mathrm{d}t'}. \tag{25}$$

The RHS of equation (25) represents a Weyl fractional integral (FI). We stop to note that Weyl's formalism follows naturally from the causal convolution, equation (25), in which the strain fields start in the distant past, so that $c \to -\infty$. On the other hand restricting the dynamics to positive times only, i.e. setting $\sigma = \epsilon \equiv 0$ for $t \leq 0$ leads to the Riemann–Liouville–formalism. This corresponds to an initial–value problem, in which the initial values must be specified [28,84].

Such a $G(t)$ can be found for hierarchical mechanical elements as discussed in [81]. If one chooses the constituting elements (springs and dashpots), to follow a scaling law in dependence of the running index k, one obtains expressions like equation (24), for the mechanical case, and similar expressions for the electromagnetic model. The basic ideas leading to memory integrals are also valid for the ones which will be encountered in anomalous diffusion theory later on.

3.2 Markovian chains and Bernoulli scaling in ion channelling

Markovian models assume that the probability distribution function for some events may be described by a sum of exponential functions

$$f_N(t) = \sum_{n=0}^{N} \frac{a_n}{\tau_n} e^{-t/\tau_n}. \tag{26}$$

Modelling dynamical processes of proteins, it is generally accepted that, according to their biological function, proteins assume (during their functional activity) a number of different conformational states. Transitions between these states are usually formulated as a kinetic or relaxation process characterised by kinetic or rate constants $k_n = \tau_n^{-1}$. The number N of time constants τ_n, $(n = 1, 2, \ldots, N)$ indicates how many conformational states of the energy landscape are taking part in the relaxation process.

As an example let us discuss patch clamp experiments for studying single ion channel gating kinetics. Such experiments provide important information on the dynamics of ion channel proteins. Generally ion channels are considered to be large integral membrane proteins which allow passive flux of ions through cell membranes. Currents recorded from ion channels show that the channels open and close their pores repeatedly during normal activity. Counting the number of channel events and plotting such a count versus duration time, one can construct a histogram which can be fitted by a probability density function $f_N(t)$ for the lifetime of the conducting state [80]. A typical measurement is shown in Fig. 4

The amplitudes a_n and the time constants τ_n, introduced in Eq. (26) are usually presumed (for the fitting procedure) to be independent. N counts the number of conformational states being active during the gating process. If the coefficients a_n and τ_n are not independent but correlated, then the Markovian property is destroyed. A simple model to introduce scaling properties by assuming correlations between the coefficients a_n and τ_n is based on Bernoulli scaling [87]. The central idea is this [59]: The probability of occurence of an event is inversely proportional to the time difference to the previous event. This scaling condition can be described mathematically by taking the choice $\tau_n = \tau\lambda^n$, $a_n = a_N p^n$, where a_N is a normalisation constant, $\lambda > 1$, $0 < p < 1$, and $p = \lambda^{-\mu}$ with $\mu = -\log p / \log \lambda > 0$. Introducing these Bernoulli scaling constraints into the series (26), leads to the result ($\xi \equiv t/\tau$):

$$f_N(t) = \sum_{n=0}^{N} g_n(\xi) \tag{27}$$

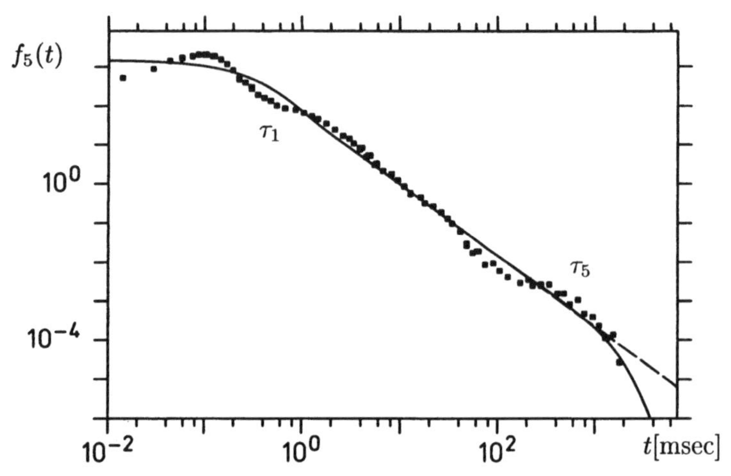

Figure 4: Measured data points from Blatz *et al* [2] of a Cl^- channel oscillate along the power–law trend (– – –). The full curve (—) approaches the power–law trend between the limits τ_1 and τ_5, and crosses over to the exponential decay for $t \geq \tau_5$.

where the g_n are given by:

$$g_n(\xi) = \begin{cases} \dfrac{a_N}{\tau} \lambda^{-(\mu+1)n} e^{-\xi\lambda^{-n}}, & 0 \leq n \leq N \\ 0 & \text{otherwise} \end{cases} \tag{28}$$

We notice that this series represents no more (and no less) than the original Markovian series (26) subjected to the constraints of Bernoulli scaling.

Making the transition to the continuum limit ($n \to x$) via the Poisson summation formula

$$\sum_{j=-\infty}^{\infty} f_j = \sum_{m=-\infty}^{\infty} \int_{-\infty}^{\infty} \mathrm{d}x \, f(x) e^{2\pi i m x} \tag{29}$$

one obtains the final result [64]

$$f_N(\xi) = \frac{a_N}{\tau \log \lambda} \sum_{m=-\infty}^{\infty} \xi^{-\nu_m} I_N(\nu_m, \xi/\lambda) + \mathcal{O}\left(e^{-\xi/\lambda}\right) \tag{30}$$

where $\xi = t/\tau$, $\nu_m = 1 + \mu - 2\pi i m/\log\lambda$, and

$$I_N(\nu_m, y) \stackrel{\text{def}}{=} \gamma(\nu_m, y) - \gamma\left(\nu_m, y\lambda^{-N+1}\right). \tag{31}$$

Here, $\gamma(\nu_m, y)$ is the incomplete Gamma function. In Fig. 5 the result (30) is graphed.

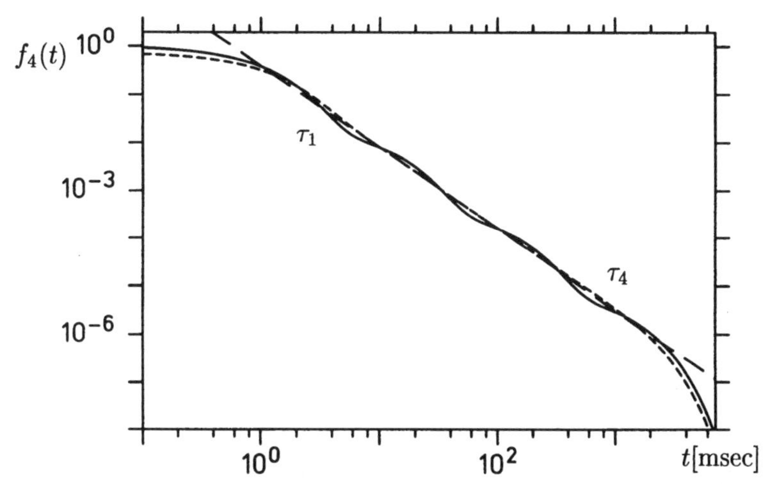

Figure 5: This plot demonstrates how the scaling region between the limits τ_1 and τ_4 emerges out of Eq. (30). The dashed curve (- - -) is calculated from Eq. (30) with only the $m = 0$ term, *i.e.* neglecting the oscillations. The full line takes several summands into account: $m = 0, \pm 1, \ldots, \pm 12$. The straight line (— —) indicates the ideal fractal behaviour without limits. We chose $\tau = 0.1$, $N = 4$, $\lambda = 10$ and $\mu = 0.7$.

3.3 Some comments on fractional relaxation equations

In principle, there are always two possible ways to generalise *ad hoc* a linear differential equation, the so-called I (integral) or the D (differential) form. In Sec. 3.1 and in Eq. (4), the I form has already been presented. On the other hand, the D form is obtained by simply replacing $\tau \mathrm{d}/\mathrm{d}t \to \tau^\beta \mathrm{d}^\beta/\mathrm{d}t^\beta$ in the standard Debye equation (1), leading to:

$$\tau^\beta D_t^\beta \Phi(t) = -\Phi(t). \tag{32}$$

An equation of this type was presented and discussed by Oldham and Spanier [69], by Tobolsky [93], by Friedrich [25], by Glöckle and Nonnenmacher [28], and was later discussed by Schiessel *et al* [82] within the context of a Weyl operator generalisation. However, we remark that (32) does not generalise the initial value problem (1). The correct fractional initial value problem is mathematically obtained via (4).

At this point, we shall not elaborate on the solution procedure. This is reported elsewhere [28,82]. It suffices to note that these equations can be transformed to purely algebraic ones in Laplace space which can be solved. The back-transform to time space now involves some awkward non-integer powers of the Laplace variable. The resulting function class is the generalised Mittag-Leffler function. For the sake of an integrated presentation, however, the Fox

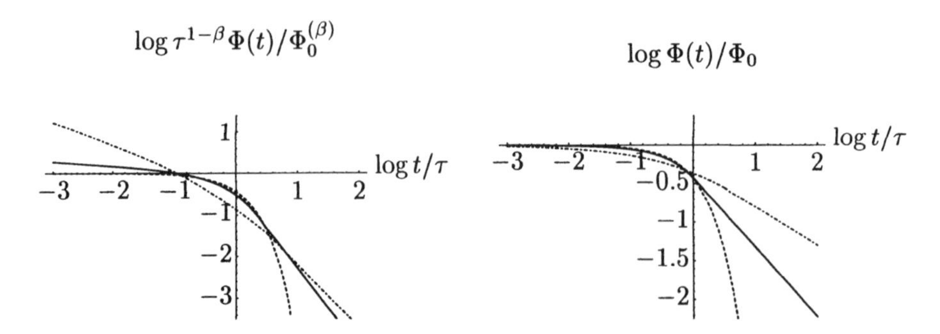

Figure 6: Relaxation function $\Phi(t)$ from *Left:* equation (32) and *Right:* equation (4) for $\beta = 1$ (Debye case) (- - -) $\beta = 0.9$ (—) and $\beta = 0.5$ (- · -).

function representation is more suitable. They also offer the advantage of easy mathematical manipulations [28,29,52,54,56,82,91].

For the two models, the final results which are graphed in Figure 6, can then be expressed in terms of Fox functions of the H^{11}_{12} type as follows:

$$\Phi(t) = \Phi_0 \beta^{-1} \tau^{\beta-1} H^{11}_{12}\left[\frac{t}{\tau}\,\middle|\, \begin{matrix}(1-\beta^{-1},\beta^{-1})\\ (1-\beta^{-1},\beta^{-1}),(0,1)\end{matrix}\right] \tag{33a}$$

for the D model (32), and

$$\Phi(t) = \Phi_0 \beta^{-1} H^{11}_{12}\left[\frac{t}{\tau}\,\middle|\, \begin{matrix}(0,\beta^{-1})\\ (0,\beta^{-1}),(0,1)\end{matrix}\right] \tag{33b}$$

for the I model (4). The asymptotic behaviour is, respectively,

$$\Phi(t) \sim \begin{cases} t^{\beta-1} & t \ll \tau \\ t^{-1-\beta} & t \gg \tau \end{cases} \tag{34a}$$

for the D model and

$$\Phi(t) \sim \begin{cases} \Phi_0 & t \ll \tau \\ t^{-\beta} & t \gg \tau \end{cases} \tag{34b}$$

for the I model. It is important to note, that D and I form exhibit some quite different behaviours. This can be seen in Fig. 6. Whilst the I form converges $\lim_{t\to 0+}\Phi(t) = \Phi_0$, the D form diverges for small times. This is a crucial point if an initial value problem is to be fulfilled. It is also worth noting, that only the

I form is consistent with above mentioned Zwanzig formalism, as shown in [30]. For completeness, we calculate the relaxation spectrum $A_\beta(k)$ of solution (33*b*), which is defined via the Laplace transform (a_0 is the normalisation constant):

$$\Phi(t) = a_0 \int_0^\infty \mathrm{d}k e^{-kt} A_\beta(k). \tag{35}$$

The result

$$A_\beta(k) = \frac{1}{(k\tau)} \times \frac{(k\tau)^\beta \sin(\pi\beta)}{(k\tau)^{2\beta} + 2(k\tau)^\beta \cos(\pi\beta) + 1} \tag{36}$$

is plotted in Fig. 7.

3.4 Fractional constitutive rheological models

To include, on an *ad hoc* phenomenological level, more subtle properties of rheological materials[c], rheological models were introduced quite early in 1868 (Maxwell's model) and 1890 (Kelvin's model). In rheology, one models the behaviour of a given system by the combination of Hookean springs and Trouton's (Newton's) dashpots, as shown in Figure 8a and b. The whole entity of above mentioned hierarchical models [81], shall be symbolised with the ladder drawn in Figure 8c, following [82]. It will be used to construct rheological models in the same spirit as before with elements a and b.

In this paragraph we limit ourselves to a brief introduction into this topic. A detailed discussion is found in Chapter VII in this volume. The point we want to focus upon here, is to demonstrate, using experimental data, that the parameters in the fractional models are physically meaningful. To this end we study the parameter dependence on the filler degree in polymers.

As a prototype example of constitutive rheological model elements, the generalised or fractional Maxwell model will be considered here. Other models are described in [82]. The fractional Maxwell model is shown in Figure 9, besides the standard analogue.

Adding up the contributions of each ladder element in the same manner as for standard elements (mirroring the serial and parallel character) one arrives at the constitutive equation:

$$\sigma(t) + \tau^{\alpha-\beta} \, {}_{-\infty}D_t^{\alpha-\beta} \sigma(t) = E\tau^\alpha \, {}_{-\infty}D_t^\alpha \varepsilon(t) \tag{37}$$

[c] *Greek.* flow. Heraklit, 495 BC, said: $\pi\alpha\nu\tau\alpha\ \varrho\varepsilon\iota$, everything flows.

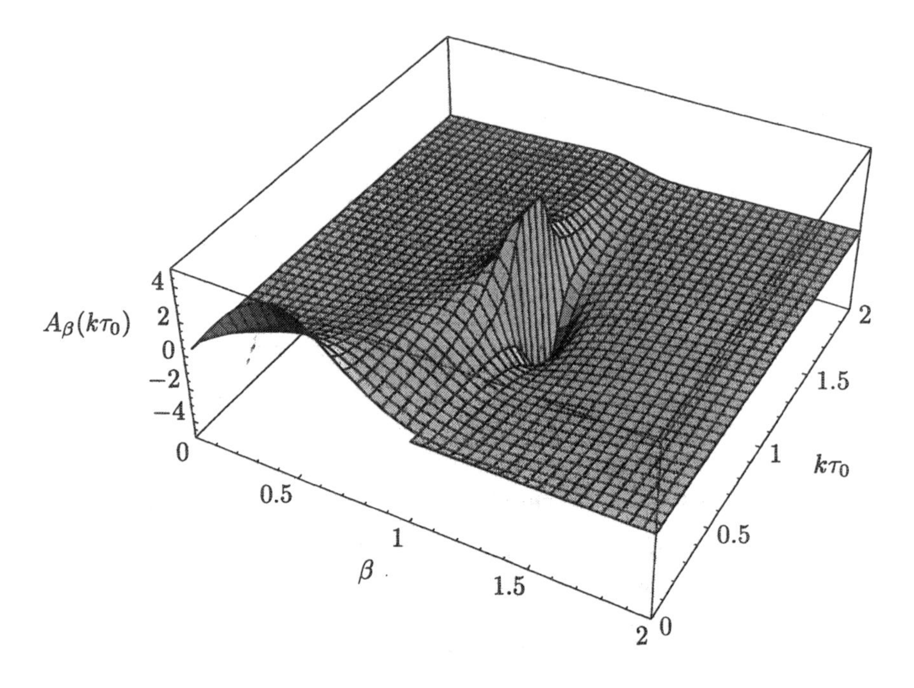

Figure 7: A three–dimensional plot of the spectral function (36) for values of β in the range $0 < \beta < 2$. For $\beta \to 1$ and $k\tau_0 \to 1$ the spectral function $A_\beta(k)$ diverges.

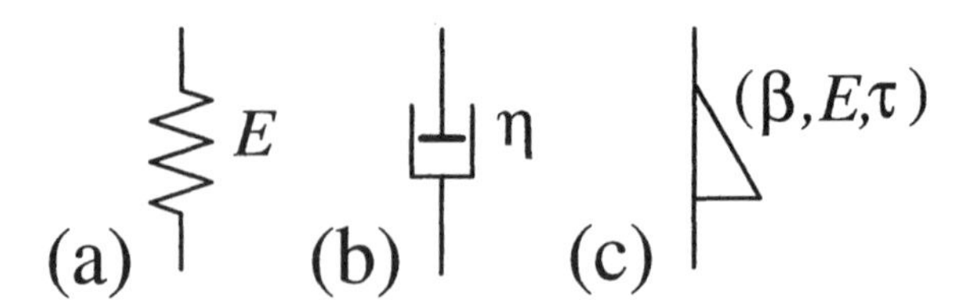

Figure 8: Basic elements: a. purely elastic, b. purely viscous and c. fractional element.

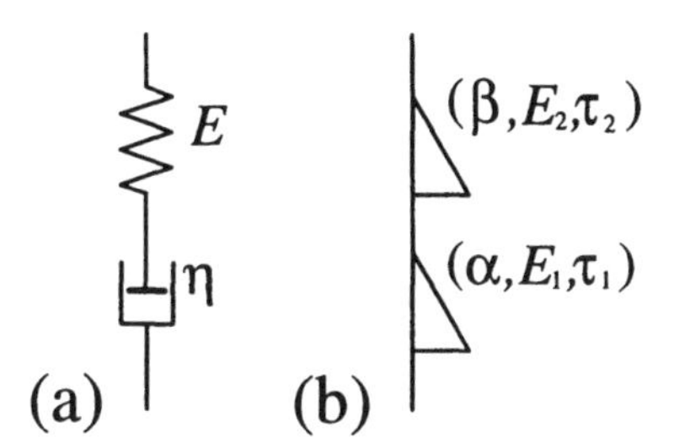

Figure 9: a. Standard and b. fractional Maxwell model.

with the abbreviations $\tau = (E_1\tau_1^\alpha/E_2\tau_2^\beta)^{1/(\alpha-\beta)}$ and $E = E_1(\tau_1/\tau)^\alpha$. From this relation, Fourier transformed ($\mathcal{F}\{{}_{-\infty}D_t^\alpha f(t)\} = (\mathrm{i}\omega)^\alpha f(\omega)$), the complex modulus $G^* = \sigma^*/\varepsilon^*$ is readily recovered

$$G^*(\omega) = \frac{E(\mathrm{i}\omega\tau)^\alpha}{1+(\mathrm{i}\omega\tau)^{\alpha-\beta}}, \tag{38}$$

Using the so-called Fourier–Mellin technique, the relaxation function can be found, after some tedious calculations [82]:

$$\begin{aligned} G(t) &= \frac{E}{\alpha-\beta} H_{12}^{11}\left[\frac{t}{\tau}\,\middle|\, \begin{matrix} \left(-\frac{\beta}{\alpha-\beta}, \frac{1}{\alpha-\beta}\right) \\ \left(-\frac{\beta}{\alpha-\beta}, \frac{1}{\alpha-\beta}\right), (0,1) \end{matrix}\right] & (39a)\\ &= \frac{E}{\alpha-\beta}\, {}_1\psi_1\left[\begin{matrix}(1,1)\\(1-\beta,\alpha-\beta)\end{matrix}; -\frac{t}{\tau}\right] & (39b)\\ &= E\left(\frac{t}{\tau}\right)^{-\beta} E_{\alpha-\beta,1-\beta}\left(-\left(\frac{t}{\tau}\right)^{\alpha-\beta}\right). & (39c) \end{aligned}$$

Here, we have also given the identification with Maitland's generalised hypergeometric function ${}_p\psi_q(z)$ and the generalised Mittag–Leffler function $E_{\alpha,\beta}(z)$. The solution (39*a*,39*c*) was obtained in a different way earlier by Glöckle and Nonnenmacher [65,28] and Friedrich [25]. This model was used for the fits in Figs. 1 and 2.

Besides the relaxation function and the complex modulus, one can also calculate the retardation function and the complex compliance. This is exactly possible by use of the Fox functions which are compiled in Appendix 5, see [82]. Here, however, the focus shall now be given to the physical interpretations of the model parameters.

3.5 Application to filled polymers

The generalised Maxwell model issued only one turnover point from one power–law to another, for $G(t)$. To describe typical data from harmonic mechanical experiments on polymers, one must employ the so–called Zener model [94]. Such a model was also discussed in [82], here we use a slightly different model according to Glöckle and Nonnenmacher, however [28]. The result reads

$$G(t) = \frac{G}{q} H^{1,1}_{1,2}\left[\frac{t}{\tau_0}\,\middle|\, \begin{matrix}(0,1/q)\\ (0,1/q),(0,1)\end{matrix}\right] + \\ + \frac{G_e}{q}\left(\frac{t}{\tau_0}\right)^{\mu} H^{1,1}_{1,2}\left[\frac{t}{\tau_0}\,\middle|\, \begin{matrix}(0,1/q)\\ (0,1/q),(-\mu,1)\end{matrix}\right] + G_\infty \tag{40}$$

where we find the intuitive identifications that G_e is the equilibrium modulus, whereas $G+G_\infty \equiv G_e+G_m+G_\infty$ is the glass modulus [30]. G_∞ is the value for G for very large times t. Glöckle and Nonnenmacher called this the fractional solid model. The complex modulus in this case is given by

$$G^*(\omega) = \frac{G + G_e(\mathrm{i}\omega\tau_0)^{-\mu}}{1+(\mathrm{i}\omega\tau_0)^{-q}} + G_\infty. \tag{41}$$

We apply this model to discuss data from filled polymers [55]. In the underlying experiments, polymers filled with carbon black and silicates were measured, for different filler degrees. Typical result for the storage and loss moduli are depicted in Figures 10 and 11, see Ref. [55]. The full line is the fit with the fractional model (40).

In general, the mechanical properties of polymer networks are massively influenced by the addition of certain filling substances (fillers) [23]. The quasistatic behaviour of filled networks is fairly well understood [41], but it is the dynamics that still lacks adequate descriptions. Here, the fractional model is shown to provide a reasonable description for the harmonic measurements.

The addition of filler enhances crosslinking and the intrinsic stress of the polymer network. In that course the involved parameters should vary with changing filler content. Reversely—if experimentally measured–the parameters should retain information on the filler content of the sample.

Results

The filler content was varied from 0 up to 60 phr (chemical mass concentration) meaning relative mass fractions in the range 0..37.5%. Via harmonic stress–strain experiments the complex modulus was measured. By construction of a master curve one obtained the modulus in a frequency window from approximately $10^{-2.5}$ to 10^{12} cyc/sec.

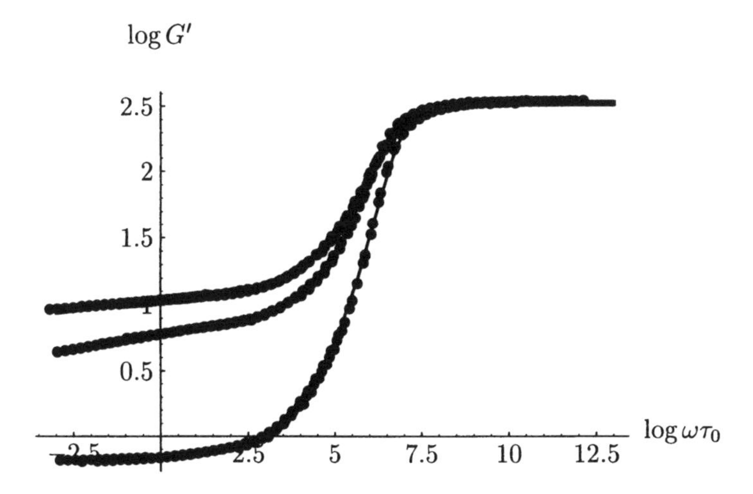

Figure 10: Storage modulus of NR32237 for the filler contents of 0, 30, and 60% (from bottom to top).

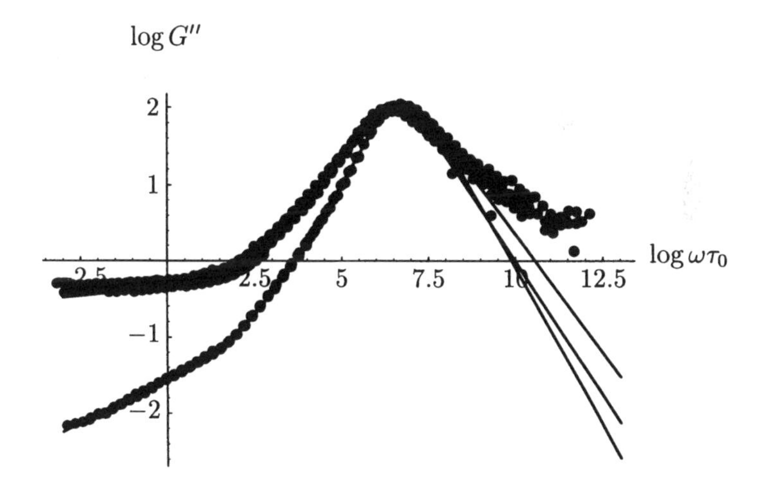

Figure 11: Loss modulus of NR32237 for the filler contents of 0, 30, and 60% (from bottom to top).

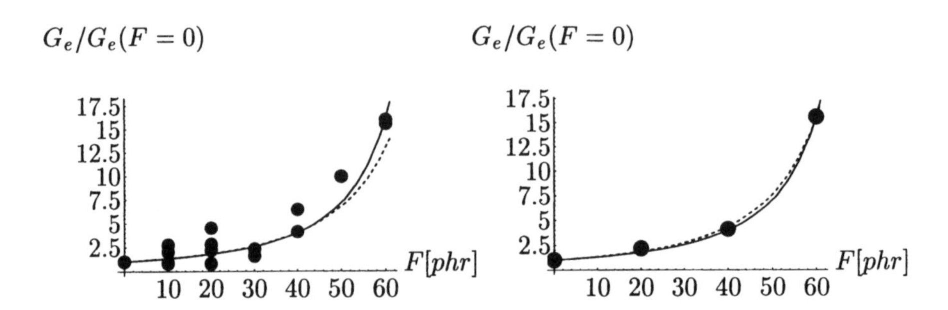

Figure 12: a. Normalised equilibrium modulus over the filler content F for all samples, b. for the most carefully measured series S10.

Equilibrium modulus

Results for the equilibrium modulus G_e are plotted in Fig. 12. The normalised data show a satisfying data collapsing indicating the independence of the actually underlying specimen's matrix. Thus it is not possible to distinguish between silica and carbon black filled systems at hands of $G(F = 0)$. That a difference of the fit parameter F_{crit} of about 10phr does not affect the fit's quality significantly is due to the nonlinear connection of F to the critical volume fraction v_{crit} that is not too much affected in this special range of variation.

Two classical models can be employed to fit these data: Brinkman's formula[8]

$$\frac{G_e}{G_0} = \left(1 - \frac{F}{F_{crit}}\right)^{-2.5} \tag{42}$$

and Eilers' & van Dijck's formula [20,72]

$$\frac{G_e}{G_0} = \left(1 + 1.25 \frac{F}{1 - F/F_{crit}}\right)^2 \tag{43}$$

where F_{crit} is a free fit parameter. Both classical descriptions do show excellent consistence with the data from the fractional model. Both models exhibit a pole for the critical filler content. This is connected to the percolation clustering of the filler particles which finally produces a cluster running through the whole sample, *i.e.* the divergence of $G_e(F)$. The critical volume fraction found from Fig. 12, about 31% shows excellent agreement with the percolation threshold for a cubic lattice.

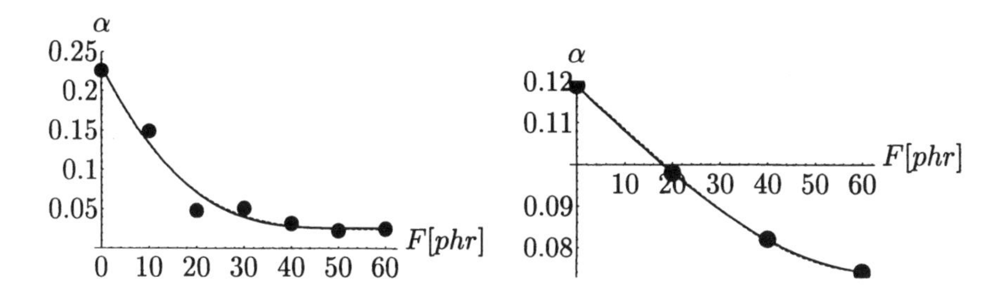

Figure 13: Homogeneity α for the series a. NR32237 and b. S10.

Fig. 13 shows another model parameter, the so-called homogeneity α. It describes the long tail of the relaxation time spectrum, see [55]. The F–dependence is very delicate and one can show that α may serve as a parameter to differentiate between the underlying matrix.

All model parameters show reasonable filler dependence and characterise a given system. The fractional Zener model is thus well–suited for the description of polymeric systems. In addition, starting off from a fit like in Figs. 10 and 11, once the parameters are determined, all the other interesting functions (relaxation *etc.*) can be calculated analytically. They show perfect agreement with the experimental measurements, as proven in [28].

Thus we can say that the fractional constitutive model (the fractional solid model), as a prototype for a fractional modelling, which was applied to filled polymers, is indeed a physically meaningful model. With only a few number parameters, the complex behaviour of the polymer could be described. Glöckle and Nonnenmacher also demonstrated that once the parameters are determined by a fit to a given function (*e.g.* the complex modulus), all other conjugate functions are determined via analytical calculations, and co–incide perfectly with the measurements on the same system [28].

3.6 Fractional protein dynamics

The fractional relaxation properties of protein dynamics for a typical process, the ligand re–binding to the heme iron of Myoglobin (Mb) after a flash dissociation, is discussed by Glöckle and Nonnenmacher [32]. $N(t)$ measures the free ligands. In equilibrium, only a negligible number of ligands should be left. The relaxation behaviour is graphed in Figs. 14 and 15. Clearly, for large times, the relaxation is power–law. The quality of the description is excellent.

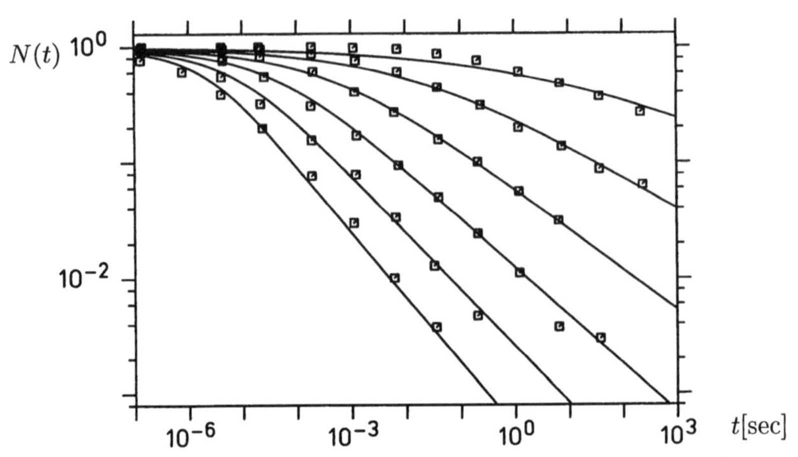

Figure 14: Three–parameter model according to equation (45) for the re–binding process of CO to Mb, after a photo dissociation. The fit parameters are $\tau_m = 8.4 \times 10^{-10}$sec, $\alpha = 3.5 \times 10^{-3}\text{K}^{-1}$ and $k = 130$. Data from Austin *et al* [1]. The temperature dependence of τ_0 follows the Arrhenius law (47).

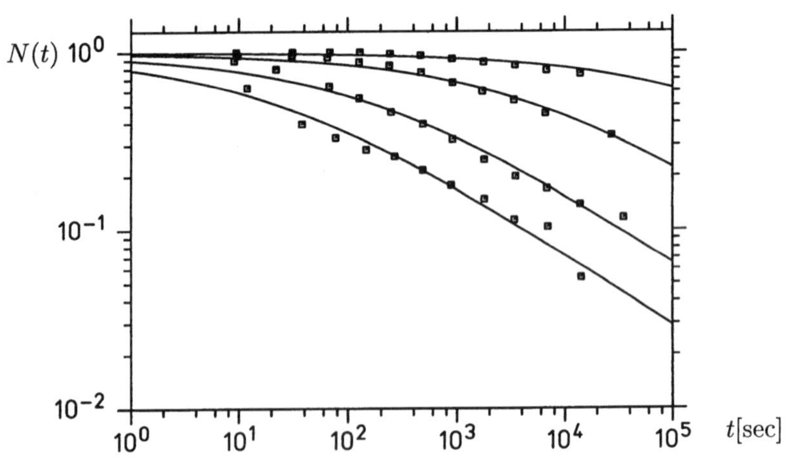

Figure 15: Relaxation of Mb after pressure release, data points from Iben *et al* [39]. The solid lines correspond to equation (45), with $\beta = 0.4$ and τ_0 according to the VTF model (48), details see Glöckle and Nonnenmacher [32].

The relative number of free ligands, $N(t)$, can be described by the fractional relaxation equation (I form), Eq. (4), with $N(t) = \Phi(t)/\Phi_0$, *i.e.*:

$$\tau_0^{-\beta} \, {}_0D_t^{-\beta} N(t) + N(t) - 1 = 0. \tag{44}$$

The solution is thus given by the Mittag–Leffler function

$$N(t) = E_\beta \left(-(t/\tau_0)^\beta \right). \tag{45}$$

The power–law exponent β which also occurs as order of the fractional integration in (4), can thereby be modelled via the proportionality relation

$$\beta(T) = \frac{0.41}{120\text{K}} T \tag{46}$$

in dependence of the temperature. The characteristic time constant τ has an Arrhenius dependence on T:

$$\tau(T) = \tau_m \exp\left\{ \frac{E^*}{T} \right\} \tag{47}$$

with an activation energy $E^* = 1470\text{K} \triangleq 12.3\text{kJ/mol}$. For the second measurement which is in a temperature range near the glass temperature, $\beta = 0.4$ is approximately constant, whereas the characteristic time can be described via a Vogel Tamann Fulcher type equation:

$$\tau(T) = \tau_0 \exp\left\{ \frac{E}{T - T_0} \right\} \tag{48}$$

with $\tau_0 = 2.5 \times 10^{-8}$sec and $T_0 = 129\text{K}$, $E = 1040\text{K}$. Ref. [32] discussed the connection to a fractal scaling model and its consequences in an energy barrier landscape picture.

4 Anomalous diffusion

Like the observed deviations from Debye relaxation patterns, many experiments reveal that diffusion processes usually do not follow the standard Gaussian behaviour. In this section, we will report some modelling considerations of *anomalous diffusion*.

4.1 Fickean diffusion

Normally, Fick's second law is used to describe standard diffusive processes. The derivation combines the continuity equation:

$$\partial_t \varrho(x,t) = -\jmath_x(x,t) \tag{49}$$

and a constitutive equation:

$$\jmath(x,t) = -K\varrho_x(x,t) \tag{50}$$

which is also called Fick's first law, analogous to Fourier's law in heat conduction. Here, $\jmath(x,t)$ denotes the flux, $\varrho(x,t)$ the distribution function of the diffusing quantity, and K the diffusion coefficient, which we assume to be constant. Combining equations (49) and (50), we arrive at the diffusion equation (Fick's second law):

$$\partial_t \varrho(x,t) = K\varrho_{xx}(x,t). \tag{51}$$

For an initial delta distribution $\varrho(x,0) = \delta(x)$ and natural boundary conditions $\varrho(|x| \to \infty, t) = 0$, one finds a typical Gaussian solution, namely

$$\varrho(x,t) = \frac{1}{\sqrt{4\pi K t}} \exp\left\{-\frac{x^2}{4Kt}\right\}. \tag{52}$$

In a Euclidean space of integer dimension d, equation (51) has to be replaced by

$$\partial_t \varrho(\mathbf{x},t) = K\nabla^2 \varrho(\mathbf{x},t), \tag{53}$$

which can be re-written as

$$\partial_t P(r,t) = K r^{1-d} \partial_r r^{d-1} \partial_r P(r,t) \tag{54}$$

in the isotropical case. Here, it is also usual to denote the propagator by P.

The root mean square distance (RMS) or variance of Fickean diffusion, *i.e.* the mean $\langle r^2(t)\rangle = \mathcal{M}\{r^2\}$, is linear in time,

$$\langle r^2(t)\rangle \propto t. \tag{55}$$

The return probability to the origin is

$$P(0,t) \propto t^{-d/2}. \tag{56}$$

4.2 Cattaneo diffusion

Regarding the solution (52), one recognises that even for very small times, there exists a finite amount of the diffusing quantity at large distances from the origin. It is therefore an intrinsic property of equation (51), that it gives rise to an infinite velocity of propagation. Mathematically speaking, this is due to the fact that equation (51) is a parabolic partial differential equation. Physically, however, the propagation velocity must be finite.

Consequently, Cattaneo proposed in 1948 a modified diffusion equation [12,14]. He replaced the constitutive equation (50) by

$$\jmath(x,t) + \tau\partial_t \jmath(x,t) = -K\varrho_x(x,t). \tag{57}$$

Thus, the flux $\jmath$ relaxes with some given characteristic time constant τ. Combining now (57) with the continuity equation (49), one is led to the so-called Cattaneo equation

$$\partial_t\varrho(x,t) + \tau\partial_t^2\varrho(x,t) = K\varrho_{xx}(x,t). \tag{58}$$

This equation is of the hyperbolic type [89]. It is a damped wave or telegrapher's equation. Now, the propagation velocity is finite, namely $v = \sqrt{K/\tau}$. In the diffusion limit $\tau \to 0$, one recovers Fick's equations with an infinite v. We note in passing that the Cattaneo equation (58) can be derived from the Boltzmann equation [63].

4.3 Properties of anomalous diffusion

Many experiments and theoretical considerations [5,37] lead to a RMS of power-law type

$$\langle r^2\rangle \propto t^{2/d_w}, \tag{59}$$

where the real-valued constant d_w is called the *anomalous diffusion exponent*. On fractals, $d_w > 2$, *i.e.* the transport is dispersive. The returning probability

$$P(0,t) \propto t^{-d_s/2} \tag{60}$$

involves an additional geometrical quantity, the fracton or spectral dimension $d_s = 2d_f/d_w$. In this relation, d_f is the familiar fractal Hausdorff dimension of the underlying geometry, on which diffusion takes place. On fractal structures, in general, $d_s < d_f < d$ (d the integer embedding dimension), so that the returning probability is larger than in normal diffusion. This is to be expected, since the probability distribution (61) has wider wings than a standard Gaussian.

	α_i	u_i
OP	0	d_w
KZB	$(d_f - d_w/2)/(d_w - 1)$	$d_w/(d_w - 1)$
M	0	$d_w/(d_w - 1)$
Gen	$(d_f - Dd_w/2)/(d_w - 1)$	$d_w/(d_w - 1)$

Table 1: Asymptotic parameters for the various models discussed, cf. Eq. (61): OP (O'Shaugnessy [70]), KZB (Klafter *et al* [44], Roman *et al* [77]), M (Metzler [54], see also Schneider and Wyss [84] and Giona and Roman [27]) and Gen.

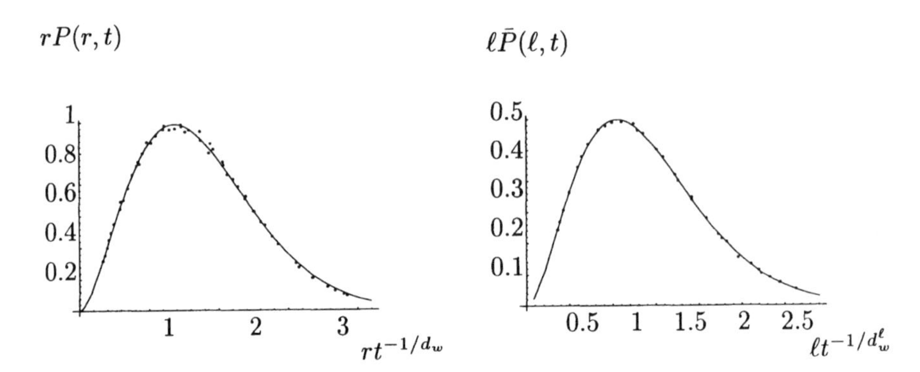

Figure 16: Probability distribution according to the exact enumeration method. a. in r space and b. in the chemical distance measure, reproduced from Havlin *et al* [36,37].

Furthermore, the Gaussian behaviour gets modified. The asymptotic shape of the propagator is usually written as

$$P(r,t) \sim t^{-d_s/2} \xi^{\alpha_i} \exp\left\{-c_i \xi^{u_i}\right\}, \tag{61}$$

where the index i refers to the different models in literature. (61) is valid for $\xi \gg 1$ and $t \to \infty$. Table 1 gives the references to the different models. $\xi = rt^{-1/d_w}$ is the similarity variable, for all models. A detailed discussion of the power–law pre–factor ξ^{α_i} may be found in [77,78]. The influence of the number of configurations, taken into account in the averages, on $P(r,t)$ is investigated in [9]. Figure 16 shows the numerical results of the exact enumeration method[37] and the fit with the model function (61) with $\alpha = 0$.

According to the procedure in relaxation modelling, the task is now to find a modified diffusion equation accounting for these generalised properties.

4.4 Fractional diffusion equations

O'Shaugnessy and Procaccia [70] presented an additional power of r in the Laplacian. Their result, however, did not comply with the meanwhile accepted asymptotic result (61). The first fractional diffusion equation was reported by Wyss [99] and soon afterwards by Schneider and Wyss [84]. Within an I formulation they replaced the first order time derivative by one of fractional order. Their equation is valid for Euclidean dimensions and include arbitrary initial conditions. Giona and Roman [27] modified both, the spatial and the time derivatives and came up with a fractional diffusion equation that did not reduce to the Gaussian case for all dimensions. This equation is nonetheless correct in the limit of equation (61).

Metzler *et al* [54] reported the generalised diffusion equation [d]

$$\frac{\partial^\gamma P(r,t)}{\partial t^\gamma} = r^{1-D}\frac{\partial}{\partial r} r^{D-1} r^{-\Theta} \frac{\partial}{\partial r} P(r,t) \tag{62}$$

which involves a fractional t–derivative of order γ, a modified Laplacian operator with fractal order D, and $r^{-\Theta}$ being a possibly non–constant diffusion coefficient. Some comments on Eq. (62) and its connection to CTRW are given by Shlesinger *et al* [88]. The solution procedure of Eq. (62), shown in Ref. [54], remains valid and the propagator is given, in terms of Fox' H–functions:

$$\begin{aligned} P(r,t) &= \frac{A}{t}\left(r^{\frac{2+\Theta}{2}}\right)^{\frac{2}{\gamma}-\frac{2d_f}{2+\Theta}} \\ &\times H^{20}_{12}\left[\frac{r^{(2+\Theta)/\gamma}}{(2+\Theta)^{2/\gamma}t}\,\middle|\, \begin{matrix}(0,1)\\ \left(1-\frac{1}{\gamma}+\frac{d_f-D}{2+\Theta},\frac{1}{\gamma}\right),\left(\frac{d_f}{2+\Theta}-\frac{1}{\gamma},\frac{1}{\gamma}\right)\end{matrix}\right], \end{aligned} \tag{63}$$

A being the normalisation constant. The occurring parameters may now be reduced by considering the conditions (59) to (61). One thus arrives at the identification of the parameters for the different models listed in Table 2. The model Gen leaves α open, according to the discussion in Roman [78], and embraces the more special models M and KZB. The single restriction for D is that it must reduce to the Euclidean dimension d if the standard Fickean case ($d_w \to 2$, $d_f \to d \in \mathbb{N}$) shall be recovered. For $\alpha = u(d_s/2 - D/2)$, as suggested in Klafter *et al* , one is led to the peculiar constraint that D must equal 1 for the standard Fickean pendant. The reason is shown below. A typical solution is displayed in Fig. 17.

The advantage of the present approach involving H–functions which may seem rather complicated is, that Eq. (62) can be solved in closed form, and

[d] Here and in the following, we re–scale the diffusion equations to absorb the (generalised) diffusion constant.

	M	KZB	Gen	OP
Θ	0	0	0	$d_w - 2$
γ	$2/d_w$	$2/d_w$	$2/d_w$	1
D	d_s	1	D	d_f

Table 2: Identification of the occurring parameters (see Eq. (62)) for the different models discussed.

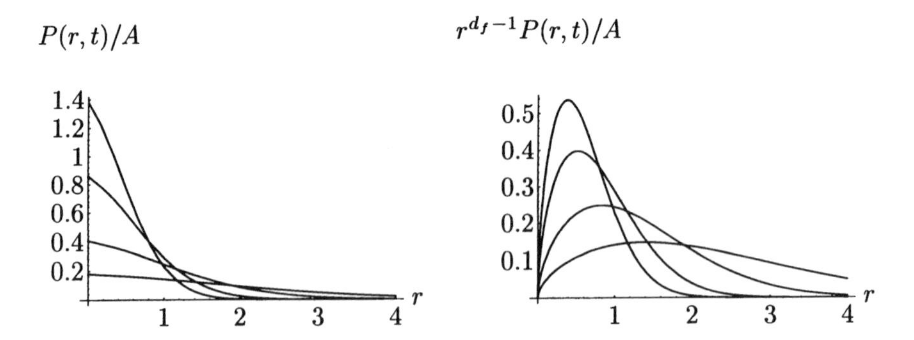

Figure 17: a. Solution $P(r,t)$ of the fractional diffusion equation (62) of the M model and b. the according density $r^{d_f-1}P(r,t)$, for the times $t = 0.5, 1, 3, 10$.

the spectral functions can also be calculated *exactly*. Thus one can find access to more than only the $\xi \gg 1$ asymptote: to the $\xi \ll 1$ and the transition region. The exact solutions for $P(r,t)$ and its spectral functions provide increased information for comparison to experimental or computer data. We do not hesitate to mention that the presented Eq. (62) remains valid, even in the intermediate ultra–diffusive régime in between standard ($d_w = 2$) and ballistic ($d_w = 1$) transport [98].

The modified Laplacian in Eq. (62) may formally be written in the form $(\mathrm{d}V_{\mathrm{eff}}/\mathrm{d}r)^{-1}\partial/\partial r\,(\mathrm{d}V_{\mathrm{eff}}/\mathrm{d}r)\partial/\partial r$, involving the *effective volume* V_{eff} as sensed by the wiggling random walker. Identifying $V_{\mathrm{eff}} \propto r^D$, $D < d_f$ suggests that the walker senses a smeared–out structure. Thus it is a matter of importance to have at hands adequate information—either by simulations or by very precise experimental measurements—on the "true" shape of $P(r,t)$, *i.e.* which D is significant for the actually underlying structure, see also Roman [78] in this connection. As our closed form solution for $P(r,t)$ includes the $\xi \ll 1$ and the *transition region* towards $\xi \gg 1$ it should be significantly better for comparisons to data than the mere stretched exponential asymptote, or the asymptotic power–laws in the transformed spaces.

4.5 Spectral transforms

For the discussion of the asymptotic probability density in the Laplace domain, let us at first consider the Gaussian position probability density

$$P(r,t) = (2\pi t)^{-d/2} \exp\left(-\frac{r^2}{4t}\right) \tag{64}$$

in $d \in \mathbb{N}$ dimensions, and its Laplace transform

$$\begin{aligned}\tilde{P}(r,s) &= \pi^{-d/2} r^{(2-d)/2} s^{(d-2)/4} K_{d/2-1}\left(rs^{1/2}\right)\\ &\sim r^{(1-d)/2} s^{(d-3)/4} \exp -rs^{1/2}\end{aligned} \tag{65}$$

which involves the Bessel function K. Guyer's result [35] in the Laplace domain

$$\tilde{P}_{Guy} \sim s^{d_s/2-1} \exp -ars^{1/d_w}, \tag{66}$$

upon which Klafter *et al* [44] base their research, does not include the pre–factor $(rs^{1/2})^{(1-d)/2}$ before the exponential in comparison to Eq. (65). But the models previously labelled OSP, M, and Gen reduce exactly to the Laplace transformed Gaussian, Eq. (65). Especially for the model Gen one finds

$$\begin{aligned}\tilde{P}(r;s) &\propto \left(rs^{1/d_w}\right)^{1-D/2} s^{d_s/2-1} K_{1-D/2}\left(rs^{1/d_w}\right)\\ &\sim \left(rs^{1/d_w}\right)^{(1-D)/2} s^{d_s/2-1} \exp\left(-rs^{1/d_w}\right)\end{aligned} \tag{67}$$

For a proper description in Laplace space Eq. (67) should therefore be preferred to Eq. (66). It is exactly the choice of (66) instead of (67) that causes the failure of the KZB model to reduce to the transformed Gaussian for any integer dimension but 1.

Physical measurements mostly reveal not $P(r,t)$ but some spectral function, *e.g.* the Fourier transformed spectral density $P^*(k,t)$. Especially for these spectral functions relatively large ranges are experimentally accessible. It is there, where the transition region described by our results, significantly enhances the accuracy of data fits, a fact well known from asymptotic fractals. Therefore we present the general integral transforms of (63) in closed form. This is made possible, again, by use of the Fox function. Below, A denotes the appropriate normalisation constant. The asymptotic behaviour of the calculated functions is summarised in Table 3. All the occurring H-functions may be represented in computable form by simply inserting the parameters into Eq. (B.7). This can be done conveniently with a `Mathematica` or `Maple` program.

A detailed discussion in connection of NMR measurements (*no* fractal but a linear Fourier transform) basing upon the M model can be found in [18].

The fractal Fourier transform (see Appendix) of $P(r,t)$, Eq. (63), is given by

$$P^*(k,t) = \tag{68}$$

$$AH_{23}^{12}\left[(2+\Theta)^{2/(2+\Theta)}kt^{\gamma/(2+\Theta)}\left|\begin{matrix}\left(\frac{D-d_f}{2+\Theta},\frac{1}{2+\Theta}\right),\left(1-\frac{d_f}{2+\Theta},\frac{1}{2+\Theta}\right)\\ \left(0,\frac{1}{2}\right),\left(0,\frac{\gamma}{2+\Theta}\right),\left(-\frac{1}{2},\frac{1}{2}\right)\end{matrix}\right.\right]. \tag{69}$$

The Laplace transform of $P(r,t)$, Eq. (63), can either be written in terms of modified Bessel functions (see Ref. [54]) or directly be expressed by the corresponding Fox function:

$$\tilde{P}(r,s) = Ar^{(2+\Theta)/\gamma-d_f}$$

$$\times H_{02}^{20}\left[\frac{r^{(2+\Theta)/\gamma}s}{(2+\Theta)^{2/(2+\Theta)}}\left|\left(1-\frac{1}{\gamma}+\frac{d_f-D}{2+\Theta},\frac{1}{\gamma}\right),\left(\frac{d_f}{2+\Theta}-\frac{1}{\gamma},\frac{1}{\gamma}\right)\right.\right]. \tag{70}$$

Finally, the Fourier–Laplace transform of Eq. (63) turns out to be

$$\tilde{P}^*(k,s) = \tag{71}$$

$$\frac{A}{s}H_{22}^{12}\left[(2+\Theta)^{2/(2+\Theta)}ks^{-\gamma/(2+\Theta)}\left|\begin{matrix}\left(\frac{D-d_f}{2+\Theta},\frac{1}{2+\Theta}\right),\left(1-\frac{d_f}{2+\Theta},\frac{1}{2+\Theta}\right)\\ \left(0,\frac{1}{2}\right),\left(-\frac{1}{2},\frac{1}{2}\right)\end{matrix}\right.\right]. \tag{72}$$

It is worth mentioning that the reduction of this function to the standard Fickean case is accomplished easily for the H-function, using its standard properties.

Consulting 3 one observes that, due to the choice of the volume element in the fractal Fourier transform, both P^* and $\tilde{P}^*$ show a horizontal asymptote, i.e. constant behaviour for $\xi \ll 1$. Thus it is a reasonable generalisation of

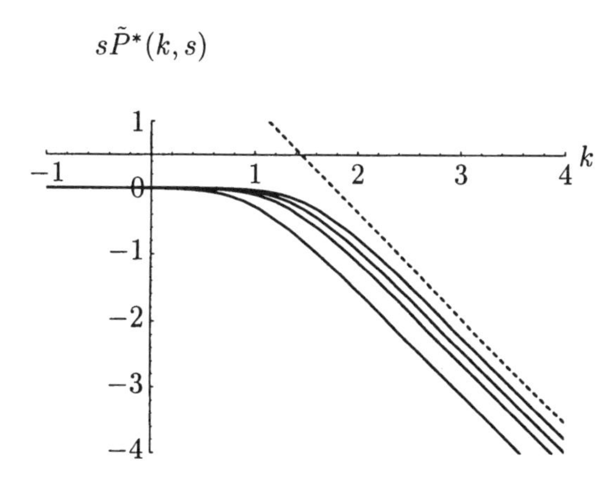

Figure 18: Fourier–Laplace transform $\tilde{P}^*(k,s)$ for $s = 100, 500, 1000, 2000$ (bottom to top) and $d_f = 1.58$, $d_w = 2.32$ in log–log representation. The dashed line indicates the long–tail power law $\sim k^{-d_f}$. Note the late inset of the asymptotic power–law behaviour.

	z	c	d
$P^*(k,t)$	$kt^{\gamma/(2+\Theta)}$	0	$-\min\left(\tilde{d},\Omega\right)$
$r^{\tilde{d}-(2+\Theta)/\gamma}\tilde{P}(r,s)$	$r^{(2+\Theta)/\gamma}s$	$\min(\Psi,\Gamma)$	$z^{\gamma(\alpha-1/2)/2}e^{-dz^{\gamma/2}}$
$s\tilde{P}^*(k,s)$	$ks^{-\gamma/(2+\Theta)}$	0	$-\min\left(\tilde{d},\Omega\right)$

Table 3: Asymptotic behaviour of the different models, in the transformed spaces. $\Omega \equiv 2+\Theta-D+\tilde{d}$, $\Psi \equiv \gamma - 1 + \gamma\frac{\tilde{d}-D}{2+\Theta}$, and $\Gamma \equiv \gamma\frac{\tilde{d}}{2+\Theta} - 1$.

the standard case and underlines the significance of the (k,s)–space. Fig. 18 shows $\tilde{P}^*$ as an example. The inserted long–tail asymptote visualises the relatively late inset of the calculated behaviour. This again underlines the great importance of the knowledge of the transition region.

Now let us turn to consider applications of fractional diffusion theory. In contrast to the simply time–dependent relaxation theory, experimental observations in space–time–coupled diffusion are very difficult. Especially there is at present no convincing measurement in both space and time which would offer a wide enough experimental window to investigate fractional diffusion. Here, two experimental techniques are discussed, where progress towards these aspects are to be expected.

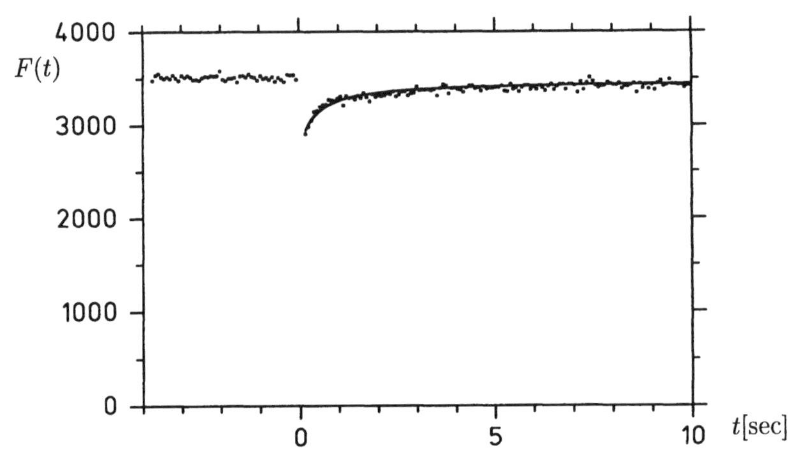

Figure 19: Data points from a typical FRAP experiment [53]. Solid line according to diffusion model.

4.6 Anomalous diffusion and fluorescence recovery

Fluorescence recovery after photobleaching (FRAP) is an experimental method to investigate transport processes of fluorescent molecules. It is often applied *in vivo* in biological systems like cells. A selective measurement of processes in a thin surface layer is achieved under the condition of total reflection. In the experiment, a short, intense laser pulse irreversibly bleaches the fluorophore in a small region. The underlying transport coefficients can be determined via the rate of fluorescence recovery, *i.e.* the exchange of bleached molecules and fluorescent ones coming from non–illuminated regions. Under total reflection conditions, due to the exponentially decaying evanescent wave, only the molecules in a thin surface layer are bleached. This enables for a selective monitoring of protein or lipid molecules labelled with fluorescent dyes in cell membranes. Via FRAP, the lateral diffusion of molecules in the membrane can be studied.

The process can be differentiated into the three stages of bleaching by a laser pulse, transport processes (diffusion) exchanging bleached and fluorescent molecules, and the detection. Here, we concentrate on the second stage, the diffusion. Details can be found in Refs. [31,53] and the references therein. Figure 19 shows a typical result. (The solid line will be explained below.) There, the fluorophore function $F(t)$ is shown versus the time t. At $t = 0$, the laser pulse is applied. Afterwards, the relaxation back to the equilibrium is visible. The parameters refer to [31]. The interesting quantity is thus the excitation function

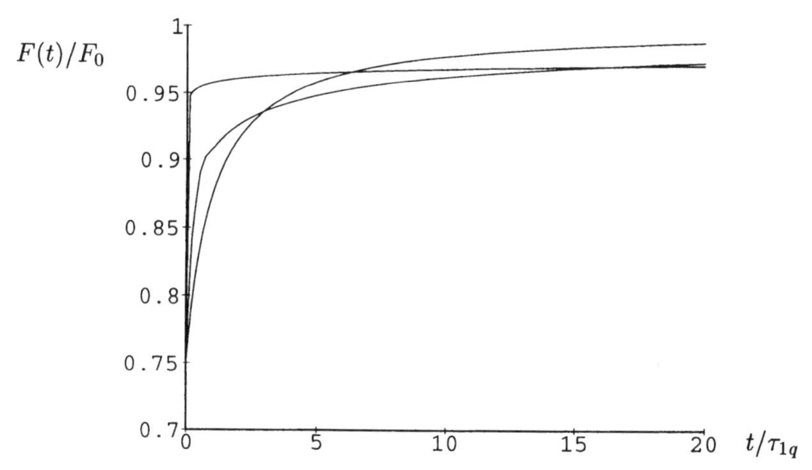

Figure 20: Fluorescence recovery for an underlying anomalous diffusion process ($q = 0.3$ and $q = 0.7$), compared with normal diffusion ($q = 1.0$). The smaller q is, the slower the relaxation towards F_0 becomes.

$\Delta F(t) = F_0 - F(t)$, where F_0 is the equilibrium value. According to [31], $F(t)$ can be expressed by

$$F(t) = QC_0 \int d\mathbf{r}' \int d\mathbf{r} \exp\left\{-\alpha T I^b(\mathbf{r}')\right\} G(\mathbf{r} - \mathbf{r}', t) I^p(\mathbf{r}) \tag{73}$$

valid for arbitrary transport processes. In equation (73), Q contains the quantum efficiencies of light absorption and emission, C_0 is the equilibrium concentration of unbleached fluorophore, T the (short) laser pulse interval, α the characteristic recovery frequency, I^b the bleaching beam profile, I^p the probing beam profile, and G the Green's function of the transport process (propagator). The nomenclature is similar to [31].

According to equation (59), the RMS, given by

$$\langle r^2 \rangle = -\nabla_k^2 \hat{G}^*(\mathbf{k}, t)_{|\mathbf{k}=0} = -\nabla_k^2 \mathcal{FT}\{G(\mathbf{x}, t)\} \propto t^q, \tag{74}$$

shows a weaker dependence on time as the normal linear behaviour. The exact calculations for $\Delta F(t)$ in [31] lead to the deviations from the standard situation shown in Figure 20. Especially for large molecules with low transport coefficients these deviations will become important.

4.7 Anomalous diffusion and NMR in biological tissue

Via the *supercon fringe field method*, see [43], a stimulated echo was measured in the fringe field of a large superconductor, where field gradients of 50 T/m

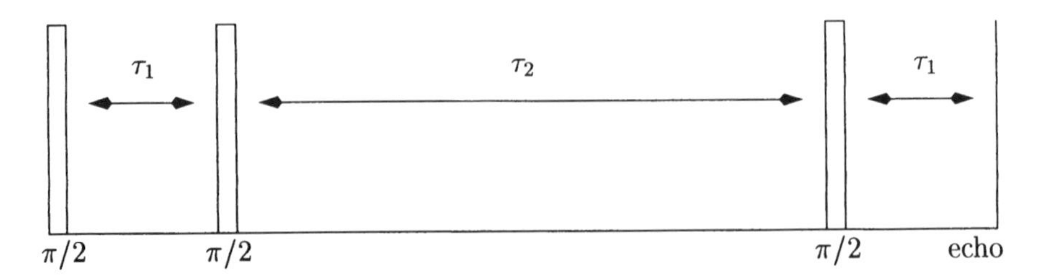

Figure 21: Pulse sequence of the NMR fringe field gradient method measuring a stimulated echo after three $\pi/2$–pulses. $\tau_2 \gg \tau_1$ guarantees the narrow pulse approximation. Note that the field gradient $\mathbf{g}$ is stationary.

can be achieved [46]. The applied pulse sequence is shown in Fig. 21. After an initial $\pi/2$ pulse (rotation from the z–axes to the x–y–plane) the spins de–phase during the time interval τ_1. Then they are rotated back by the second $\pi/2$ pulse. Due to diffusion processes additional de–phasing occurs. The third $\pi/2$ pulse brings the spins back to the x–y–plane where re–phasing occurs. Due to the diffusion during τ_2 the re–phasing is incomplete and allows for a quantitative analysis of the proton diffusion. $\tau_2 \gg \tau_1$ ensures the narrow pulse approximation. For each sample two series were measured. At first τ_1 was held fixed (40, 70 and 100 μsec, respectively). For the second series, τ_2 was kept constant at logarithmically equidistant intervals (from 2 to 320 msec). These are called τ_1– and τ_2–measurements, respectively.

According to the theory of NMR field gradient methods, the measured echo amplitude $E(\mathbf{q}, \tau_2)$, depending on the relaxation time τ_2 and the wave vector $\mathbf{q}$ is

$$E(\mathbf{q}, \tau_2) = \int \rho(\mathbf{r}) \int P_S(\mathbf{r}|\mathbf{r}', \tau_2) \exp\left(2\pi i \mathbf{q}(\mathbf{r}' - \mathbf{r})\right) \mathrm{d}\mathbf{r}'\mathrm{d}\mathbf{r} \tag{75}$$

where $\mathbf{q} \equiv (2\pi)^{-1}\gamma_0\tau_1\mathbf{g}$, with the gyromagnetic ratio γ_0, the pulse separation τ_1 and the field gradient vector $\mathbf{g}$ [10] Thus it is exactly the occurrence of a field *gradient* that enables for a spatial (or Fourier spatial) resolution. In Eq. (75), $\rho(\mathbf{r})P_S(\mathbf{r}|\mathbf{r}', \tau_2)$ is the probability density for a random walker (here: a spin) that started in $\mathbf{r}$ and arrived in $\mathbf{r}'$ after the time τ_2. Therefore, Eq. (75) is the spatial Fourier transform of the probability density of proton diffusion in the sample. The mathematical treatment is analogous to elastic scattering. Keeping either τ_1 constant and varying τ_2 or vice versa, one can resolve the temporal or the wave vector dependence of $E(\mathbf{q}, \tau_2)$, respectively. With this interpretation it is useful to drop the τ_2 notation and use the ordinary t for the diffusion time. Substituting $\mathbf{R} \equiv \mathbf{r} - \mathbf{r}'$, one can re–write Eq. (75) using spherical co–ordinates, in the form

$$E(\mathbf{q},t) = \int \bar{P}_S(\mathbf{R},t)\exp(2\pi i\mathbf{qR})\,\mathrm{d}\mathbf{R}. \qquad (76)$$

Here $\bar{P}_S(\mathbf{R},t)$ is the mean probability density for a random walker that started in the origin, after time t. If, for example, $\mathbf{g}$ is parallel to the z–axes of the co–ordinate system, the final form for Eq. (75) is

$$E(q,t) = 2\int_0^\infty \bar{P}_S(Z,t)\cos(\gamma_0\tau_1 GR)\,\mathrm{d}Z = 2\mathcal{F}_C\left\{\bar{P}_S(Z,t);q\right\} \qquad (77)$$

with $\mathcal{F}_C$ denoting the Fourier cosine transform. In the above Eq. (77) a probability density symmetric to the origin was assumed. In the case of standard diffusion with a Gaussian shape of the propagator the Fourier transform $E(q,t)$ is also of Gaussian shape. For anomalous diffusion processes more complicated functions come into play. As was shown in Section 4.5, $E(q,t)$ will have a long power–law tail [18,45,46].

The results of the experiments are displayed exemplarily in Figs. 22 to 26 employing a decadic double–logarithmic scale. The abscissa is scaled in the dimensionless variable $4\pi^2q^2Dt \equiv Db$ in arbitrary units, as the diffusion constant D is not explicitly specified. Clearly the measured data do not show an exponential decay.

In a first phenomenological approach two different functions were fitted to the data sets, these being a stretched exponential function

$$E(q_0,t) \propto \exp\left(-[t/T]^{\alpha_1}\right) \qquad (78.a)$$

$$E(q,t_0) \propto \exp\left(-\left[(q/Q)^2\right]^{\alpha_2}\right) \qquad (78.b)$$

and a Rigaut–type asymptotic fractal of the form:

$$E(q_0,t) \propto (1+t/T)^{-n_1} \qquad (79.a)$$

$$E(q,t_0) \propto \left(1+[q/Q]^2\right)^{-n_2} \qquad (79.b)$$

for either the τ_1 or the τ_2 experiment, *i.e.* for fixed q or fixed t, respectively. T and Q are constants with suitably chosen physical dimensions. The dashed lines in Fig. 22 indicate the stretched exponential, the full line corresponds to the Rigaut–type formula. The relevant fit parameters are listed in Table 4.7 for some typical specimens. Note that formulae (79.a) and (79.b) differ somewhat from Rigaut's original ones. Our investigations have shown, however, that the above listed functions do reproduce the slope in the given window more appropriately. One immediately realises from Table 4.7 that the stretched exponential function does not provide an appropriate tool for characterisation as

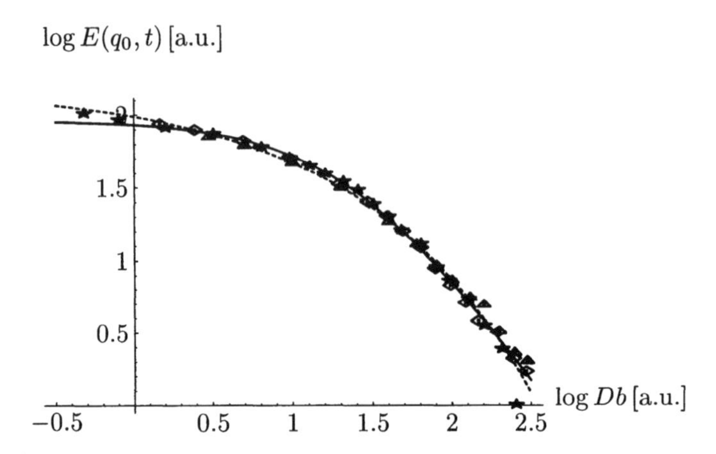

Figure 22: Result of the τ_1–measurement of the liver tissue S10. The symbols correspond to the following time pre–sets for τ_1: star 40 μsec; diamond 70 μsec; triangle 100 μsec. (—) asymptotic fractal, (– – –) KWW according to Eqs. (78.a–79.b).

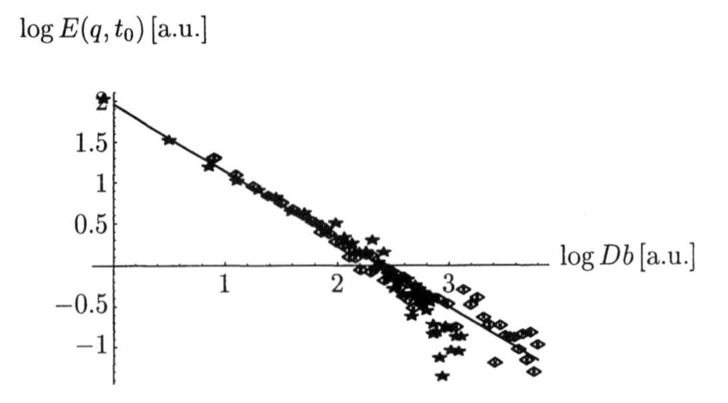

Figure 23: Result of the complete τ_2–measurement of the tumourous tissue E2552. The symbols correspond to the following pre–sets for τ_2: star 5 msec; diamond 50 msec.

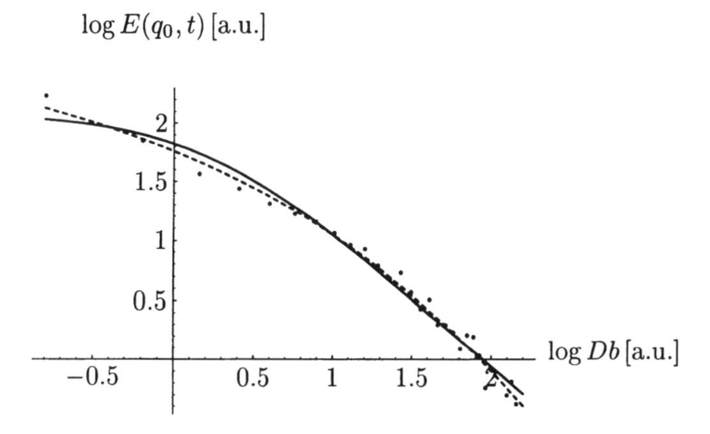

Figure 24: Result of the first ($\tau_2 = 5$msec) τ_2–measurement of the liver tissue S10.

many of the samples could not be fitted reasonably with it. On the other hand, using the Rigaut–type asymptotic fractal approach, all of the experimental results could be fitted with a common `Simplex` algorithm on `Fortran 77`. The single exception is the measurement of free water which reveals a probability distribution of approximate Gaussian shape, as it should be.

Let us briefly discuss some of the features of the plots. In Fig. 22 the representative τ_1–measurements for the liver S10 are depicted. Numerically one cannot differentiate between the quality of either fit function, stretched exponential or Rigaut–type. However, inspection of the initial data points of all investigated tissues clearly gives preference to the Rigaut–type formula (or the Lévy distribution discussed below). For the τ_1–experiments the data points show a satisfactory collapsing if one uses the Gaussian similarity variable $b \propto \tau_2 \tau_1^2$. Quite a different situation is found for the τ_2–measurements. There, no collapsing of the data sets for different parameters τ_2 can be obtained. Instead one observes a transition from a stretched exponential or asymptotically fractal to a "pure" power–law behaviour from low to large values of τ_2. In the case of the mastopathy not plotted here the power–law behaviour is only found for the largest value of τ_2. The liver specimens both show power–laws for the two larger τ_2, as it is shown in the three splitted plots of Figs. 25 and 26, whereas Fig. 24 was fitted by a stretched exponential. Whereas, for both of the tumourous samples one observes the power–law over the whole window of measurement for all the three values of τ_2, see Fig. 23. Thus the richer the sub–structure (the higher the cellular ramification), the earlier (in terms of $t \equiv \tau_2$) is the inset of pure power–law behaviour. (In other words, the turnover region

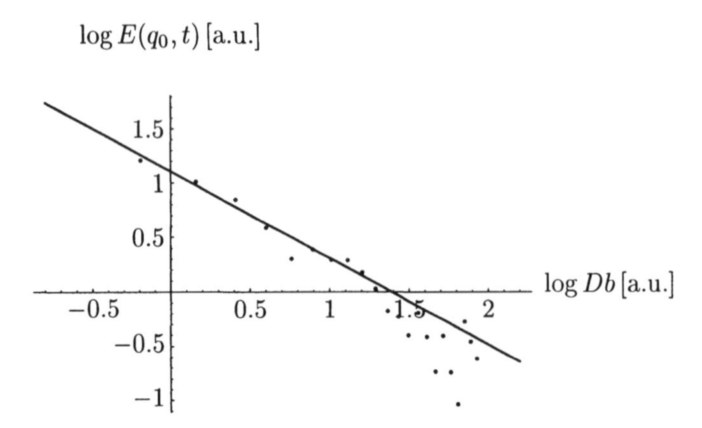

Figure 25: Result of the second ($\tau_2 = 20$msec) τ_2–measurement of the liver tissue S10.

Specimen	α_1	α_2	n_1	n_2
Fib. Mastopathy E3008	0.10	*	0.62	0.65
Liver E3716	0.27	*	1.22	0.70
Liver S101	0.40	*	1.54	0.71
Tumour E5598	0.48	*	2.70	0.88
Tumour E2552	0.20	*	1.19	0.78
Free H_2O	0.91	0.98	*	*

Table 4: Stretching exponents α_i and Rigaut–type power–law exponents n_i for some specimens. The asterisks indicate that no reasonable fit with the corresponding formula was possible.

in the asymptotic fractal model where the constant behaviour turns over to the power–law region, gets shifted to the left, the more ramified the structure becomes.)

Due to the limited accuracy of the measurement and the relatively small width of the experimental window, we do not try to over–interpret the data in the spirit of our analytical anomalous diffusion model. Thus, despite not being a proof for our anomalous diffusion theory developed in Section 4.4, the NMR measurements clearly show a power–law decay in Fourier space ($E(\mathbf{q}, t)$ which corroborates the expected anomalous behaviour in wave vector space (Fourier space), the clear consequence of a stretched exponential in (r, t) space, see also the discussion in Refs. [18,56].

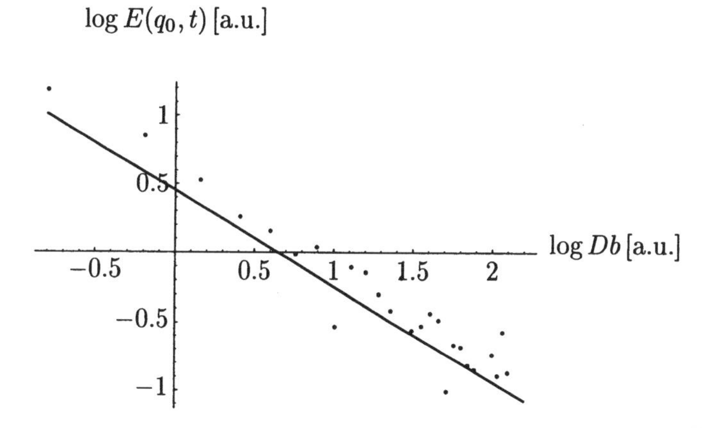

Figure 26: Result of the third ($\tau_2 = 80$msec) τ_2–measurement of the liver tissue S10.

4.8 Generalised Cattaneo equations (GCEs)

Three different generalisations of the Cattaneo equation are discussed in [16], and physical motivations are given. Here the GCE basing upon a non–local transport theory is reviewed.

Following [7,40,90], in media with memory the flux $\jmath$ is related to the previous history of the density ϱ through a relaxation function $K(t)$ as

$$\jmath\,(x,t) = -\int_0^t K(t-t')\varrho_x\,(x,t')\,\mathrm{d}t'. \tag{80}$$

For a suitable choice for $K(t)$, the standard Cattaneo equation is obtained. Introducing (80) into the LHS of equation (57), it follows:

$$\jmath\,(x,t) + \tau\partial_t\jmath\,(x,t) = -\,(\tau\partial_t + 1)\int_0^t K\,(t-t')\,\varrho_x\,(x,t')\,\mathrm{d}t'. \tag{81}$$

By using the Leibniz's formula for the differentiation of an integral this results in

$$\begin{aligned}\jmath\,(x,t) + \tau\partial_t\jmath\,(x,t) &= -\tau K\,(0)\,\varrho\,(x,t) \\ &\quad -\int_0^t \left[\tau\partial_t K\,(t-t') + K\,(t-t')\right]\varrho_x\,(x,t')\,\mathrm{d}t'. \end{aligned}\tag{82}$$

Hence by comparing with Eq (57) it appears clear that $\tau K\,(0) = D$ and $\tau\partial_t K\,(t) + K\,(t) = 0$. Solving this differential equation, the relaxation function that makes the non-local theory of transport compatible with the Cattaneo equation, is obtained:

$$K(t) = \frac{D}{\tau} \exp\left(-\frac{t}{\tau}\right). \tag{83}$$

The objective now is to embed by a similar computation a generalised Cattaneo equation within a theory of transport with memory kernels. In [16] it is shown that a kernel of the form

$$K(u) = \frac{D}{\tau^\gamma} \frac{u^{-1}}{1 + \tau^{-\gamma} u^{-\gamma}} \tag{84}$$

in the Laplace domain, or

$$K(t) = \frac{D}{\tau^\gamma} E_{\gamma,1}\left[-\left(\frac{t}{\tau}\right)^\gamma\right], \tag{85}$$

in term of a generalised Mittag–Leffler function, in t domain, gives rise to a GCE of the form:

$$\partial_t \varrho(x,t) + \tau^\gamma \partial_t^{\gamma-1} \partial_t^2 \varrho(x,t) = D \partial_t^{\gamma-1} \varrho_{xx}(x,t). \tag{86}$$

The corresponding generalised constitutive equation reads:

$$\jmath(x,t) + \tau^\gamma \partial_t^\gamma \jmath(x,t) = -D \partial_t^{\gamma-1} \varrho_x(x,t). \tag{87}$$

Thus, a flux relaxation function of the form (85), which for long times has a power-law behaviour $K(t) \sim t^{-\gamma}$ with $0 < \gamma < 1$, can be associated with a GCE of the form (86).

Using the standard Fourier–Laplace transform technique, one calculates

$$\varrho(k,u) = \frac{\tau^\gamma u^{-1} + u^{-(\gamma+1)}}{\tau^\gamma + u^{-\gamma} + D u^{-2} k^2}. \tag{88}$$

In (x,t)-space, this leads to a modified Gaussian behaviour (stretched Gaussian), due to the occurrence of k^2 in the denominator. All the even spatial moments of $\varrho(x,t)$ exist, whereas all uneven moments vanish. For the MSD $\langle x^2(t)\rangle = -\left.\varrho_{kk}(k,t)\right|_{k=0}$, it results

$$\langle x^2 \rangle \sim \frac{2Dt^{2-\gamma}}{\Gamma(3-\gamma)} \qquad t \gg \tau \tag{89}$$

which corresponds to super–diffusion since $0 < \gamma < 1$. In the short-time limit, it is a ballistic transport

$$\langle x^2 \rangle \sim \frac{D}{\tau^\gamma} t^2, \qquad t \ll \tau \tag{90}$$

which is recovered. Thus, the GCE preserves the standard Cattaneo behaviour for short times $t \ll \tau$. This can be interpreted as the most sensible behaviour for times much shorter than the mean collision time, since we then have a cloud of particles advancing ballistically independently of each other before the diffusion mechanism sets in. As defined in [12], assuming a plane-wave solution and some tedious calculations, the phase velocity $v_{\mathrm{Ph}} \stackrel{\mathrm{def}}{=} \omega/\mathrm{Re}k$ is:

$$v_{\mathrm{Ph}} = \frac{\sqrt{2D}\omega^{\gamma/2}}{\sqrt{\sqrt{1 - 2\tau^\gamma\omega^\gamma \cos\frac{\pi\gamma}{2} + \tau^{2\gamma}\omega^{2\gamma}} + \tau^\gamma\omega^\gamma - \cos\frac{\pi\gamma}{2}}} \tag{91}$$

The asymptotic behaviours are

$$v_{\mathrm{Ph}} \sim \sqrt{\frac{D}{\tau^\gamma}} \qquad \tau\omega \gg 1 \tag{92}$$

$$v_{\mathrm{Ph}} \sim \frac{\sqrt{2D}\omega^{\gamma/2}}{\sqrt{1 - \cos\frac{\pi\gamma}{2}}} \qquad \tau\omega \ll 1. \tag{93}$$

They reduce, for $\gamma \to 1$, to the standard results $v_{\mathrm{Ph}} \sim \sqrt{D/\tau}$ for $\tau\omega \gg 1$, and $v_{\mathrm{Ph}} \sim \sqrt{2D\omega}$ for $\tau\omega \ll 1$.

5 Conclusions

Fractional model equations for relaxation and diffusion processes in complex systems have been reviewed. The discussion of some recent applications show the importance, to have at hands a relatively simple phenomenological theory to describe the typical features of such complex systems: power–law memory and relaxation dynamics, modified Gaussian laws in diffusion.

Acknowledgements

RM would like to thank Albert Compte, Universitat Autónoma de Barcelona for discussions. Financial support from the Deutsche Forschungsgemeinschaft (SFB 239) is gratefully acknowledged. RM is an Alexander von Humboldt Feodor–Lynen fellow.

Appendix. Stable laws

Following Paul Lévy [48], one says that a probability distribution is stable ('loi stable', stable law), if X_1 and X_2 are two independent probability variables dependent on the same probability distribution and their sum

$$c_1 X_1 + c_2 X_2 = cX, \tag{A.1}$$

depends on the same law. c is a positive constant suitably chosen as a function of the c_i. In terms of the characteristic function $\varphi(z) = \mathcal{M}\{e^{izx}\}$, $\mathcal{M}\{\}$ denoting the mean value, this property can be expressed as follows:

$$\varphi(c_1 z)\varphi(c_2 z) = \varphi(cz) \tag{A.2}$$

with the special case

$$\varphi^2(z) = \varphi(qz). \tag{A.3}$$

The characteristic function can be expressed in terms of a stretched exponential:

$$\psi(z) = \log \varphi(z) = \imath\gamma z - c|z|^\alpha \left\{1 + \imath\beta \frac{z}{|z|}\omega(z,\alpha)\right\}, \tag{A.4}$$

where $0 < \alpha \leq 2$, $-1 \leq \beta \leq 1$, $\gamma \in \mathbb{R}$ and $c \geq 0$. Furthermore,

$$\omega(z,\alpha) = \begin{cases} \tan \frac{\pi}{2}\alpha, & \text{falls} \quad \alpha \neq 1 \\ \frac{2}{\pi} \log|z|, & \text{falls} \quad \alpha = 1 \end{cases}. \tag{A.5}$$

α is called characteristic exponent or Lévy index. For $\alpha = 2$ one recovers a Gaussian, for $\alpha = 1$ a Cauchy or Lorentz, and for $\alpha = 1/2$ a Holtsmark distribution.

Appendix. Fox' H–functions

In literature, the H–function is known as generalised Mellin–Barnes function, generalised (Meijer's) G–function or as Fox' H–function. Originally applied in statistics, it arises in the physical sciences or engineering naturally in the solution of linear differential equations of fractional order. The importance of the Fox function is the fact that it includes nearly all special functions occurring in applied mathematics and statistics as its special cases. Even sophisticated functions like Wright's generalised Bessel functions, Maijer's G–function or Maitland's generalised hypergeometric function are embraced by the H-function.

In 1961 Fox defined the H-function in his studies of symmetrical Fourier kernels as the Mellin–Barnes type path integral

$$\begin{aligned} H^{mn}_{pq}(x) &= H^{mn}_{pq}\left[x \,\middle|\, \begin{matrix} (a_p, A_p) \\ (b_q, B_q) \end{matrix} \right] \\ &= H^{mn}_{pq}\left[x \,\middle|\, \begin{matrix} (a_1, A_1), (a_2, A_2), \dots, (a_p, A_p) \\ (b_1, B_1), (b_2, B_2), \dots, (b_q, B_q) \end{matrix} \right] \\ &= \frac{1}{2\pi i} \int_L \mathrm{d}z\, \chi(z) x^z \end{aligned} \tag{B.1}$$

with the integral density:

$$\chi(z) = \frac{\prod_1^m \Gamma(b_j - B_j z) \prod_1^n \Gamma(1 - a_j + A_j z)}{\prod_{m+1}^q \Gamma(1 - b_j + B_j z) \prod_{n+1}^p \Gamma(a_j - A_j z)}. \tag{B.2}$$

(For constraints on the occurring parameters see e.g. [28,52,91]). Note that the path integral in equation (B.1) represents just the inverse Mellin transform of $\chi(s)$.

We list now some very convenient properties of Fox' H-functions [52,91]:

PROPERTY 1: The H-function is symmetric with respect to the permutations of $(a_1, A_1), \dots, (a_n, A_n)$, of $(a_{n+1}, A_{n+1}), \dots, (a_p, A_p)$, of $(b_1, B_1), \dots,$ (b_m, B_m), and of $(b_{m+1}, B_{m+1}), \dots, (b_q, B_q)$.

PROPERTY 2: For $n \geq 1$ and $q > m$ one has

$$\begin{aligned} &H^{mn}_{pq}\left[x \,\middle|\, \begin{matrix} (a_1, A_1), (a_2, A_2), \dots, (a_p, A_p) \\ (b_1, B_1), (b_2, B_2), \dots, (b_{q-1}, B_{q-1}), (a_1, A_1) \end{matrix} \right] \\ &= H^{mn-1}_{p-1q-1}\left[x \,\middle|\, \begin{matrix} (a_2, A_2), \dots, (a_p, A_p) \\ (b_1, B_1), \dots, (b_{q-1}, B_{q-1}) \end{matrix} \right] \end{aligned} \tag{B.3}$$

PROPERTY 3:

$$H^{mn}_{pq}\left[x \,\middle|\, \begin{matrix} (a_p, A_p) \\ (b_q, B_q) \end{matrix} \right] = H^{nm}_{qp}\left[\frac{1}{x} \,\middle|\, \begin{matrix} (1 - b_q, B_q) \\ (1 - a_p, A_p) \end{matrix} \right] \tag{B.4}$$

PROPERTY 4: For $k > 0$

$$H^{mn}_{pq}\left[x \,\middle|\, \begin{matrix} (a_p, A_p) \\ (b_q, B_q) \end{matrix} \right] = k H^{mn}_{pq}\left[x^k \,\middle|\, \begin{matrix} (a_p, kA_p) \\ (b_q, kB_q) \end{matrix} \right] \tag{B.5}$$

PROPERTY 5:

$$x^\sigma H^{mn}_{pq}\left[x \,\middle|\, \begin{matrix} (a_p, A_p) \\ (b_q, B_q) \end{matrix} \right] = H^{mn}_{pq}\left[x \,\middle|\, \begin{matrix} (a_p + \sigma A_p, A_p) \\ (b_q + \sigma B_q, B_q) \end{matrix} \right] \tag{B.6}$$

An H–function can be expressed as a computable series in the form

$$H_{pq}^{mn}(z) = \sum_{h=1}^{m}\sum_{\nu=a}^{\infty} \frac{\prod\limits_{j=1,j\neq h}^{m} \Gamma\left(b_j - B_j(b_h+\nu)/B_h\right)}{\prod\limits_{j=m+1}^{q} \Gamma\left(1-b_j + B_j(b_h+\nu)/B_h\right)}$$

$$\times \frac{\prod\limits_{j=1}^{n} \Gamma\left(1-a_j + A_j(b_h+\nu)/B_h\right)}{\prod\limits_{j=n+1}^{p} \Gamma\left(a_j - A_j(b_h+\nu)/B_h\right)} \times \frac{(-1)^{\nu} z^{(b_h+\nu)/B_h}}{\nu! B_h} \tag{B.7}$$

which is an alternating series and thus shows slow convergence.

The asymptotic expansion of a Fox function are given by Braaksma [6] and are also compiled in [52,91]:

$$H_{pq}^{mn}(x) = \mathcal{O}\left(|x|^{c}\right), \tag{B.8}$$

for small x, where $\mu \geq 0$ and $c = \min \mathrm{Re}(b_j/B_j)$; $j = 1, \ldots, m$. For large x one finds

$$H_{pq}^{mn}(x) = \mathcal{O}\left(|x|^{d}\right), \tag{B.9}$$

where now $\mu > 0$, $n \geq 1$ and $|\arg x| < \alpha\pi/2$, $d = \max \mathrm{Re}((a_j - 1)/A_j)$, $j = 1, \ldots, n$. Is $n = 0$, an exponential decay is recovered for large x:

$$H_{pq}^{m,0}(x) \sim \mathcal{O} \exp\left(-\mu x^{1/\mu} \beta^{1/\mu}\right) x^{(\gamma+1/2)/\mu} \tag{B.10}$$

under the condition $|\arg x| < \lambda\pi/2$ and $\mu > 0$. The parameters are defined as follows:

$$\mu = \sum_{j=1}^{q} B_j - \sum_{j=1}^{p} A_j \tag{B.11a}$$

$$\alpha = \sum_{j=1}^{n} A_j - \sum_{j=n+1}^{p} A_j + \sum_{j=1}^{m} B_j - \sum_{j=m+1}^{q} B_j \tag{B.11b}$$

$$\gamma = \sum_{j=1}^{q} b_j - \sum_{j=1}^{p} a_j + \frac{p}{2} - \frac{q}{2} \tag{B.11c}$$

$$\lambda = \sum_{j=1}^{m} B_j - \sum_{j=m+1}^{q} B_j - \sum_{j=1}^{p} A_j \tag{B.11d}$$

Appendix. Fractal Fourier transform

To get an explicit expression for the Fourier transform on fractals one can re-write the definition of a spherical N-dimensional ($N \in \mathbb{N}$) Fourier transform by use of the surface of the corresponding unit-hypersphere

$$\mathcal{F}_{(N)}\{f(r);q\} = (2\pi)^{-N/2} S_N(1) \int \mathrm{d}r\, r^{N-1} f(r) \frac{\sin qr}{qr} \tag{C.1}$$

where $S_N(R) = 2\pi^{N/2} R^{N-1}/\Gamma(N/2)$ is well-known from statistical mechanics. For a fractal hypersphere this relation is now—heuristically—generalised (see Ref. [92]) and usually written as $S_\mathrm{f}(R) = 2\pi^{d_f/2} R^{d_f-1}/\Gamma(d_f/2)$ so that a fractal volume element $\mathrm{d}V_\mathrm{f} = 2\pi^{d_f/2} r^{d_f-1}/\Gamma(d_f/2)\,\mathrm{d}r$ is recovered. Thus the fractal Fourier transform is defined as

$$\mathcal{F}_\mathrm{f}\{f(r);q\} = (2\pi)^{-N/2} \frac{2\pi^{d_f/2}}{\Gamma(d_f/2)} q^{-1} \mathcal{F}_S \left\{ r^{d_f-2} f(r); q \right\}. \tag{C.2}$$

The fractal Fourier transform can be expressed by an ordinary Fourier sine transform. In the main text we assume $\mathrm{d}V_\mathrm{f} \propto r^{\tilde{d}-1}$.

References

1. R.H. Austin, K.W. Beeson, L. Eisenstein, H. Frauenfelder and I.C. Gunsalus, *Biochemistry* **14** 5355 (1975)
2. A.L. Blatz and K.L. Magleby, *J. Physiol.* (London) **378** 141 (1986)
3. A. Blumen, J. Klafter and G. Zumofen in *Optical Spectroscopy of Glasses*, Ed. I. Zschokke (Reidel, Dordrecht, 1986)
4. A. Blumen, G. Zumofen and J. Klafter, *Phys. Rev.* A **40**, 3964 (1989)
5. J.-P. Bouchaud and A. Georges, *Phys. Rep.* **195** 127 (1990)
6. B.L.J. Braaksma, *Compos. Math.* **15** 239 (1964)
7. J.F. Brady and D.L. Koch, *Disorder and Mixing* Ed. E Guyon *et al* (Kluver, Doordrecht, 1988)
8. H.C. Brinkman, *J. Chem. Phys.* **20** 571 (1952)
9. A. Bunde and J. Draeger, *Phys. Rev.* E **52** 53 (1995)
10. P. Callaghan, *Principles of Nuclear Magnetic Resonance Microscopy* (Clarendon Press, Oxford, 1991)
11. J. Casas-Vázquez and D. Jou ed., *Rheological Modeling: Thermodynamical and Statistical approaches* (Springer, Berlin, 1991)
12. J. Casas-Vázquez, D. Jou and Lebon, *Extended Irreversible Thermodynamics* (Springer, Berlin, 1996)

13. M. Caputo and F. Mainardi, *Riv. Nuovo Cimento* (ser. 2) **1** 161 (1971)
14. G. Cattaneo, *Atti. Sem. Mat. Fis. Univ. Modena* **3** 83 (1948)
15. A. Compte, R. Metzler and J. Camacho, *Phys. Rev.* E **56** 1445 (1997)
16. A. Compte and R. Metzler, *Generalised Cattaneo equation for the description of anomalous transport processes*, *J. Phys. A: Math. Gen.* in press (1997)
17. J. Crank, *The Mathematics of Diffusion* (Clarendon Press, Oxford, 1970)
18. R.A. Damion and K.J. Packer, *Proc. Roy. Soc.* **453** 205 (1997)
19. J. Dollinger, R. Metzler and T.F. Nonnenmacher, *Bi–asymptotic fractals: Fractals between lower and upper bounds*, subm.
20. H. Eilers, *Kolloid–Z.* **97** 313 (1941)
21. K. Falconer, *Fractal Geometry—Mathematical Foundations and Applications* (John Wiley, Chichester, 1990)
22. J. Feder, *Fractals* (Plenum Press, New York, 1988)
23. D. Ferry, *Viscoelastic Properties of Polymers* (John Wiley, New York, 1970)
24. H. Frauenfelder, *Am. NY Acad. Sci.* **504** 151 (1987)
25. C. Friedrich, *Rheol. Acta* **30** 151 (1991) and 1992 *Phil. Mag. Lett.* **66** 287 (1992)
26. M. Ghosh and B.K. Chakrabarti, *Indian J. Phys.* **65A** 1 (1991)
27. M. Giona and H. E. Roman, *Physica* **185A** 87 (1992)
28. W.G. Glöckle and T.F. Nonnenmacher, *Macromolecules* (1991)
29. W.G. Glöckle and T.F. Nonnenmacher, *J. Stat. Phys.* **71** 741 (1993)
30. W.G. Glöckle and T.F. Nonnenmacher, *Rheol. Acta* **33** 337 (1994)
31. W.G. Glöckle and T.F. Nonnenmacher in [67]
32. W.G. Glöckle and T.F. Nonnenmacher *Biophys. J.* **68** 46 (1995)
33. R. Gorenflo and S. Vessela, *Abel Integral Equations: Analysis and Application* Lecture Notes in Mathematics, Vol. **1461** (Springer Verlag, Berlin, 1991)
34. R. Gorenflo and Y. Luchkow, *Int. Transf. Spec. Funcs.* **5** 47 (1997)
35. R.A. Guyer, *Phys. Rev.* A **29** 2751 (1984)
36. S. Havlin, D. Movshovitz, B. Trus and G.H. Weiss, *J. Phys. A: Math. Gen.* **18** L719 (1985)
37. S. Havlin and D. Ben–Avraham, *Adv. Phys.* **36** 695 (1987)
38. R. Hilfer and L. Anton, *Phys. Rev.* A **35** R848 (1995)
39. I.E.T. Iben, D. Braunstein, W. Doster, H. Frauenfelder, M.K. Hong, J.B. Johnson, S. Luck, P. Ormos, A. Schulte, P.J. Steinbach, A.H. Xie and R.D. Young, *Phys. Rev. Lett.* **62** 1916 (1989)
40. D.L. Koch and J.F. Brady, *Chem. Eng. Sci.* **42** 1377 (1987)
41. H.–G. Kilian, M. Strauss and W. Hamm, *Rubber Chem. Technol.* **67** 1

(1994)
42. H.-G. Kilian, R. Metzler and B. Zink *Aggregate model of liquids*, *J. Chem. Phys.* in press
43. R. Kimmich, W. Unrath, G. Schnur and E. Rommel 1991 *J. Magn. Res.* **91** 136 (1991)
44. J. Klafter, G. Zumofen and A. Blumen, *J. Phys. A: Math. Gen.* **25** 4835 (1991)
45. M. Köpf, C. Corinth, O. Haferkamp and T.F. Nonnenmacher, *Biophys. J.* **70** 2950 (1996)
46. M. Köpf, R. Metzler, O. Haferkamp and T.F. Nonnenmacher, *NMR studies of anomalous diffusion in biological tissues: Experimental observation of Lévy stable processes* in press
47. C. Kumar and S.R. Shenoy, *Solid State Commun.* **57** 927 (1986)
48. P. Lévy, *Théorie de l'addition des variables aléatoires* (Masson, Paris, 1954)
49. G. Losa and T.F. Nonnenmacher, *Mod. Pathology* **9** 174 (1997)
50. F. Mainardi, *J. Alloy and Comp.* **211/212** 534 (1994)
51. B.B. Mandelbrot, *The fractal geometry of nature* (Freeman, New York, 1983)
52. A.M. Mathai and R.K. Saxena, *The H-function with Applications in Statistics and Other Disciplines* (Wiley Eastern Ltd., New Delhi, 1978)
53. T. Mattfeld in [67]
54. R. Metzler, W.G. Glöckle and T.F. Nonnenmacher, *Physica* **211A** 13 (1994)
55. R. Metzler, W. Schick, H.-G. Kilian and T.F. Nonnenmacher, *J. Chem. Phys.* **103** 7180 (1995)
56. R. Metzler and T.F. Nonnenmacher, *J. Phys. A: Math. Gen.* **30** 1089 (1997)
57. R. Metzler, W.G. Glöckle, T.F. Nonnenmacher and B.J. West, *"Fractional tuning" of the Riccati equation*, *Fractals* in press
58. K.S. Miller and B. Ross, *An introduction to the fractional calculus and fractional differential equations* (Wiley, New York, 1993)
59. E.W. Montroll and M.F. Shlesinger, *Proc. Natl. Acad. Sci. USA* **79** 3380 (1982)
60. K.L. Ngai in [73]
61. K.L. Ngai and R.W. Rendell in *Relaxation in Complex Systems and Related Topics*, Ed. I.A. Campbell and C. Giovanella *NATO ASI Series* B (Plenum Press, New York, 1989)
62. K.L. Ngai, K.A. Rajagopal and S. Teitler, *J. Chem. Phys.* **88** 5086 (1988)
63. T.F. Nonnenmacher, *J. Non-Equil. Thermod.* **9** 171 (1984)

64. T.F. Nonnenmacher and D.J.F. Nonnenmacher, *Phys. Lett.* A **140** 323 (1989)
65. T.F. Nonnenmacher in [11]
66. T.F. Nonnenmacher in [67]
67. T.F. Nonnenmacher, G.A. Losa and E.R. Weibl *Fractals in Biology and Medicine* (Birkhäuser, Basel, 1993)
68. T.F. Nonnenmacher and R. Metzler, *Fractals* **3** 557 (1995) and in C.J.G. Evertsz, H.–O. Peitgen and R.F. Voss Ed., *Fractal Geometry and Analysis, The Mandelbrot Festschrift, Curaçao 1995* (World Scientific, Singapore, 1996)
69. K.B. Oldham and J. Spanier, *The fractional calculus* (Academic Press, New York, 1974)
70. B. O'Shaugnessy and I. Procaccia, *Phys. Rev. Lett.* **54** 455 (1985)
71. E. Pezron, A. Ricard and L. Leibler, *J. Polym. Sci.* **28** 2445 (1990)
72. C. van der Poel, *Rheol. Acta* **1** 158 (1958)
73. T.V. Ramakrishnan and L. Raj Lakshmi Ed., *Non–Debye Relaxation in Condensed Matter* (World Scientific, Singapore, 1987)
74. R. Rammal and G. Toulouse, *J. Phys. Lett.* (Paris) **44** L–13 (1983)
75. M. Reiner, *Rheologie in elementarer Darstellung* (VEB Fachbuchverlag, Leipzig, 1969)
76. J.P. Rigaut, *J. Microsc.* **133** 41 (1984)
77. H.E. Roman and P. Alemanyi, *J. Phys. A: Math. Gen.* **25** 2107 (1994)
78. H.E. Roman, *Phys. Rev.* E **51** 5422 (1995)
79. B. Ross Ed., *The Fractional Calculus and its Application*, *Lecture Notes in Mathematics* **457** (Springer Verlag, Berlin, 1975)
80. B. Sakmann and E. Neher, *Single–channel recording* (Plenum, New York, 1983)
81. H. Schiessel and A. Blumen, *J. Phys. A: Math. Gen.* **26** 5057 (1993)
82. H. Schiessel, R. Metzler, A. Blumen and T.F. Nonnenmacher, *J. Phys. A: Math. Gen.* **28** 6567 (1995)
83. W.R. Schneider in *Stochastic Processes in Classical and Quantum Systems*, S. Albeverio, G. Casati and D. Merlini editors (Springer, Berlin, 1986)
84. W.R. Schneider and W. Wyss, *J. Math. Phys.* **30** 134 (1989)
85. R.K. Schofield and W.G. Scott Blair, *Proc. Roy. Soc.* A **132** 707 (1932)
86. M.F. Shlesinger and J. Klafter in *Fractals in Physics*, L. Pietronero and E. Tossati editors (Elsevier, Amsterdam, 1986)
87. M.F. Shlesinger, *Ann. Rev. Phys. Chem.* **39** 629 (1988)
88. M.F. Shlesinger, *Fractals* **3** (1995)
89. V.I. Smirnow, *Lehrgang der höheren Mathematik* (VEB Deutscher Verlag

der Wissenschaften, Berlin, 1989)
90. S.L. Sobolev, *Sov. Phys. Usp.* **34** 217 (1991)
91. H.M. Srivastava, K.C. Gupta and S.P. Goyal, *The H-function of one and two variables with applications* (South Asian Publ., New Delhi, 1982)
92. H. Takayasu, *Fractals in the physical sciences* (Manchester University Press, Manchester, 1990)
93. A.V. Tobolsky and E. Catsiff, *J. Polym. Sci.* **19** 111 (1956)
94. N.W. Tschoegl, *The Phenomenological Theory of Linear Viscoelastic Behavior* (Springer, Berlin, 1989)
95. M.O. Vlad, R. Metzler, T.F. Nonnenmacher and M.C. Mackey, *J. Math. Phys.* **37** 2279
96. I.M. Ward, *Mechanical Properties of Solid Polymers* (Wiley, Chichester, 1983)
97. B. J. West and W. Deering, *Phys. Rep.* **246** 1 (1994)
98. B.J. West, P. Grigolini, R. Metzler and T.F. Nonnenmacher, *Phys. Rev.* E **55** 99 (1997)
99. W. Wyss, *J. Math. Phys.* **27** 2782 (1986)

CHAPTER IX

FRACTIONAL CALCULUS AND REGULAR VARIATION IN THERMODYNAMICS

R. HILFER
ICA-1, Universität Stuttgart, 70569 Stuttgart, Germany
Institut für Physik, Universität Mainz, 55099 Mainz, Germany

Contents

1 Introduction

Almost all applications of fractional calculus to physics propose fractional equations of motion for various complex nonequilibrium phenomena (see the other chapters in this book). Despite this major direction of research within "fractional physics" there is also a somewhat different application of fractional calculus to equilibrium thermodynamics and statistical mechanics where time plays no role [1,2,3].

Many analytical and numerical investigations of critical points in fluids, magnets and other systems have led to a deepened understanding of phase transitions [4,5,6,7,8,9,10,11,22]. An important objective in these studies is the determination of the phase diagram for the systems of interest [5,11]. It is interesting that contemporary scaling and renormalization theories have difficulties predicting critical points or phase boundaries for continuous phase transitions [10,12]. On the other hand the Clausius-Clapeyron formula from the last century [13] allows to determine such nonuniversal features. Remarkably, fractional calculus seems to provide tools for extending some of the classical results for first order transitions to continuous transitions [1,2,3]. Examples of such extensions are discussed in this chapter. Many concepts and results from the classical theory appear to have a fractional counterpart that provides a deepened perspective on the phenomenological scaling theory of phase transitions.

Discontinuities in the first derivatives of thermodynamic potentials correspond to thermodynamic phase transitions with macroscopic phase coexistence [22]. Ehrenfest advanced the idea that discontinuities in higher order derivatives may be used to characterize critical behaviour [14]. It is natural to extend Ehrenfests idea using derivatives of noninteger (fractional) order [1,2,3].

Given the general objective of extending Ehrenfests classification scheme by employing fractional calculus, and given that the presentation should be self contained, it is useful to discuss first fractional calculus, then the traditional thermodynamic theories, and only then the application of the former to the latter. Let me therefore begin by establishing the notation and some basic facts for fractional calculus. Of particular interest will be the behaviour of fractional integrals and derivatives near their lower limit. Regular variation appears as a natural concept in this context. Its discussion is followed by recalling some elements from thermodynamics, scaling theory, and the classification scheme of Ehrenfest. After these preparations the fractional generalization of Ehrenfests classification scheme is carried out, and it leads to interesting universal and nonuniversal predictions for critical behaviour. My presentation concludes by showing that the generalized order of the classical van der Waals (mean field) critical point is 4/3 rather than 2.

2 Mathematical Background

2.1 Fractional Integrals

Definition 2.1 (Riemann-Liouville fractional integral). *The right-sided Riemann-Liouville fractional integral of order $\alpha > 0$ is defined for a function $f \in L^1_{\text{loc}}(a,\infty)$ as*

$$(\mathrm{I}^\alpha_{a+} f)(x) = \frac{1}{\Gamma(\alpha)} \int_a^x (x-y)^{\alpha-1} f(y)\, \mathrm{d}y \tag{1}$$

for $x > a$, the left-sided Riemann-Liouville fractional integral is defined as

$$(\mathrm{I}^\alpha_{a-} f)(x) = \frac{1}{\Gamma(\alpha)} \int_x^a (y-x)^{\alpha-1} f(y)\, \mathrm{d}y \tag{2}$$

for $x < a$.

The domain of definition is taken as the space of locally integrable functions [18], $L^1_{\text{loc}}(a,\infty)$ consisting of functions that are integrable on every compact subset of $[a,\infty[$ (see e.g. [15]). As an operator on Lebesgue spaces I^α_{a+} maps $L^p(a,\infty)$ continuously into $L^q(a,\infty)$ for $1 \le p,q \le \infty, \alpha > 0$ if and only if $0 < \alpha < 1, 1 < p < 1/\alpha$ and $q = p/(1-\alpha p)$ [16,17]. For $\alpha = 0$ the operator $\mathrm{I}^0_{a+} = \mathbf{1}$ is defined as the identity. The additivity of exponents (or semigroup property)

$$\mathrm{I}^\alpha_{a+}\mathrm{I}^\beta_{a+} = \mathrm{I}^{\alpha+\beta}_{a+}, \tag{3}$$

valid for all $\alpha, \beta > 0$, is basic to the subject. Riemann-Liouville integrals with lower limit $a = -\infty$ are frequently called Weyl fractional integrals and denoted as $\mathrm{I}^\alpha_{+} = \mathrm{I}^\alpha_{-\infty+}$.

The following simple observation is a special case of a more general theorem. It says that $(\mathrm{I}^\alpha_{0+} f)(x)$ varies essentially as a power law when $x \to 0$. The concept of regular variation will be discussed in section 2.3.

Theorem 2.1 (Regular variation of fractional integrals). *Let $f : \mathbb{R} \to \mathbb{R}$ be a continuous and bounded function with $\lim_{x\to 0} f(x) = f(0) \neq 0$, and let $0 < \alpha < 1$. Then the limit*

$$\lim_{x\to 0} \frac{(\mathrm{I}^\alpha_{0+} f)(\lambda x)}{(\mathrm{I}^\alpha_{0+} f)(x)} = g(\lambda) \tag{4}$$

exists and is finite for all $0 < \lambda < 1$.

Proof. Substituting $z = y/(\lambda x)$ gives

$$\begin{aligned}
\lim_{x\to 0} \frac{(\mathrm{I}^{\alpha}_{0+}f)(\lambda x)}{(\mathrm{I}^{\alpha}_{0+}f)(x)} &= \lim_{x\to 0} \frac{\int_0^{\lambda x}(\lambda x - y)^{\alpha-1}f(y)dy}{\int_0^{x}(x - y)^{\alpha-1}f(y)dy} \\
&= \lim_{x\to 0} \frac{(\lambda x)^{\alpha}\int_0^{1}(1-z)^{\alpha-1}f(\lambda xz)dz}{x^{\alpha}\int_0^{1}(1-z)^{\alpha-1}f(xz)dz} \\
&= \lambda^{\alpha}. \qquad (5)
\end{aligned}$$

Thus $g(\lambda) = \lambda^{\alpha}$. The interchange of the limits in the last step is justified because $f(x)$ can be majorized as $|f(x)| \le cx^{\beta-1}$ for suitable $c > \sup_{x\in[0,1]}|f(x)|$ and $0 < \beta < 1$, the resulting Euler integral being finite under the assumed conditions. □

2.2 Fractional Derivatives

There are various definitions for fractional derivatives. The most important of them are discussed in detail in Chapter I of this book. A common construction represents fractional derivatives through a combination of integer order derivatives and fractional integrals. The following definition generalizes this approach. It was already found to be useful in chapter II .

Definition 2.2 (Fractional derivatives). *The (right-/left-sided) fractional derivative of order* $0 < \alpha < 1$ *and type* $0 \le \mu \le 1$ *with respect to* x *is defined by*

$$\mathrm{D}^{\alpha,\mu}_{a\pm}f(x) = \left(\pm \mathrm{I}^{\mu(1-\alpha)}_{a\pm}\frac{\mathrm{d}}{\mathrm{d}x}(\mathrm{I}^{(1-\mu)(1-\alpha)}_{a\pm}f)\right)(x) \qquad (6)$$

for functions for which the expression on the right hand side exists.

The most commonly used fractional derivatives are those of type $\mu = 0$. For $a > -\infty$ they are named Riemann-Liouville fractional derivatives and denoted by the symbols $\mathrm{D}^{\alpha}_{a\pm} := \mathrm{D}^{\alpha,0}_{a\pm}$. For $a = \mp\infty$ they are named Weyl fractional derivatives and denoted by $\mathrm{D}^{\alpha}_{\pm} = \mathrm{D}^{\alpha,0}_{\mp\infty\pm}$. The notation $\mathrm{D}^{\alpha,\mu}_{x,a\pm}f(x)$ is used to indicate the variable of differentiation explicitly.

Definition 2.3 (Fractional derivatives at the limit). *For* $-\infty < a < \infty$ *the fractional derivative of order* $0 < \alpha < 1$ *and type* $0 \le \mu \le 1$ *at the limit* a *is defined by*

$$\mathrm{D}^{\alpha,\mu}f(a) = \pm \lim_{x\to a\pm} \mathrm{D}^{\alpha,\mu}_{a\pm}f(x) \qquad (7)$$

whenever the two limits exist and are equal. If $\mathrm{D}^{\alpha,\mu}f(a)$ *exists the function is called fractionally differentiable at the limit* a.

While fractional derivatives are nonlocal operators the fractional derivative at the limit is a local operator. Two fractional derivative operators $\mathrm{D}^{\alpha,\mu}_{a+}$ and $\mathrm{D}^{\beta,\mu}_{a+}$ do not commute, and the semigroup property for exponents, eq. (3), does not hold. This sets them apart from integer order derivatives. Note also, that the constant function cannot be Weyl-differentiated, which is sometimes overlooked in physical applications. The differences to the behaviour of ordinary derivatives highlight the necessity of modifying and extending the definition in various directions (see Chapter I and III). For the purposes of the present chapter the definition in eq. (6) will be sufficient.

The relation between fractional derivatives of the same order but different types $\mu < \lambda$ depends upon the behaviour of f near the limit.

Proposition 2.1. *Let* $f : [a, A] \to \mathbb{R}$ *where* $a < A < \infty$ *be absolutely continuous*[a]. *If* $(\mathrm{I}^{\beta}_{a+}f)(a)$ *is finite for some* β *with* $0 \le \beta \le 1-\alpha$ *then*

$$(\mathrm{D}^{\alpha,\mu}_{a+}f)(x) = (\mathrm{D}^{\alpha,\lambda}_{a+}f)(x) \tag{8}$$

whenever $0 \le \mu < \lambda < \frac{1-\alpha-\beta}{1-\alpha}$, *and*

$$(\mathrm{D}^{\alpha,\mu}_{a+}f)(x) = (\mathrm{D}^{\alpha,(1-\alpha-\beta)/(1-\alpha)}_{a+}f)(x) + \frac{(\mathrm{I}^{\beta}_{a+}f)(a)}{\Gamma(1-\alpha-\beta)(x-a)^{\alpha+\beta}} \tag{9}$$

holds for all $\mu < \lambda = \frac{1-\alpha-\beta}{1-\alpha}$.

Proof. The relation

$$(\mathrm{D}\mathrm{I}^{1-\alpha}_{a+}f)(x) = \frac{f(a)}{\Gamma(1-\alpha)(x-a)^{\alpha}} + (\mathrm{I}^{1-\alpha}_{a+}\mathrm{D}f)(x) \tag{10}$$

holds under the assumed conditions on f [18]. Setting temporarily $g(x) = (\mathrm{I}^{(1-\lambda)(1-\alpha)}_{a+}f)(x)$ and using eq.(3) one finds

$$\begin{aligned}
\mathrm{D}^{\alpha,\lambda}_{a+}f(x) &= (\mathrm{I}^{\lambda(1-\alpha)}_{a+}\mathrm{D}\mathrm{I}^{(1-\lambda)(1-\alpha)}_{a+}f)(x) \\
&= (\mathrm{I}^{\mu(1-\alpha)}_{a+}\mathrm{I}^{(\lambda-\mu)(1-\alpha)}_{a+}\mathrm{D}g)(x) \\
&= \mathrm{I}^{\mu(1-\alpha)}_{a+}\left((\mathrm{D}\mathrm{I}^{(\lambda-\mu)(1-\alpha)}_{a+}g)(x) - \frac{(x-a)^{(\lambda-\mu)(1-\alpha)-1}}{\Gamma((\lambda-\mu)(1-\alpha))}g(a)\right) \\
&= (\mathrm{D}^{\alpha,\mu}_{a+}f)(x) - \frac{(x-a)^{\lambda(1-\alpha)-1}}{\Gamma(\lambda(1-\alpha))}g(a)\,,
\end{aligned} \tag{11}$$

[a] If $A = \infty$ let f be continuously differentiable and such that it vanishes together with its derivative as $|x|^{\alpha-1-\varepsilon}$ for some $\varepsilon > 0$.

In the last step the fractional integral

$$\mathrm{I}_{a+}^{\alpha}(x-a)^{\beta} = \frac{\Gamma(\beta+1)}{\Gamma(\alpha+\beta+1)}(x-a)^{\alpha+\beta} \tag{12}$$

was used. For $\lambda < \frac{1-\alpha-\beta}{1-\alpha}$ eq. (8) follows because $g(a) = 0$, and eq. (9) follows by setting $\lambda = \frac{1-\alpha-\beta}{1-\alpha}$. □

The proposition highlights the important role played by initial and boundary values when dealing with fractional derivatives.

Proposition 2.2. *Let $f : [a, A] \to \mathbb{R}$ where $a < A < \infty$ be a differentiable and monotonously increasing function. Then $(\mathrm{D}_{a+}^{\alpha,1} f)(x) \geq 0$ for all α for which it exists.*

Proof. $f'(x) \geq 0$ by monotonicity. Also $(x-y)^{-\alpha} \geq 0$ holds for all y with $a < y < x$. Therefore $\int_a^x (x-y)^{-\alpha} f'(y) \mathrm{d}y \geq 0$ and the result follows from the definitions. □

2.3 Regular Variation

The theory of regularly varying functions is based on the following definition [19,20,21].

Definition 2.4 (Regular variation). *A positive and measurable function $f : \mathbb{R} \to \mathbb{R}$ is called regularly varying at infinity if*

$$\lim_{x\to\infty} \frac{f(bx)}{f(x)} = g(b) \tag{13}$$

for all $b > 0$ and an arbitrary function g. A function f is called slowly varying at infinity if

$$\lim_{x\to\infty} \frac{f(bx)}{f(x)} = 1 \tag{14}$$

for all $b > 0$. A function $f(x)$ is called regularly (slowly) varying at $a \in \mathbb{R}$ if $f(1/(x-a))$ is regularly (slowly) varying at infinity.

The importance of regular variation for physics and other disciplines results from the following two theorems.

Theorem 2.2 (Characterization theorem). *If $f > 0$ is measurable and*

$$\lim_{x\to\infty} \frac{f(\lambda x)}{f(x)} = g(\lambda) \in (0, \infty) \tag{15}$$

holds for all λ in a set of positive measure, then

1. *the limit exists for all $\lambda > 0$,*
2. *there exists a number $\alpha \in \mathbb{R}$ such that $g(\lambda) = \lambda^\alpha$ for all $\lambda > 0$,*
3. *$f(x) = x^\alpha \Lambda(x)$ with $\Lambda(x)$ a slowly varying function, i.e.*

$$\lim_{x\to\infty} \frac{\Lambda(bx)}{\Lambda(x)} = 1 \tag{16}$$

   for all $b > 0$.

4. *for $x \to \infty$*

$$\lim_{x\to\infty} f(x) = \begin{cases} \infty & \text{if} \quad \alpha > 0 \\ 0 & \text{if} \quad \alpha < 0 \end{cases} \tag{17}$$

Partial Proof. To prove 2. note that if b, c are chosen such that (15) holds then

$$g(bc) = \lim_{x\to 0} \frac{f(bcx)}{f(x)} \frac{f(cx)}{f(cx)} = \lim_{x\to 0} \frac{f(bcx)}{f(cx)} \lim_{x\to 0} \frac{f(cx)}{f(x)} = g(b)g(c) \; . \tag{18}$$

Setting $h(x) = \log(g(\mathrm{e}^x))$ this becomes

$$h(x+y) = h(x) + h(y) \tag{19}$$

whose only measurable solution is $h(x) = \alpha x$. Thus

$$g(x) = \exp(h(\log x) = \exp(\alpha \log x) = x^\alpha \tag{20}$$

as claimed in 2. Now let $\Lambda(x) = f(x)/x^\alpha$. Then

$$\lim_{x\to 0} \frac{\Lambda(bx)}{\Lambda(x)} = \lim_{x\to 0} \frac{f(bx)x^\alpha}{(bx)^\alpha f(x)} = g(b)b^{-\alpha} = 1 \tag{21}$$

and this proves 3. The last statement 4. follows directly from $f(x) = x^\alpha \Lambda(x)$. For the proof of 1. see [19,20,21]. □

The number α is called index of regular variation. The set of all regularly varying functions of index α is denoted as RV^α. Functions in RV^0 are slowly varying. Note that the set RV^0 is closed under pointwise addition, multiplication, and composition of functions. This is not true for RV^α for $\alpha \neq 0$. If $f \in RV^\alpha$ for $\alpha \neq 0$ then $\lim_{x\to\infty} f(x) = 0$ if $\alpha > 0$ while $\lim_{x\to\infty} f(x) = \infty$ if $\alpha < 0$.

The characterization theorem reveals that the index of regular variation resembles the so called scaling exponent in the theory of critical phenomena. There one usually writes [22,10]

$$f(x) \sim x^{\alpha} \tag{22}$$

for $x \to \infty$. The notation $\sim$ means that there exists a finite α such that

$$\lim_{x\to\infty} \frac{\log f(x)}{\log x} = \alpha \tag{23}$$

holds. The number α is then called a scaling exponent. A related concept from mathematics is the (upper) [b] order $\alpha(f)$ of a positive function f. It is defined by

$$\alpha(f) = \limsup_{x\to\infty} \frac{\log f(x)}{\log x} \tag{24}$$

which may become infinite. While there are many positive functions for which the scaling exponent in eq. (23) does not exist, the order defined in eq. (24) exists for all positive functions f. To exclude those cases in which the scaling exponent does not exist it is usually argued that they play no role for the particular problem at hand (such as scaling theory in thermodynamics). It is therefore of some general interest that the need for such assumptions can be partially eliminated by another theorem from the theory of regularly varying functions whose proof can be found in [21].

Theorem 2.3 (Approximation theorem). *Let $g > 0$ be a positive function of finite order α. Then there exists a function $f \in RV^{\alpha}$ such that*

$$\limsup_{x\to\infty} \frac{f(x)}{g(x)} = 1 \tag{25}$$

Thus every positive function which needs not even to be measurable can be approximated by a regularly varying function. Moreover $f(x) \geq g(x)$ for large x, and there is sequence of points $x_n \to \infty$ such that $|f(x_n) - g(x_n)| = 0$. The two theorems together provide the connection between physical scaling theories and the theory of regular variation.

Fractional calculus is closely related to the theory of regularly varying functions [19,20] although this connection is rarely appreciated [1,3]. Equation (4) shows that the fractional Riemann-Liouville integral of a continuous and bounded function varies regularly upon approaching its lower limit. By definition (6) this property carries over to fractional derivatives.

[b] Replacing lim sup with lim inf defines the lower order of a function f.

Theorem 2.4 (Regular variation of fractional derivatives). *Let the function $f : [0,\infty[\to \mathbb{R}$ be monotonously increasing with $f(x) \geq 0$ and $f(0) = 0$, and such that $(\mathrm{D}_{0+}^{\alpha,\lambda} f)(x)$ with $0 < \alpha < 1$ and $0 \leq \lambda \leq 1$ is also monotonously increasing on a neighbourhood $[0,\delta]$ for small $\delta > 0$. Let $0 \leq \beta < \lambda(1-\alpha)+\alpha$, let $C \geq 0$ be a constant and $\Lambda(x) \in RV^0$ a slowly varying function for $x \to 0$. Then*

$$\lim_{x\to 0} \frac{f(x)}{x^{\beta}\Lambda(x)} = C \tag{26}$$

holds if and only if

$$\lim_{x\to 0} \frac{(\mathrm{D}_{0+}^{\alpha,\lambda} f)(x)}{x^{\beta-\alpha}\Lambda(x)} = C\frac{\Gamma(\beta+1)}{\Gamma(\beta-\alpha+1)} \tag{27}$$

holds.

Proof. By the Hardy-Littlewood-Karamata theorem [19,20,21] eq. (26) for f is equivalent with the Laplace transform $f(u)$ obeying

$$\lim_{u\to\infty} \frac{f(u)}{u^{-\beta}\Lambda(1/u)} = C\Gamma(\beta+1) \tag{28}$$

and hence also with

$$\lim_{u\to\infty} \frac{f(u) - u^{\lambda(\alpha-1)-\alpha}c_0}{u^{-\beta}\Lambda(1/u)} = C\Gamma(\beta+1) \tag{29}$$

for an arbitrary constant c_0 by virtue of $\beta < \lambda(1-\alpha)+\alpha$. It follows that

$$\lim_{u\to\infty} \frac{u^{\alpha}f(u) - u^{\lambda(\alpha-1)}c_0}{u^{\alpha-\beta}\Lambda(1/u)} = C\Gamma(\beta+1)\frac{\Gamma(\beta-\alpha+1)}{\Gamma(\beta-\alpha+1)}. \tag{30}$$

Applying the Hardy-Littlewood-Karamata Theorem in the reverse direction and recognizing the numerator as the Laplace transform of $(\mathrm{D}_{0+}^{\alpha,\lambda} f)(x)$ gives eq. (27). □

2.4 Convexity

Convex functions play a central role in thermodynamics and it is useful to collect various of their properties. A real valued function $f : \mathbb{V} \to \mathbb{R}$ on a vector space $\mathbb{V}$ is called convex if and only if

$$f(\lambda x + (1-\lambda)y) \leq \lambda f(x) + (1-\lambda)f(y) \tag{31}$$

for all $x, y \in \mathbb{V}$ and $0 \leq \lambda \leq 1$. Convexity has a number of useful implications for thermodynamic functions.

Theorem 2.5 (Consequences of convexity).

1. *A twice continuously differentiable function $f \in C^2(\mathbb{R})$ is convex if and only if $f'' \geq 0$.*

2. *If f is convex on $\mathbb{R}$ then*

   (a) *at every point x the left and right sided derivative*

   $$(\partial^{\pm} f)(x) = \lim_{\varepsilon \to 0+} \frac{f(x \pm \varepsilon) - f(x)}{\pm \varepsilon} \tag{32}$$

   exists.

   (b) *The derivative is monotonously increasing, i.e. for $x < z$ one has $(\partial^- f)(x) \leq (\partial^+)(z) \leq (\partial^- f)(z) \leq (\partial^+ f)(z)$.*

   (c) *$(\partial^- f)(x) = (\partial^+ f)(x)$ for all but countably many x.*

3. *For any convex function f on $\mathbb{R}$ and all x there exists a number a such that $f(z) - f(x) \geq a(z - x)$ holds for all z.*

4. *A convex function f on $\mathbb{R}^n$ is continuous. It fulfills a Hölder (Lipschitz) condition, namely for fixed y*

   $$|f(x) - f(y)| \leq C|x - y| \tag{33}$$

   holds for all x with $|x - y| \leq 1$ and $C = 2\sup_{|z-y|\leq 1} |f(z) - f(y)|$.

5. *If f is convex on $\mathbb{R}^n$ then for all $a, x_0 \in \mathbb{R}^n$ the directional derivative*

   $$(\partial_a f)(x_0) = \lim_{\varepsilon \to 0+} [f(x_0 + \varepsilon a) - f(x_0)]/\varepsilon \tag{34}$$

   exists, and is monotonously increasing, i.e. $(\partial_{-a} f)(x_0) \geq -(\partial_a f)(x_0)$. f is differentiable at x_0 if and only if $(\partial_{-a} f)(x_0) = -(\partial_a f)(x_0)$ holds for all a. Moreover, the map $a \mapsto (\partial_a f)(x_0)$ is convex in a.

For proofs see [23].

3 Phase Transitions in Thermodynamics

3.1 Equilibrium States

The thermodynamic equilibrium states $X = (X_1, \ldots, X_{n+1})$ are specified in terms of their extensive coordinates $X_i, i = 1, \ldots, n+1$, such as entropy S, volume V, and particle number N [24]. For solids and magnets further coordinates, such as mechanical strains and magnetization [c], are needed. The set of equilibrium states $\mathbb{E} \subset \mathbb{R}^{n+1}$ forms a convex cone [24]. This means that (1) if $X_1, X_2 \in \mathbb{E}$ and $0 \leq \alpha \leq 1$ then $\alpha X_1 + (1-\alpha)X_2 \in \mathbb{E}$, and (2) if $X \in \mathbb{E}$ then $\alpha X \in \mathbb{E}$ for all $\alpha > 0$ and if $X_1, X_2 \in \mathbb{E}$ then $X_1 + X_2 \in \mathbb{E}$.

3.2 General Laws

The first law of thermodynamics finds its mathematical expression in the existence of a potential function $U(X) : \mathbb{E} \to [0, \infty[$, that is called the (total) internal energy. Its gradient $\nabla U = (p_1, \ldots, p_{n+1})$

$$p_i = \frac{\partial U}{\partial X_i} \tag{35}$$

defines a conservative vector field of conjugate forces. For a simple fluid $n = 2$, the variables are $X_1 = S, X_2 = V, X_3 = N$, and the conjugate variables are $p_1 = T(S, V, N)$ temperature, $p_2 = -p(S, V, N)$ pressure, and $p_3 = \mu(S, V, N)$ chemical potential. The second law is equivalent to the convexity of $U(X)$, i.e. to

$$U(\lambda X_1 + (1-\lambda)X_2) \leq \lambda U(X_1) + (1-\lambda)U(X_2) \tag{36}$$

for all $X_1, X_2 \in \mathbb{E}$ and $0 \leq \lambda \leq 1$. Finally, the zeroth law states that U must be homogeneous of degree 1, i.e.

$$U(\lambda X) = \lambda U(X) \tag{37}$$

for all $X \in \mathbb{E}$ and $\lambda > 0$. The positivity of temperature implies that U must be a monotonously increasing as function of entropy S, i.e.

$$\left.\frac{\partial U}{\partial S}\right|_{V,N} \geq 0. \tag{38}$$

Differentiating Equation (37) with respect to λ and setting $\lambda = 1$ gives Eulers homogeneity relation

$$U = \sum_{i=1}^{n} p_i X_i \tag{39}$$

which reduces to $U = TS - pV + \mu N$ for simple fluids.

---

[c] For simple magnetic systems $z = (S, V, N, \mathbf{M})$ where $\mathbf{M}$ denotes the total magnetic dipole moment, and where one should remember that the sample must be of ellipsoidal shape to ensure magnetic homogeneity.

3.3 Densities and Legendre Transforms

In infinite systems the extensive quantities diverge ($U, X_i \to \infty$), and it becomes convenient to pass to densities or to specific (resp. molar) quantities. Setting $\lambda = 1/X_{n+1}$ in Equation (37) gives

$$X_{n+1}u(x_1, \ldots, x_n) := X_{n+1}U\left(\frac{X_1}{X_{n+1}}, \ldots, \frac{X_n}{X_{n+1}}, 1\right) = U(X_1, \ldots, X_{n+1}) \tag{40}$$

which suggests to define the densities

$$u = \frac{U}{X_{n+1}} \tag{41}$$

$$x_i = \frac{X_i}{X_{n+1}} \tag{42}$$

for $i = 1, \ldots, n$. While the function $u(x_1, \ldots, x_n)$ remains convex (and a potential), it is no longer homogeneous of first degree. Note also that the set $\mathbb{E}_\infty$ of equilibrium states $(x_1, \ldots, x_n)$ of an infinite system need no longer form a convex cone.

If $f : \mathbb{E} \to \mathbb{R}$ is a convex function then $f^* : \mathbb{E}^* \to \mathbb{R}$ denotes its **conjugate function**. f^* is defined as [23]

$$f^*(Y) = \sup_{X \in \mathbb{E}} (YX - f(X)) \tag{43}$$

with domain $\mathbb{E}^* = \{Y \in \mathbb{R}^{n+1} : f^*(Y) < \infty\}$. Here YX denotes the scalar product. The mathematical definition of duality in Eq. (43) is equivalent to the Legendre transform up to a minus sign. Thus

$$G(Y) = -U^*(Y) = -\sup_{X \in \mathbb{E}} (YX - U(X)) = \inf_{X \in \mathbb{E}} (f(X) - YX) \tag{44}$$

is obtained. Dividing by X_{n+1} in Eq. (39) yields

$$g(p_1, \ldots, p_n) = \inf_{x_i} \left[u(x_1, \ldots, x_n) - \sum_{i=1}^{n} p_i x_i \right] \tag{45}$$

for the Legendre transform.

3.4 Response Functions

Expanding the fundamental potential $u = u(x_1, \dots, x_n) = u(x)$ around a point $y = (y_1, \dots, y_n)$ gives

$$u(x) = u(y) + \sum_{i=1}^{n} p_i(y)\delta x_i + \frac{1}{2}\sum_{i=1}^{n}\sum_{j=1}^{n} u_{ij}(y)\delta x_i \delta x_j + \cdots \tag{46}$$

where $\delta x_i = x_i - y_i$ represents a small deviation from y. The corresponding deviation of the conjugate forces p_i is $\delta p_i = p_i(x) - p_i(y)$. Hence

$$p_i(x) = \frac{\partial u}{\partial x_i} = p_i(y) + \delta p_i \tag{47}$$

and differentiation gives a system of linear equations

$$\delta p_i = \sum_{j=1}^{n} u_{ij}(y)\delta x_j \tag{48}$$

for $i = 1, \dots, n$. The fundamental response coefficients

$$u_{ij}(x) = u_{ji}(x) = \frac{\partial^2 u(x)}{\partial x_i \partial x_j} = \frac{\partial p_i(x)}{\partial x_j} = \frac{\partial p_j(x)}{\partial x_i} \tag{49}$$

describe the response of the system against a small perturbation of the coordinates x. They may be viewed as generalized stiffness moduli. In particular they characterize the stability of the equilibrium state. The state x is stable against perturbations if the stiffness matrix $\mathbf{U} = (u_{ij})$ of response coefficients is positive definite. In this case the state is extremal and represents a pure phase. A pure phase cannot coexist with other phases without walls or restrictions.

It is frequently desirable to change to other independent variables. In particular one would like to know how the densities x respond to perturbations when the forces p are considered to be the indpendent variables. If $\det(\mathbf{U}) \neq 0$, which holds when x is stable, the linear system (48) can be be inverted and in this case

$$\delta x_i = \sum_{i=1}^{n} \chi_{ij}\delta p_j \tag{50}$$

where $\chi_{ij} = U_{ij}/\det(\mathbf{U})$ and U_{ij} is the cofactor of the matrix element u_{ik}. The physical meaning of the new response coefficients χ_{ij} is that of generalized susceptibilities or compliances. One has

$$\chi_{ij}(x) = \chi_{ji}(x) = -\frac{\partial^2 g(p)}{\partial p_i \partial p_j} = \frac{\partial x_i(p)}{\partial p_j} = \frac{\partial x_j(p)}{\partial p_i} \tag{51}$$

and

$$\frac{\partial g(p)}{\partial p_i} = -x_i. \tag{52}$$

The generalized susceptibility matrix $\chi = (\chi_{ij}) = \mathbf{U}^{-1}$ is the inverse of the generalized stiffness matrix $\mathbf{U}$.

For later convenience recall the case of the simple fluid whose equilibrium states are given by the specific entropy $x_1 = s$ and specific volume $x_2 = v$,

$$s(T,p) = -\left.\frac{\partial g}{\partial T}\right|_p \tag{53a}$$

$$v(T,p) = \left.\frac{\partial g}{\partial p}\right|_T . \tag{53b}$$

The susceptibilities are then obtained as

$$\chi_{11}(T,p) = \frac{c_p(T,p)}{T} = \left.\frac{\partial s}{\partial T}\right|_p = -\left.\frac{\partial^2 g}{\partial T^2}\right|_p \tag{54a}$$

$$\chi_{22}(T,p) = v\kappa_T(T,p) = -\left.\frac{\partial v}{\partial p}\right|_T = -\left.\frac{\partial^2 g}{\partial p^2}\right|_T \tag{54b}$$

$$\chi_{12}(T,p) = -v\alpha_p(T,p) = -\left.\frac{\partial v}{\partial T}\right|_p = \left.\frac{\partial s}{\partial p}\right|_T = -\frac{\partial^2 g}{\partial p \partial T} \tag{54c}$$

where c_p is the heat capacity at constant pressure, α_p is the isobaric coefficient of thermal expansion, and κ_T is the isothermal compressibility. Using the general thermodynamic relation with the specific heat at constant volume $c_v = c_p - Tv\alpha_p^2/\kappa_T$ the stiffnes matrix $\mathbf{U}$ is given as

$$\mathbf{U} = \begin{pmatrix} \dfrac{T}{c_v} & \dfrac{\alpha_p T}{c_v \kappa_T} \\ \dfrac{\alpha_p T}{c_v \kappa_T} & \dfrac{c_p}{v c_v \kappa_T} \end{pmatrix} \tag{55}$$

in terms of these quantities.

3.5 Phase Transitions

Equilibrium phase transitions can be viewed as as equilibrium states at which the stability or analyticity of the thermodynamic potentials is lost. A loss of stability is indicated by the stiffness matrix $\mathbf{U}$ ceasing to be positive definite.

In particular a critical point is signalled by **U** becoming positive semidefinite, i.e. singular, or equivalently by the susceptibility matrix becoming divergent. The loss of analyticity is indicated by such divergences or by the fact that macroscopic phase coexistence is associated with discontinuities in the first derivatives of thermodynamic potentials.

Let $p = (p_1, \ldots, p_n)$ denote the intensive field variables of an equilibrium state for a thermodynamic system described by the potential g.

Definition 3.1 (Phase Transitions). *An equilibrium state p_c is called a phase transition of the thermodynamic system described by the potential $g(p)$ if the function $g(p)$ is singular at p_c in the sense that $g(p)$ cannot be expanded into a convergent power series with finite and nonzero radius of convergence.*

A concrete example, that is also technically important, is the van der Waals theory for nonideal gases or fluids whose free enthalpy reads

$$g(T,p) = pv - \frac{3}{2}RT\log T - RT\log(v-b) - \frac{a}{v} - RTf_0 \tag{56}$$

where $v = v(T,p)$ is the solution of van der Waals' equation of state

$$p + \frac{a}{v^2} = \frac{RT}{v-b}. \tag{57}$$

The van der Waals constants a, b are material parameters determined mainly by the size and interaction energies of the fluid molecules. The universal gas constant R is the product of Avogadro's number and Boltzmann's constant, and f_0 is a free constant normalizing the energies. The van der Waals enthalpy exhibits a line of phase transitions ending in a critical point given by

$$T_c = \frac{8a}{27Rb}, \qquad p_c = \frac{a}{27b^2}, \qquad v_c = 3b. \tag{58}$$

Introducing new dimensonless variables $\widehat{T} = T/T_c, \widehat{p} = p/p_c$ and $\widehat{v} = v/v_c$ leads to the dimensionless equation of state

$$(3\widehat{v} - 1)\left(\widehat{p} + \frac{3}{\widehat{v}^2}\right) = 8\widehat{T}. \tag{59}$$

Denoting the dimensionless free enthalpy $g/(p_c v_c)$ as $\widehat{g}$ gives

$$\widehat{g}(\widehat{T},\widehat{p}) = \widehat{p}\widehat{v} - 4\widehat{T}\log\widehat{T} - \frac{8\widehat{T}}{3}\log\left(\widehat{v} - \frac{1}{3}\right) - \frac{3}{\widehat{v}} - \widehat{T}g_0 \tag{60}$$

where $g_0 = 8f_0/3 + \log(v_c^{8/3}T_c^4)$ is a new constant. The critical point in the dimensionless variables is $\widehat{p} = \widehat{v} = \widehat{T} = 1$.

For fixed dimensionless $\widehat{T} < 1$ the equation of state has three solutions $\widehat{v}_l(\widehat{T},\widehat{p}) < \widehat{v}^*(\widehat{T},\widehat{p}) < \widehat{v}_g(\widehat{T},\widehat{p})$. Two of them, $\widehat{v}_l, \widehat{v}_g$, are stable and correspond to the liquid and gas phase, the third, $\widehat{v}^*$, is unstable. The solutions are related to each other and to the coexistence pressure and temperature through Maxwells construction giving

$$\widehat{p} = \frac{8\widehat{T}}{3(\widehat{v}_g - \widehat{v}_l)} \log\left|\frac{3\widehat{v}_g - 1}{3\widehat{v}_l - 1}\right| - \frac{3}{\widehat{v}_l\widehat{v}_g} \tag{61}$$

as the relation between them. Using equations (53a)–(54c) one finds

$$\widehat{s}(\widehat{T},\widehat{p}) = \frac{T_c}{p_c v_c} s(\widehat{T},\widehat{p}) = \frac{8}{3}\log\left(\widehat{v} - \frac{1}{3}\right) + 4\log\widehat{T} + 4 + g_0 \tag{62}$$

$$\widehat{c}_p(\widehat{T},\widehat{p}) = \frac{T_c}{p_c v_c} c_p(\widehat{T},\widehat{p}) = \frac{64\widehat{v}^2\widehat{T}}{(3\widehat{v}-1)(9\widehat{p}\widehat{v}^2 - 2(\widehat{p}+8\widehat{T})\widehat{v}+9)} + 4 \tag{63}$$

$$\widehat{\kappa}_T(\widehat{T},\widehat{p}) = p_c\kappa_T(\widehat{T},\widehat{p}) = \frac{\widehat{v}(3\widehat{v}-1)}{9\widehat{p}\widehat{v}^2 - 2(\widehat{p}+8\widehat{T})\widehat{v}+9} \tag{64}$$

$$\widehat{\alpha}_p(\widehat{T},\widehat{p}) = T_c\alpha_p(\widehat{T},\widehat{p}) = \frac{8\widehat{v}}{9\widehat{p}\widehat{v}^2 - 2(\widehat{p}+8\widehat{T})\widehat{v}+9} \tag{65}$$

where $\widehat{v} = \widehat{v}(\widehat{T},\widehat{p})$ is the solution of eq. (59). Note that all three reponse functions $c_p(1,1) = \kappa_T(1,1) = \alpha_p(1,1) = \infty$ diverge at the critical point.

Figure 1 illustrates the phase diagram of the van der Waals theory. The solid line is the phase boundary obtained from the Maxwell construction, and the dashed lines delimit the region inside which the equation of state has multiple solutions.

3.6 Scaling Theory

Numerous experimental and theoretical studies suggest that the response functions diverge algebraically near a critical point [22,10,12]. Equivalently $g(p)$ is found to vary as a power near a critical point p_c. More precisely, for almost all directions of approach to p_c one finds

$$g(p) - g(p_c) \sim |p - p_c|^{1+1/\delta} \tag{66}$$

where $|p - p_c| = \sqrt{(p_1 - p_{c1})^2 + \cdots + (p_n - p_{cn})^2}$ and δ is called the equation of state exponent. In nature the basic exponent δ seems to obey the inequality

$$\delta \geq 1. \tag{67}$$

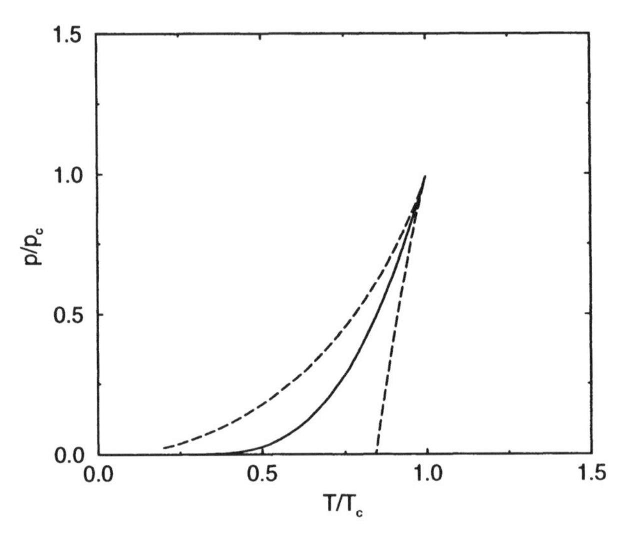

Figure 1: Vapour pressure curve (solid) of the van der Waals gas. The dotted lines delimit the region inside which equation (59) has multiple solutions.

While $\delta < 1$ cannot be ruled out theoretically, it would suggest an anomalous phase transition [25,3]. Such anomalous transitions can occur in systems with many-body interactions [25,11].

Although the scaling relation (66) holds for almost all directions there exist special directions such as the phase boundary along which different exponents are observed. These findings are traditionally summarized by the general homogeneity hypothesis [22,26,27]. It postulates that the singular part g_s of g behaves as

$$\lambda \tilde{g}_s(t, h_1, \dots, h_{n-1}) = \tilde{g}_s\left(t\lambda^{1/(2-\alpha)}, h_1\lambda^{1/(2-\alpha_1)}, \dots, h_{n-1}\lambda^{1/(2-\alpha_{n-1})}\right) \tag{68}$$

for all $\lambda > 0$ in the limit $t, h_i \to 0$. The homogeneity relation introduces a complete set of thermodynamic fluctuation exponents $\alpha, \alpha_1, \dots, \alpha_{n-1}$. [d] Without loss of generality they may be taken to be partially ordered as

$$\alpha_1 \geq \alpha_2 \geq \dots \geq \alpha_{n-1}. \tag{69}$$

[d] Unfortunately the notation α_i conflicts slightly with α_p for the thermal expansion coefficient. What is meant should become clear from the context.

The so called nonlinear scaling fields

$$\begin{aligned} t &= t(\widetilde{p}_1, \dots, \widetilde{p}_n) \\ h_i &= h_i(\widetilde{p}_1, \dots, \widetilde{p}_n) \end{aligned} \tag{70}$$

for $i = 1, \dots, n-1$ are defined such that $t(0, \dots, 0) = 0$ and $h_i(0, \dots, 0) = 0$ in terms of the dimensionless variables $\widetilde{p}_i$. These are given by $\widehat{p}_i = p_i/p_{ci} = 1 + \widetilde{p}_i$ in terms of p_i. The potential $g(p) = g_r(p) + g_s(p)$ is decomposed into a singular part g_s and a regular analytic background piece g_r, and it is made nondimensional through $\widehat{g} = g/g(p_c) = 1+\widetilde{g}$. The temperature-like scaling field $t = h_0$ and the corresponding exponent α are singled out notationally, because temperature and entropy play a distinguished role. This is also reflected by setting $\lambda = |t|^{\alpha-2}$ which leads to the familiar form of the scaling hypothesis [22,26,27]

$$\widetilde{g}_s(t, h_1, \dots, h_{n-1}) = |t|^{(2-\alpha)} Y_\pm(h_1|t|^{-\phi_1}, \dots, h_{n-1}|t|^{-\phi_{n-1}}) \tag{71}$$

with scaling function $Y_\pm(y_1, \dots, y_{n-1}) = \widetilde{g}_s(\pm 1, y_1 \dots, y_{n-1})$ for $t > 0$ resp. $t < 0$, and crossover exponents

$$\phi_i = \frac{2-\alpha}{2-\alpha_i}. \tag{72}$$

The scaling functions $Y_\pm$ for $t > 0$ resp. $t < 0$ must join smoothly to fulfill the analyticity requirements for g. The exponent δ in eq. (66) may be identified by setting $\lambda = r^{\alpha_i - 2}$ in eq. (68) where

$$r = \sqrt{t^2 + h_1^2 + \dots + h_{n-1}^2} = |h_i|\sqrt{(t/h_i)^2 + \dots + 1 + \dots (h_{n-1}/h_i)^2}$$

and demanding that the arguments of $\widetilde{g}_s(y_0, y_1, \dots, 1, \dots, y_{n-1})$ remain finite in the limit $t, h_i \to 0$. Then eq. (69) requires $i = 1$ and thus

$$\delta = \frac{1}{1-\alpha_1}. \tag{73}$$

The derivative of eq. (68) with respect to h_1 gives the homogeneity relation

$$\begin{aligned} \frac{\partial \widetilde{g}_s}{\partial h_1} &= \Psi_s(t, h_1, \dots, h_{n-1}) \\ &= \lambda^{\frac{\alpha_1 - 1}{2-\alpha_1}} \Psi_s\left(t\lambda^{1/(2-\alpha)}, h_1\lambda^{1/(2-\alpha_1)}, \dots, h_{n-1}\lambda^{1/(2-\alpha_{n-1})}\right) \end{aligned} \tag{74}$$

for the singular part Ψ_s of the so called order parameter Ψ of the transition. Setting $\lambda = |h_1|^{\alpha_1 - 2}$ and $t = h_2 = \ldots = h_{n-1} = 0$ yields

$$\Psi_s(0, h_1, 0, \ldots, 0) = \Psi_s(0, \pm 1, 0, \ldots, 0)\ |h_1|^{1/\delta}. \tag{75}$$

Identifying $t = h_0$ one may define exponents δ_i for $i = 0, \ldots, n-1$ through

$$\frac{\partial \widetilde{g}_s}{\partial h_i}(0, \ldots, h_i, \ldots, 0) \sim |h_i|^{1/\delta_i} \tag{76}$$

and finds thence that the two sets of exponents α_i and δ_i are equivalent to each other through the relation

$$\delta_i = \frac{1}{1 - \alpha_i} \tag{77}$$

with $\delta_1 = \delta$, and $\delta_0 = 1/(1-\alpha)$. It is customary to introduce also exponents for various derivatives. Specifically one defines β_j through

$$\frac{\partial \widetilde{g}_s}{\partial h_i}(t, 0, \ldots, 0) \sim (-t)^{\beta_i} \tag{78}$$

for $t < 0$. For $i = 1$ the exponent $\beta_1 = \beta$ is known as the order parameter exponent. Similarly one can define cross-susceptibility exponents γ_{ij} through

$$\frac{\partial^2 \widetilde{g}_s}{\partial h_i \partial h_j}(t, 0, \ldots, 0) \sim |t|^{-\gamma_{ij}} \tag{79}$$

for $h_1 = h_2 = \ldots = h_{n-1} = 0$. The exponent $\gamma_{11} = \gamma$ is called (order parameter-) susceptibility exponent, while $\gamma_{00} = \alpha$ is the specific heat exponent, if, as before, t is identified with h_0.

Differentiating eq. (71) allows to evaluate the exponents β_i and γ_{ij} as

$$\beta_i = 2 - \alpha - \phi_i \tag{80}$$

$$\gamma_{ij} = \phi_i + \phi_j - (2 - \alpha) \tag{81}$$

from which one obtains the relation

$$2 - \alpha = \beta_i + \beta_j + \gamma_{ij} \tag{82}$$

by eliminating the crossover exponents. For $i = j = 1$ this reduces to the celebrated scaling law $2 - \alpha = 2\beta + \gamma$ between the thermodynamic exponents α, β and γ. The crossover exponents themselves may be written as

$$\phi_i = \beta_i + \gamma_{ii} = \beta_i \delta_i \tag{83}$$

recovering another well known scaling law (in case $i = 1$).

Practical application of the scaling hypothesis (71) requires knowledge of the thermodynamic potential $g(p)$, and of the scaling fields in eq. (70). The scaling fields are defined as those fields for which the renormalization group recursion relations become exactly linear [28,10]. Thus they cannot be determined from thermodynamic potentials alone. The problem of determining the fields t and h_i in practice is closely related to the problem of finding the order parameter [9,27]. Most authors avoid the problem by assuming "Ising symmetry". This implies that the field $t = (T - T_c)/T_c$ becomes the temperature. In general the order parameter should be related to the "most strongly" fluctuating microscopic density [9,27]. This idea is embodied in the inequalities (69) leading to the identification (73).

The scaling and renormalization group theory of phase transitions concentrates on the so called "universality" of critical exponents. The concept of universality originates from the observation that for systems with short range interactions the critical exponents do not depend on how the system is realized (as a magnet or fluid, on a lattice or in the continuum). By contrast the classical thermodynamics of phase transitions concentrates on general (but nonuniversal) properties of phase transitions, such as the location of the critical point. An example is the Clausius-Clapeyron relation giving a general rule for the slope of coexistence curves. The modern theory of critical phenomena seems of considerably lower practical importance in applications than the classical theory [12]. This may be partly so, because it applies only in a very narrow region around a critical point [12], and provides few results for determining critical lines or phase boundaries. For this reason phase boundaries are mostly determined with the help of classical theory through simulation or experiment. It is therefore useful to search for analogues of the classical results, and to elucidate general relations for continuous phase transitions.

4 The Classification Scheme of Ehrenfest

4.1 Jump Singularities

Let $f : \mathbb{R}^n \to \mathbb{R}$ be a real-valued function, and let ∂^m be a shorthand notation for the mixed partial differential operator

$$\partial^m = \frac{\partial^{m_1}}{\partial x_1^{m_1}} \cdots \frac{\partial^{m_n}}{\partial x_n^{m_n}} \tag{84}$$

with multiindex $m = (m_1, ..., m_n) \in \mathbb{N}^n$. The number $|m| = m_1 + ... + m_n$ is called the order of ∂^m. The space of k times continously differentiable functions is denoted by $C^k(\mathbb{R}^n)$. Let $\mathbb{S}^{n-1} := \{x \in \mathbb{R}^n : |x| = 1\}$ denote the unit sphere.

Definition 4.1 (Sectorwise Continuity). *A function $f : \mathbb{R}^n \to \mathbb{R}$ is called sectorwise continuous at a point x_0 if there exists*

1. *a decomposition $\mathbb{S}^{n-1} = \bigcup_{i=1}^{P} \mathbb{S}_i$ of the unit sphere $\mathbb{S}^{n-1}$ into a finite number $2 \leq P < \infty$ of mutually disjoint singly connected compact subsets (sectors) $\mathbb{S}_i \subset \mathbb{S}^{n-1}$ and*
2. *a set of P values $f_i(x_0) < \infty, 1 \leq i \leq P$ with $f_i(x_0) \neq f_j(x_0)$ for $i \neq j$*

such that

$$\lim_{\substack{\varepsilon \to 0+ \\ e \in \mathbb{S}_i}} f(x_0 + \varepsilon e) = f_i(x_0) \tag{85}$$

for all unit vectors $e \in \mathbb{S}_i$.

The simplest singularities are those of a piecewise continuous function. This motivates the

Definition 4.2 (Discontinuities). *Let $m = (m_1, ..., m_n) \in \mathbb{N}^n$. A function $f \in C^{k-1}(\mathbb{R}^n)$ is said to have a discontinuity of order $k \in \mathbb{N}, k \geq 1$ at a point $x_0 \in \mathbb{R}^n$ if all partial derivatives $(\partial^m f)(x)$ with $|m| = k$ are sectorwise continuous at the point x_0.*

At a discontinuity of order k the k-th order partial derivatives of f are multivalued, and they can have up to P different values. The requirement of connectedness for the sectors $\mathbb{S}_i$ in the decomposition of $\mathbb{S}^{n-1}$ ensures that for each $\mathbb{S}_i$ there exists at least one $\mathbb{S}_j$ such that $(-\mathbb{S}_i) \cap \mathbb{S}_j \neq \emptyset$ where $(-\mathbb{S}_i)$ is defined as $(-\mathbb{S}_i) = \{e \in \mathbb{S}^{n-1} : -e \in \mathbb{S}_i\}$. Therefore also the directional derivatives (see eq. (34)) of order k in the directions e and $-e$ are in general discontinuous at a discontinuity of order k. Thus, if $e \in \mathbb{S}_i$, then there exists at least one $\mathbb{S}_j$ such that

$$\partial_e^k f(x) \neq (-1)^k \partial_{-e}^k f(x) \tag{86}$$

holds for all $e \in (-\mathbb{S}_i) \cap \mathbb{S}_j$.

4.2 Ehrenfest's order of a phase transition

Ehrenfests classification scheme [14] is based on the experimental observation of the density contrast between coexisting phases and latent heat at a phase transition. Equations (53a) and (53b) show that these observations correspond to discontinuities in the first derivatives of the free enthalpy. Ehrenfest proposes that all phase transtions correspond to discontinuities, although possibly in derivatives of higher order. He studies explicitly first and second order derivatives of $g(T,p)$ for a simple thermodynamic system corresponding to the case $n = 2$ and $P = 2$. Generalizing his idea slightly leads to the following definition.

Definition 4.3 (Ehrenfest classification). *An equilibrium state p_c is called a phase transition of order $k \in \mathbb{N}$ in the sense of Ehrenfest if and only if the free enthalpy $g(p)$ has a discontinuity of order k at p_c.*

The best known examples of phase transitions are first order transitions. A concrete case is the coexistence line, given by eq. (61), in the van der Waals theory. The critical point in van der Waals mean field theory is commonly presented as an example of a second order phase transition in the sense of Ehrenfest (see [29] p. 548). Although this terminology has already found its way into reference tables [30] it will be seen below that the transition is of order 4/3 in a sense to be made precise.

Ehrenfests classification while allowing for jump discontinuities does not classify divergences (e.g. algebraic or logarithmic) in thermodynamic quantities. For this reason it has been frequently criticized as being incomplete [10,31,32]. Note also that more complicated jump discontinuities could arise in Definition 4.2 if more general decompositions of $\mathbb{S}$ are admitted.

4.3 Thermodynamic Consequences

In his original paper Ehrenfest derived analogues of the Clausius-Clapeyron equation for transitions of second order based on the traditional concept of metastability. He assumes implicitly that the free enthalpy is multivalued in the vicinity of a critical point. More precisely he assumes the existence of two functions $g_1(p), g_2(p) \in C^k(\mathbb{E}_\infty)$ such that the equilibrium free enthalpy is given by

$$g(p) = \min\{g_1(p), g_2(p)\} \tag{87}$$

for $p \in \mathbb{B}(p_c, \varepsilon)$, where $\mathbb{B}(p_c, \varepsilon)$ denotes a spherical neighbourhood of radius ε around a phase transition point p_c. The metastable branches are given by $g^\dagger(p) = \max\{g_1(p), g_2(p)\}$. The existence of such metastable extensions has meanwhile become controversial among theorists [9,11,33,34,35] due to a weak essential singularity at the coexistence line, and it is now believed that the time honoured concept of metastable states "lacks fundamental justification in statistical mechanics" [5].

For $n = 2$ with variables $p_1 = T$ and $p_2 = p$ Ehrenfest derives the Clausius-Clapyron relation starting from the equality of the equilibrium free enthalpies of the two phases

$$g_1(T,p) = g_2(T,p) \tag{88}$$

whose solution $p(T)$ is the vapour pressure curve $\mathcal{C}$. Let (T_c, p_c) denote a point on the curve $\mathcal{C}$ parametrized as

$$\begin{aligned} \mathcal{C} : [0,1] &\to \mathbb{R}^2 \\ \lambda &\mapsto (T(\lambda), p(\lambda)). \end{aligned} \tag{89}$$

Differentiating eq. (88), valid on $\mathcal{C}$, with respect to λ gives

$$\left(\left.\frac{\partial g_2}{\partial T}\right|_p - \left.\frac{\partial g_1}{\partial T}\right|_p \right) \frac{dT}{d\lambda} + \left(\left.\frac{\partial g_2}{\partial p}\right|_T - \left.\frac{\partial g_1}{\partial p}\right|_T \right) \frac{dp}{d\lambda} = 0, \tag{90}$$

and from this follows with the help of eqs. (53a) and (53b)

$$\frac{dp}{dT}(T_c) = \frac{s_2(T_c,p_c) - s_1(T_c,p_c)}{v_2(T_c,p_c) - v_1(T_c,p_c)}. \tag{91}$$

This is the Clausius-Clapeyron equation for the slope of the vapour pressure curve.

Ehrenfest observes that for a jump discontinuity of second order the right hand side of eq. (91) degenerates into $\frac{0}{0}$. Considering a line of such second order points he repeats the argument for s and v to obtain

$$\frac{dp}{dT}(T_c) = -\frac{\dfrac{\partial^2 g_2}{\partial T^2} - \dfrac{\partial^2 g_1}{\partial T^2}}{\dfrac{\partial^2 g_2}{\partial T \partial p} - \dfrac{\partial^2 g_1}{\partial T \partial p}} = \frac{c_{p2} - c_{p1}}{T_c v_c(\alpha_{p2} - \alpha_{p1})} \tag{92}$$

$$\frac{dp}{dT}(T_c) = -\frac{\dfrac{\partial^2 g_2}{\partial p \partial T} - \dfrac{\partial^2 g_1}{\partial p \partial T}}{\dfrac{\partial^2 g_2}{\partial p^2} - \dfrac{\partial^2 g_1}{\partial p^2}} = \frac{\alpha_{p2} - \alpha_{p1}}{\kappa_{T2} - \kappa_{T1}} \tag{93}$$

at every point (T_c, p_c) on the second order line. Here $v_2(T_c,p_c) = v_1(T_c,p_c) = v_c$. Equating the two last equations yields the result

$$\frac{(c_{p2} - c_{p1})(\kappa_{T2} - \kappa_{T1})}{(\alpha_{p2} - \alpha_{p1})^2 T_c v_c} = 1 \tag{94}$$

known as the Ehrenfest relation for a second order phase transition. Similar relations for third order derivatives can be found for a line of third order phase transitions.

Note that the Ehrenfest relations are not applicable for the van der Waals critical point because equations (63)–(65) show that c_p, κ_T and α_p diverge at the critical point. This implies that the mean field critical point is not of second order in the sense of Ehrenfest.

The usual derivation of the Clausius-Clapeyron and Ehrenfest relations assumes the analytic continuation of the metastable branches g_1, g_2 across the vapour pressure curve $\mathcal{C}$. Nowadays this assumption has become controversial [9,11,33,34,35] and it is therefore worthwile to derive the results without assuming metastable extensions. A second reason to look for an alternative is that the derivation assumes the vapour pressure curve to be differentiable which can be violated e.g. at triple points. A third reason is that the derivation cannot be used at a critical point.

Consider a single-valued equilibrium potential $g(p)$ (i.e. one without metastable branches) which may, however, have creases or other singularities.

Proposition 4.1 (Generalized Clausius-Clapeyron Equation). *Let p_c be a phase transition in the single valued potential $g : \mathbb{R}^n \to \mathbb{R}$ at which $M < \infty$ macroscopic phases coexist. Then the tangential hyperplane at the point p_c to the phase boundary hypersurface separating two pure phases j and k has the equation*

$$\sum_{i=1}^{n}(g_i^j - g_i^k)p_i = 0 \tag{95}$$

where

$$g_i^j = \frac{\partial g^j}{\partial p_i}(p_c) \tag{96}$$

and $g^j(p)$ is the restriction of $g(p)$ to the equilibrium states representing the pure phase $j (j = 1, \ldots, M)$.

Proof. Introduce spherical coordinates $p = p_c + re$ with $e = (e_1, \ldots, e_n)$ denoting a unit vector in the unit sphere $\mathbb{S}^{n-1}$. Convexity guarantees the existence of the one sided radial derivatives

$$\lim_{r\to 0} \frac{\partial g}{\partial r}(p_c + re) = \lim_{r\to 0} \sum_{i=1}^{n} e_i \frac{\partial g}{\partial p_i}(p_c + re) \tag{97}$$

and therefore also the limits of the partial derivatives on the right hand side exist for each e. For each $j = 1, \dots, M$ the set

$$\mathbb{S}_j^{n-1} = \left\{ e \in \mathbb{S}^{n-1} : \lim_{r \to 0} \frac{\partial g}{\partial p_i}(p_c + re) = g_i^j \text{ for all } i = 1, \dots, n \right\} \tag{98}$$

defines the directions pointing into the pure phase j. By Theorem 2.5 the map

$$e \mapsto (\partial_e g)(p_c) = \lim_{r \to 0} \frac{\partial g}{\partial r}(p_c + re) \tag{99}$$

is convex and hence continuous in e. Let $e^i \in \mathbb{S}_j^{n-1}$ and $f^i \in \mathbb{S}_k^{n-1}$ be two sequences of unit vectors converging to the same unit vector $e = \lim_{i \to \infty} e^i = \lim_{i \to \infty} f^i$. Such sequences exist because the phases j and k are assumed to coexist at p_c. Then continuity in e implies that the limits

$$\lim_{e^i \to e} \lim_{r \to 0} \frac{\partial g}{\partial r}(p_c + re^i) = \lim_{f^i \to e} \lim_{r \to 0} \frac{\partial g}{\partial r}(p_c + rf^i) \tag{100}$$

can be interchanged and are equal. Using eqs. (96) and (97) now gives

$$\sum_{i=1}^{n} g_i^j e_i = \sum_{i=1}^{n} g_i^k e_i \tag{101}$$

showing that the vector $g^j - g^k$ is normal to the tangential hyperplane plane to the phase boundary at p_c. □

For $n = 2$ and $M = 2$ the Clausius-Clapeyron equation (91) is recovered by setting $e = (\cos\varphi, \sin\varphi)$, writing eq. (101) as

$$\tan\varphi = -\frac{g_1^2 - g_1^1}{g_2^2 - g_2^1}, \tag{102}$$

and recognizing $\tan\varphi$ as the slope of the phase boundary. Inserting $g_1^i = -s_i$ and $g_2^i = v_i$ from eqs. (53a) and (53b) this reduces to eq. (91). The considerations show that the Clausius Clapeyron equation follows from the second law (convexity) and does not require the existence of metastable analytic continuations.

5 A Generalized Classification Scheme

5.1 Fractional Singularities

Let $f : \mathbb{R}^n \to \mathbb{R}$ be a real-valued function. Introduce dimensionless spherical coordinates (r, e) around a point $x_0 \in \mathbb{R}^n$ through

$$x = x_0 + re \tag{103}$$

where $0 \leq r = |x - x_0| < \infty$. [e]

Continuity of f implies $\lim_{r\to 0} f(x_0 + re) = f(x_0)$ for all e. A k-times continuously differentiable function $f \in C^k(\mathbb{R}^n)$ is also $k+1$-times continuously differentiable at x_0 if and only if

$$\partial_e \partial^m f(x_0) = \frac{\partial}{\partial r}\partial^m f(x_0 + re)\bigg|_{r=0} = -\frac{\partial}{\partial r}\partial^m f(x_0 - re)\bigg|_{r=0} = -\partial_{-e}\partial^m f(x_0) \tag{105}$$

holds for all $m = (m_1, ..., m_d)$ with $|m| = k$, and for all $e \in \mathbb{S}^{n-1}$. This can be generalized to define fractional differentiability of order $k + \alpha$ in several variables by using radial fractional derivatives. In the following the notation $[\alpha] = \max\{i \in \mathbb{Z} : i \leq \alpha\}$ is used for the integer part of a real number $\alpha \in \mathbb{R}$. The notation $\mathrm{D}^{\alpha,\mu}_{r,0+}$ and $\mathrm{I}^{\alpha}_{r,0+}$ will be used for fractional derivatives and integrals with respect to r.

Definition 5.1 (Fractional Differentiability at a Point). *A function $f : \mathbb{R}^n \to \mathbb{R}$ is called fractionally differentiable of order $k + \alpha$ with $0 < \alpha < 1$ at the point $x_0 \in \mathbb{R}^n$ in the direction $e \in \mathbb{S}^{n-1}$ if f is k times differentiable at x_0, and the limit*

$$\begin{aligned} f^{(m,\alpha)}(x_0, e) &= \lim_{r\to 0} f^{(m,\alpha)}(x_0 + re) \\ &= \lim_{r\to 0} \left(\mathrm{D}^{\alpha,1}_{r,0+}\partial^m f\right)(x_0 + re) \\ &= \lim_{r\to 0} \left(\mathrm{I}^{1-\alpha}_{r,0+}\frac{\partial}{\partial r}\partial^m f\right)(x_0 + re) < \infty \end{aligned} \tag{106}$$

exists and is finite for all multiindices $m = (m_1, ..., m_n)$ with $|m| = k$. The function is called fractionally differentiable of order $k + \alpha$ at the point x_0 if the condition (106) holds for all $e \in \mathbb{S}^{n-1}$. The function is called continuously fractionally differentiable of order $k + \alpha$ at the point x_0 in the direction e if

$$f^{(m,\alpha)}(x_0, e) = -f^{(m,\alpha)}(x_0, -e) \tag{107}$$

[e] Explicitly the unit vector $e \in \mathbb{S}^{n-1}$ is given by

$$\begin{aligned} e_1 &= \cos\varphi_1 \\ e_2 &= \sin\varphi_1 \cos\varphi_2 \\ e_3 &= \sin\varphi_1 \sin\varphi_2 \cos\varphi_3 \\ &\dots \\ e_{n-1} &= \sin\varphi_1 \sin\varphi_2 ... \sin\varphi_{n-2} \cos\varphi_{n-1} \\ e_n &= \sin\varphi_1 \sin\varphi_2 ... \sin\varphi_{n-2} \sin\varphi_{n-1} \end{aligned} \tag{104}$$

with $0 \leq \varphi_k \leq \pi, k = 1, ..., n-2$, and $0 \leq \varphi_{n-1} < 2\pi$.

holds for all m with $|m| = k$. f is called continuously fractionally differentiable of order $k(x_0) + \alpha(x_0)$ at the point x_0 if (107) holds for all $e \in \mathbb{S}^{n-1}$. The set of continuously fractionally differentiable functions at the point x_0 will be denoted $C^{k,\alpha}(x_0)$.

The definition is readily extended to radial Riemann-Liouville derivatives $\mathrm{D}_{0+}^{\alpha,\beta}$ of order α and type $\beta \neq 1$. One advantage of $\mathrm{D}_{0+}^{\alpha,1}$ is that it obviates the necessity to extract the singular part from the free energies [10,36] in applications to scaling theory of critical phenomena.

Definition 5.2 (Fractional singularity). *A function $f \in C^k(\mathbb{R}^n)$ is said to have a fractional singularity of order $\omega = k + \alpha$ $(0 < \alpha < 1)$ at a point $x_0 \in \mathbb{R}^n$ if $k = [\omega]$ and*

$$\alpha(x_0) = \sup\left\{0 < \beta < 1 : f \in C^{k,\beta}(x_0)\right\}. \tag{108}$$

The fractional singularity is called regular if for all directions e the k-th radial derivative

$$f^{(k)}(x_0 + re) = \frac{\partial^k f}{\partial r^k}(x_0 + re) \tag{109}$$

varies regularly with index $\gamma(x_0, e) \geq \alpha(x_0)$ as $r \to 0$.

Note that there exist functions having singularities that are not fractional singularities. Many singularities may however be approximated as fractional singularities by virtue of Theorem 2.3.

5.2 Phase Transitions of Fractional Order

Ehrenfest's classification of phase transitions was based on discontinuities as defined above in Definition 4.2. A much more general classification scheme based on fractional derivatives along arbitrary curves in the thermodynamic state space was introduced in Refs. [1,2,37,38]. In this section a related classification scheme is introduced that is based on fractional discontinuities. It is more general than Ehrenfest's scheme; but may be seen as a special case of the generalized classification in [1,2,37,38] in the sense that it restricts attention to a special class of curves, namely rays. The new scheme allows to identify the order parameter field and hence the order parameter within thermodynamics. Secondly the new classification allows the calculation of nonuniversal properties such as phase boundaries or field mixing coefficients for scaling fields.

Definition 5.3 (Fractional phase transition). *An equilibrium state p_c is called a phase transition of fractional order $\omega \in \mathbb{R}$ if the free enthalpy $g(p)$ has a regular fractional singularity of order ω at p_c.*

Because fractional singularities are defined in Definition 5.2 by requiring fractional differentiability up to order ω the fractional classification scheme is a natural generalization of Ehrenfests idea. Compared with previous classification schemes [1,2,3] the present scheme does not depend on the path of approach to the critical point. This has the advantage that it singles out the most strongly diverging/fluctuating observable, i.e. the order parameter. In this way the critical exponents are naturally coupled to the identification of an order parameter.

6 Thermodynamic Consequences

6.1 Scaling Exponents

The generalized thermodynamic classification of phase transitions implies that thermodynamic potentials exhibit algebraic scaling near a critical point.

Proposition 6.1. *Let $g(p_1, \dots, p_n)$ be the single valued equilibrium free enthalpy of a thermodynamic system having a regular fractional phase transition of order $1 < \omega < 2$ at p_c. Then for each $e = (e_1, \dots, e_n) \in \mathbb{S}^{n-1}$ there is number $0 \le g^{(\omega)}(p_c, e) < \infty$ such that*

$$\lim_{r\to 0} \frac{g(p_c + re) - g(p_c)}{r^\omega} = \frac{g^{(\omega)}(p_c, e)}{\Gamma(\omega + 1)}. \tag{110}$$

Proof. Definitions 5.2 and 5.3 together with eq. (106) guarantee the existence of numbers

$$0 \le g^{(m,\omega-1)}(p_c, e) = \lim_{r\to 0} (\mathrm{D}^{\omega-1,1}_{r,0+} \partial^m g) < \infty \tag{111}$$

with multiindices $m = (m_1, \dots, m_n)$ and $|m| = 1$. The notation $g_i^{(\omega-1)} = g^{(m,\omega-1)}$ is used whenever $m = (m_1, \dots, m_n)$ with $m_i = 1$ and $m_j = 0$ for $j \neq i$. Applying Theorem 2.4 to $\partial^m g$ shows that for each $e \in \mathbb{S}^{n-1}$

$$\lim_{r\to 0} r^{1-\omega} \frac{\partial g}{\partial p_i}(p_c + re) = \frac{g_i^{(\omega-1)}(p_c, e)}{\Gamma(\omega)} \tag{112}$$

holds for all $i = 1, \dots, n$. Multiplying with e_i and summing over i gives

$$\lim_{r\to 0} r^{1-\omega} \frac{\partial g}{\partial r}(p_c + re) = \frac{1}{\Gamma(\omega)} \sum_{i=1}^{n} e_i g_i^{(\omega-1)}(p_c, e). \tag{113}$$

Integration of this equation using Karamatas theorem [21] now yields eq. (110) with $g^{(\omega)}(p_c, e) = \sum_{i=1}^n e_i g_i^{(\omega-1)}(p_c, e)$. □

Writing $r = |p - p_c|$ the result (110) may be written as

$$g(p) - g(p_c) \approx \frac{g^{(\omega)}(p_c, e)}{\Gamma(\omega + 1)} |p - p_c|^{\omega} \tag{114}$$

for $p \to p_c$. If the scaling amplitudes $g^{(\omega)}(p_c, e)$ are nonzero for almost all directions e then this algebraic scaling behaviour corresponds to eq. (66) from conventional scaling theory. Comparing the exponents yields the identification $\omega = 1 + 1/\delta$, and using relation (73) one arrives at the

Corollary 6.1 (Identification of critical exponents). *Let $g(p_1, \dots, p_n)$ be the single valued equilibrium free enthalpy of a thermodynamic system having a regular fractional phase transition of order $1 < \omega < 2$ at p_c. Let $g^{(\omega)}(p_c, e) \neq 0$ for almost all e be as in (110). Then the order ω of the phase transition is*

$$\omega = 1 + \frac{1}{\delta} = 2 - \alpha_1 \tag{115}$$

where δ and α_1 are the critical exponents from eqs. (66) and (73).

Convexity restricts the order of a transition at equilibrium states that are not boundary points.

Proposition 6.2. *Let $g : \mathbb{G} \to \mathbb{R}$ be the free enthalpy of a thermodynamic system defined on a singly connected subset $\mathbb{G} \subset \mathbb{R}^n$. If p_c is a fractional singularity of g of order ω, and p_c is an interior point of $\mathbb{G}$, then $\omega \geq 1$.*

Proof. The function $-g$ is convex. Theorem 2.5 then guarantees the existence of one sided derivatives of first order, and hence $\omega \geq 1$. □

The inequality $\omega \geq 1$ implies $\delta \geq 0$. Note that the stronger inequality (67) implies $\omega \leq 2$. The assumption, that p_c is an interior point, is important because it excludes anequilibrium phase transitions [3] that may exist at boundary points.

6.2 Fractional Clausius-Clapeyron Relation

This section investigates some thermodynamic consequences of the fractional classification scheme introduced in Definition 5.3. Consider again a single valued equilibrium free enthalpy $g(p_1, \dots, p_n)$ of a thermodynamic system having a regular fractional phase transition of order $1 < \omega < 2$ at p_c. From the definitions or from eq. (110) it follows that

$$\mathrm{D}^{\omega-1,1}_{r,0+} \frac{\partial g}{\partial r}(p_c + re) = g^{(\omega)}(p_c, e) \tag{116}$$

holds asymptotically for $r \to 0$. The scaling amplitudes are thus recognized as radial fractional derivatives in the limit $r \to 0$. Equation (108) implies that $g^{(\zeta)}(p_c, e) = 0$ for $\zeta < \omega$. The order ω of the phase transition is the smallest order for which the radial fractional derivative at the origin is nonzero. Assume now that also $g^{(\omega)}(p_c, e) \neq 0$ for almost all $e \in \mathbb{S}^{n-1}$ and that the amplitude $g^{(\omega)}(p_c, e)$ is continuous as a function of e. Then the directions such that

$$\lim_{r \to 0} \mathrm{D}^{\omega-1,1}_{r,0+} \frac{\partial g}{\partial r}(p_c + re) = g^{(\omega)}(p_c, e) = 0 \tag{117}$$

are of special interest. It follows from the definitions that in these directions the free enthalpy has either a higher index of regular variation than ω or else there is a slowly varying correction. The directions e solving eq. (117) therefore correspond to the phase boundaries or scaling axes for continuous phase transitions. Equation (117) may be seen as an analogue of the Clausius-Clapeyron equation for continuous transitions.

Fractional differentiation of order $\omega_i > \omega$ along other directions e_i such that $g^{(\omega)}(p_c, e_i) = 0$ yields additional scaling exponents. Repeating the argument leads to a hierarchy $\alpha_i = 2 - \omega_i$ of exponents as in eq. (69). Higher orders of differentiation correspond to larger values of the critical exponent δ_i and to smaller values of α_i. Note however that the generalized classification allows also for continuously varying exponents.

6.3 Application to van der Waals theory

The previous results may be applied to van der Waals theory. Shifting the critical point to the origin by changing coordinates in eq. (59) through the replacement $\widehat{T} = \widetilde{T} + 1, \widehat{p} = \widetilde{p} + 1, \widehat{v} = \widetilde{v} + 1$, leads to the cubic equation

$$3(\widetilde{p} + 1)\widetilde{v}^3 + 8(\widetilde{p} - \widetilde{T})\widetilde{v}^2 + (7\widetilde{p} - 16\widetilde{T})\widetilde{v} + 2(\widetilde{p} - 4\widetilde{T}) = 0 \tag{118}$$

for v. Introducing polar coordinates $\widetilde{T} = r\cos\varphi, \widetilde{p} = r\sin\varphi$ yields for $\tan\varphi \neq 4$

$$\widetilde{v}^3 = \frac{2r(4\cos\varphi - \sin\varphi)}{3(1 + r\sin\varphi)} \left[1 + \frac{16\cos\varphi - 7\sin\varphi}{8\cos\varphi - 2\sin\varphi}\widetilde{v} + \frac{4\cos\varphi - 4\sin\varphi}{4\cos\varphi - \sin\varphi}\widetilde{v}^2\right]. \tag{119}$$

Taking the third root and iterating the equation gives the asymptotic expansion

$$\widetilde{v}(r, \varphi) \approx \left(\frac{2r}{3}(4\cos\varphi - \sin\varphi)\right)^{1/3} \left[1 + \frac{16\cos\varphi - 7\sin\varphi}{(8\cos\varphi - 2\sin\varphi)^{2/3}}\left(\frac{r}{3}\right)^{1/3} + \ldots\right] \tag{120}$$

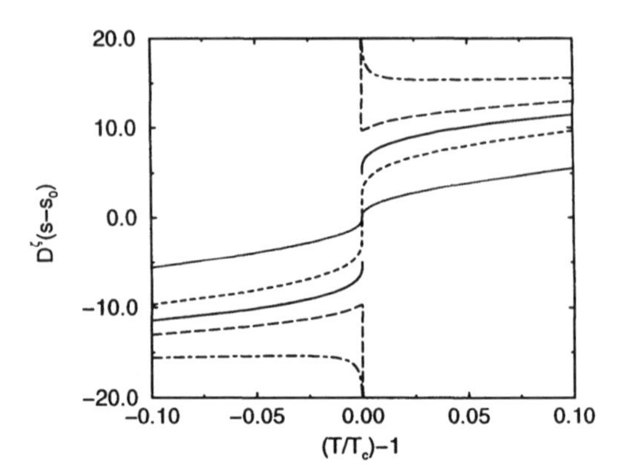

Figure 2: Asymptotic behaviour of the fractional derivatives of order $\zeta = 0$ (dotted), $\zeta = 1/4$ (dashed), $\zeta = 1/3$ (solid), $\zeta = 0.4$ (long dashed), and $\zeta = 1/2$ (dashdotted) for the entropy $\widetilde{s}(\widetilde{T}, \widetilde{p} = 0)$ of the van der Waals gas calculated from eq. (121). The curves for $\zeta = 0$ and $\zeta = 1/4$ are continuous at 0. The curves for $\zeta = 0.4$ and $\zeta = 1/2$ are divergent at 0. The discontinuity $16\Gamma(4/3)3^{-1/3}$ in the curve for $\zeta = 1/3$ is the fractional latent heat.

valid for $r \to 0$ and $\varphi \neq \arctan 4$. Using equation (62) it follows from this that

$$\widetilde{s}(r,\varphi) \approx 4\left(\frac{2r}{3}(4\cos\varphi - \sin\varphi)\right)^{1/3}\left[1 + \frac{10\cos\varphi - (11/2)\sin\varphi}{(8\cos\varphi - 2\sin\varphi)^{2/3}}\left(\frac{r}{3}\right)^{1/3} + \ldots\right] \tag{121}$$

where $\widetilde{s} = s - s_0$ with $s_0 = (8/3)\log(2/3) + 4 + g_0$. These results imply that

$$\lim_{r\to 0} \mathrm{D}^{\zeta,1}_{r,0+}\frac{\partial g}{\partial r}(r,\varphi) = \begin{cases} 0 & \text{for } \zeta < \frac{1}{3} \\ -\Gamma\left(\frac{4}{3}\right)\left(\frac{2}{3}\right)^{1/3}(4\cos\varphi - \sin\varphi)^{4/3} & \text{for } \zeta = \frac{1}{3} \\ \infty & \text{for } \zeta > \frac{1}{3} \end{cases} \tag{122}$$

and hence that the order of the van der Waals phase transition is $\omega = 4/3$. Equation (115) then gives $\delta = 3$ and $\alpha_1 = 2/3$ for the critical exponents. Figure 2 illustrates the fractional derivatives $\mathrm{D}^{\zeta,1}_{r,0+}\widetilde{s}(r,\varphi)$ of the entropy for $\varphi = 0, \pi$ and $\zeta = 0, 0.25, 1/3, 0.4, 0.5$ as calculated for $r \approx 0$ from eq. (121). The curve is continuous at T_c for $\zeta < 1/3$ but becomes progressively steeper as ζ approaches $1/3$. It becomes discontinuous at T_c for $\zeta = 1/3$ and divergent for $\zeta > 1/3$. The discontinuity of magnitude $16\Gamma(4/3)3^{-1/3} \approx 9.9$ appearing for $\zeta = 1/3$ may be viewed as a fractional analogue of the latent heat. Note that the slope is infinite at the discontinuity for $\zeta = 1/3$. Hence one does not expect hysteresis or metastability at phase transitions of order $\omega > 1$ in agreement with theoretical and experimental observations.

Next, Equation (117), the fractional analogue of the Clausius-Clapeyron equation, predicts $\tan\varphi = 4$ for the slope of the phase boundary at the critical point by virtue of eq. (122). This prediction is indeed borne out as may be seen from Figure 1 above. The fractional classification yields simultaneously the scaling exponents together with the scaling axes. Thereby it provides a method to identify the thermodynamic field conjugate to the order parameter. The order parameter is identified as the most strongly fluctuating observable.

Finally the analysis is completed by considering the free enthalpy along the direction of the scaling axis, i.e. for $\tan\varphi = 4$ or equivalently for $\widetilde{p} = 4\widetilde{T}$. In this case eq. (118) reduces to a quadratic equation for $\widetilde{v}$. As a consequence $\widetilde{v}(r, \varphi = \arctan 4) \sim r^{1/2}$ and the order parameter exponent becomes $\beta = \beta_1 = 1/2$. Of course this is equivalent to the identification of the scaling exponents $\alpha = 0$ and $\delta_0 = 1$ corresponding to order $\omega_0 = 1 + 1/\delta_0 = 2$ along the scaling axis.

In summary the critical point of the van der Waals free enthalpy is classified as a phase transition of order 4/3 in the generalized classification scheme. The radial fractional derivatives of order 4/3 exhibit discontinuities at the critical point upon approach from all except one direction. The exceptional direction is the scaling axis given by $\varphi = \arctan 4$. The second order derivative of the free energy becomes discontinuous when approaching the critical point along the exceptional direction $\varphi = \arctan 4$. This reflects the familiar discontinuity in the specific heat capacity at a mean field critical point.

References

1. R. Hilfer. Thermodynamic scaling derived via analytic continuation from the classification of Ehrenfest. *Physica Scripta*, 44:321, 1991.
2. R. Hilfer. Multiscaling and the classification of continuous phase transitions. *Phys. Rev. Lett.*, 68:190, 1992.
3. R. Hilfer. Classification theory for anequilibrium phase transitions. *Phys. Rev. E*, 48:2466, 1993.
4. A. Aharony. Multicritical points. In F.J.W. Hahne, editor, *Critical Phenomena*, page 210, Berlin, 1983. Springer Verlag.
5. K. Binder. Theory of first-order phase transitions. *Rep. Prog. Phys.*, 50:783, 1987.
6. K. Binder. Some recent progress in the phenomenological theory of finite

size scaling and application to Monte Carlo studies of critical phenomena. In V. Privman, editor, *Finite Size Scaling and Numerical Simulation of Statistical Systems*, page 173, Singapore, 1990. World Scientific.

7. K. Binder. Finite size effects at phase transitions. In H. Gausterer and C.B. Lang, editors, *Computational Methods in Field Theory*, page 59, Berlin, 1992. Springer Verlag.
8. J.L. Cardy. Conformal invariance. In C. Domb and J.L. Lebowitz, editors, *Phase Transitions and Critical Phenomena*, volume 11, page 55, London, 1987. Academic Press.
9. M.E. Fisher. The theory of crtical point singularities. In M.S. Green, editor, *Critical Phenomena*, page 1, New York, 1971. Academic Press.
10. M.E. Fisher. Scaling, universality and renormalization group theory. In F.J.W. Hahne, editor, *Critical Phenomena*, page 1, Berlin, 1983. Springer Verlag.
11. M.E. Fisher. Phases and phase diagrams: Gibbs's legacy today. In G. Mostow and D. Caldi, editors, *Proceedings of the Gibbs Symposium*, page 39, Providence, 1990. American Mathematical Society.
12. J.V. Sengers and J.M.H. Levelt-Sengers. Thermodynamics behaviour of fluids near the critical point. *Ann.Rev.Phys.Chem.*, 37:189, 1986.
13. R. Clausius. Über die bewegende Kraft der Wärme und die Gesetze, welche sich daraus für die Wärmelehre selbst ableiten lassen. *Poggendorff's Annalen*, 79:368, 1850.
14. P. Ehrenfest. Phasenumwandlungen im üblichen und erweitereten Sinn, classifiziert nach den entsprechenden Singularitäten des thermodynamischen Potentiales. *Suppl. Mitteilungen aus dem Kamerlingh-Onnes Institut, Leiden*, 75b:153, 1933.
15. H.W. Alt. *Lineare Funktionalanalysis*. Springer Verlag, Berlin, 1992.
16. G.H. Hardy and J.E. Littlewood. Some properties of fractional integrals. I. *Math. Zeitschr.*, XXVII:565, 1928.
17. G.H. Hardy and J.E. Littlewood. Some properties of fractional integrals. II. *Math. Zeitschr.*, XXXIV:403, 1932.
18. S.G. Samko, A.A. Kilbas, and O.I. Marichev. *Fractional Integrals and Derivatives*. Gordon and Breach, Amsterdam, 1993.
19. W. Feller. *An Introduction to Probability Theory and Its Applications*, volume II. Wiley, New York, 1971.
20. E. Seneta. *Regularly Varying Functions*. Springer Verlag, Berlin, 1976.
21. N.H. Bingham, C.M. Goldie, and J.L Teugels. *Regular Variation*. Cambridge University Press, Cambridge, 1987.
22. H.E. Stanley. *Introduction to Phase Transitions and Critical Phenomena*. Oxford University Press, New York, 1971.

23. A.W. Roberts and D.E. Varberg. *Convex Functions.* Academic Press, New York, 1973.
24. N. Straumann. *Thermodynamik.* Lecture Notes in Physics, Vol. 265. Springer Verlag, Berlin, 1986.
25. M.E. Fisher and G.W. Milton. Classifying first-order phase transitions. *Physica A*, 138:22, 1986.
26. A. Hankey and H.E. Stanley. Systematic application of generalized homogeneous functions to static scaling, dynamic scaling, and universality. *Phys.Rev.B*, 6:3515, 1972.
27. M.E. Fisher. General scaling theory for crtical points. In B. Lundqvist and S. Lundqvist, editors, *Proc. Nobel Symposium XXIV on Collective Properties of Physical Systems*, page 16, New York, 1974. Academic Press.
28. F.J. Wegner. The critical state, general aspects. In C. Domb and M.S. Green, editors, *Phase Transitions and Critical Phenomena*, volume 6, page 7, London, 1976. Academic Press.
29. J. Zinn-Justin. *Quantum Field Theory and Critical Phenomena.* Oxford University Press, Oxford, 1989.
30. H. Stöcker, editor. *Taschenbuch der Physik.* Verlag Harri Deutsch, Thun/Frankfurt, 1998.
31. K. Huang. *Statistical Mechanics.* Wiley, New York, 1987.
32. R. Becker. *Theorie der Wärme.* Springer Verlag, Berlin, 1985.
33. A.F. Andreev. Singularities of thermodynamic quantities at a first order phase transition point. *Soviet Physics JETP*, 18:1415, 1964.
34. A.S. Wightman. Convexity and the notion of equilibrium state in thermodynamics and statistical mechanics. In *R.B. Israel, Convexity in the Theory of Lattice Gases*, page ix, Princeton, 1979. Princeton University Press.
35. S.N. Isakov. Nonanalytic features of the first order phase transition in the ising model. *Commun. Math. Phys.*, 95:427, 1984.
36. A. Aharony and M.E. Fisher. Nonlinear scaling fields and corrections to scaling near criticality. *Phys.Rev.B*, 27:4394, 1983.
37. R. Hilfer. Scaling theory and the classification of phase transitions. *Mod. Phys. Lett. B*, 6:773, 1992.
38. R. Hilfer. Classification theory for phase transitions. *Int.J.Mod.Phys.B*, 7:4371, 1993.

14256352R00270

Made in the USA
Lexington, KY
17 March 2012